HOMOLOGIES IN ENZYMES AND METABOLIC PATHWAYS

METABOLIC ALTERATIONS IN CANCER

homologies in enzymes and metabolic pathways

metabolic alterations in cancer

Proceedings of the Miami Winter Symposia, January 19–23, 1970
Organized by the Department of Biochemistry, University of Miami, and the Papanicolaou Cancer Research Institute

W. J. WHELAN and J. SCHULTZ, Editors

1970

NORTH-HOLLAND PUBLISHING COMPANY – AMSTERDAM · LONDON

ISBN: 0 7204 4081 5

Published by:

NORTH-HOLLAND PUBLISHING COMPANY – AMSTERDAM
NORTH-HOLLAND PUBLISHING COMPANY, LTD. – LONDON

Printed in The Netherlands

FOREWORD

Symposia on biochemical topics have been arranged by the Department of Biochemistry and the Program in Cellular and Molecular Biology of the University of Miami for a number of years. In January 1969 the Department of Biochemistry joined with the University-affiliated Papanicolaou Cancer Research Institute to stage consecutive symposia, entitled respectively, "Control Mechanisms in Intermediary Metabolism"and "Membrane Function and Electron Transfer to Oxygen". The interest aroused in the symposia, and the high standards of the presentations, prompted the sponsors to continue the venture and to commence the publication of the symposia in full, so that the information, much of it original, could be shared with a wider audience. As a beginning, a selection of papers from the second of the above symposia was published under the title "Biochemistry of the Phagocytic Process".*

The present volume contains the full report of the 1970 meeting and will be the first volume of a continuing series under the title *Miami Winter Symposia.* The two symposia were entitled "Homologies in Enzymes and Metabolic Pathways" and "Metabolic Alterations in Cancer".

Associated with the symposia is an opening lecture, now named in honor of the University of Miami's distinguished Visiting Professor, Professor Feodor Lynen. The first Lynen Lecture, given by Dr. George Wald, is reproduced here. Earlier lectures in the symposia series have been published elsewhere.**

It is hoped that the method of organization of the symposia will ensure their publication as rapidly as possible. The speakers are strictly enjoined to present their finished manuscripts at the time of the meeting, and our publishers use a rapid, highly attractive photo-offset printing process. We thank them, the speakers, and the many local helpers, faculty and administrative staff, coordinated by Dr. K. Savard, who have made this venture possible. We also acknowledge with gratitude the financial assistance of the National

* "Biochemistry of the Phagocytic Process." Ed. by J. Schultz, North-Holland Publishing Co., Amsterdam, 1969.

** F. Lynen, "Life, Luck and Logic in Biochemical Research", *Perspectives in Biology and Medicine,* Winter 1969, 204.
H.A. Krebs, "The History of Tricarboxylic Acid Cycle", *Perspectives in Biology and Medicine,* in the press.

Science Foundation, the Howard Hughes Medical Institute, the Departments of Anesthesiology, Medicine, and Ophthalmology of the University of Miami School of Medicine, Eli Lilly Inc., Merck, Sharpe and Dohme Inc., and E.R. Squibb and Company, Inc.

W.J. Whelan
J. Schultz

CONTENTS

PART 1

HOMOLOGIES IN ENZYMES AND METABOLIC PATHWAYS

THE FIRST FEODOR LYNEN LECTURE

VISION AND THE MANSIONS OF LIFE

George WALD
Biological Laboratories, Harvard University,
Cambridge, Massachusetts, USA

Abstract: Wald, George, Vision and the Mansions of Life, *Miami Winter Symposia,* 1, pp. pp. 1–32. North-Holland Publishing Company, Amsterdam, 1970.

The theory of evolution is founded primarily upon morphological data. In its approach to such data it is guided always by the fundamental distinction between analogous organs, which, whatever their basic composition, serve similar functions; and homologous organs, which, whatever their functions, share deep-seated correspondences in structure, origin, and position in the organism. Only homologies are accepted as a proper indication of biological relationship.

As the biologist penetrates to smaller and smaller elements of the organism, he reaches finally, and as we now know with no essential discontinuity, the molecular level. Here physiology fuses with reaction patterns and mechanisms; and morphology becomes one with chemical structure. On this level again one can readily distinguish analogies and homologies. The physical chemistry of biological systems, taken by itself, leads only to analogies. Their chemical structure and origins provide the stuff of homologies. As such they open the possibility of pursuing biological relationship in the realm of molecular dimensions. In this sense one can speak of biochemical evolution. This enterprise has a long history, beginning on the level of gross molecular relationships and now reaching its culmination in the detailed structures of proteins and nucleic acids. Compared with classic morphological approaches it lacks, and will always lack, a detailed paleontology, and for this it will continue to lean on the classic morphological data. On the other hand it has interesting roots in biochemical embryology. It opens also new vistas for defining phylogenetic and systematic relationships on levels of discrimination not available heretofore. These necessitate a reexamination of our older concepts of species, analogy, homology and recapitulation.

1. Introduction

One cannot go deeply into any aspect of biology without encountering evolution. I came into biochemical evolution, not by design, but by necessity. It was something that happened to me in the course of my work. I think it is much better that way, for then one is not an advocate, and can be a judge.

I was trying to learn as much as I could of the chemistry of vision, first in one animal, then in many others. As the work developed, it began to form a pattern; and gradually it emerged that that was the pattern of evolution. So I found myself in biochemical evolution, tying up with other aspects of biochemical evolution. One can do that without becoming an "evolutionist", just as one can pursue one's work into the ocean without becoming an "oceanographer", or become absorbed with the environment of organisms without becoming an "ecologist". One remains a biologist, doing his work, trying to understand. If one is to raise a banner, let it be to life, which includes all these things, and to which one can give oneself wholeheartedly and without thought of special pleading.

I took my degree at Columbia University in 1932, as the culmination of five years in the laboratory of Selig Hecht. During those five years Hecht called me his "assistant". It was his way of helping me to live, and study, and do my first research. It was a bad time outside, the period of the Great Depression. I was very fortunate at the end in being awarded one of five National Research Council Fellowships.

That brought me to the laboratory of Otto Warburg in Dahlem. During the five years with Hecht I had worked up a great hunger to lay hands on the molecules for which we had only symbols. That in itself was symptomatic. I had taken almost every graduate course in physical chemistry offered at Columbia, yet no biochemistry. I had done that in the conviction, to which Hecht's laboratory subscribed, that the molecules themselves didn't matter very much; the important thing was their behavior. So coming to Otto Warburg was for me entering a new universe.

In a first conversation with Warburg he asked me, did I think that rhodopsin is an *Atmungsferment*? I said with a little embarrassment that I had no idea; whereupon he said, "Try it!"

He said something else that relieved me enormously. There I was coming from the principal vision laboratory in the world, presumably knowing all about the eye. Yet strange as it may seem, all that time in Hecht's laboratory I had never seen a retina. Warburg, on principle, assumed that I knew nothing. To my huge relief, he said to me, "Kubowitz will show you how to take out a retina". Kubowitz did; and I have been taking out retinas that way ever since.

So I did my first experiment to see whether rhodopsin is an *Atmungsferment*. I put a batch of dark-adapted retinas into a Warburg respirometer, measured their oxygen uptake in the dark, and then at a certain moment, turned on a bright light so as to bleach them. To my surprise and joy, the rate of oxygen consumption suddenly began to leap upward and kept going up and up. I went back and told Warburg, and set up the next experiment. That next time when I turned on the light, nothing whatever happened; nor did it ever happen again. That first time apparently bacteria had got into the act, waiting as usual to make the most dramatic entrance. "You see, Herr Wald," said Warburg, "The better one's technique, the harder it is to make discoveries."

The respirometry of retinas and rhodopsin solutions never did turn very interesting. Meanwhile, I had begun to wonder what sort of molecule rhodopsin might be. I was reading old Kühne assidously, Kühne, who after Boll had discovered rhodopsin in frog retinas in 1877, in two furious years of work together with Ewald, learned virtually everything known about that pigment until a half century later. Ewald and Kühne's work made me think that rhodopsin must be a protein; but what could its color group be? Visual pigments are pretty widespread; so one would think that that chromophore probably belongs to one of the large and prevalent types of animal pigment. But what? A porphyrin? Its spectrum didn't look right for that. An anthocyanin? That didn't seem very likely. A carotenoid? That sounded possible. Carr and Price in England had just described a simple general test for carotenoids. On mixing carotenoids or vitamin A with antimony trichloride in chloroform, one obtained a blue color. That sounded worth a try. So I extracted some frog retinas with chloroform, and mixed the extract with antimony chloride in chloroform. There was the blue color. That was all it took. It opened the door to what was to become my life, indeed what was to become for me my door to life.

Pretty soon I had measured the absorption spectrum of the substance that gave the blue color with antimony chloride, and realized that it must be vitamin A. Then, a little late in the day, I began to read in the literature that vitamin A has a connection with vision, the history of which went back to the ancient Egyptian medical papyri. Under conditions of food shortage, known even to the ancient Egyptians, a curious disease occurred called night blindness. With its first description one was offered the correct prescription for its cure – the eating of liver. Late in World War I, Bloch in Copenhagen identified the factor in liver that cures night blindness to be the then newly discovered vitamin A. It all began to make sense: for there was vitamin A, in the retina itself, as though acting directly. That was a new thought for vitamins. which had been supposed to act by some mysterious interaction. Yet in that year, 1932, Warburg's laboratory was already engaged in experiments that would show riboflavin, and shortly afterward another B vitamin, nicotinamide, working directly as the prosthetic groups of enzymes of energy metabolism.

When I told Warburg that I had found vitamin A in the retina, and its association with night blindness, he said, "In that case you must go to Karrer." Paul Karrer in Zurich, had just isolated and worked out the correct structure of vitamin A. Indeed he had reported this at the International Physiological Congress in Rome that summer. I was at the Congress, but not at all interested in vitamin A; and so did not go to Karrer's lecture. "Karrer is vitamin A, vitamin A is Karrer", said Warburg. "What do you suppose? Should Karrer begin to work on *Atmungsferment*?"

So I went to Karrer. During the next four months my wife and I received the entire output of pig and cattle eyes from the Zurich city abattoir. We took the retinas out of 6000 pig and 5000 cattle eyes. Four months of dissecting – and a little of the Swiss Alps, and my first skiing – and then Karrer was satisfied that it was indeed vitamin A.

With that I pulled up stakes again and went off to Heidelberg and Otto Meyerhof. That brought me back to a changed Germany. The very day on which I had left Warburg, Adolf Hitler became *Reichskanzler*. Four months later I returned to what was now Nazi Germany. It was a strange and horrifying experience. Meyerhof's laboratory was a little island of sanity, not spared an occasional invasion. One day our *Diener* who, I was told, had been a red street fighter in earlier times, disappeared.

You might think that, having just found vitamin A in the retina that was what I was working on. Through some strange perversity, I had decided to do something else in Meyerhof's laboratory, something more characteristically Meyerhof. I spent a few months fractionating the organic phosphate compounds of the retina. Nothing very interesting emerged – nothing that I ever published, no more than I published the results of all the respirometry done in Warburg's laboratory. I was just learning more biochemistry.

Then the Summer vacation came, and Meyerhof and all his assistants went away, leaving me alone, waiting to go back to America. Then something fortunate happened.

Shortly after the Nazis took over, the Society of Animal Friends, the president of which appropriately was a retired General, succeeded in having a law passed forbidding the killing of frogs in the state of Baden. That is, German frogs. It was still all right to kill non-Aryan frogs, which we imported from Hungary. Just after the laboratory went on vacation a large shipment of Hungarian frogs arrived. The *Diener* was about to throw them into the Neckar, but was glad to give them instead to me. Working with those frogs I discovered retinene – later shown by R.A. Morton in Liverpool to be vitamin A aldehyde – and found it to be an intermediate in the bleaching of rhodopsin to vitamin A.

And then I came back to America. Though I wanted very much to stay with Meyerhof, and Meyerhof was kind enough to try to keep me there, I had to go back, because the Rockefeller Foundation in Paris thought Germany had become too dangerous for Americans. So I went to the University of Chicago.

Why there? Because, with that strange perversity still working, I had respirometry and organic phosphates on my mind; and there in Chicago was Ralph Gerard, who knew all about such things. This time a wise old man, Anton Carlson, straightened me out.

"You leave the respirometry and the phosphates to somebody else", he said, "That vitamin A is your baby. You do that." So that's what I did. I finally stopped putting off my fate, and instead embraced it.

Let me tell you now of some of the paths down which it led me. Let me stop this personal history, which is trivial, and go to life, which is important: and say as best I can, where the story comes out.

This particular story had an odd beginning. I worked out the rhodopsin system first on frogs, the classic animal; with excursions into a variety of mammals. Many years ago, however, two German workers, Köttgen and Abelsdorff [1] extracted the visual pigments from a variety of vertebrate eyes into water solution, and measured their difference spectra. In frogs, mammals and owls they found ordinary red rhodopsin; but from the retinas of eight species of fish they obtained a purple visual pigment with a different absorption spectrum.

Clearly one had to look into the visual pigments of fishes. That brought me to Woods Hole. To my great surprise however, the first fishes I worked on at Woods Hole, all had ordinary rhodopsin, vitamin A, and retinal. Either Köttgen and Abelsdorff or the fishes were wrong. On re-reading Köttgen and Abelsdorff, I saw something strange. When they had reached out for fishes, in the interior of Germany so long ago, they had come up entirely with freshwater fishes; whereas my fishes in Woods Hole were all marine. That raised a strange and almost incredible possibility: that somehow freshwater fishes differ in this regard from fishes out of the sea. So I got myself a freshwater fish; and promptly found that new visual pigment, which I called porphyropsin. It had a new chromophore, and bleached to yield a new kind of vitamin A. That marked the discovery of retinal_2 and vitamin A_2, the substances that replace retinal and vitamin A in the visual systems of freshwater fishes, and numbers of other freshwater vertebrates. That began the story which I should now like to summarize.

2. Freshwater and saltwater fishes

Two kinds of visual system are found in the rods of vertebrate retinas. One is based upon the red visual pigment, rhodopsin, formed by the combination of the protein opsin with retinal, the aldehyde of vitamin A. The other is based

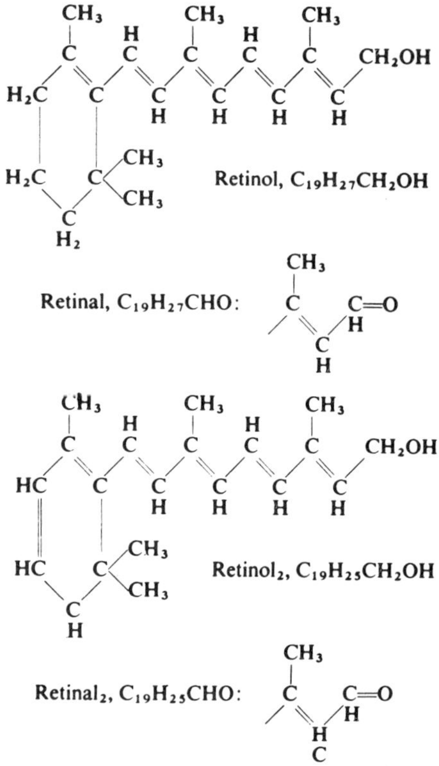

Retinol, $C_{19}H_{27}CH_2OH$

Retinal, $C_{19}H_{27}CHO$:

Retinol₂, $C_{19}H_{25}CH_2OH$

Retinal₂, $C_{19}H_{25}CHO$:

Fig. 1. Structures of vitamin A (retinol), retinal, vitamin A_2 (retinol$_2$), and retinal$_2$.

upon the purple pigment, porphyropsin, formed from the same type of opsin combined with retinal$_2$, the aldehyde of vitamin A_2. Retinal$_2$ and vitamin A_2 differ from retinal and vitamin A only in possessing an added double bond in the ring (fig. 1) [2–5].

The porphyropsin system was first discovered in fresh-water fishes. Marine fishes and land vertebrates characteristically possess the rhodopsin system (figs. 2 and 3) [6,7].

What of the fishes that are neither freshwater nor marine, but migrate between both environments? It would be well before discussing them to clarify somewhat their biological position.

Most fishes are restricted throughout their lives to narrow ranges of salinity. Such forms are called "stenohaline" and are of two kinds, fresh-water and marine. A much smaller group of fishes can live as adults in a wide range of

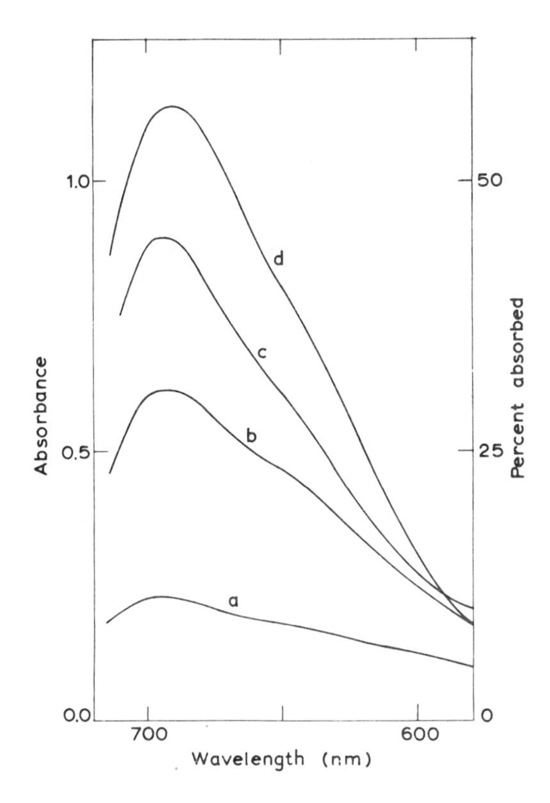

Fig. 2. The vitamin A_2 of the retinas of fresh-water fishes. Antomony chloride tests with extracts of wholly bleached retinas display only the absorption band maximal at 690 to 696 nm characteristic of vitamin A_2. This result has been obtained invariably in about 12 widely distributed species of fresh-water teleost. In almost all cases the visual pigment has also been extracted and this pigment has been found to be porphyropsin, with λ_{max} about 522 nm. (Republished by permission of the Journal of General Physiology [6].)

salinities. They are called "euryhaline," and, again, are of two kinds, anadromous and catadromous, meaning "upstream" and "downstream". These terms refer to the direction of the spawning migrations. Salmon, for example, are typically anadromous forms, coming upstream to spawn, whereas the "fresh-water" cells are catadromous, going downstream to the sea on their spawning migration.

It is probably true, however, that no euryhaline fish has to leave its spawning environment to complete a normal life cycle. Many instances are known in which anadromous fishes remain permanently in fresh water. The

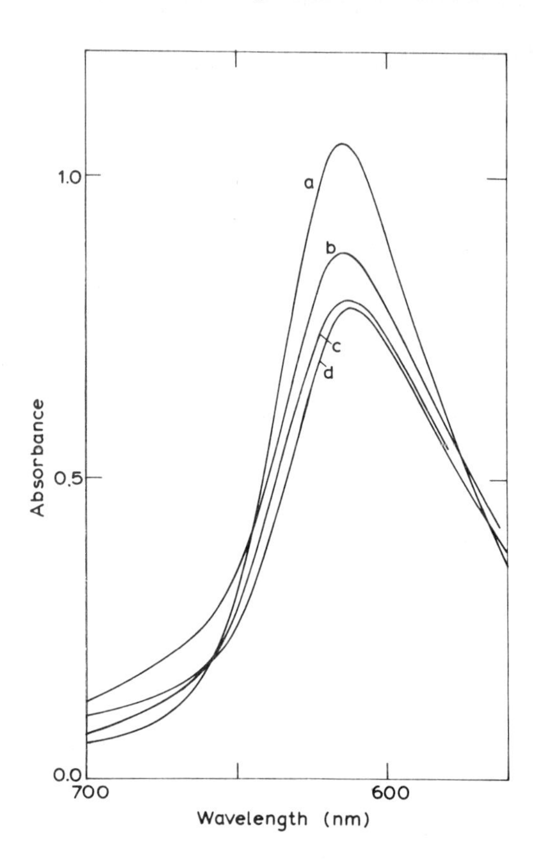

Fig. 3. The vitamin A_1 of the retinas of marine fishes. Spectra of the antimony chloride tests with extracts of bleached retinas reveal the λ_{max} at 615 to 620 nm characteristic of vitamin A_1. This result has been obtained with a great variety of bony, and a few elasmobranch, fishes and is characteristic also of land vertebrates. Two wrasse fishes (Labridae), however, the cunner and tautog, though wholly marine, are exceptional in having a predominance of vitamin A_2 in their retinas. (Republished by permission of the Journal of General Physiology [6].)

same is true of such an anadromous cyclostome as the sea lamprey, which has recently colonized the Great Lakes and virtually destroyed the fresh-water fisheries there.

So far as we know, the spawning environment is always fixed. The eggs, the sperms, or the embryos, perhaps sometimes all three, are stenohaline. Euryhalinity develops later in life and permits, though does not compel, these animals to migrate to the other environment. Migration is only a potentiality,

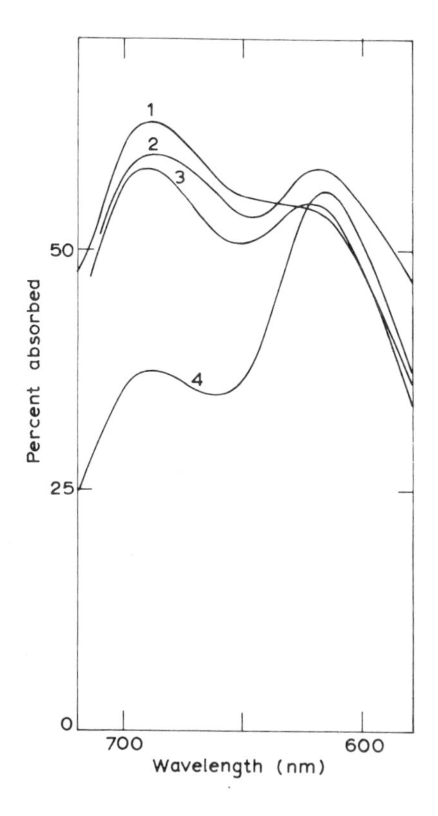

Fig. 4. Retinal vitamins A in euryhaline fishes. Spectra of antimony chloride tests with extracts of bleached retainas from (1) chinook salmon, (2) rainbow trout, (3) brook trout, and (4) the American "fresh-water" eel. All these tissues contain both vitamins A_1 and A_2, the anadromous salmonids a predominance of vitamin A_2, the catadromous eel a higher proportion of vitamin A_1. (Republished by permission of the Journal of General Physiology [6].)

which some of these forms exploit regularly and others rarely. The salmons are essentially fresh-water fishes with the privilege of going to sea as adults; the fresh-water eels are marine fishes with the capacity of coming as adults into fresh water.

The significant biological statement concerning such fishes is not that they migrate but that, being fixed in spawning environment, they are euryhaline as adults. I should like on this basis to redefine the terms applied to them. An anadromous fish is a euryhaline form which spawns in fresh water; a catadromous fish, one which spawns in the sea [6].

On examining the visual systems of several genera of salmonids, I found that all of them possess mixtures of the rhodopsin and porphyropsin systems, yet primarily the latter, characteristic of the spawning environment. Conversely, the American fresh-water eel possesses a mixture of both visual pigments, in which rhodopsin-again the spawning type-predominates (fig. 4) [7]. Certain other anadromous fishes–alewife, white perch-possess porphyropsin almost alone. All the euryhaline fishes examined follow a simple rule: all of them possess, either predominantly or exclusively, the type of visual system characteristically associated with the spawning environment [6,8].

To a first approximation these patterns are genetic and independent of the immediate environment. The salmonids which were found to possess mixtures of both visual systems had spent their entire lives in fresh-water. Alewives just in from the sea on their spawning migration possess porphyropsin almost exclusively. Most striking of all, the cunner and tautog, members of the wholly marine family of Labridae, the wrasse fishes, possess porphyropsin [6,8].

Since the distribution of visual systems among fishes is genetic, one may ask whether it fits into some evolutionary pattern. Many paleontologists are convinced that the vertebrate stock originated in fresh water. It is from such fresh-water ancestors that our fresh-water fishes were ultimately derived.* The observation that these animals characteristically have the porphyropsin system suggests that this may have been the ancestral vertebrate type – a view that receives some support from the observation that the sea lamprey, *Petromyzon marinus*, a member of the most primitive living group of vertebrates, possesses porphyropsin as an adult [10] (see below). Subsequently vertebrates undertook two great evolutionary migrations, one into the sea, the other to land. Both led them to the use of rhodopsin in rod vision, for this is the pigment we find characteristically in marine fishes and land vertebrates. The euryhaline fishes are intermediate between fresh-water and marine forms both in life history and in the composition of their visual systems. In this regard one can arrange the fishes in such an ordered sequence as shown in fig. 5.

* The "ultimately" here conceals a thorny problem. Fish evolution has probably involved numerous interchanges between fresh-water and marine existence, and many present-day fresh-water fishes may have had marine forms in their ancestry. In that case one should have to assume that the complex of genetic changes that has brought stocks into fresh water has regularly carried with it the property of using vitamin A_2 and porphyropsin in vision. I cannot suggest a genetic mechanism for this association; nor is it absolute. Numbers of fresh-water fishes are now known to possess rhodopsins, some apparently exclusively [9].

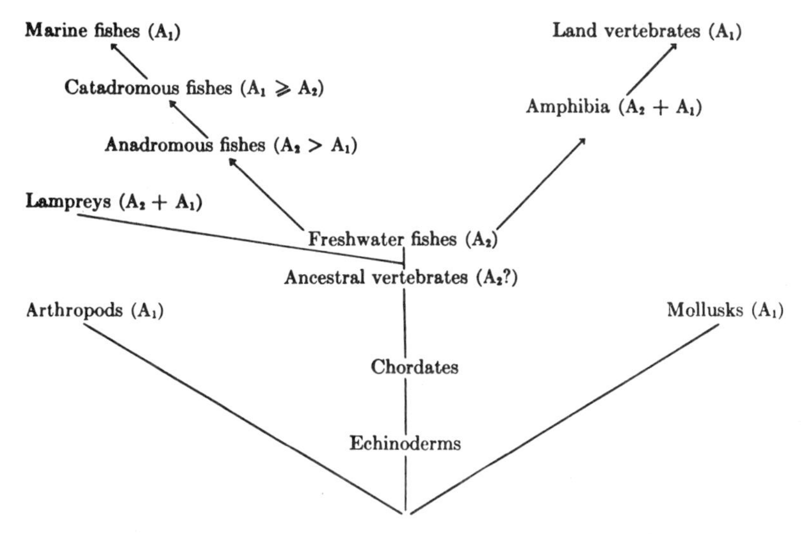

Fig. 5. Distribution of vitamins A_1 and A_2 in vertebrate and invertebrate retinas. The observations, all made on contemporary animals, are here correlated with the present ecology. They may also, however, represent evolutionary sequences, and in that case they convey the suggestion that primitive vertebrate vision was based upon vitamin A_2.

3. Euryhaline fishes and amphibia

One can hardly develop such an argument without raising questions regarding the amphibia. These animals come between fresh-water fishes and land vertebrates, just as the euryhaline fishes do between fresh-water and marine fishes. Most amphibia, like most euryhaline fishes, spawn in fresh water. Indeed, the life cycle of the common frog runs strikingly parallel with that of such an anadromous fish as the salmon. Both originate and go through a larval period in fresh water. Both, after undergoing deep-seated anatomical and physiological changes which can in both be described as metamorphosis, migrate for the growth phase, the salmon to sea, the frog to land. Both return to fresh water at sexual maturity to spawn. What land is to the frog, the sea is to the salmon. *The euryhaline fishes are the amphibia among the fishes.*

In this sense one might speak of almost all amphibia as "anadromous", meaning that they spawn in fresh water and are free, as adults, to go back and forth between fresh water and the land. A few amphibia (red-backed and tree salamanders) have developed special devices for living permanently ashore. I know of no "catadromous" amphibian – that is, one that spawns on land and

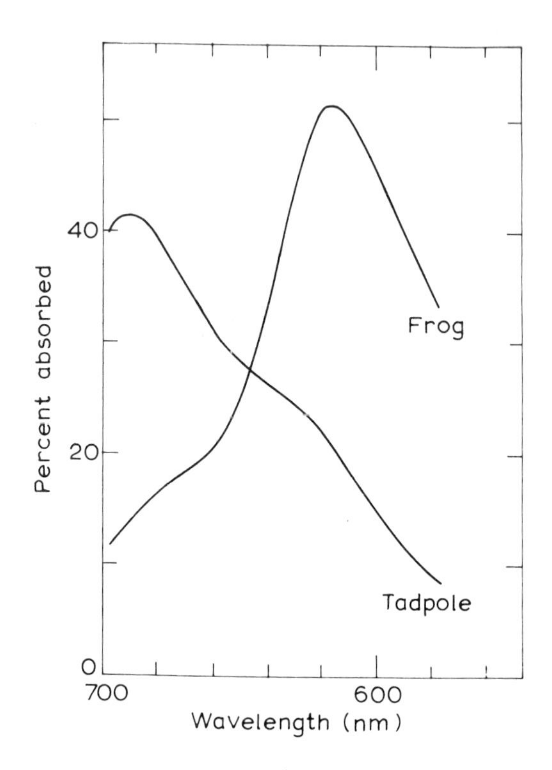

Fig. 6. Biochemical metamorphosis of visual systems in the bullfrog, *Rana catesbiana*. The tadpole just entering the metamorphic climax has in its retina vitamin A_2 (that is, pophyropsin) with only a trace of vitamin A_1 and rhodopsin, whereas the newly emerged froglet has just the reverse pattern. (Republished by permission of the Harvey Lectures [12].)

goes through its growth phase in the water. A number of aquatic reptiles, however (alligators, fresh-water snakes, and turtles), fulfill this description nicely.

If these are substantial parallels, and if the spawning environment decides the pattern of visual pigments, then one should expect such an "anadromous" amphibian as the common frog to possess mainly pophyropsin, like a salmon. Yet rhodopsin was originally discovered in the rods of frogs, and for a long period all that we knew of this pigment was learned with frogs.

In this dilemma I turned to a tailed emphibian with the thought that it might display more primitive properties than the tailles types. Adults of the common New England spotted newt, *Diemyctylus* (formerly *Triturus*) *viridescens*, were found to possess porphyropsin exclusively [11]. This brought a first amphibian into the same fold with certain anadromous fishes but left the frogs in a more aberrant position than ever.

On examining bullfrogs in metamorphosis, however, I found that tadpoles just entering the metamorphic climax possess porphyropsin almost entirely, whereas newly emerged frogs have changed almost entirely to rhodopsin [12]. The anatomical metamorphosis, which in this species takes about 3 weeks, is accompanied by this biochemical metamorphosis of visual systems. The bullfrog enters metamorphosis with porphyropsin, like a fresh-water fish, and emerges with rhodopsin, like a land vertebrate (fig. 6).

These observations have recently been confirmed [13,14]; and Wilt [13] has shown that the metamorphosis of visual pigments in the bullfrog can be stimulated to occur prematurely by treatment with thyroxine. A similar metamorphosis has also been observed in the Pacific tree frog, *Hyla regilla* [14]. On the other hand, several instances have been recorded in which the visual pigment does not appear to change at anatomical metamorphosis: the frogs, *Rana esculenta* and *temporaria,* [15] and the toad, *Bufo boreas halophilus* [14].

Certain amphibia, therefore, like euryhaline fishes, display both the rhodopsin and porphyropsin systems. It seemed for a time that one difference between both groups might be that in euryhaline fishes the patterns of visual system are fixed, whereas in amphibia they change abruptly with metamorphosis. I shall have more to say of this later.

4. **Biochemistry of metamorphosis**

At the time the metamorphosis of visual systems was discovered in the bull-frog, another such change in the same species had already been described. McCutcheon [16] had found that the properties of hemoglobin in this animal change markedly at metamorphosis. The oxygen equilibrium curve of hemoglobin, measured at one temperature and pH, goes through a remarkable transition between tadpoles and adults. The hemoglobin of tadpoles has a high affinity for oxygen, and it seemed from McCutcheon's measurements that the shape of its oxygen equilibrium curve might be hyperbolic, whereas the hemoglobin of young adults has a relatively low affinity for oxygen, and its equilibrium curve is distinctly S-shaped.

Riggs [17] re-examined this situation in our laboratory. He confirmed McCutcheon's finding of a striking loss of oxygen affinity at metamorphosis. He found, however, that the shape of the oxygen equilibrium curve does not alter at metamorphosis; it is equally sigmoid throughout development. He found another important change: tadpole hemoglobin exhibits almost no loss of oxygen affinity on acidification (that is, no Bohr effect), whereas frog hemoglobin has a very large Bohr effect.

It is clear therefore that hemoglobin, like the pigment of rod vision, metamorphoses in the bullfrog at the time of anatomical metamorphosis. Both substances are conjugated proteins. In the rod pigment, it is the prosthetic group, retinene, which changes; the protein opsin, so far as known, remains unaltered. In hemoglobin it is the protein, globin, which changes; the prosthetic group, heme, is the same always.*

Frieden et al. [22] have described a third change in the proteins of this species at metamorphosis. In the bullfrog tadpole the predominant proteins of the blood plasma are globulins. At metamorphosis, the protein concentration of the plasma doubles, and albumins become predominant. These changes can be induced prematurely, just as can anatomical metamorphosis, by administering tri-iodothyronine.

Still another type of biochemical change has been shown to accompany metamorphosis in frogs and salamanders. Fishes excrete most of their nitrogen as ammonia, whereas land vertebrates excrete their nitrogen primarily as urea or uric acid. Munro [23] showed some years ago that whereas the tadpoles of the frog, *Rana temporaria,* excrete the great bulk of their nitrogen as ammonia, at the metamorphic climax this animal goes over to excrete its nitrogen primarily as urea. At this time, also, arginase, the last in the chain of enzymes that forms urea, makes its first appearance in the liver.†

Munro [25] has demonstrated similar changes accompanying metamorphosis in the toad *Bufo Bufo*, the salamanders *Triturus vulgaris* and *T. cristatus*, and the axolotl *Siredon mexicanum*.

* Such changes in hemoglobin seem to present a fundamental property of vertebrates, for they penetrate to the most primitive forms. Adinolfi, Chieffi and Siniscalco [18] have reported that larvae of the lamprey, *Petromyzon planeri*, possess 2 hemoglobins, and on metamorphosis change to 2 others. Sometimes they found all 4 hemoglobins together in metamorphosing animals. Manwell [19] has reported finding different hemoglobins in postlarval and adult California sculpins (*Scorpaenichthys marmoratus,* a marine teleost). He states also that preliminary experiments on a live-bearing surf perch (*Embiotica lateralis*) reveal a fetal form of hemoglobin. Evidence is beginning to emerge also that certain euryhaline fishes have mixed hemoglobin patterns as adults that parallel their possession of mixed visual systems; for 2 to 3 hemoglobins have been found in each of a variety of salmons, trout and shad; whereas a number of permanently marine and fresh-water fishes each possessed a single hemoglobin [20,21]

† Brown and Cohen [24] have analyzed in detail the appearance at the time of metamorphosis in the bullfrog, *Rana catesbiana*, of the enzyme systems concerned with urea synthesis.

These observations make a beginning with the biochemistry of metamorphosis. They show that just as animals in metamorphosis undergo radical alterations in anatomy, so their biochemistry is fundamentally revised at the same time. Indeed both kinds of change, anatomical and biochemical, herald an ecological transition, for they are followed by radical changes of habitat. They mark also an evolutionary transition, for these changes offer the most striking instances we know of recapitulation. The amphibian in metamorphosis seems to repeat in rapid summary the changes which accompanied the emergence of vertebrates from fresh water onto land. The transformations of visual systems and of the patterns of nitrogen excretion seem to provide clear instances of *biochemical recapitulation*. The changes in hemoglobin also seem to involve aspects of recapitulation [11,26]. Whether the changes in serum proteins have this character, it is too early to say. In any case, in metamorphosis the anatomy, the biochemistry, and, shortly afterward, the ecology all are transformed, and frequently in some degree of accord with the animal's evolutionary history.

It is interesting to realize how closely these patterns hold together. An aberration in one of them seems to call forth appropriate aberrations in the others. The mud puppy, *Necturus maculosus*, for example, remains to some degree a permanent larva, never losing its external gills and never emerging from the water. Some years ago I found that adult mud puppies have porphyropsin alone, like a fresh-water fish [11].

The clawed toad, *Xenopus laevis*, a member of the peculiar family Aglossa, which possesses neither tongue nor teeth, is a purely aquatic form, which, though it metamorphoses, ordinarily never emerges from the water. Adults of this species possess in their retinas vitamin A_2 and porphyropsin almost exclusively [27,28]. Underhay and Baldwin [29] have shown that this species also exhibits peculiar changes in the pattern of its nitrogen excretion. As a tadpole it excretes nitrogen primarily as ammonia. At metamorphosis, like other amphibia, it begins to change over toward urea excretion, so that at the height of metamorphosis it excretes a little more nitrogen as urea than as ammonia. Toward the end of metamorphosis, however, it swings back again, so that the adult excretes about 3 times as much ammonia as urea nitrogen. It is as though this animal, having got ready to leave the water, changed its mind; and both the getting ready and the change of mind are reflected in the nitrogen excretion. Indeed, *Xenopus* can change its mind again; for if kept moist while yet out of water it accumulates huge amounts of urea, perhaps as a device for conserving water such as is practiced by the elasmobranch fishes. Its return to water is attended by a massive excretion of urea accompanied by very little ammonia [30].

We see therefore, that even the aberrations of amphibian metamorphosis, anatomical and ecological, are paralleled closely by the biochemistry. It is probably true that in all cases in which the anatomy or the ecology changes, the biochemistry also changes. Indeed the biochemistry may have a primary status; the visible alterations in anatomy and ecology may only reflect prior biochemical changes.

5. Second metamorphosis

The first requirement of a life cycle is that it be *circular.* An organism that leaves its natal environment to explore, or grow up in, another must return at maturity to reproduce its kind. The spawning environment is fixed, whatever excursions animals may take as adults, and it is a truism that all animals must return to their natal environment to spawn.

For this reason, any animal that undergoes profound changes preparatory to migrating from its natal environment is likely to undergo a second series of changes *in the reverse direction* before returning. *Every metamorphosis invites a second metamorphosis.*

Let us begin with the common spotted newt mentioned above. This animal begins its life as an olive-green, gilled larva, living wholly in the water. After several months it metamorphoses to a lung-breathing, land-dwelling eft. The color changes to a brilliant orange-red, the skin becomes rough and dry, the lateral-line organs recede. The newt now lives 2 to 3 years wholly on land, growing meanwhile almost to full size. Then it undergoes a second metamorphosis: the color returns approximately to that of the larva, and the newt regains the wet, shiny, mucus-covered skin, the keeled tail, the functional lateral-line organs, though not, of course, gills. In this mature state it re-enters the water to spawn and live out the remainder of its life [31,32].

Many anatomical and behavioral aspects of the second metamorphosis can be induced prematurely in red efts by injection or implantation of anterial pituitary preparations [32,33]; and a significant part of this complex of changes – the drive to re-enter the water, and the moult to a smooth, wet-skin – is stimulated in hypophysectomized red efts by injections of prolactin, the lactogenic hormone of the anterior pituitary [34].

I have already said that the mature animal possesses porphyropsin, like a fresh-water fish. These animals, however, had already undergone the second metamorphosis. Red efts on examination were found to possess mixtures of rhodopsin and porphyropsin, predominantly rhodopsin (fig. 7). The second metamorphosis in this species is accompanied therefore by the biochemical

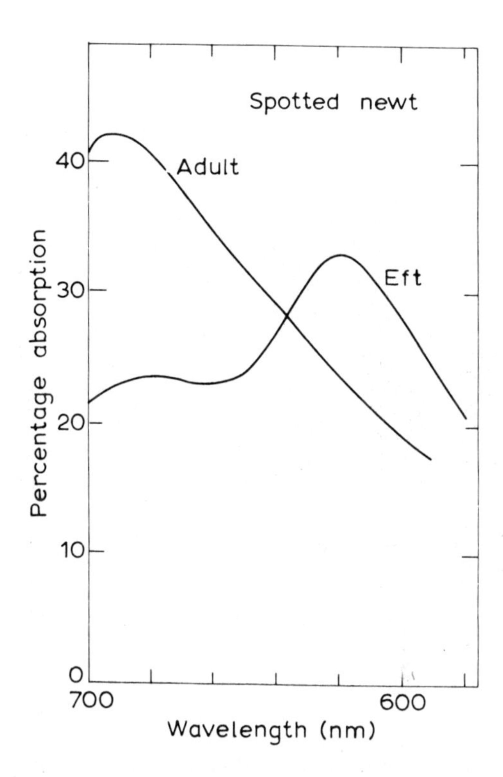

Fig. 7. Second metamorphosis of visual systems in the New England spotted newt, *Diemyctylus viridescens*. Retinas of the land-living red eft contain a preponderance of vitamin A_1, with a minor admixture of A_2; retinas of water-phase, sexually mature adults contain vitamin A_2 predominantly or exclusively. (See Wald [11].)

metamorphosis of its visual system from a predominantly land type to that characteristic of fresh-water types [11].

In extension of the present argument. Nash and Frankhauser [35] have examined the nitrogen excretion of this animal. Like other amphibia already mentioned, the larval newt excretes about 90 per cent of its total nitrogen as ammonia. At the first metamorphosis, it goes over to excreting urea, and the red eft excretes almost 90 per cent of its nitrogen in this form. Then, at the second metamorphosis, it turns back again, so that in the adults, about one-fourth of the total nitrogen is excreted again as ammonia.

This, in turn, brings us back to the sea lamprey. This animal has a life cycle much like that of a salmon (fig. 8). After passing 4 to 5 years as a blind am-

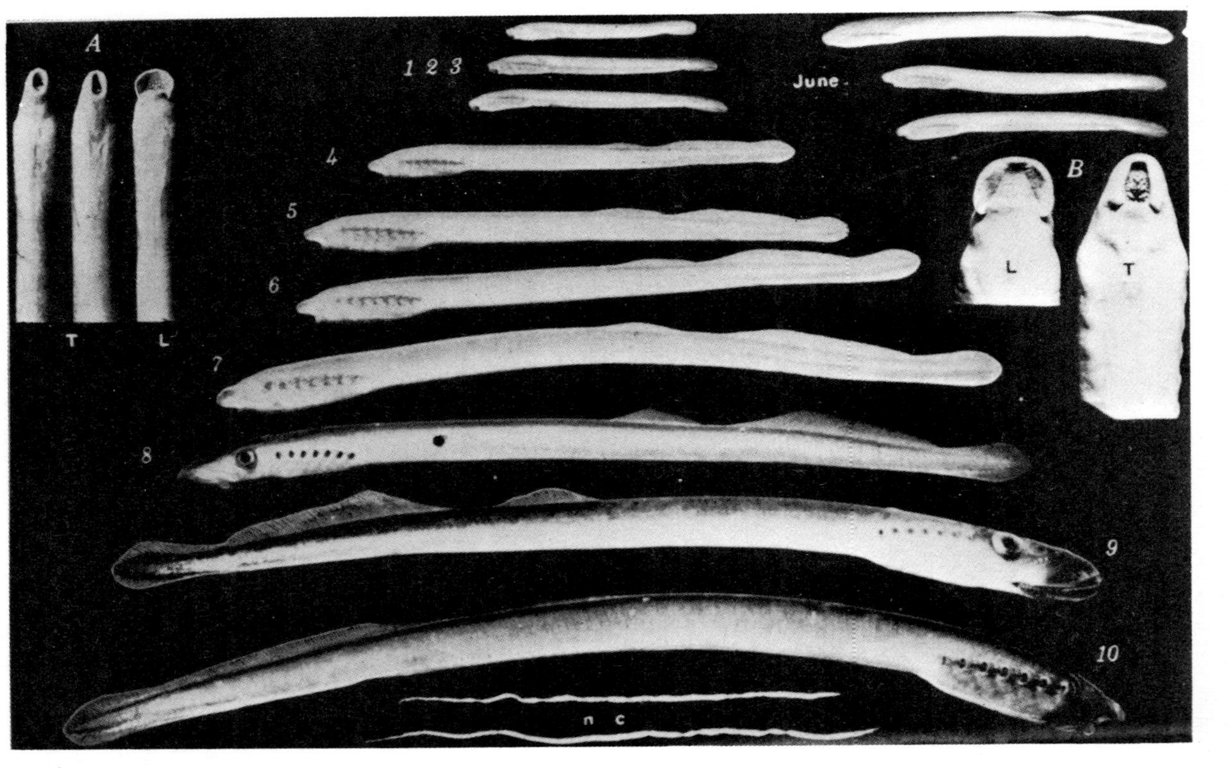

Fig. 8. Development of the sea lamprey, *Petromyzon marinus*. This animal begins its life in streams as a blind larva, buried in mud or sand (stages 1 to 7), Then it undergoes a first metamorphosis while still in this position (stages 8 to 10), preparatory to migrating downstream. Several years later it undergoes a second metamorphosis, to the sexually mature adult, and migrates upstream again to spawn and die. (A,B) Transformation of the mouth at first metamorphosis from the larval, hooded form (L) to the contracted circular form (T). In (c) Notochords of decayed adults found in streams after spawning. (Republished by permission of the New York State Conservation Department [36].)

mocoete larva, living buried in the sand or mud of its natal stream, it undergoes, while still in that position, a profound metamorphosis, preparatory to migrating downstream to the ocean or a lake for its growth phase. This lasts 1½ to 3½ years. Then the sea lamprey undergoes a second metamorphosis, to the sexually mature adult. The sexes differentiate visibly for the first time: the gonads mature, secondary sex characteristics appear, the male develops a rope-like ridge along the back, and either sex may assume golden mating tints. Then these animals migrate upstream to spawn [36].

Some years ago I found that the sea lamprey, taken on its spawning migration, has almost exclusively vitamin A_2 in its retina, and I concluded from this that it probably possesses porphyropsin [10]. Since lampreys are members of the ancient class Agnatha, which includes the most primitive living vertebrates, I took this observation to support the view that porphyropsin is the ancestral type of visual pigment in vertebrates.

Then, however, Crescitelli [37] reported that he had extracted *rhodopsin* from the retinas of this species and pointed out that this goes better with the opposed view, that rhodopsin is the primitive vertebrate pigment.

The specimens of sea lamprey examined by Crescitelli had just metamorphosed from the larval condition and had begun to migrate downstream, whereas the ones I had examined were at the other end of their life cycle, migrating upstream to spawn. On obtaining downstream migrants like Crescitelli's, I confirmed his observations exactly (fig. 9). The retinas of such animals contain vitamin A and rhodopsin alone. The upstream migrants, however, possess vitamin A_2 and porphyropsin virtually alone [38]. We find therefore in this most primitive group of vertebrates another biochemical example of second metamorphosis, like that previously observed in the newt.

Such second metamorphoses expose fundamental characteristics of the metamorphic process:

1) Both the first and second metamorphoses *anticipate* changes in environment. Ordinarily they occur in the old environment and are completed. There are *preparations* for the new environment, not responses to it.

2) Striking hormonal relationships are associated with these events. It has been known for many years that the first metamorphosis is stimulated in many instances by the thyroid hormone. A number of instances are now known in which phenomena associated with the second metamorphosis are stimulated by hormones of the anterior pituitary, including specifically the lactogenic hormone, prolactin [32–34].

3) Just as the first metamorphosis prepares the animal to leave its natal environment, so the second metamorphosis prepares it to return, completing the life cycle. It is of the essence of a second metamorphosis to reverse in part

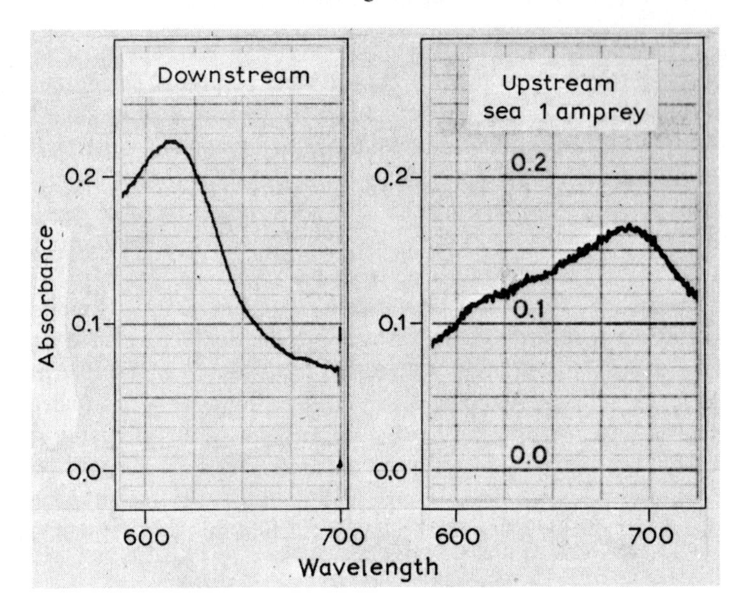

Fig. 9. Spectra of antimony chloride tests with extracts of bleached retinas from (left) lampreys that have just completed their first metamorphosis, and are migrating downstream; and (right) sexually mature lampreys migrating upstream to spawn [38].

the changes which accompanied the first metamorphosis. The 2 metamorphoses tend to be opposed in direction, anatomically and biochemically.

4) Just as the changes in the first metamorphosis tend to have the character of recapitulations – that is, to coincide somewhat with the animal's evolutionary history – so the changes which occur in a second metamorphosis are likely to be antirecapitulatory, to reverse in direction the sequence of changes that accompanied the animal's evolution.

The last consideration involves a potential source of confusion. As I have already said, a life cycle is circular. If one section of it runs parallel with the course of evolution, another section is likely to run counter to that course. Just as every metamorphosis invites a second metamorphosis, so every associated recapitulation invites a subsequent antirecapitulation. This is only proper, provided it occurs at the point in the animal's history when it is being prepared for the return to the natal environment.

6. Deep-sea fishes; eels

Heretofore I have discussed only changes in the visual pigments that involve their prosthetic groups. I should like now to discuss another type of change, involving the other component of a visual pigment, the protein opsin.

Denton and Warren [39,40] reported that the visual pigments of deep-sea fishes, instead of having absorption maxima (λ_{max}) near 500 nm, as do the rhodopsins of surface forms, have λ_{max} near 480 nm. In consequence, they are orange in color rather than red. and Denton and Warren proposed that they be called chryopsins, or visual gold. For reasons which appear below, I prefer to call them deep-sea rhodopsins.

This observation has since been confirmed by Munz [41] and by Wald, Brown and Brown [42]. It makes good ecological sense; for the surface light that penetrates most deeply into clear sea water is blue, and made up of wavelengths near 480 nm, and the rhodopsins of deep-sea fishes are more effective through having their maximal absorption in this region of the spectrum [39,41].

As might be expected, the transition from surface to deep-sea rhodopsin is not sudden. A preliminary exploration shows that the absorption spectra of the rhodopsins shift more or less systematically with depth from the surface to about 200 fathoms. We find that throughout such a series the prosthetic group – the retinene – remains the same. It is the opsin which alters [42]. We have here a relationship comparable with that familiar in the hemoglobins, all of which possess the same heme joined with a variety of globins, different in every species.

Disregarding the relatively few rhodopsins and porphyropsins which lie in exceptional positions, one sees, therefore, a major transition from λ_{max} 480 to λ_{max} 500 nm in the rhodopsins of marine fishes, correlated with depth, and depending on a systematic change of opsins; this connects with a further major transition from rhodopsin to porphyropsin (from λ_{max} 500 to λ_{max} 522 nm) correlated with the transfer to fresh water, and depending on the change of chromophore from retinene$_1$ to retinene$_2$.

With this we can return to the "fresh-water" eel (*Anguilla*) (fig. 11). Carlisle and Denton [43] have confirmed our observation that this animal, when taken in fresh water, ordinarily possesses the mixture of rhodopsin and porphyropsin described earlier; but they find that toward the beginning of its spawning migration it goes over to deep-sea rhodopsin (fig. 10). Whereas the absorption peak of its usual mixture of visual pigments, when the eel is in fresh water, lies at about 500 nm, that of the animal about to migrate lies close to 485 nm. Indeed, the rhodopsin of such a "fresh-water" eel prepara-

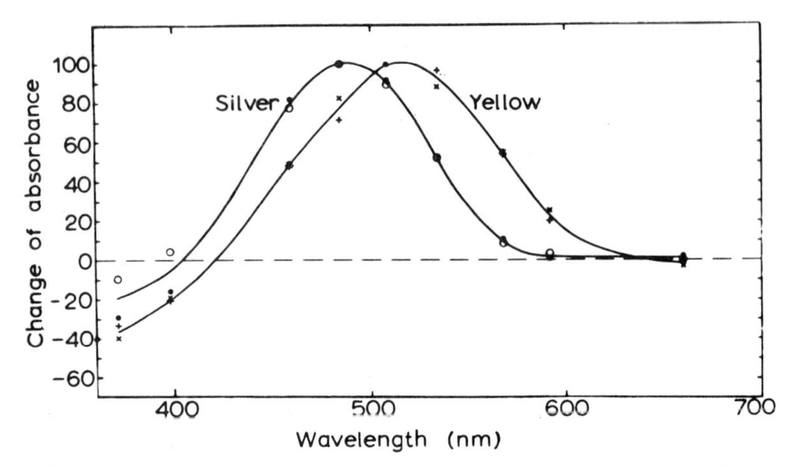

Fig. 10. Second metamorphosis of visual pigments in the European "fresh-water" eel, *Anguilla anguilla.* The yellow eel, prior to second metamorphosis, has a mixture of rhodopsin and porphyropsin (net λ_{max} 500 to 505 nm) corresponding to the mixture of vitamins A shown in fig. 4. The sexually mature silver eel, about to begin its spawning migration to the Sargasso Sea, has changed over to deep-sea rhodopsin (λ_{max} about 487 nm). (Republished by permission of the Journal of the Marine Biological Association of the United Kingdom [43].)

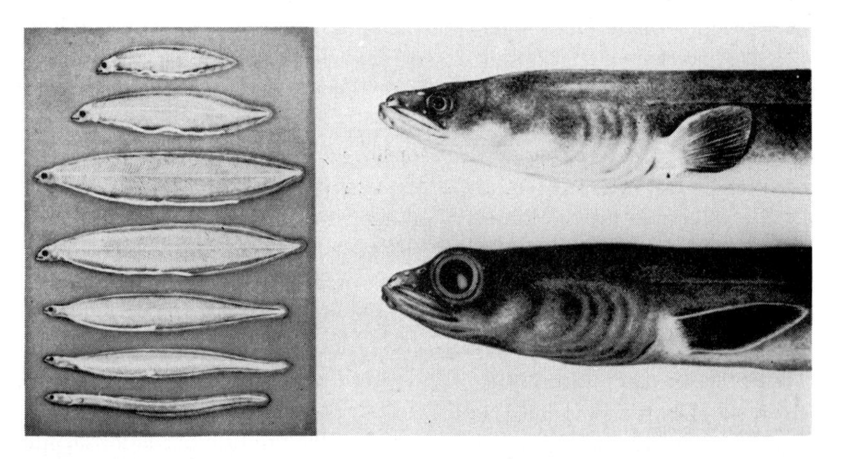

Fig. 11. Development of the American eel, *Anguilla rostrata. Left*: top to bottom, three larvae, or leptocephali, of various sizes; then3 larvae undergoing the first metamorphosis; and the metamorphosed "glass eel". The uppermost 3 pictures are enlarged 2.7 times relative to the others. *Right*: the second metamorphosis: (top) American eel in the "green" stage; (bottom) a European eel in the "silver" stage, ready for its spawning migration. Note particularly in the latter the approximate doubling in diameter of the eye. This large-eyed stage has not yet been observed in the American species. (Republished by permission of the Department of Fisheries, Province of Quebec [45].)

tory to migration is virtually identical in spectrum with that of the permanently deep-sea conger eel [44].

This is another instance of a second metamorphosis (fig. 11). The eel, having been spawned in the depths of the Sargasso Sea, journeys as a larva (leptocephalus) to the shores of America or Europe [46]. There it metamorphoses to the adult from and usually, though probably not always, migrates into fresh water for its growth phase. Eventually it metamorphoses again: its color changes, the eyes approximately double in diameter, the digestive system deteriorates. As though getting ready for its return to the Sargasso Sea, it changes also to deep-sea rhodopsin.

My co-workers, Paul and Patricia Brown, have examined such animals at the Stazione Zoologica in Naples.*

The European eel about to migrate seaward seems first to lose its retinal vitamin A_2 and retinene$_2$ and then to begin to combine vitamin A_1 and retinene$_2$ with a new, deep-sea opsin. The animal has already progressed far with this, as with the anatomical changes of the second metamorphosis, while still in fresh water.

Such observations can tell us something concerning the larval condition. To my knowledge no one has yet examined the visual pigment of the leptocephalus larva, but the foregoing discussion suggests strongly that the pigment is deep-sea rhodopsin. Similarly, though no one seems as yet to have examined the retinal pigment of the larval New England newt, our observation that the adult at maturity metamorphoses to porphyropsin implies that this is also the larval pigment. Again, since the blind ammocoete larva of the sea lamprey metamorphoses to an eyed adult possesssing rhodopsin, this is the first visual pigment to appear in this species. Yet the fact that in the second metamorphosis the pigment changes to porphyropsin implies that the latter represents the true, albeit missing, larval type. That is, since the second metamorphosis involves some measure of return to the larval condition, it can tell us something of the larval state, even of larval properties which have been lost in the course of evolution.

* Experiments in our laboratory and at the Stazione Zoologica in Naples by P.K. Brown and P.S. Brown have demonstrated considerable variation in the proportions of vitamins A_1 and A_2 in the retinas of individual eels of both the American and the European species. These proportions vary between 65:35 and 25:75, with a mean value of approximate equality. The visual pigments, rhodopsin and porphyropsin, are present in approximately the same *molar* ratios; but since rhodopsin possesses a higher specific extinction than porphyropsin, extinction-wise rhodopsin tends to predominate.

7. Land vertebrates

Land vertebrates still pursue their embryogeny in water, but they have brought the water ashore. In a sense they are erstwhile amphibia which have carried water ashore in which their embryos go through the larval stages and first metamorphosis. They have developed special devices for this: the boxed-in or cleidoic egg, and viviparity. Amphibia still experiment with both. Certain of them for example, the American red-backed, slimy, and worm salamanders lay eggs on land within which the larvae complete their entire devellopment. Others such as the European black salamander, *Salamandra atra*, retain the eggs in the body until the young are fully formed. The European spotted salamander, *S. maculosa*, ordinarily lays its eggs in streams, but if it cannot reach water, permits them to develop internally.

One might hope, therefore, to find residues of metamorphosis in the embryogeny of land vertebrates, and in this one is not disappointed. Anatomical residues abound; they were the original source of the idea of recapitulation and were principally responsible for its nearly overexuberance. The embryo of land vertebrate undergoes an anatomical metamorphosis approaching that of an amphibian. Unlike a larval amphibian, it never has functional gills; but, for a time, it does of course have gill slits, as well as other evidences of earlier aquatic life.

One finds biochemical metamorphosis also in the embryos of land vertebrates, and it includes some of the same changes with which metamorphosis in amphibia has already made us familiar.

So, for example, the chick embryo developing in the egg displays a changing pattern of nitrogen excretion which mimics to a degree the evolutionary sequence. Measurements by Fisher and Eakin [47,48] in the whole egg (fig. 12) show that the ammonia content remains almost constant, perhaps even falling slightly, throughout incubation. Ammonia seems never to be "excreted". Urea increases regularly in amount, being not only formed, but excreted selectively into the allantois after the fifth to seventh day. Urea is not formed from added ammonia, ornithine or uric acid; and this seems to be true also of the adult chicken [49]. That is, the chicken seems at no time to possess the enzymatic apparatus for synthesizing urea from ammonia and carbon dioxide. Of the enzymes of the ornithine cycle, it possesses only the terminal member, arginase. The arginase activity per unit weight of tissue declines rapidly, reaching very low values in the fifth to seventh day of incubation [50,51], owing in part to a decrease in the concentration of the enzyme, in part to the formation of an arginase-inhibitor [51].

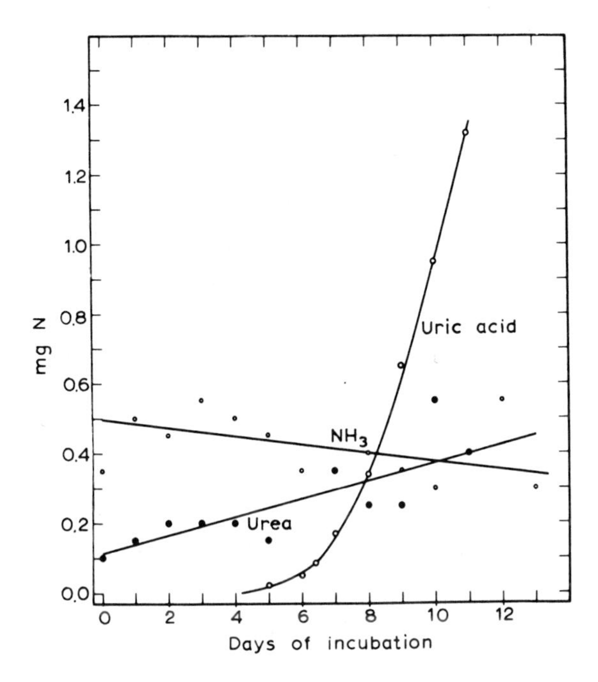

Fig. 12. Ammonia, urea and uric acid content of the incubating chick egg. The ammonia content of the whole egg remains almost constant, perhaps declining slightly. The urea content slowely rises, and after a time (days 9 to 11) about half of it is excreted into the allantois. Uric acid appears in the allantois on about the fifth day, and thereafter accumulates rapidly. (Drawn from measurements by Fisher and Eakin [47].)

At about the same time, uric acid synthesis and excretion begin, and are maintained throughout the further life of the organism.

A second example: In general, vertebrates hold the osmotic pressures of their body fluids within narrow limits. In the various groups of vertebrates the blood osmotic pressure takes characteristic values, correlated to a degree with the ecology, and perhaps also with the phylogeny (see [11]). Thus, the freezing-point depression (Δ F.P.) of the plasma in fresh-water fishes and amphibia lies at 0.45 to 0.55°, whereas adult birds and mammals exhibit values of 0.55 to 0.65°. Measurements in the developing chick [52] show that the fluids of the early embryo have freezing-point depressions of about 0.47°. (fig. 13). The fluid osmotic pressure rises throughout development, with a final spurt at hatching that brings it to the adult value (Δ F.P. = 0.58°). That is, the embryo begins with a fluid osmotic pressure characteristic of fresh-

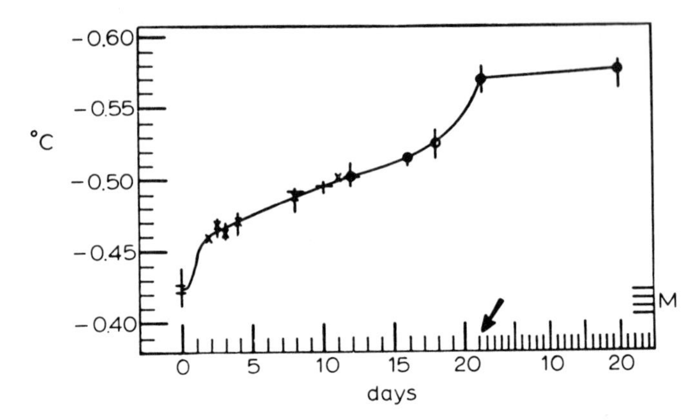

Fig. 13. Biochemical metamorphosis of fluid osmotic pressure in the developing chick. *Ordinates*: freezing-point depression, a measure of osmotic pressure. *Abscissae*: days of incubation, to hatching on the 21st day (arrow), and days thereafter; (=) unincubated egg white, (X) subgerminal fluids, (–) amniotic fluids, (o) bloods. The embryo begins with osmotic pressures characteristic of fresh-water fishes and amphibia and ends with the much higher osmotic pressure characteristic of mature birds and mammals. The duration of functional activity of the mesonephros (M) is also indicated. Mesonephros is indicated by an arrow. (Republished by permission of the Journal of Cellular Physiology [52].)

water fishes and amphibia and ends with that characteristic of mature birds.*

A third example: Hall [56] has shown that during the embryonic development of the chick its hemoglobin changes radically, continuously losing affinity for oxygen, so that in an adult chicken more than twice as much oxygen pressure is needed for half-saturation as is needed in a ten-day-old chick (fig. 14). These changes persist for some time after hatching. They are similar in direction to the change in hemoglobin that accompanies metamorphosis in the bullfrog.

Comparable changes in hemoglobin accompany the embryonic development of all mammals so far examined. It is now well recognized that in mam-

* This may be one example of a much more general phenomenon. In the frog *Rana temporaria*, the ovarian eggs have an osmotic pressure like that of adult blood (Δ F.P. = 0.41°). Within a few hours after fertilization this has fallen to about 0.33°, and in the gastrula stage reaches the extraordinarily low minimum of 0.275°. Then it rises again, so that toward the end of the first week of development it again approaches the adult level [53–55]. I hardly know whether these changes in frogs and chicks are properly to be described as "metamorphoses". They may come too early in development and may be too continuous for that. I include them tentatively in this discussion in the hope that further examination will clarify their status.

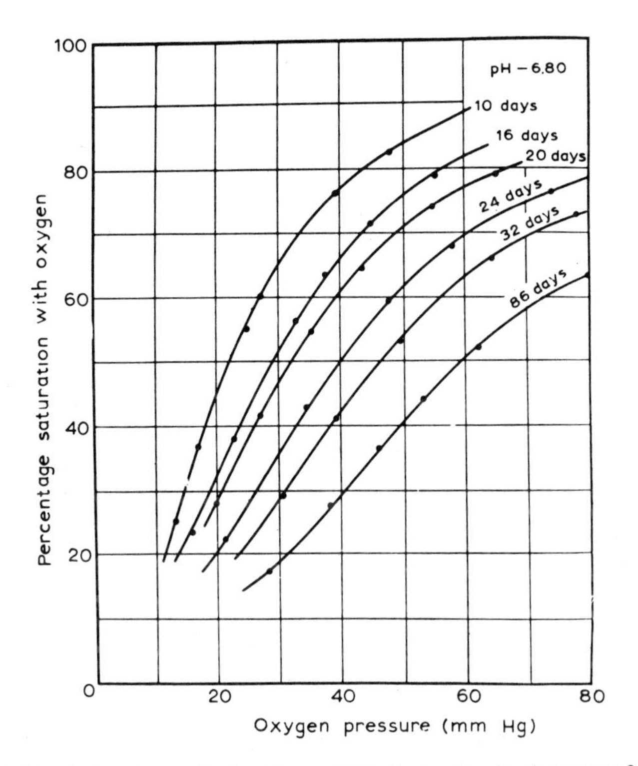

Fig. 14. Biochemical metamorphosis of hemoglobin during the development of the chick. Measurements on dilute solutions of hemoglobin, buffered at pH 6.80, and equilibrated with oxygen at 37°. The affinity for oxygen decreases regularly from the tenth day of incubation, and this change continues for some time after hatching. (Republished by permission of the Journal of Physiology [57].)

mals generally, man included, fetal hemoglobin is a different species of molecule from maternal or adult hemoglobin (see [57]). Always – with the possible exception of man – the change in oxygen affinity is in the same direction, a loss of affinity as development progresses (fig. 15). The fetal and adult hemoglobins of mammals differ also in many other ways: in electrophoretic mobility, sedimentation rate, resistance to alkali, immunological specificity, solubility, crystal shape, and amino acid composition (for references see [11]). All these changes involve the globin moiety of hemoglobin; the heme is the same always.

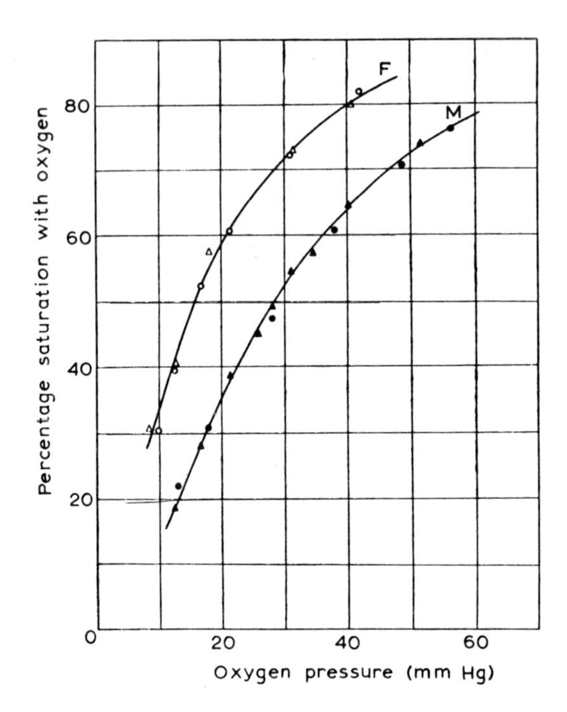

Fig. 15. Biochemical metamorphosis of hemoglobin in a placental mammal. Oxygen equilibrium curves of hemoglobin from a goat fetus (F) and from the mother (M). Both hemoglobins were obtained at 15 weeks (triangles) and 18 weeks (circles) gestation, and measured in solution at 37° and pH 6.8. The fetal hemoglobin has almost twice as high an affinity for oxygen as the maternal hemoglobin. (Republished by permission of the Journal of Physiology [57].)

The phenomenon of metamorphosis, biochemical as well as anatomical, extends therefore beyond the amphibia and fishes to include the land vertebrates, both egg-laying and placental.

Do land vertebrates exhibit also vestiges of a second metamorphosis? I suppose that *puberty* is so to be regarded. To be sure, this does not prepare a land vertebrate to migrate, for the natal environment is now segregated, and puberty prepares the animal only to mate. Here only one representative cell – the spermatozoon – completes the return to the natal environment; and this, of course, undergoes a profound metamorphosis before being launched upon a migration as formidable, relative to its size, as that of any salmon.

8. L'Envoi

I have just gone over with you one of the most exciting experiences of my scientific life; one that excites me again as I review it. I stopped doing such experiments some years ago, in order to go on to other things. After I stopped, something began to happen that happens often enough, and is important enough, so that I should like to say a word about it.

The entire point of science is not to amass facts, but to achieve generalizations. What one is preparing in science is not a catalog or handbook, but a fabric of relationships. Facts are only the raw material of science; the generalizations to which they had lead, the relationships in which they are involved, are all that matter.

If one is fortunate in one's work, it begins to fall into a pattern; and out of that pattern comes a generalization. As soon however as one has stated a generalization that becomes a target, there is little fame or fortune or even elation in confirming it. All the fun now is in destroying the generalization. To make that more worthwhile, one begins by dignifying it: one calls it "classic". That means it's wrong, but old enough to be forgiven. Then one announces that one has found an exception, or better still a number of exceptions to the generalization. So one has brought down big game, and can display the horns, or even blow them.

One of the nicest things that ever happened to me was to find my own exception. I have already told about it. It was certain Labrid or Wrasse fishes, which in spite of being altogether marine, possess the fresh-water visual pigment porphyropsin. That was a fine thing, not only because it was an exception, but because it said something, as every exception does. It demonstrated that I was dealing with a genetic character, relatively independent of the immediate environment.

Lately, as mentioned above, a number of other exceptions have been found to the general distribution of vitamins A_1 and A_2 among the fishes. Numbers of fresh-water fishes, for example, have now been found whose vision is based upon vitamin A_1 and rhodopsin, sometimes exclusively so. Such exceptions are very interesting, very important; but not, as sometimes said, because they show that "Wald was wrong", and that his work, sadly, has become "classic". My being wrong is of no interest at all. But every exception is trying to say something, trying indeed to point the way to a new and more complete generalization. To destroy a generalization is only a violence. The job is to find the new generalization. That is the only real point in exceptions: every exception invites a new generalization.

I am sure that that will prove to be true in this case also. Indeed, I have an idea where I would look. Since I cannot be sure of doing the experiments myself; let me say what I would try first, and why.

You see, we found another exception to our generalization. We found a fresh-water turtle with porphyropsin and vitamin A_2, though fresh-water turtles like other reptiles spawn on land. Add at once that marine turtles, equally of terrestrial origin, use rhodopsin and vitamin A_1 in vision. What could that be about?

There is something curious about fresh-water turtles, those that have been measured. They have extraordinarily low blood osmotic pressures, as low as those of fresh-water fishes. Marine turtles and land tortoises have much higher blood osmotic pressures, like those of marine fishes and land vertebrates. That made me wonder some time ago whether the blood osmotic pressure might possibly be the independent variable that sets the pattern of vitamins A_1 and A_2 among vertebrates. The few measurements I know of the blood osmotic pressures of Labrid fishes also seem to put them rather low among marine fishes.

So, given the opportunity, what I would do is to make accurate measurements of blood osmotic pressure, and see how they correlate with the use of vitamins A_1 and A_2 in vision. That would be interesting to do whichever way it came out.

Meanwhile, that old "classic" generalization is not doing so badly. The chances are somewhere between 0.8 and 0.9 that if you draw a fish out of fresh water, knowing nothing else about it, it has exclusively vitamin A_2 and porphyropsin in its retina; and if out of the ocean, it has vitamin A_1 and rhodopsin. That's what it is to be "classic". Yet the new excitement is in the exceptions, and their promise of that new generalization, for which all of us are waiting.

References

[1] E. Köttgen and G. Abelsdorff, Z. Physcol. u. Physiol. 12 (1896) 161.
[2] G. Wald, Am. J. Ophthalmol. 40 (1955) 18.
[3] G. Wald, in: Enzymes: Units of Biological Structure and Function, ed. O.H. Gaebler (Academic Press, New York, 1956) p. 355.
[4] R.A. Morton and G.A.J. Pitt, Fortschr. Chemie Org. Naturstoffe 14 (1957) 244.
[5] G. Wald, Exptl. Cell Res., Suppl. 5 (1958) 389.
[6] G. Wald, J. Gen. Physiol. 22 (1939) 391.
[7] G. Wald, J. Gen. Physiol. 22 (1939) 775.
[8] G. Wald, J. Gen. Physiol. 25 (1941) 235.
[9] C.D.B; Bridges, Vision Res. 5 (1969) 223, 229; S.A. Schwanzara, Vision Res. 7 (1967) 121.

[10] G. Wald, J. Gen. Physiol. 25 (1942) 331.
[11] G. Wald, in: Modern Trends in Physiology and Biochemistry, ed. E.S.G. Barron (Academic Press, New York, 1952) p. 337.
[12] G. Wald, Harvey Lectures 41 (1945) 117.
[13] F.H. Wilt, Develop. Biol. 1 (1959) 199.
[14] F. Crescitelli, Ann. N.Y. Acad. Sci. 74 (1958) 230; and, in: Photobiology (Proc. 19th Ann. Biol. Colloq., Oregon State College, 1958) p. 30.
[15] F.D. Collins, R.M. Love and R.A. Morton, Biochem. J. 53 (1953) 632.
[16] F.H. McCutcheon, J. Cell. Comp. Physiol. 8 (1936) 63.
[17] A.F. Riggs, J. Gen. Physiol. 35 (1951) 23.
[18] M. Adinolfi, G. Chieffi and M. Siniscalco, Nature 184 (1959) 1325.
[19] C. Manwell, Science 126 (1957) 1175.
[20] D.R. Buhler and W.E. Shanks, Science 129 (1959) 899.
[21] K. Hashimoto and F. Matsuura, Nature 184 (1959) 1418.
[22] E. Frieden, A.E. Herner, L. Fish and E.J.C. Lewis, Science 126 (1957) 559.
[23] A.F. Munro, Biochem. J. 33 (1939) 1957.
[24] G.W. Brown, Jr. and P.P. Cohen, in: Chemical Basis of Development, eds. W.D. McElroy and B. Glass (Johns Hopkins Press, Baltimore, 1958) p. 495.
[25] A.F. Munro, Biochem. J. 54 (1953) 29.
[26] G. Wald and D.W. Allen, J. Gen. Physiol. 40 (1957) 593.
[27] H.J.A. Dartnall, J. Physiol. 125 (1954) 25.
[28] G. Wald, Nature 175 (1955) 390; H.J.A. Dartnall, J. Physiol. 134 (1956) 327.
[29] E.E. Underhay and E. Baldwin, Biochem. J. 61 (1955) 544.
[30] M.M. Cragg: Ph.D. Thesis. University of London, 1953. Cited in: Underhay and Baldwin [29].
[31] G.K. Noble, American Museum Novitates 228 (1926); 348 (1929).
[32] A.B. Dawson, J. Exptl. Zool. 74 (1936) 221.
[33] E.E. Reinke and C.S. Chadwick, J. Exptl. Zool. 83 (1940) 224.
[34] W.C. Grant, Jr. and J.A. Grant, Biol. Bull. 114 (1958) 1.
[35] G. Nash and G. Frankhauser, Science 130 (1959) 714.
[36] S.H. Gage, in: Biological Survey of the Oswego River System (Suppl. 17th Annual Report, New York State Conservation Department, 1927) p. 158.
[37] F. Crescitelli, J. Gen. Physiol. 39 (1956) 423.
[38] G. Wald, J. Gen. Physiol. 40 (1957) 901.
[39] E.J. Denton and F.J. Warren, Nature 178 (1956) 1059.
[40] E.J. Denton, J. Marine Biol. Assoc. U.K. 36 (1957) 651.
[41] F.W. Munz, Science 125 (1957) 1142.
[42] G. Wald, P.K. Brown and P.S. Brown, Nature 180 (1957) 969.
[43] D.B. Carlisle and E.J. Denton, J. Marine Biol. Assoc. U.K. 38 (1959) 97.
[44] E.J. Denton and M.A. Walker, Proc. Roy. Soc. London Ser. B 148 (1958) 257.
[45] V.D. Vladykov, in: Fishes of Quebec (Album no. 6, Department of Fisheries, Province of Quebec, 1955).
[46] J. Schmidt, Ann. Rep. Smithsonian Inst. (1924) 279; Phil. Trans. Roy. Soc. London Ser. B 211 (1922) 179.
[47] J.R. Fisher and R.E. Eakin, J. Embryol. Exptl. Morphol. 5 (1957) 215.
[48] R.E. Eakin and J.R. Fisher, in: Chemical Basis of Development, eds. W.D. McElroy and B. Glass (Johns Hopkins Press, Baltimore, 1958) p. 514.

[49] J. Needham, J. Brachet and R.K. Brown, J. Exptl. Biol. 12 (1935) 222.
[50] J. Needham and J. Brachet, Compt. Rend. Soc. Biol. Paris 18 (1935) 840.
[51] M. Ceska and J.R. Fisher, Biol. Bull. 117 (1959) 611.
[52] E. Howard, J. Cell. Comp. Physiol. 50 (1957) 451.
[53] A. Krogh. K. Schmidt-Nielsen and E. Zeuthen, Z. Vergleich. Physiol. 26 (1938) 230.
[54] E.L. Backman and J. Runström, Biochem. Z. 22 (1909) 290.
[55] K. Bialaszewicz, Arch. Entwicklungsmech. Organ. 34 (1912) 489.
[56] F.G. Hall, J. Physiol. 83 (1934) 222.
[57] F.G. Hall, J. Physiol. 82 (1934) 33.

THE EVOLUTIONARY INFORMATION CONTENT OF PROTEIN AMINO ACID SEQUENCES

E. MARGOLIASH and W.M. FITCH

Department of Molecular Biology, Abbott Laboratories, North Chicago, Illinois 60064, and Department of Physiological Chemistry, University of Wisconsin, Madison, Wisconsin 53706, USA

Abstract: Margoliash, E. and Fitch, W.M. The Evolutionary Information Content of Protein Amino Acid Sequences. *Miami Winter Symposia* 1, pp. 33–51. North-Holland Publishing Co., Amsterdam, 1970.

The comparative study of amino acid sequences of proteins can lead to knowledge of evolutionary, developmental and genetic mechanisms and relations. Illustrations are taken from eukaryotic cytochromes *c*, and the generalizations are applied to other proteins.

Definitions of evolutionary analogy and homology, as applied to protein structure, and a description of statistical methods which determine homologous proteins and phylogenetic trees from orthologous proteins are given.

Such trees lead to the development of the complete mutation pathways relating the proteins of extant species back to various intermediate ancestral forms at the different branch points of the phylogenetic trees, and thus, in turn, to the amino acid sequence of the protein in the ancestral form common to all extant species. Statistical tests on model cases indicate that the procedure which accounts for the proteins of descendents in the fewest number of nucleotide replacements does indeed give good estimates of the mutation pathways followed and of the structure of ancestral forms of the protein. If this is done for two sets of orthologous sequences, it is possible to distinguish between convergent and divergent evolutionary processes.

From statistical phylogenetic trees for cytochrome *c* and fibrinopeptides it appears that in the descent of these proteins the nuclotide replacement frequencies are not random. Rather, there is a marked preponderance of guanine to adenine evolutionary fixations over and above the frequency expected on the assumption that all nucleotide replacements are equiprobable.

It is also possible to estimate the number of residues in the set of proteins examined which must remain strictly invariant for biological reasons. For cytochrome *c*, over the entire range of species examined to date, this number is near 30. However, as the calculation is done on a narrower and narrower taxonomic range, the number of invariant residues increases to a limit of about 90% of the peptide chain for present day mammalian cytochromes *c*. Thus only about 10 of the residues in cytochrome *c* are concomitantly variable which explains why this protein is so evolutionarily conservative; a single fixation occurring in two diverging lines of descent, on the average, every 26 million years. Fibrinopeptides, at the other extreme, vary very rapidly, evolutionary fixations occurring every million years, and 18 of the 19 residues are concomitantly variable. The rate of fixation of evolutionary variations, calculated per concomitantly variable residue, appears to be the same for both cytochrome *c* and fibrinopeptides, within the error of such estimates, which suggests that there exists a fundamental rate of protein evolution, related only to the number of residues capable of undergoing changes.

1. Introduction

It has long been obvious that because of the processes by which life perpetuates itself, living organisms are an excellent repository of the evidence of their own evolutionary history. Every material of which an organism is composed and every phase of its activities are results of that history. However, as pointed out by Zuckerkandl and Pauling [1], some biological substances retain the traces of the past in a relatively easily identifiable form, while for others the relation to evolution is much more difficult to discern. There is in this regard a very fundamental difference between so-called "informational macromolecules", DNA, RNA and proteins, and the other substances found in living organisms. The former are simple images of each other in which the linear sequence of chemical building blocks carries the biological information, so that whether one determines the amino acid sequence of a protein chain or the structure of a transfer RNA molecule, one is merely examining the fine structure of a very small segment of the genome at its simplest molecular level. This simplicity is the crucial advantage. All other biological substances which are elaborated by the organism represent a far more complex interplay of sources of biological information. For example, chemically simple micromolecular substances, such as flavinoids or any of the intermediates of metabolic energy cycles, are the products of whole assembly lines of enzymes, each derived from one or more structural genes, each of which is in turn likely to be controlled by one or more so-called regulatory genetic influences. Thus, though the number of genes controlling the synthesis of a micromolecular biological entity may be in the order of 100, far less than the possibly 10^6 genes which may affect a complex morphological character, such as the shape of the human nose, the genetic complexity of the micromolecule is more than enough to give it the same status as that of the ordinary morphological characters employed for classical evolutionary appraisals. The advantage of chemistry, represented by the understanding of the structure of substances at the molecular level, has been entirely lost.

This is not the case with the amino acid sequences of proteins, since the chemical structure is itself an expression of the structure of a gene. Thus, with information on the primary structure of a sufficient number of different proteins from a sufficient number of different and properly chosen species, it may eventually be possible, independently of any other knowledge, to read directly the record of the evolutionary history of these species encoded in the proteins they synthesize. An obvious attraction of the molecular taxonomy of proteins is the possibility of reconstituting today the temporal order of long past evolutionary changes in terms of unit mutational events. This could possibly lead to an estimate of the structures of informational macromolecules, proteins and nucleic acids, as they occurred further and further back to that shadowy point in biological history when chemical evolution ended and replicating biological systems took over. In this process one can expect to obtain a wealth of information concerning evolutionary mechanisms as they relate to protein structure and function. This short review attempts to summarize the present status of the endeavour.

2. The significance of amino acid sequence similarities

Similarities between different proteins in the same species or between ostensibly similar proteins of different species are apparent by any of the large variety of techniques which define their structural and functional parameters. These extend from similarities in tissue and cellular localizations, similarities in physiological and physico-chemical modes of function, all the way to precise details of amino acid sequence and of three-dimensional spatial structure. Since the primary structures are the direct expression of the organism's store of biological information, this paper will concern itself solely with amino acid sequences. However, proteins were classified before their primary structures were known and the search for similarities is still to a large extent limited to groups defined by criteria other than primary structures. This will necessarily exclude descendants of the ancestral form which have varied to the extent of acquiring new functions and the physico-chemical attributes which fit the new functions. It is only when primary structures will have become available for a large proportion of all proteins that it will be possible to discuss relations of proteins which are no longer apparent in their functions. In the meantime, expected similarities in function of proteins that have undergone relatively small divergences, as in the case of the digestive proteolytic enzymes [2,3], or quite unexpected similarities, as in the case of lysozyme and α-lactalbumin [4], have already provided vivid illustrations of the evolutionary shaping at the molecular level of new functions from old structures.

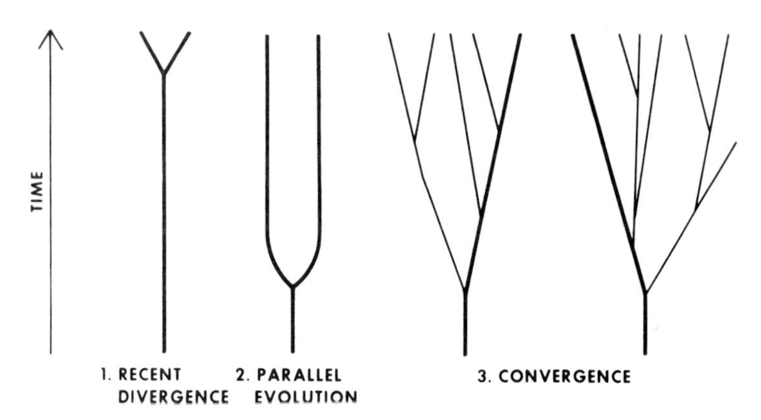

Fig. 1. Possible evolutionary reasons for similarity of polypeptide chains.

However, common ancestry is not the only possible basis for similarity in amino acid sequence. Indeed, two proteins may be similar at the time they are examined not only because they diverged from a common origin relatively recently in their evolutionary history or because having diverged a long time ago they have followed largely parallel pathways, but also because having arisen from different ancestral origins they have tended to evolve to similar or identical functions in different lines of evolutionary descent, and have therefore acquired the degree of similarity of structure required by this similarity of function. These possibilities are diagrammed in fig. 1. Thus, before one can conclude that a set of proteins of apparently similar amino acid sequence has a common evolutionary origin, i.e. are *homologous* in the ordinary biological usage of the term, one must answer two questions, as follows:

1) *Are the similarities of primary structure greater than could occur by chance?*

A systematic approach to this question [5] requires in essence the ability to calculate the probability of random similarity. This can be done by comparing all possible pairs of segments of a fixed length (such as 20 or 30 residues long) between the two protein chains under consideration. For example, in a comparison of two sequences 100 residues long, for segments of 20 residues there are (100-20+1) (100-20+1) or 6561 possible pairs. One can calculate the minimal number of single nucleotide changes required to transform the gene segment coding for one member of each pair into that coding for the other ("mutation" or "replacement distance"), and plot the total number of times each particular replacement distance occurs in all the comparisons as a function of the replacement distance. Such a plot is given in fig. 2 for human

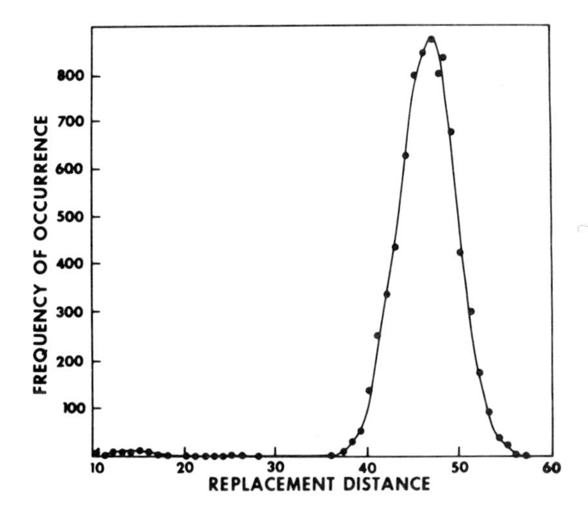

Fig. 2. Comparison of replacement distances for all possible 30 residue segments of human cytochrome *c* [6] and bakers' yeast iso-1- cytochrome *c* [7] by the procedure of Fitch [5]. The number of times various replacement distances occur in the comparisons are given on the ordinate (frequency of occurrence).

and the iso-1-cytochrome *c* of bakers' yeast. The average replacement distance for any randomly chosen pair of amino acids is 1.5, so that for a pair of random 30-residue segments it would be 45. In fig. 2, the random comparisons are given by the Gaussian portion of the curve, very nearly centered, as expected, on a replacement distance of 45. The long tail of the curve to the left represents those comparisons for which the replacement distances are smaller than would be expected on a random basis, and which therefore indicate that the degree of similarity between human and yeast cytochromes *c* is greater than could be accounted for by chance.

Data of the type given in fig. 2 can be recalculated to give cumulative distributions. When these are plotted on probability paper, the Gaussian portion of the curve becomes a straight line, and non-random comparisons are detected as deflections from the linear curve towards the lower left [5]. Such a probit plot is shown for the human-yeast cytochrome *c* comparison in fig. 3. To supplement the graphic comparisons, particularly in cases in which the degree of non-randomness is not as obvious as in the example given in figs. 2 and 3, an arbitrary statistic having the characteristic distribution of χ^2 when unrelated amino acid sequences are compared, can be employed. This permits determination of the probability that a given departure from linearity would

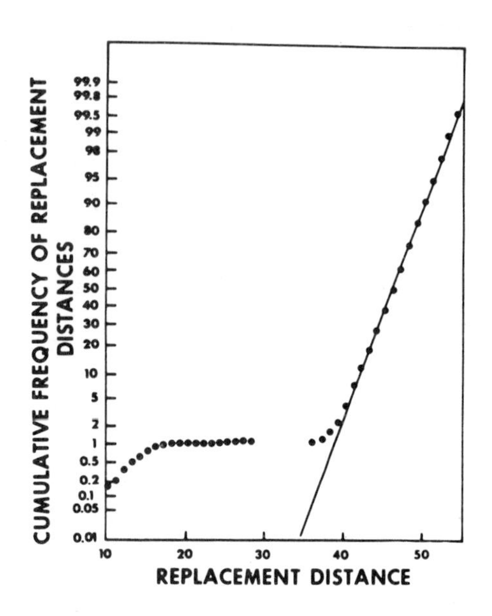

Fig. 3. Probability plot of the data from fig. 2. The random part of the distribution is given by the linear portion of the curve. The comparisons for which the replacement distances are smaller than expected for a random distribution are represented by the points which deviate from the straight line at the lower left. The probability that such a distribution occurred by chance is less than 10^{-80} [8].

occur by chance [8]. Thus, for example, the data in fig. 3 indicate that the probability that such a distribution would occur by chance is less than 10^{-80}.

The proper alignment of two amino acid sequences, for which a degree of similarity greater than random has been established, is to a large extent obtained merely by considering those pairs of segments which yielded the non-random portion of the distribution curve. Furthermore, a method has been devised to locate the gaps required to align two primary structures so as to minimize the total number of nucleotide replacements, deletions and insertions necessary to account for the differences between the sequences [9].

The search for significant similarities need not be limited to different proteins, but may also be usefully conducted with portions of a single polypeptide chain. If such are found, one may reasonably infer that partial internal duplications have occurred during the evolution of the corresponding structural genes. Such phenomena could result from unequal crossing over within one gene, as is considered to account for the remarkable similarity between the first and last 26 amino acids of bacterial ferredoxins [10–14], the variable and constant segments, respectively, of the light and heavy chains of γ-immunoglobulins [15–19]. Moreover, equal crossing over can take place between adjoining genes, a phenomenon which presumably accounts for the non-α chains of the abnormal human Lepore hemoglobins, hybrids of δ and β chains [20–24]. It also may occur between two alleles in a heterozygote, as must have been the case for the 2-α chain of human haptoglobin, the 142 residues of which are derived from the amino-terminal and carboxyl-terminal segments of the 83-residue 1Fα and 1Sα common haptoglonin allelic chains [25].

2) *Are significant similarities of primary structure due to common ancestry or to functional convergence?*

Statistical answers to this question require the techniques employed in estimating evolutionary relations from amino acid sequence information (phylogenetic trees), and the assessment of the structures of ancestral forms of the protein under consideration (reconstructed ancestral sequences). Since both these topics are considered below, any discussion of the distinction between divergence from a common ancestral form and convergence from different phylogenetic origins is best postponed till after these procedures have been considered.

3. Statistical phylogenetic trees

If the amino acid sequences for a set of proteins have been shown to possess similarities greater than random, and one further assumes that this is due to evolutionary homology, one can then set out to attempt to determine the phylogenetic relations of the species carrying these proteins purely on the basis of their structures. However, it must not be overlooked that not all homologous relationships justify such a procedure. If, in the common ancestor of all the species considered, the protein was represented by a single gene, then the descendent genes can be called *orthologous* (from ortho, meaning exact) [26], and precisely reflect, in a one-to-one fashion, the lineage of the species. As long as the evolutionary variations of this protein represent a statistically valid sample of the overall evolutionary variations of species carrying it, then one may expect to extract proper phylogenetic information from the corresponding amino acid sequences. However, homologous genes may have undergone duplication and remained side by side in all or many of the species descending from the earliest ancestor in which the duplication occurred. These may be te.med *paralogous* (from para, meaning in parallel) [26], and clearly cannot be utilized indiscriminately to ascertain phylogenetic relations. For example, in most vertebrates, hemoglobins are tetrameric and have at least two types of chains, α and β. Moreover, there often are other types of non-α chains, such as the γ and δ human chains. Vertebrates also carry another protein of the same homologous series, the monomeric myoglobin. There is general agreement that the genes for all these proteins are homologous [27–29], but if one were to utilize for the construction of a vertebrate phylogeny the amino acid sequences of the α chains of some species, those of the β chains of others and those of the myoglobins of still others, the result would be an absurdity. Indeed, the species would be mainly segregated into 3 groups, one each for those species for which the α, β or myoglobin chains were used for the analysis. This is because the gene duplications which gave rise to the three varieties of chains had occurred before the evolutionary appearance of the common ancestor of the species examined, and these genes had since evolved more or less independently. Each gene separately would be orthologous and the species variations of α, or β, or myoglobin chain structures could in principle provide data for three independent assessments of vertebrate evolutionary relations. (In the phylogenetic tree for eukaryotic cytochromes *c* shown in fig. 4, all the proteins are orthologous, except for the iso-1 and iso-2 cytochromes *c* of bakers' yeast which are paralogous. Since this is the only such relationship, it does not introduce any errors in the rest of the tree.)

Just as for the polypeptide segments utilized to establish the random or non-random nature of the similarities between two amino acid sequences (see above), it is possible to calculate the minimal replacement distances between any two orthologous amino acid sequences. For a set of n sequences, there are $n(n\text{-}1)/2$ such distances, which can be used to construct a phylogenetic tree, such as that shown in fig. 4 for the eukaryotic cytochromes *c* of 29 species [26,30,31]. Initially, each protein is assigned to a separate subset. Those two subsets which show the lowest replacement distance are joined and are henceforth treated as a single subset. The procedure is repeated until all proteins have been joined to provide the initial phylogenetic tree, which is merely a graphical representation of the order in which the subsets were joined. The replacement distances between various branch points of the tree can be calculated, and the distance between the proteins of any two species can be reconstructed by summing the appropriate branch lengths to give an "output" replacement distance. Such output distances will differ from so-called "input" distances, namely, those calculated directly from the amino acid sequences. This is because, after the first two subsets are joined, the distances from the other proteins to the new joined subset can only be calculated in terms of the average of the distances from every other protein to those in

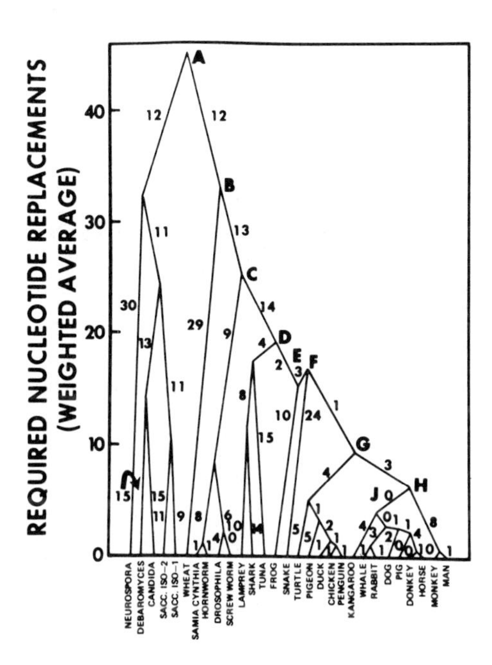

Fig. 4. Statistical phylogenetic tree based on the replacement distances between the cytochromes *c* of the species listed, as obtained by the procedure of Fitch and Margoliash [30]. Each number on the figure is the replacement distance along the line of descent as counted by the procedure of Fitch [36]. Each apex is placed at an ordinate value which is the weighted average of the sums of all nucleotide replacements in the lines of descent from that apex. References to the amino acid sequences of the cytochrome *c* are given in references [28 and 45].

the first subset, and the utilization of average distances necessarily continues throughout the computation. Therefore, the initial tree constructed need not necessarily represent the best utilization of the data. One procedure for seeking an optimal tree is to calculate a percent standard deviation between the distances reconstructed from the tree and the original input replacement distances. Alternative trees are examined, and that which shows the smallest percent standard deviation is considered to be the best.

However, this is not the only criterion that can be used in seeking an optimal tree. One could, for example, choose the tree for which the total number of mutations is the least. Moreover, it is not possible to examine all possible trees since there are a very large number of such trees, and there are no known alogorithms which can choose the one best tree, by either of the above criteria, without examining too many trees to be practicable. Thus, for n species there are $(2n-3)\{(2n-5)!/(n-3)!\ 2^{n-3}\}$ trees [31]. For 29 species this corresponds to more than 10^{36} different trees. Several variations of common numerical taxonomic methods are therefore used by different authors [30, 32,33] to pick "reasonable" trees for examination.

Whatever criteria are utilized, it is remarkable that the resulting phylogenies are generally in good accord with ordinary biological classifications, even though the amino acid sequences of the set of orthologous proteins, the genetic code and a simple set of statistical calculations were strictly the only information employed. The phylogenetic tree derived from the structures of eukaryotic cytochromes *c* (fig. 4) is not by any means perfect. Some of the relations depicted are certainly erroneous. Thus, primates branch off the ancestral mammalian line before marsupials, the turtle is nearer the birds than the other reptile (the rattlesnake) in the set, and the shark appears to relate more closely to the lamprey than to the tuna. Nevertheless, before this type of procedure was available, a phylogeny as accurate as this could not be derived from a single trait, let alone a single gene. Clearly, this must be because an examination of the number of mutations fixed in the course of the evolution of a single gene yields a considerably more precise estimate of the extent of evolutionary divergence than that from a single morphological trait. Indeed,

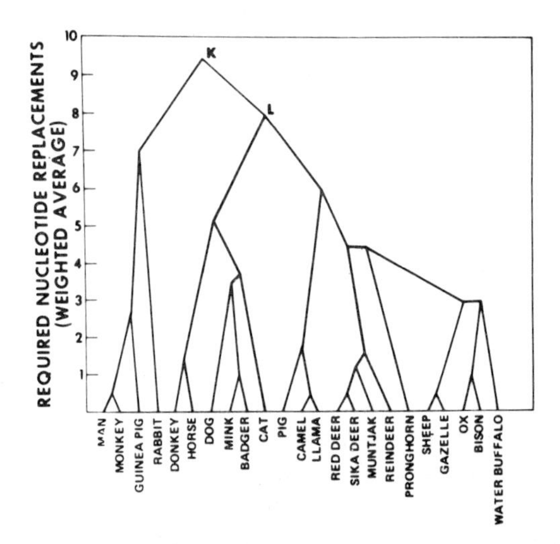

Fig. 5. Statistical phylogenetic tree based on the replacement distances between the fibrinopeptides A of the species listed. The topology of the tree was obtained by the procedure of Fitch and Margoliash [30]. The nucleotide replacements were counted by the procedure of Fitch [36]; Other markings as for fig. 4. References to the amino acid sequences of the fibrinopeptides are given in [34 and 35].

one can expect that when sufficient amino acid sequence data for various sets of proteins become available, precise phylogenies will be readily obtainable by such procedures.

Of the other proteins for which structural information has accumulated, fibrinopeptide A, cleaved by thrombin from the amino-terminal segment of fibrinogen in blood clotting, has led to a satisfactory phylogenetic tree (fig. 5) for a set of 23 species much more closely related than those represented in the cytochrome *c* tree (fig. 4). This segment of fibrinogen varies rather rapidly during evolution, which together with its small size (19 residues) makes it most useful in examining a narrow taxonomic span of species. As shown by Mross and Doolittle [34,35], the structures of the fibrinopeptides from 19 artiodactyls fit very well the classical phyletic relations of these species. Other such relatively small groups can surely be studied as effectively on this basis.

An insufficient number of structures of orthologous proteins in the hemoglobin-myoglobin and in the ferredoxin series are as yet available to lead to useful phylogenetic trees.

4. Reconstruction of ancestral amino acid sequences and the distinction between divergent and convergent evolutionary processes

The reconstruction of the amino acid sequence of the ancestral form of the protein at each of the branching points of the phylogenetic tree can be carried out, following certain rules, from a phylogenetic tree and the amino sequences of the present day proteins [30,31,36]. An example of the result of such a procedure is the ancestral cytochrome *c* sequence (fig. 6) corresponding to the structure derived for the cytochrome *c* of the ancestral species at the topmost apex of the phylogenetic tree. The ambiguities result from the lack of sufficient data to decide unequivocally what is the codon for every residue position.

Similar procedures can be utilized to distinguish between divergent and convergent evolutionary processes [36]. Consider two sets of orthologous proteins which are to be tested for homology; it is possible to reconstruct the probable nucleotide (or nucleotides where some ambiguity may exist) for every position of the two ancestral genes for the two ancetral genes for the two sets. If the same nucleotide is present in a certain position in both ancestral genes, then any differences in present day sequences are of a divergent character. If, on the other hand, a different nucleotide occurs in a given position in the two ancestral genes, then any similarity between the present day proteins of the two sets is of a convergent character for that particular posi-

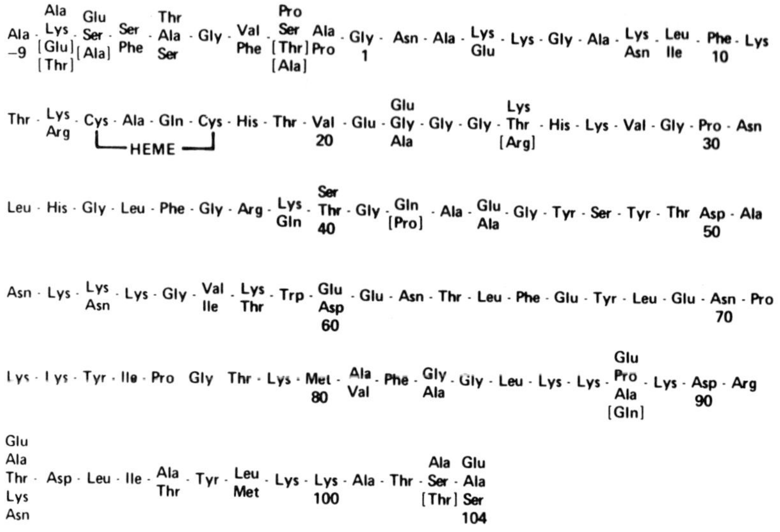

Ala Glu Thr Pro
Ala · Lys · Ser · Ser · Ala · Gly · Val · Ser Ala · Gly · Asn · Ala · Lys · Lys · Gly · Ala · Lys · Leu · Phe · Lys
−9 [Glu] [Ala] Phe Ser Phe [Thr] Pro 1 Glu Asn Ile 10
[Thr] [Ala]

Glu Lys
Thr · Lys · Cys · Ala · Gln · Cys · His · Thr · Val · Glu · Gly · Gly · Gly · Thr · His · Lys · Val · Gly · Pro · Asn
Arg HEME 20 Ala [Arg] 30

Ser
Leu · His · Gly · Leu · Phe · Gly · Arg · Lys · Thr · Gly · Gln · Ala · Glu · Gly · Tyr · Ser · Tyr · Thr Asp · Ala
Gln 40 [Pro] Ala 50

Asn · Lys · Lys · Lys · Gly · Val · Lys · Trp · Glu · Glu · Asn · Thr · Leu · Phe · Glu · Tyr · Leu · Glu · Asn · Pro
Asn Ile Thr Asp 70
60

Glu
Lys · Lys · Tyr · Ile · Pro Gly Thr · Lys · Met · Ala · Phe · Gly · Gly · Leu · Lys · Lys · Pro · Lys · Asp · Arg
80 Val Ala Ala 90
[Gln]

Glu
Ala Ala Glu
Thr · Asp · Leu · Ile · Ala · Tyr · Leu · Lys · Lys · Ala · Thr · Ser · Ala
Lys Thr Met 100 [Thr] Ser
Asn 104

Fig. 6. Amino acid sequence of the ancestral form of cytochrome *c* at the topmost apex of the phylogenetic tree. Any of the amino acids shown would permit the evolution of the 29 descendent cytochromes *c* in the maximum number of 366 nucleotide replacements, assuming the topology shown in fig. 4. Amino acids in brackets have not yet been observed in any present day cytochrome *c*.

tion. To consider the complete structural genes, one can, assuming that the descendent nucleotide sequences are completely unrelated, estimate how many of the some 300 ancestral nucleotide comparisons (for proteins of 100 residues) would be expected to be of a divergent and how many of a convergent type. A significant excess of one or the other type would make it possible to decide whether the two sets were divergently related or only similar because of convergence. Typical results are shown in fig. 7. The abcissa plots the number of different species in the two trees being compared. The horizontal line is the line of mean expectation. The curve fluctuating about it was obtained using random sequences of 100 amino acids, showing that when the proteins are entirely unrelated there is no excess of either divergent or convergent relationships. The ordinate gives the standard deviation from expectation, so that points above the line of mean expectation indicate excess of divergent over convergent comparisons, and points below the line the opposite situation. The lower curve was obtained from two sets of amino acid sequences that were made to simulate a convergent evolutionary process by a

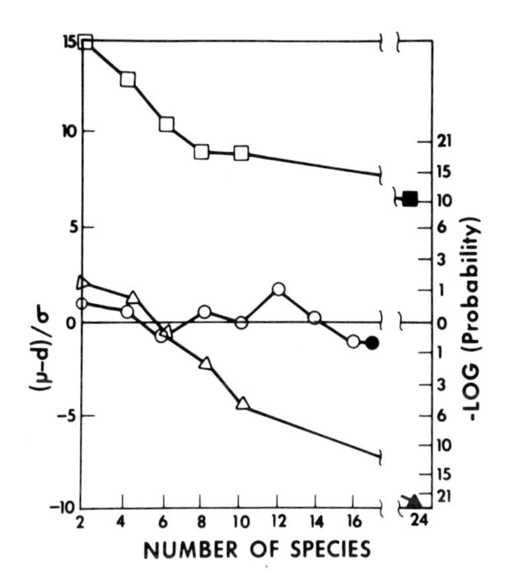

Fig. 7. Convergence and divergence as a function of the number of species. The abcissa gives the total number of sequences examined. Open symbols indicate that equal numbers of sequences were present in the two groups compared, closed symbols that they were divided unequally between the two groups. The ordinate gives the deviation (δ) from expectation (μ−d), in standard deviation units on the left, and the equivalent probability of a result being due to chance is given as negative powers of 10 on the right. Points above the zero line represent an excess of divergent comparisons, below the line. an excess of convergent comparisons. Random sequences of amino acids are shown by circles (○——○), convergent sequences by triangles (△——△), and divergent sequences of squares (□——□). The convergent sequences were obtained by computer simulation. The divergent sequences compare fungal to non-fungal cytochromes *c*. According to Fitch [36].

computer. The curve above the line is for two sets of eukaryotic cytochromes *c* composed of fungal and non-fungal proteins. The result clearly shows that fungal and non-fungal eukaryotic cytochromes *c* had a common evolutionary origin. It should be noted that orthology within each of the two groups is the only required assumption.

5. **Invariant codons and covarions**

Possibly the most useful of all present applications of statistical phylogenetic trees is the estimation of the number of invariant codons in the structural gene for the protein considered [37]. These represent positions in the polypeptide chain for which only one particular amino acid can fulfill the required function satisfactorily, so that the probability of a line of evolutionary descent surviving the fixation of a mutation in these codons is essentially nil. All mutations in such codons are termed *malefic* [37].

The phylogenetic tree based on cytochrome *c* structures (fig. 4) prescribes the distribution of codons in the structural gene which have undergone 0, 1, 2, 3 or more replacements in their descent from the common ancestral form. That distribution can be accounted for if one assumes that there are three classes of codons. One class is invariant. The other two vary in a random fashion according to two different rates, one, the "hypervariable" set of codons, changing much more rapidly than the other [37–39]. There are probably more than two rates of variation, but two rates are sufficient to fit the presently available data [38]. All codons belonging to the same variable set are equally likely to fix the next nucleotide replacement, and for each, therefore, the number of codons that have undergone 1, 2, 3 ... replacements will follow a Poisson distribution. Fitting such distributions to the data obtained from the cytochrome *c* phylogenetic tree in fig. 4 makes it possible to estimate the size of the tree sets. The best fit is for an invariant set of 32 residues, a normally variable set of 65 residues and a "hypervariable" set of 16 residues [38]. This last appears to fix mutations in the course of evolution some 3.2 times faster than the normally variable codons [38].

The above calculation employed 29 different cytochromes *c* of species ranging from fungi to vertebrates (see fig. 4), and yielded an estimate of the percent of the cytochrome *c* gene that was invariant of about 25% [38]. A similar estimate was made earlier using the cytochromes *c* of only 20 species, but covering the same taxonomic range [37]. However, if one selectively and gradually excludes the proteins of the more remote groups of species from the calculation, the resulting percent of the gene found to be invariant increases. If these values are plotted as a function of the average replacement distance for all the species taken into account for each recalculation (fig. 8), a roughly linear regression is obtained. On extrapolation, a value showing over 90% of the gene to be invariant is obtained when the replacement distance is zero [38]. This demonstrates that in any one mammalian cytochrome *c* at the present time, only about 10 residues can undergo changes without leading to a lethal or malefic change [38]. Moreover, it seems reasonable that if enough data were available to make similar extrapolations towards fungal cytochromes *c* or insect cytochromes *c*, for example, an essentially similar result would be obtained. The codons corresponding to those amino acid positions which, in any one species and at any one time in the course of evolution, are free to fix mutations may be termed *concomitantly variable codons* or *covarions* [38].

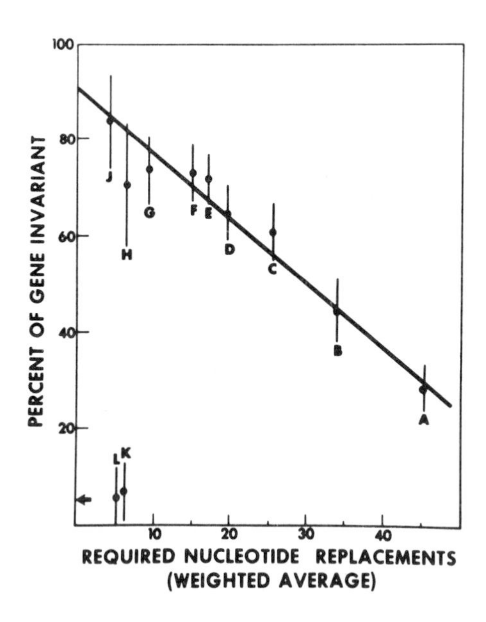

Fig. 8. Concomitantly variable codons. The percent of the gene found to be invariant is plotted as a function of the weighted average of required nucleotide replacements (height of peaks in figs. 4 and 5 for all the species in each comparison). Letters A to J represent the groups of cytochromes *c* indicated in fig. 4, letters L and K the groups of fibrinopeptides indicated in fig. 5. The arrow on the ordinate is the position equivalent to one invariant residue out of the 19 residues of fibrinopeptide. The line at each point is an estimate of the standard deviation of the ordinate value of the point. A weighted least squares fit to the results for cytochrome *c* is extrapolated to the abcissa to estimate the fraction of the gene for which all mutations are lethal or malefic. According to Fitch and Markowitz [38].

This conclusion is particularly important as it demonstrates that in the cytochrome *c* of any one species only a very small proportion of all residue positions that have varied, as among the cytochromes *c* of the more than 30 species investigated, are in fact variable. This very stringent limitation on evolutionary change in protein structure must be due to the complex interplay of structural-functional requirements. In cytochrome *c*, in addition to the types of residue interaction common to other proteins, the relatively short peptide chain must essentially wholly enclose the evolutionarily invariant heme, in a way that requires a relatively large number of internal residues to be in contact with the prosthetic group [40–42]. Moreover, provision must be made to adapt the outer surface to specific interactions with three different macromolecular surfaces, those of cytochrome oxidase, cytochrome reductase and the mitochondrial membrane binding site for cytochrome *c*. These contacts could well involve a major proportion of the surface of the protein.

In order to account for the observed variation of over two thirds of the residues of cytochrome *c* in the proteins of a wide taxonomic range of species, one must assume that when a mutation is fixed in a particular covarion, it may also change some of the members of the set of covarions. Thus, over extended periods of evolutionary history, more than 70 residue positions have shown substitutions.

The number of covarions, obviously an expression of the extent and tightness of structural-functional requirements, appears to represent a fundamental parameter which is nothing else than a quantitative expression of the effect of function on the evolutionary behavior of proteins. Though the number of covarions may well vary somewhat for the same protein in different species, it nevertheless appears to impose the average rate of evolutionary change so characteristic of every protein.

An excellent example is provided by the comparison of cytochromes *c* and fibrinopeptides A [38]. As depicted in fig. 8, 18 of the 19 residues of fibrinopeptides A are variable if one considers the fibrinopeptides of all the species listed in fig. 5, and remarkably, this number does not appear to change as the range of species is decreased. The number of covarions for fibrinopeptide A thus appears to be 18. (Because of the relatively small range of species for which the data are available, this estimate is probably not as accurate as that for cytochrome *c*, and the correct value could be 17.) Since the time of the common ancestor of the horse and the pig, the phylogenetic tree for cytochrome *c* indicates that 5 nucleotide replacements were fixed in the 104 codons in both lines of descent to the present day genes, while the tree for fibrinopeptide A shows 13 nucleotide replacements for 19 codons. This corresponds to 0.048 and 0.684 fixations/codon, as expected from the known slow conservative nature of evolutionary changes in cytochromes *c* [40–43] and the very rapid changes of fibrinopeptides [34]. However, such calculations include not only the codons which can undergo changes, namely covarions, but also all the codons for which variations are either lethal or malefic. If the latter is excluded the values become 5/10 = 0.50 for cytochrome *c*, and 13/18 = 0.72 for fibrinopeptide A in fixations/covarion [38]. Considering the probable error of such estimates, these two values cannot be significantly different and should be thought of as in remarkably good concordance.

If similar values are obtained for other proteins when a sufficiency of primary structures become available it will be possible to conclude that the overall rate of evolutionary change of a protein is determined by the number of its covarions. Such a result would also be compatible with the contention that for those positions that are amenable to change the occurrence of variations is governed by random processes, namely, that the mutations fixed are selectively neutral. The pros and the cons of such an interpretation have been discussed elsewhere [26, 38, 43–49] and it would not be practicable to reproduce the entire, rather inconclusive, argument here. However, regardless of whether the mutations are or are not neutral, wide applicability of the relation between covarions and evolutionary rate of change would provide, given a suitable paleontological reference point in time, a method for dating evolutionary events, such as speciations in the case of species phylogenies and gene duplications in the case of gene phylogenies. This procedure would be a statistically elaborated counterpart of earlier crude attempts to date evolutionary events directly from the average numbers of residue or nucleotide differences for the eukaryotic cytochromes c of different taxonomic groups of species and a paleontological reference point [41–43, 50].

6. Early evolutionary history of replicating macromolecules

The most fascinating possibility opened up by the statistical approach to the evolutionary information contained in the structures of present-day macromolecules, is an approach to the distant past. Our ability to extrapolate backward in biological history is indeed not limited to the latest common ancestor of present species. By examining numerous orthologous sets of proteins and nucleic acids, it will be possible to reconstruct the structure of the ancestral form for each set. Having in this way eliminated the mutations that were accumulated between the time when these early species existed and the present, one may well observe similarities between the ancestral sequences which are not obvious in their descendants. This procedure would thus identify paralogous relationships when none could be observed in present day macromolecules and make it possible to establish an extensive gene phylogeny. One could work out a tree based on the ancestral sequences and thus estimate the structure of the ancestral sequence of these ancestral sequences. This could in principle be done both with proteins and nucleic acids, independently. The second order ancestral sequences, if these can be attained, may well take us back to very early stages at the borders of chemical and biological evolutions at which the genetic code and machinery actually originated. The earliest "protein" and "nucleic acid" should indeed correspond one to the other if our present speculative concepts of the evolution of the genetic code [51–54] are factually correct.

References

[1] E. Zuckerkandl and L. Pauling, J. Theoret. Biol. 8 (1965) 357.
[2] H. Neurath, K.A. Walsh and W.P. Winter, Science 158 (1967) 1638.
[3] B.S. Hartley, this volume, p. 191.
[4] K. Brew, T.C. Vanaman and R.L. Hill, J. Biol. Chem. 242 (1967) 3747.
[5] W.M. Fitch, J. Mol. Biol. 16 (1966) 9.
[6] H. Matsubara and E.L. Smith, J. Biol. Chem. 238 (1963) 2732.
[7] Y. Yaoi, K. Titani and K. Narita, J. Biochem. 59 (1966) 247.
[8] W.M. Fitch, J. Mol. Biol. 49 (1970) 1.
[9] W.M. Fitch, Biochem. Gen. 3 (1969) 99.
[10] M. Tanaka, T. Nakashima, A.M. Benson, M.F. Mower and K.T. Yasunobu, Biochemistry 5 (1966) 1666.
[11] A.M. Benson, H.F. Mower and K.T. Yasunobu, Arch. Biochem. Biophys. 121 (1967) 563.
[12] W.M. Fitch, J. Mol. Biol. 16 (1966) 17.
[13] R.V. Eck and M.O. Dayhoff, Science 152 (1966) 363.
[14] T.H. Jukes, in: Molecules and Evolution (Columbia Univ. Press, New York, 1966).
[15] S.J. Singer and R.F. Doolittle, Science 153 (1966) 13.
[16] R.L. Hill, R. Delany, R.E. Fellows and H.E. Lebowitz, Proc. Natl. Acad. Sci. 56 (1966) 1762.
[17] R.L. Hill, H.E. Lebowitz, R.E. Fellows and R. Delaney, in: Gamma-Globulins, ed. J. Killander (Interscience, New York, 1967) p. 109.
[18] U. Rutishauser, B.A. Cunningham, C. Bennett, W.H. Konigsberg and G.M. Edelman, Proc. Natl. Acad. Sci. 61 (1968) 1414.
[19] G.M. Edelman, B.A. Cunningham, W.E. Gall, P.D. Gottlieb, U. Rutishauser and M.J. Waxdal, Proc. Natl. Acad. Sci. U.S. 63 (1969) 78.
[20] G. Baglioni, Proc. Natl. Acad. Sci. U.S. 48 (1962) 1880.
[21] C. Baglioni, Biochim. Biophys. Acta 97 (1965) 37.
[22] C.C. Curtain, Australian J. Exptl. Biol. Med. Sci. 42 (1964) 89.
[23] J. Barnabas and C.J. Muller, Nature 194 (1962) 931.
[24] D. Labie, W.A. Schroeder and T.H.J. Huisman, Biochim. Biophys. Acta 127 (1966) 428.
[25] J.A. Black and G.H. Dixon, Nature 218 (1968) 736.
[26] W.M. Fitch and E. Margoliash, Evolutionary Biol. 4 (1970) 67.
[27] H.A. Itano, Advan. Protein Chem. 12 (1957) 215.
[28] V.M. Ingram, Nature 189 (1961) 704.
[29] V.M. Ingram, in: The Hemoglobins in Genetics and Evolution (Columbia University Press, New York) 1963.
[30] W.M. Fitch and E. Margoliash, Science 155 (1966) 279.
[31] W.M. Fitch and E. Margoliash, Brookhaven Symp. Biol. 21 (1968) 217.
[32] R.R. Sokal and P.H.A. Sneath, in: Principles of Numerical Taxonomy (Freeman, San Francisco, 1963).
[33] M.O. Dayhoff, Atlas of Protein Sequence and Structure (National Biomedical Research Foundation, Silver Spring, Md., 1969).
[34] G.A. Mross and R.F. Doolittle, Arch. Biochem. Biophys. 122 (1967) 674.
[35] R.F. Doolittle, personal communication.

[36] W.M. Fitch, Systematic Zool. 19 (1970) 99.
[37] W.M. Fitch and E. Margoliash, Biochem. Gen. 1 (1967) 65.
[38] W.M. Fitch and E. Markowitz, Biochem. Gen. in press.
[39] E. Markowitz, Biochem. Gen., in press.
[40] E. Margoliash and A. Schejter, Advan. Protein Chem. 21 (1966) 113.
[41] E. Margoliash and W.M. Fitch, Trans. N.Y. Acad. Sci. 151 (1968) 359.
[42] C. Nolan and E. Margoliash, Ann. Rev. Biochem. 37 (1968) 727.
[43] E. Margoliash, W.M. Fitch and R.E. Dickerson, Brookhaven Symp. Biol. 21 (1968) 259.
[44] M. Kimura, Nature 217 (1968) 624.
[45] J.M. Smith, Nature 219 (1968) 1114.
[46] J.L. King and T.H. Jukes, Science 164 (1969) 788.
[47] K.W. Corbin and T. Uzzell, American Naturalist, 140 (1970) 37.
[48] P. O'Donald, Nature 221 (1969) 815.
[49] N. Arnheim and C.E. Taylor, Nature 223 (1969) 900.
[50] E. Margoliash and E.L. Smith, in: Evolving Genes and Proteins, ed. V. Bryson and H.J. Vogel (Academic Press, New York, 1965) p. 221.
[51] T.H. Jukes, Biochem. Biophys. Res. Commun. 19 (1965) 391.
[52] C.R. Woese, in: The Genetic Code (Harper and Row, New York, 1967).
[53] F.H.C. Crick, J. Mol. Biol. 38 (1968) 367.
[54] L.E. Orgel, J. Mol. Biol. 38 (1968) 381.

AMINOACYL - tRNA SYNTHETASES OF *ESCHERICHIA COLI* AND OF MAN

K.H. MUENCH, A. SAFILLE, M.-L. LEE, D.R. JOSEPH, D. KESDEN and J.C. PITA, Jr.

Departments of Medicine and Biochemistry, University of Miami School of Medicine, Miami, Florida 33136, USA

Abstract: Muench, K.H., Safille, A., Lee, M.-L., Joseph, D.R., Kesden, D. and Pita, J.C. Jr. Aminoacyl-tRNA Synthetases of *Escherichia Coli* and of Man. *Miami Winter Symposia,* 1, pp. 52–65. North-Holland Publishing Company, Amsterdam, 1970.

Even within the single species *Escherichia coli* the aminoacyl - tRNA synthetases exhibit considerable diversity in quaternary structure. For most of these enzymes molecular weights near 100,000 have been reported. Two types can now be defined: those synthetases with and without subunits. Synthetases may sometimes be isolated as active aggregates with molecular weights near 200,000.

We have purified the prolyl-tRNA synthetase of *E. coli* 500-fold partially characterized it. The active form of the enzyme is a dimer of molecular weight near 100,000, stable at 37° or in concentrations of glycerol 15% or greater. It can be reversibly converted into an inactive monomer of molecular weight near 50,000 by storage at 0° in the absence of glycerol. If multiple active sites are present on the dimer, they are kinetically independent, because the enzyme exhibits linear kinetic plots with all three substrates. The enzyme exists entirely as the dimer in growing cells.

We have purified tryptophanyl-tRNA synthetase from *E. coli* to homogeneity according to the criteria of gel filtration and disc gel electrophoresis. The former method gives a molecular weight of 74,000 for this enzyme, and sucrose density gradient centrifugation gives a molecular weight of 76,000. Dodecyl sulfate-polyacrylamide gel electrophoresis reveals subunits of molecular weight 37,000. Again, the enzyme exhibits linear kinetic plots with all three substrates. In this instance conversion of the monomeric units into an active dimer has not been achieved. Another enzyme in this category is tyrosyl-tRNA synthetase of *E. coli*, of molecular weight 95,000 with subunits of molecular weight near 44,000.

We have purified the leucyl-tRNA synthetase of *E. coli* to an estimated 90% according to disc gel electrophoresis. According to dodecyl sulfate-polyacrylamide gel electrophoresis the enzyme has a molecular weight of 93,000. In sucrose density gradient centrifugation the enzyme behaves as a protein of molecular weight 90,000.

Experience in the study of the enzymes from *E. coli* has permitted an initial enzymological attack on the lymphocytes of human chronic lymphocytic leukemia. These cells are rich in lysyl-, tryptophanyl- alanyl-, and arginyl-tRNA synthetases, and we have purified the lysyl-tRNA synthetase 90-fold and started its characterization. The enzyme is capable of charging lysine to the tRNA of *E. coli* and of yeast, a property facilitating characterization.

1. Introduction

Because of the role of tRNA as adapter in protein synthesis, its structural requirements are narrowly defined, and tRNAs throughout the evolutionary sequence from bacteria to mammals are similar. Thus all of the tRNAs whose primary sequence has been described contain 75-85 nucleotidyl residues and can be arranged in the cloverleaf secondary structure [1–4] and in the same tertiary structure [4].

Apparently, greater structural latitude is allowed for the aminoacyl-tRNA synthetases, which catalyze the reactions:

$$\text{Amino acid} + \text{ATP} + \text{Synthetase} \rightleftharpoons \text{Aminoacyl-AMP}\cdot\text{Synthetase} + \text{PP}_i$$

$$\text{Aminoacyl-AMP}\cdot\text{Synthetase} + \text{tRNA} \rightleftharpoons \text{Aminoacyl-tRNA} + \text{Synthetase} + \text{AMP}$$

On the basis of quaternary structure even in the single species *Escherichia coli* at least two classes of these enzymes are now emerging: those with and those without subunits. In the former category are the aminoacyl-tRNA synthetases for phenylalanine [5,6], methionine [7–9], alanine [10], serine [11], proline [12], tryptophan [13] and probably tyrosine [14]. In the latter category are those for isoleucine [15], valine [16] and leucine. Whereas the latter enzymes have similar molecular weights, near 100,000, the enzymes with subunits exhibit considerable variation in molecular weight. For example, the molecular weights of the subunits of tryptophanyl-tRNA synthetase are 37,000 and of phenylalanyl-tRNA synthetase are 100,000.

Catalytic activity has never been unequivocally demonstrated for a subunit. However, the subunits (molecular weight 48,000) of methionyl-tRNA synthetase have a binding site for each of the three substrates, L-methionine, ATP and tRNA^{Met} [8], and unless they dimerize under assay conditions, these subunits may be active [9]. The methionyl-tRNA synthetase tends to aggregate. It is isolated from *E. coli* as a dimer or tetramer, depending on the purification procedure used [7,17]. Subunits (molecular weight 70,000) of alanyl-tRNA synthetase from a temperature-sensitive strain of *E. coli* possibly activate alanine and attach it to tRNA^{Ala}, but dimerization during the assay remains a strong alternative explanation for the catalysis observed [10]. In no case have amino acid activation and transfer been localized to separate polypeptide chains.

In what follows we shall report some of our own experiments on the prolyl-, tryptophanyl-, leucyl- and tyrosyl-tRNA synthetases of *E. coli,* and we shall conclude with some preliminary work on synthetases from human leukemic lymphocytes.

2. Experimental

Blue dextran was purchased from Pharmacia; yeast alcohol dehydrogenase, *E. coli* alkaline phosphatase, rabbit muscle aldolase, and trypsin from Worthington; ovalbumin from Sigma and Mann. Crystalline rabbit muscle phosphorylase was a gift from Dr. Adel Yunis, and crystalline human placental alkaline phosphatase was a gift from Dr. Donald Harkness. Purified tyrosyl-tRNA synthetase was a gift from Dr. Richard Calendar. Other materials were obtained from sources previously noted [12,18,19].

Enzyme assays were performed as previously described [12,18,19]. One unit forms 1 nmole of aminoacyl-tRNA in 10 min. Protein was measured by the method of Lowry [20] or by absorbance [21]. One A_{260} unit of tRNA has an absorbance at 260 mμ of 1.0 in a 1.0-cm light path when dissolved in 1.0 ml of 5mM KH_2PO_4–5 mM K_2HPO_4 buffer.

3. Results and discussion

3.1. *Prolyl-tRNA synthetase*

The prolyl-tRNA synthetase of *E. coli* [12] undergoes the conversions shown in the schematic equations:

Active dimer $\rightleftharpoons$ Inactive monomer $\rightarrow$ Denatured monomer.

The molecular weight of the dimer is 110,000, calculated as an average from glycerol density gradient centrifugations [12,22] with human placental alkaline phosphatase (molecular weight 125,000) [23], *E. coli* alkaline phosphatase (molecular weight 90,000) [24] and *E. coli* isoleucyl-tRNA synthetase (molecular weight 112,000) [15] as reference enzymes. The reversible monomer-dimer relationship of the prolyl-tRNA synthetase has made possible the demonstration that the monomer is inactive in both prolyl-tRNA formation and in proline activation. When placed at 0° in the absence of glycerol over a wide range of pH and ionic conditions, the active dimer becomes the inactive monomer with first-order kinetics [12]. The monomer rapidly changes into the dimer with second-order kinetics during warming at 37° [12]. The monomer dimerizes slowly during storage at 0° in concentrations of glycerol greater than 15%. Again, the dimerization occurs under a variety of ionic conditions and pH. Because the dimer catalyzes the formation of prolyl-tRNA at a measurable rate at 0°, the relative amounts of monomer and dimer are easily determined by assay at 0° before and after warming for 10 min at 37° [12].

Böck [6] has found that phenylalanyl-tRNA synthetase of *E. coli* has inactive subunits of molecular weight 100,000. The subunits appear in 1.5 M guanidine. After removal of the guanidine, concentrating the enzyme in presence of 20% glycerol or warming at 37° under assay conditions produces the active dimer. Only for prolyl- and phenylalanyl-tRNA synthetases has reconstitution of active dimer from inactive monomer been demonstrated.

The prolyl-tRNA synthetase monomer is relatively unstable and becomes irreversibly inactivated at 0° with first-order kinetics and a half-life of about 3 days. The presence of 20 mM 2-mercaptoethanol increases the half-life to 6 days, and 1 mM dithiothreitol increases the half-life to 10 days, but 0.4 mM L-proline, 0.5 mM ATP or tRNA (50 A_{260} units/ml, about 20 pmoles $tRNA^{Pro}/A_{260}$ unit) have no effect. The monomer and dimer exist in an apparent equilibrium, the proportion of dimer being a function of temperature and maximal at 37° [12]. The rate of equilibration is also a function of temperature. The relationship of dimer formation to temperature and the relative unimportance of ionic conditions suggest hydrophobic bonding as the force joining the monomers. ATP specifically stabilizes the dimer with a protection constant of 10^{-4} M at 0° [12]. Under identical conditions the K_m for ATP is 4×10^{-5} M. The values are close enough to indicate that the means of stabilizing the dimer may be the binding of ATP at the active site. If the dimer has two active sites, they function in an entirely independent manner, for we have obtained linear kinetic plots (1/rate against 1/substrate concentration) with all three substrates.

The monomer and dimer are separable by hydroxylapatite chromatography [12], by sucrose density gradient centrifugation [12], and by gel filtration (fig. 1). Unfortunately, none of these techniques has provided pure enzyme in high yield, because the monomer is unstable.

Because prolyl-tRNA synthetase exists in two quaternary structures, we considered the state of the enzyme *in vivo*. In one example described by Neidhardt and colleagues [25] a change in the quaternary structure of an aminoacyl-tRNA synthetase occurs *in vivo*: the valyl-tRNA synthetase present before infection in *E. coli* undergoes a bacteriophage-induced increase in mole-

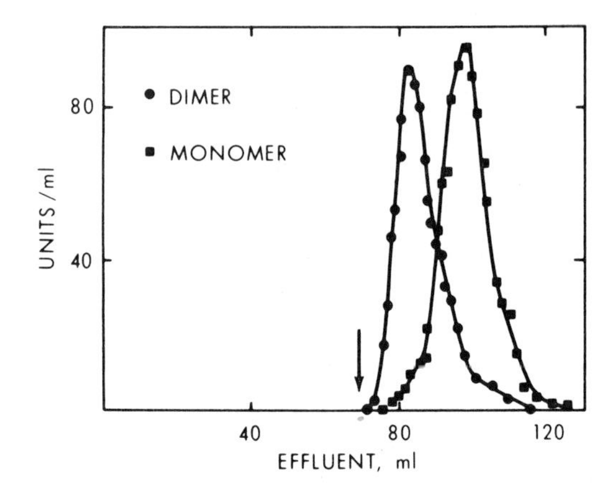

Fig. 1. Separation of the active dimer (•) and inactive monomer (▪) of prolyl-tRNA synthetase by gel filtration on Sephadex G-100. Dimer or monomer prepared as described [12] was placed over a 45 × 2.4 cm column of Sephadex G-100 equilibrated with 12.5 mM KH_2HPO_4-12.5 mM K_2HPO_4 and 20 mM 2-mercaptoethanol and, for the dimer only, 20% glycerol. The void volume (arrow indicates foot of peak) as determined by blue dextran was not changed by the presence of 20% glycerol.

cular weight from 100,000 to 200,000 with a concomitant decrease in specific activity. Prolyl-tRNA synthetase exists entirely in the dimeric form in growing cells: when growing *E. coli* cells are rapidly cooled to 0° and mechanically broken, the extract contains no monomer. The same extract permits formation of monomer during standing at 0° and subsequently permits reformation of dimer at 37°.

3.2. *Tryptophanyl-tRNA synthetase*

In contrast to the situation with prolyl-tRNA synthetase we have purified the tryptophanyl-tRNA synthetase of *E. coli* to homogeneity [26]. As shown in fig. 2 the enzyme gives a single band on disc gel electrophoresis [27] with as much as 250 μg of protein. In Sephadex G-200 gel filtration (fig. 3) the enzyme gives a peak of constant specific activity and migrates precisely with *E. coli* alkaline phosphatase, which has an apparent molecular weight of 74,000 by that method [28]. A standardized Sephadex G-150 column gives the same result. Because glycerol stabilizes the enzyme, the gel column was equilibrated with a buffer containing 10% glycerol. Glycerol up to 20% does not change the swelling or void volume of Sephadex columns. Tryptophanyl-tRNA synthetase has a molecular weight of 76,000, calculated as an average

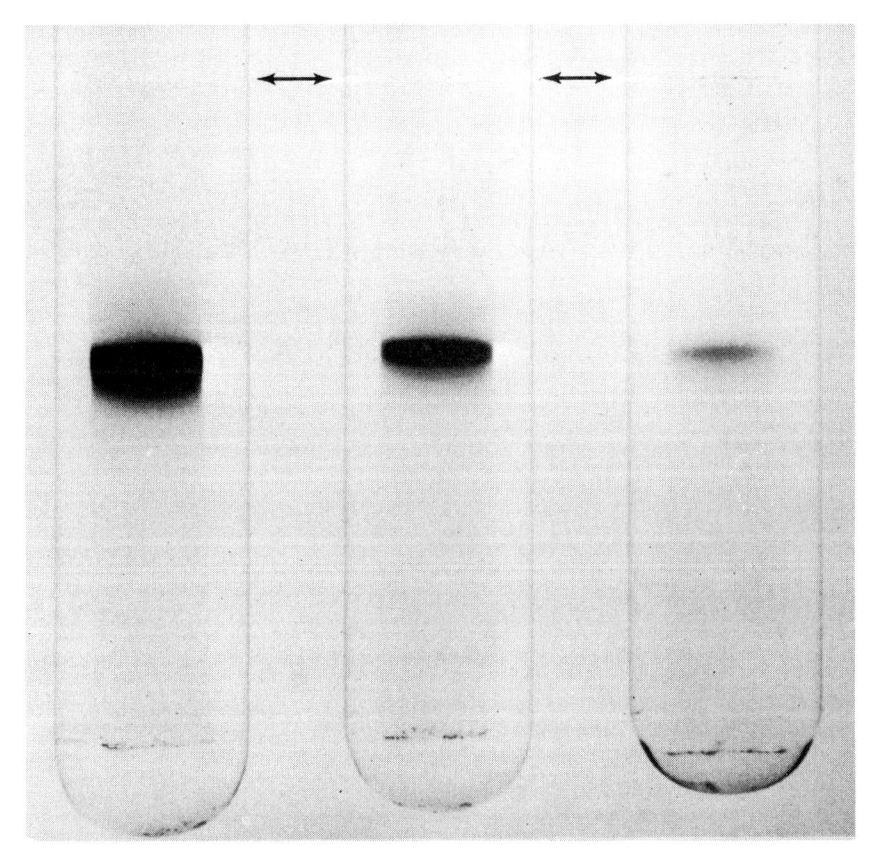

Fig. 2. Polyacrylamide disc gel electrophoresis of tryptophanyl-tRNA synthetase. The enzyme, 10 μg (right), 40 μg (center), 100 μg (left) and 250 μg (not shown) was subjected to disc gel electrophoresis [26] in the system described by the Canalco brochure (April, 1965) with the standard 7.0% polyacrylamide gel, stacking at pH 8.9 and resolving at pH 9.5. Tris-glycine buffer, pH 8.5, containing bromphenol blue was at the anode and cathode. The arrows indicate the origin. The 6.3 × 0.5 cm gels were developed at 3 milliamps per gel in a Canalco Model 12 apparatus at room temperature until the dye front reached the bottom of the gel. The gels were stained with 0.1% amido-schwarz and washed with 10% acetic acid.

from sucrose density gradient centrifugations [22] with the three standards previously mentioned for glycerol density gradient centrifugation.

The enzyme exhibits typical linear kinetics with all three of its substrates. Therefore, its active sites are independent, presumably one to each enzyme molecule. To determine the number of binding sites per enzyme molecule we

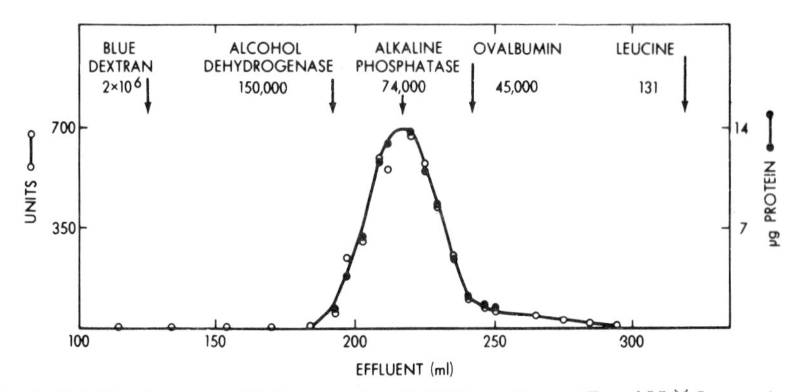

Fig. 3. Gel filtration or purified tryptophanyl-tRNA synthetase. To a 108 X 2 cm column of Sephadex G-200, equilibrated with 12.5 mM KH_2PO_4-12.5 mM K_2HPO_4, 10% glycerol, and 20 mM 2-mercaptoethanol and standardized by the substances indicated, was added 0.48 mg of tryptophanyl tRNA-synthetase in 2 ml of buffer. The column was developed at 5 ml/hr with collection of 5 ml fractions. Fractions were assayed for tryptophanyl-tRNA synthetase (○), and for protein [20] (•). Recovery of each was 90%.

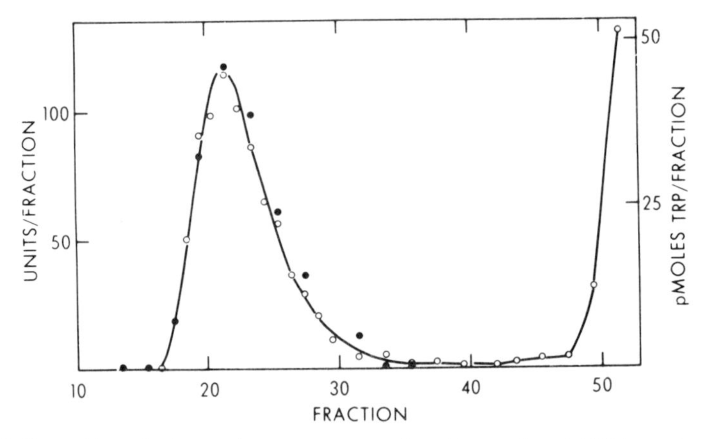

Fig. 4. Formation of a tryptophan-enzyme complex. Tryptophanyl-tRNA synthetase (23 μg) was incubated for 5 min at 23° in 0.21 ml containing 20 μmoles potassium bicine buffer, pH 8.8, 1.0 μmole $MgCl_2$, 0.8 μmole GSH, 0.20 μmole ATP, and 20 nmoles L-^{14}C-tryptophan (4.4×10^4 cpm/nmole); 0.025 ml of 0.1 M EDTA, pH 7, was added, and 0.20 ml of the solution was applied to a 32 X 1 cm column of Sephadex G-75 equilibrated with 20 mM sodium cacodylate buffer, pH 7.0, 10 mM 2-mercaptoethanol and 1 mM EDTA. Development was at 7.8 ml/hr with collection of 0.45 ml fractions. L-^{14}C-tryptophan (○) in each fraction was determined on 0.10 ml aliquots placed directly on glass fiber discs, dried, immersed in scintillation fluid, and counted. Tryptophanyl-tRNA synthetase (•) in each fraction was assayed with the result shown. Recovery of enzymatic activity was 70%.

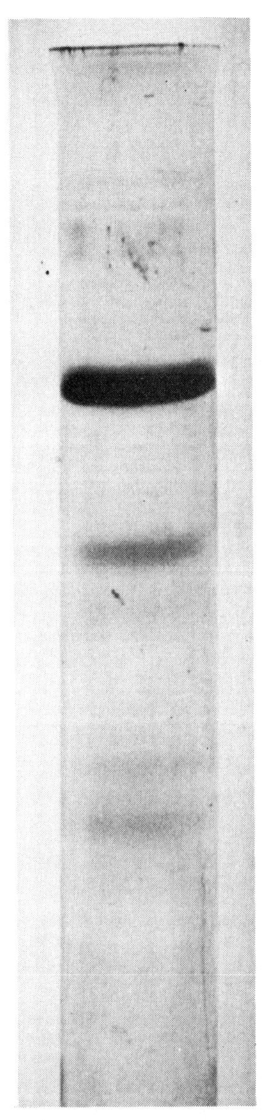

Fig. 5. Sodium dodecyl sulfate-polyacrylamide gel electrophoresis of tryptophanyl-tRNA synthetase. The enzyme (12 μg) after exposure to 1% sodium dodecyl sulfate and 1% 2-mercaptoethanol in 10 mM sodium phosphate buffer, pH 7, for 2 hr at 37° was added (0.005 ml) to 0.05 ml of 50% glycerol, 10% 2-mercaptoethanol, 0.034% sodium dodecyl sulfate and 3.4 mM sodium phosphate buffer, pH 7, layered on a 10% polyacrylamide gel (origin at top of figure), electrophoresed and stained, as described [32]. The gel was destained in 5% methanol, 7.5% acetic acid and stored in 7.5% acetic acid.

have studied ATP and tryptophan binding by the method of gel filtration [29, 30]. As shown in fig. 4, L-^{14}C-tryptophan emerges from a Sephadex G-100 column with the peak of enzyme activity and with a constant ratio of tryptophan to enzyme throughout the peak. During the preliminary incubation of tryptophan with the enzyme, ATP must be present for the binding to occur. Similarly, ^{14}C-ATP binds to the enzyme in the presence but not in the absence of tryptophan. One mole of either substrate is bound per mole of enzyme of molecular weight 74,000. When pure leucyl-tRNA synthetase is substituted for the tryptophanyl-tRNA synthetase, there is no binding of tryptophan.

When subjected to sodium dodecyl sulfate-polyacrylamide gel electrophoresis [31,32] the tryptophanyl-tRNA synthetase migrates as a single major polypepide (fig. 5) of molecular weight 37,000 (fig. 6). The precision of this method is demonstrated by the polypeptide molecular weights, all between 36,000 and 38,000, determined on six gels made on four separate occasions.

In the same procedure [32] the major band of a preparation of *E. coli* tyrosyl-tRNA synthetase judged more than 75% pure (fraction VII, table 1, ref. [14]) migrates as a polypeptide chain of molecular weight 44,000. Since the enzyme has a molecular weight of 95,000 by sedimentation studies [14], it can be placed tentatively in the dimer category.

In summary, the active form of tryptophanyl-tRNA synthetase appears to be a dimer of molecular weight near 75,000 composed of polypeptide chains

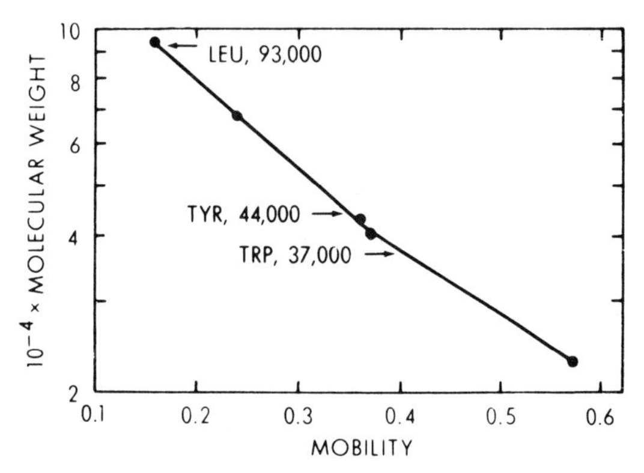

Fig. 6. Molecular weight of polypeptide chains as a function of mobility on sodium dodecyl sulfate-polyacrylamide gel. The procedure followed was that described in fig. 5. Standards (•, from left top to right bottom) were 10 μg samples of rabbit muscle phosphorylase *a* (molecular weight of polypeptide chain 94,000), bovine serum albumin (68,000), ovalbumin (43,000), rabbit muscle aldolase (40,000) and trypsin (23,000). The arrows indicate positions of tryptophanyl-tRNA synthetase (6 gels), tyrosyl-tRNA synthetase (3 gels) and leucyl-tRNA synthetase (6 gels).

of molecular weight near 37,000. The dimer possesses one binding site for ATP and tryptophan. Our experiments do not exclude the possibility of another binding site, because significant dissociation of enzyme and substrate can occur during gel filtration. We have completed the amino acid analysis, and we are preparing for *N*-terminal and peptide analysis to determine whether or not the monomers are identical. Bruton and Hartley [8] have reported the single *N*-terminal amino acid sequence, alanine, glycine, glycine, threonine, for the subunits (48,000 in molecular weight) of *E.coli* methionyl-tRNA synthetase.

3.3. *Leucyl-tRNA synthetase*

The leucyl-tRNA synthetase of *E. coli* presents a different picture. We have purified the enzyme 560-fold, which is at least 90% pure by disc gel electrophoresis. As shown in fig. 7 sodium dodecyl sulfate-polyacrylamide gel electrophoresis reveals one major polypeptide. In six separate runs with comparison to standards the molecular weight of the major polypeptide chain averaged 93,000 (fig. 6). In sucrose density gradient centrifugations [22] with the three reference enzymes previously mentioned, the leucyl-tRNA behaves as a protein of molecular weight 90,000. From these data we conclude the active form of leucyl-tRNA synthetase is a single polypeptide chain. In this respect it differs from the prolyl-, tryptophanyl-, and tyrosyl- and is similar to the valyl- and isoleucyl-tRNA synthetases. The valyl-tRNA synthetase of *E.coli* consists of a single polypeptide chain of molecular weight 110,000, containing a single binding site for valine, ATP and $tRNA^{Val}$ as shown by Yaniv and Gros [16,33]. The isoleucyl-tRNA synthetase of *E. coli*, as described by Berg and his colleagues, has a molecular weight of 112,000 [15] and contains one particular sulfhydryl group essential for isoleucyl-AMP formation [34]. The enzyme cannot be split into smaller polypeptide chains [35], bears the single *N*-terminal amino acid, threonine [35], and has an amino acid and tryptic peptide composition dictating against its being composed of identical subunits [15].

3.4. *Aminoacyl-tRNA synthetases of human leukemic lymphocytes*

The study of aminoacyl-tRNA synthetases of human origin is obstructed at the outset by the relative unavailability of tissue. The difficulty is increased by the need for human tRNA to be used as a substrate. Matthaei and Schoech [36] used the human placenta, and Anderson [37] used a human leukemic spleen as sources for both the aminoacyl-tRNA synthetases and for the tRNA's. We have been using the small lymphocytes of human chronic lymphocytic leukemia as our source of enzymes and tRNA's. These cells present several advantages from the enzymologist's point of view alone. Chronic lympho-

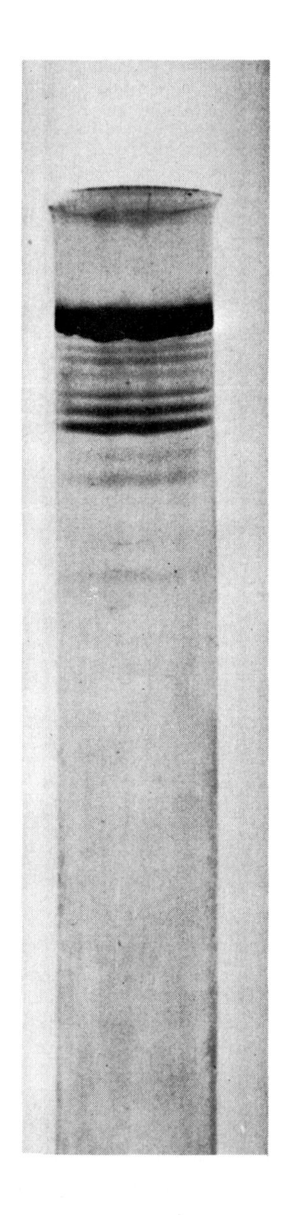

Fig. 7. Sodium dodecyl sulfate-polyacrylamide gel electrophoresis of leucyl-tRNA synthetase. The procedure was that described in fig. 5. The origin is at the top.

cytic leukemia is the most benign and most prevalent of human leukemias, and patients with this disease can exist without treatment for long periods of time. The number of small lymphocytes per volume of blood may exceed even 200 times the normal level, and by simple centrifugation techniques we have prepared 400 g of these cells, relatively free of other blood cells, from a single patient in less than 10 hr. The enzymes are stable during storage of the cells in liquid nitrogen. Lymphocytes are an extraordinarily rich source, being endowed nearly as well as rapidly dividing *E. coli* cells. Thus in a rough comparison under arbitrary assay conditions, extracts of *E. coli* contain from 200 to 2,800 units of the 20 aminoacyl-tRNA synthetases per gram of wet cells [18], whereas extracts of human lymphocytes contain 400 units of lysyl- and 250 units of tryptophanyl-tRNA synthetase per gram. However, the tRNA content of human lymphocytes and of other mammalian cells [37,38], about 6 A_{260} units/g, is only one-tenth that of *E. coli*. Therefore, availability of tRNA for routine enzyme assays becomes the limiting factor in this work. With tRNA from *E. coli* and from brewer's yeast we have screened lymphocyte extracts for all 20 aminoacyl-tRNA synthetases. The four most active enzymes are those for lysine and tryptophan with brewer's yeast tRNA and alanine and arginine with *E. coli* tRNA. The enzymes for proline and leucine are not detectable with these substrate tRNA's. Anderson [37] has proposed conservation of the co-recognition sites of $tRNA^{Ala}$ and alanyl-tRNA synthetase throughout evolution on the basis of the complete crossreactivity of these enzymes and substrates for *E. coli*, lobster tail, rabbit liver, and human leukemic spleen. However, yeast $tRNA^{Ala}$ may be an exception, for it is not a good substrate for the alanyl-tRNA synthetase of human leukemic lymphocytes.

Our studies of the human enzymes are now just beginning. We have purified the lysyl-tRNA synthetase 90-fold by chromatography on DEAE-cellulose and hydroxylapatite, and we have determined optimal assay conditions. Both the lysyl- and the tryptophanyl-tRNA synthetases are stable, and we anticipate no difficulty in purifying them to homogeneity for the characterizations we have reported for *E. coli* enzymes.

Acknowledgement

This work was supported by United States Public Health Service (National Institutes of Health) Grant 5-PO1-AM-09001 and American Cancer Society Grant IN-5IJ, Subgrant 1. K.H.M. is a Markle Scholar in Academic Medicine and was supported as a Faculty Research Associate of the American Cancer Society, Grant PRA-21. M.L. and D.K. were partly supported by National Institutes of Health Training Grant HE-5463. D.R.J. was partly supported by National Institutes of Health Training Grant GM-02011. J.C.P., Jr., is a Medical Scientist Fellow of the Life Insurance Medical Research Fund, Grant MS-69-15.

References

[1] J.T. Madison, Ann. Rev. Biochem. 37 (1968) 131.
[2] M. Staehelin, H. Rogg, B.C. Baguley, T. Ginsberg and W. Wehrli, Nature 219 (1968) 1363.
[3] G.R. Philipps, Nature 223 (1969) 374.
[4] M. Levitt, Nature 224 (1969) 759.
[5] M.P. Stulberg, J. Biol. Chem. 242 (1967) 1060.
[6] A. Böck, European J. Biochem. 4 (1968) 395.
[7] C.J. Bruton and B.S. Hartley, Biochem. J. 108(1968) 281.
[8] C.J. Bruton and B.S. Hartley, Biochem. J. In press.
[9] D. Cassio and J.P. Waller, European J. Biochem. 5 (1968) 33.
[10] M. Lazar, M. Yaniv and F. Gros, C.R. Acad. Sci. Paris 266 (1968) D531.
[11] J.R. Katze and W. Konigsberg, J. Biol. Chem. 245 (1970) 923.
[12] M.-L.Lee and K.H. Muench, J. Biol. Chem. 244 (1969) 223.
[13] D.R. Joseph and K.H. Muench, Fed. Proc. 29 (1970) 467.
[14] R. Calendar and P. Berg, Biochemistry 5 (1966) 1681.
[15] A.N. Baldwin and P. Berg, J. Biol. Chem. 241 (1966) 831.
[16] M. Yaniv and F. Gros, J. Mol. Biol. 44 (1969) 1.
[17] R.L. Heinrikson and B.S. Hartley, Biochem. J. 105 (1967) 17.
[18] K.H. Muench and P. Berg, in: Procedures in Nucleic Acid Research, eds. G.L. Cantoni and D.R. Davies (Harper and Row, New York, 1966) p. 375.
[19] K.H. Muench, Biochemistry 8 (1969) 4872.
[20] O.H. Lowry, N.J. Rosebrough, A.L. Farr, and R.J. Randall, J. Biol. Chem. 193 (1951) 265.
[21] E. Layne, in: Methods in Enzymology, vol. III, eds. S.P. Colowick and N.O. Kaplan (Academic Press, New York 1957) p. 447.
[22] R.G. Martin and B.N. Ames, J. Biol. Chem. 236 (1961) 1372.
[23] D.R. Harkness, Arch. Biochem. Biophys. 126 (1968) 503.
[24] A.J. Gottlieb and H.H. Sussman, Biochim. Biophys. Acta 160 (1968) 167.
[25] M.J. Chrispeels, R.F. Boyd, L.S. Williams, and F.C. Neidhardt, J. Mol. Biol. 31 (1968) 463.
[26] D.R. Joseph and K.H. Muench, Fed. Proc. 28 (1969) 865.
[27] L. Ornstein, Ann. N.Y. Acad. Sci. 121 (1964) 321.
[28] P. Andrews, Biochem. J. 91 (1964) 222.
[29] A.T. Norris and P. Berg, Proc. Natl. Acad. Sci. U.S. 52 (1964) 330.
[30] U. Lagerkvist, L. Rymo, and J. Waldenstrom, J. Biol. Chem. 241 (1966) 5391.
[31] J.V. Maizel, Science 151 (1966) 988.
[32] K. Weber and M. Osborn, J. Biol. Chem. 244 (1969) 4406.
[33] M. Yaniv and F. Gros, J. Mol. Biol. 44 (1969) 17.

[34] M. Iaccarino and P. Berg, J. Mol. Biol. 42 (1969) 151.
[35] D. Arndt, and P. Berg, J. Biol. Chem. 245 (1970) 665.
[36] J.H. Matthaei and G.K. Schoech, Biochem. Biophys. Res. Commun. 27 (1967) 638.
[37] W.F. Anderson, Biochemistry 8 (1969) 3687.
[38] N. Delihas, and M. Staehelin, Biochim. Biophys. Acta 119 (1966) 385.

EVOLUTION OF PYRIDINE NUCLEOTIDE LINKED DEHYDROGENASES

N.O. KAPLAN

Department of Chemistry, University of California at San Diego,
La Jolla, California 92037, USA

Abstract: Kaplan, N.O. Evolution of Pyridine Nucleotide Linked Dehydrogenases. *Miami Winter Symposia,* 1, pp. 66–88. North-Holland Publishing Company, Amsterdam, 1970.

Most pyridine nucleotide linked enzymes have been found to consist of subunits. Recent evidence has suggested that there may be similarities in the overall subunit structure of a number of dehydrogenases. However, the interaction between subunits appears to differ with the various dehydrogenases. A summary of these structural relationships is considered. The evolution of a particular dehydrogenase has resulted in the conservation of a given portion of the molecule. In the case of glyceraldehyde-3-phosphate dehydrogenase, the sequence around the active site cysteine has been almost completely maintained. In the case of lactate dehydrogenase, the essential sulfhydryl peptide has also undergone little change during evolution. The variation in homologies between the two main types (M and H) of lactate dehydrogenase are discussed. Relationships of the primary structures among the different dehydrogenases will be surveyed and a summary are presented on the nature of interaction between subunits in dehydrogenases, as well as the role of the pyridine coenzymes in maintaining the quaternary structures of this group of proteins. Allosteric factors which are involved in the regulation of catalytic activity in the dehydrogenases are discussed. Concepts of the mechanism of action of dehydrogenase activity relating to evolutionary change are proposed. Suggestions are made on the significance of the structures of the coenzymes to the origin of the A and B types of dehydrogenase.

1. Types of pyridine nucleotides

Before discussing the various pyridine nucleotide linked dehydrogenases, it is of interest to examine the two types of pyridine coenzymes themselves, namely DPN(NAD) and TPN(NADP). The two coenzymes are ubiquitously distributed in nature and appear to have distinct functional roles, DPNH being utilized for the generation of ATP, whereas TPNH is the source of reductive power in synthetic reactions. It would therefore seem reasonable to assume that both types of coenzyme originated at a very early time in evolution and that the TPN and DPN linked enzymes underwent early separate evolutionary changes. If this assumption is correct, we would expect to find the properties of TPN dehydrogenases differing somewhat from the DPN linked proteins. Although there is not as yet any extensive information available concerning this thesis, there is some data suggesting that the characteristics of the two types of dehydrogenases are different.

2. Restrictions on evolution

In considering the evolution of the pyridine nucleotide dehydrogenases, it is of importance to stress that changes occurring in a particular enzyme are limited to the extent that the alterations cannot be constituted in any manner that would inhibit the coenzymes from binding to the protein. The fact that there has not been any evolution of the coenzymes is a reflection of the large number of enzymes that require either DPN or TPN. The acetyl pyridine analogue of DPN (where a $\overset{\overset{O}{\|}}{C}CH_3$ replaces a $\overset{\overset{O}{\|}}{C}NH_2$ group) can be substituted or act as a more efficient acceptor in a number of dehydrogenases. However, as illustrated in table 1, some rabbit dehydrogenases do not react, or have little activity with the analogue, whereas other enzymes such as liver alcohol dehydrogenase show a higher V_{max} for the analogue as compared to the natural coenzyme. These observations indicate that changes in the coenzyme are not possible since this would require mutations of a large number of genes

Table 1
Reaction of 3-acetylpyridine-*DPN with different rabbit dehydrogenases.

Dehydrogenase	Rate of reaction compared with DPN (%)
Liver alcohol	450
Liver glutamic	150
Heart mitochondrial malic	125
Muscle lactic	22
Muscle triosephosphate	10
Heart lactic	4
Liver β-hydroxy butyric	<1
Muscle α-glycerophosphate	0

that control the synthesis of the many dehydrogenases. Therefore, any mutational event involving a dehydrogenase cannot result in a type of sequence change which will not allow for interaction of the coenzyme with the protein.

In order to approach the problem of dehydrogenase evolution, it is of value to compare the physical, chemical, and catalytic properties of the various catalysts; therefore a brief survey of some of these characteristics will be presented below.

3. Subunits in dehydrogenases

Most dehydrogenases have been found to consist of subunits. Some information relating to the molecular weights of dehydrogenases and number of subunits are summarized in table 2. It is of interest to note that glyceraldehyde phosphate dehydrogenase (GPD) [1-4] and the animal L-lactate dehydrogenase (LDH) [1,5–8] are of approximately the same molecular weight and contain four subunits of approximately 35,000. Yeast alcohol dehydrogenase (ADH) is also a tetramer [1,9,10] whereas the liver ADH has only two subunits [1,11,12]. Both alcohol dehydrogenases appear to have subunits of 35,000 to 40,000. The DPN linked α-glycerol phosphate dehydrogenase also appears to contain subunits with weights of roughly 35,000 [13].

Most malate dehydrogenases have been found to have molecular weights between 60,000 and 70,000 [14,15]. There are two subunits per molecule. In the case of the *Bacillus* group, and several bacteria closely related to this

Table 2
Molecular weights and number of subunits of some dehydrogenases.

Dehydrogenase	Molecular weight	Number of subunits
Glutamate	320,000	6
Glyceraldehyde phosphate	140,000	4
Lactate	140,000	4
Yeast alcohol	130,000	4
Liver alcohol	84,000	2
Malate from most species	70,000	2
Malate from *Bacillus*	125,000	4
Glycerol phosphate	65,000	2
Steroid (*Pseudomonas*)	42,000	?
Mannitol 1-phosphate	40,000	?
Dihydrofolic reductase	20,000	?

group, the molecular weight of the malate dehydrogenase is roughly twice that which is found in most organisms; there are four subunits in the *Bacillus* malate dehydrogenase [16,17].

In general, the data which has been accumulated suggests that most dehydrogenases exist either as dimers or tetramers and that the basic subunit weight is approximately 35,000. This may imply a close evolutionary relationship between the dehydrogenases, although there are a number of enzymes which have been reported not to have these general characteristics. For example, the dihydrofolic reductase from animal tissues has been found to have a molecular weight of 20,000 [18].* The animal glutamate dehydrogenase in the oligomeric form has a molecular weight of about 320,000 [1,19-21]. In the case of the beef liver enzyme, however, when high concentrations are in solution, there is association into a very high molecular weight aggregate. It is of interest to note that the molecular weight of the dogfish liver glutamate dehydrogenase [22] remains constant (320,000) over a long range of concentrations and does not associate; this non-association also applies to the enzyme from rat liver [23]. The subunit weight of the glutamate dehydrogenase has been established to be about 53,000. Hence, the enzyme is a hexamer. The hexameric structure has been verified by electron microscopic studies by the late Dr. R.C. Valentine [24]. The subunit size of the glutamate dehydrogenase appears to be significantly larger than the primary units of most other dehydrogenases. Whether this enlargement of the polypeptide chain is related to the striking allosteric characteristics of the glutamate enzyme remains to be clarified. It is of interest to note that a related enzyme, alanine dehydrogenase from *Bacillus subtilis*, also has a higher molecular weight and appears to be a hexamer [25]. From the information available at the present time, the different pyridine nucleotide linked dehydrogenases appear mainly as polymers and have been found as dimers, tetramers, and hexamers.

The difference in length of the polypeptide chain may also reflect a difference in function. The glutamate dehydrogenase promotes a dehydrogenation at a C-N bond whereas the other dehydrogenases have a C-O function [27]. The possibility exists that the glutamate enzyme separated early in evolution from an enzyme with a generalized C-O activity. In this connection, it is worthwhile noting that microbial glutamate dehydrogenases are also of high molecular weight (300,000) although the enzymes from *Neurospora* and

* A value of 40,000 was obtained for the mannitol 1-phosphate dehydrogenase from *Aerobacter aerogenase* in our laboratory in 1962 [26]. No studies on subunits were carried out at that time; this question is now being examined, and it will be of some interest to ascertain whether this enzyme exists as a single subunit.

yeast do not respond to GTP or ADP in an allosteric fashion. The microbial enzymes are either specific for DPN or TPN; the aminal glutamate dehydrogenases have the capacity to react with both types of pyridine coenzymes.

A large number of experiments have been carried out in an attempt to determine whether the subunits of LDH and GPD are enzymatically active. Most of these studies strongly suggest that the tetramers are the significantly enzymatically active entities [28-30]. There has, as yet, been no definite proof of native enzymatic activity inherent in the monomers or dimers of GPD or LDH.

Bacillus subtilis MDH is a tetramer and can be dissociated under controlled conditions into a dimer which has little or no catalytic activity. In contrast to most other MDHs, which are dimeric, the *Bacillus* malate dehydrogenase appears to require the interaction of four subunits for the generation of normal enzymatic activity [17]. It appears very likely that not only is the proper folding of the subunit of importance, but also that the spatial relationship of the subunits to each other is critical in determining the normal catalytic characteristics of a dehydrogenase.

4. Evolutionary significance of subunits

Ther are a number of factors which might be considered as evolutionary advantages for most of the dehydrogenases occurring as polymeric enzymes. Since most of the dehydrogenases have a significant number of essential sulfhydryl groups, the quaternary structure resulting from the interaction of the subunits may afford protection to such groups which is not inherent in the subunit itself.

Perhaps the most impelling reason for an active polymeric form is the potentiality of having multisites which can have negtive or positive cooperativity. Conway and Koshland [31] have made an extensive study of DPN binding to rabbit muscle GPD. In their studies, evidence was obtained indicating negative cooperativity, that is the binding of each molecule of DPN decreases the affinity for coenzyme for the recent sites of the neighbouring subunits. The first DPN is bound tightly whereas the fourth mole of nucleotide is bound loosely and has a high dissociation constant; the second and third molecules appear to be intermediate in their affinities between the first and fourth sites (table 3). The first mole of DPN bound was shown to induce a conformation change in all of the subunits; this is indicated by an increase in the reactivity of sulfhydryl groups as well as by viscosity changes. The fourth molecule of DPN bound induces little or no change in the conformation of GPD. Furthermore, it is of interest to note that as successive molecules of DPN are bound, the specific activity per DPN site increases significantly.

Table 3
Average dissociation constants for individual steps in dissociation of DPN from glyceraldehyde-3-phosphate dehydrogenase from rabbit muscle (from Conway & Koshland) [31].

	Average value of *K*
$E.DPN_1 \rightarrow E + DPN$	10^{-11}
$E.DPN_2 \rightarrow E.DPN + DPN$	10^{-9}
$E.DPN_3 \rightarrow E.DPN_2 + DPN$	3×10^{-7}
$E.DPN_4 \rightarrow E.DPN_3 + DPN$	2.6×10^{-5}

Earlier reports by deVijlder and Slater [32] as well as those of Listowsky et al. [33] suggested a unique role for the first mole of DPN bound. This uniqueness is rather surprising since preliminary X-ray crystallographic data [34] shows that the four identical subunits are engaged at least in a two-fold symmetrical axis. Although information on the other GPDs is not complete, there appears to be variation in the extent of the "negative cooperativity effect". In fact, there is some evidence from temperature jump experiments by Kirschner et al. [35] for a positive cooperative effect in yeast GPD. Recently Koshland et al. [36] have reported both negative and positive cooperative effects with DPN for the yeast enzyme. Jaenicke [37] has also observed modulation of the yeast enzyme.

As suggested by Conway and Koshland, the fact that the fourth mole of DPN is loosely bound may indicate an evolutionary advantage. This is reflected by the fact that there is a greater turnover number per site, when four moles of DPN are bound than when two are bound. The loosely bound DPN may be more easily reduced, and the resulting DPNH may dissociate readily to react with other enzymes such as LDH. From the studies summarized above, it is evident that DPN concentration may play a sensitive role in controlling the GPD reaction.

Another example of the interaction of subunits regulating the activity of dehydrogenase is a recent study of the lobster muscle LDH (38). This enzyme, when purified, shows sigmoid kinetics in the oxidation of lactate by DPN (see fig. 1). Low concentrations of DPNH have been found to activate the oxidation of lactate; (AcPy)DPNH has been found to be more effective than DPNH in promoting the activation (see fig. 2). Lactate concentrations are also factors in determining the types of kinetics observed with the lobster LDH (see fig. 3). The kinetic characteristics of the oxidation of lactate by DPN have the general features of a second order autocatalysed reaction. In order to

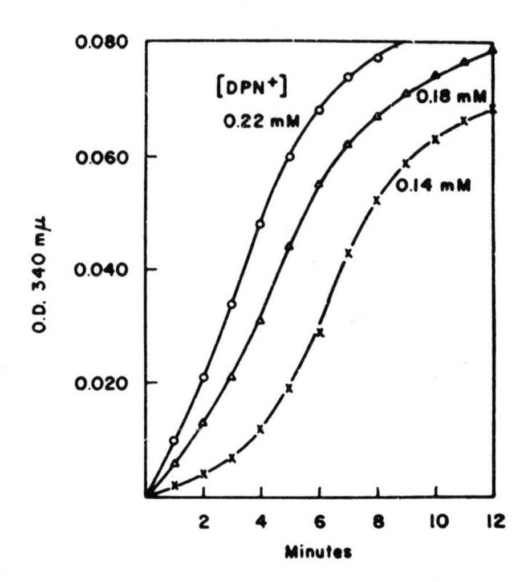

Fig. 1. Progress curves of forward reaction catalysed by lobster lactate dehydrogenase. Enzyme and L-lactate concentrations are identical for each of the curves. L-Lactate concentration was 20 mM. DPN^+ concentrations were as indicated in the diagram [38].

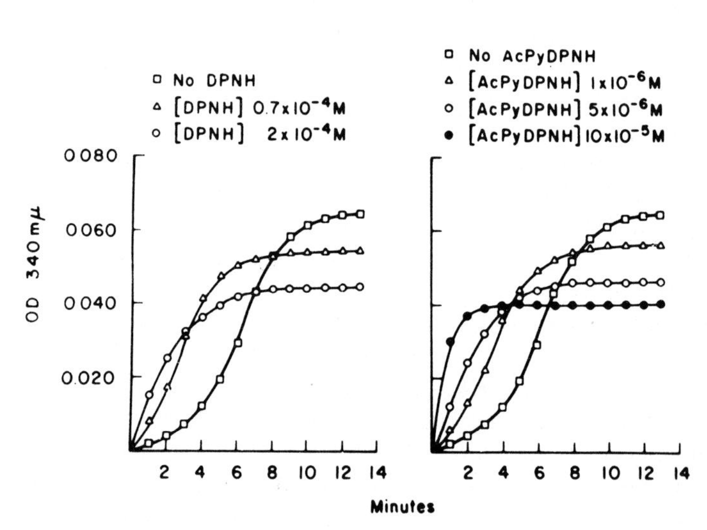

Fig. 2. Comparison of activating effects of (AcPy) DPNH and DPNH on lobster muscle lactate dehydrogenase [38].

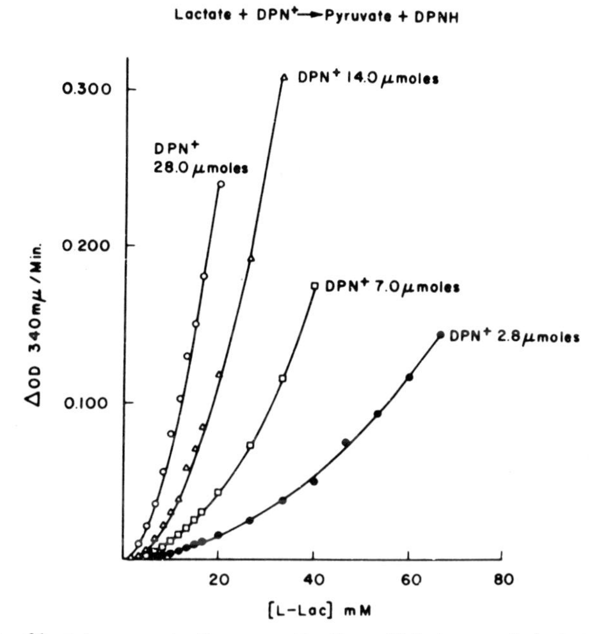

Fig. 3. Effects of lactate concentrations on activation of lobster muscle lactate dehydrogenase [38].

elucidate the unusual kinetic findings of the lobster LDH, we have proposed a model based on the concept of subunit interaction to interpret the DPNH activation. We have suggested that oxidized and reduced forms of the coenzyme occupy the same sites on the enzyme, and that the binding of the reduced nucleotide to one or more of the four binding sites, causes a structural change in the protein so that the binding of the oxidized DPN to the remaining available sites is enhanced. In a sense, this is analogous to what has been observed with rabbit muscle glyceraldehyde phosphate dehydrogenase (see above). We have also proposed that no specific regulating site is present in the lobster LDH.

Fig. 4 schematically illustrates the mechanism postulated for the regulation of the lobster LDH, the implication of this proposal being that there are five possible species of the enzyme. The binding of DPNH to one site would allow for an increase in affinity for DPN in the other sites. This would allow for the rate of lactate oxidation to be a variable of DPN and DPNH concentration, and to each of the several binding constants shown in fig. 4. As a consequence of this scheme, the initial oxidation of lactate will be slow and progressively

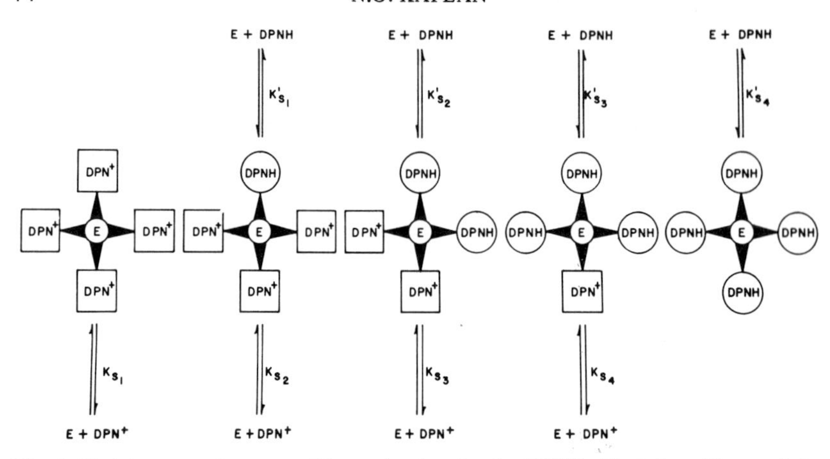

Fig. 4. Model proposed as a possible mechanism for the DPNH-effected positive modulation of the rates of the forward reaction in lobster muscle lactate dehydrogenase [38].

increase as the DPNH concentration rises. When the available sites are filled with DPNH, the reaction then reaches equilibrium. Hence, low concentrations of DPNH would induce a stimulation of lactate oxidation, whereas high concentrations of the reduced coenzyme would result in a lowering of the rate of oxidation of the lactate because of the competition of DPN and DPNH for the same site.

The above considerations suggest that there is modulation for lactate oxidation and thereby allows for a build-up in the regulating system whereby the product, DPNH, controls its own production. At low concentrations of DPNH, the reduction of DPN is favored and at high concentration of DPNH, it is inhibited. Such regulation would assure optimal DPNH concentrations under changing physiological conditions. Factors controlling carbohydrate metabolism and ATP generation in lobsters and other crustaceans are not well understood. It is known, however, that carbohydrate metabolism and muscular activity undergo changes during the different intermolt stages and the periodic ecdysis.

The DPNH activation of lactate oxidation has not been observed with the vertebrate LDHs. Other crustaceans show some of the characteristics exhibited by the lobster enzyme. There is a possibility that lactate formed in lobster muscle is not transported to the liver and metabolized by the Cori cycle to the same extent as is found in vertebrates. Lactate oxidation in voluntary muscle therefore may be of considerably more significance in crustaceans than in higher animals. The lobster muscle LDH appears to have evolved as a mechanism for regulation of lactate metabolism within the muscle itself. One might therefore speculate that following activity in lobster muscle, the accumulated lactate might be oxidized aerobically. This would lead to a decrease in DPNH concentration to the point at which the inhibition of LDH no longer occurs; the concentration of reduced coenzymes would then reach a level which would facilitate the oxidation of lactate. It is thus reasonable to assume that the unique properties of the lobster muscle LDH are particularly geared to the metabolism of lobster and that this enzyme has evolved from changes during the amino acid sequence, which allowed for a distinct type of subunit interaction.

Non-equivalent binding of DPNH to bovine M_4 LDH has been reported by Weber and Anderson [39]. Their results suggest that binding of the first molecule of reduced nucleotide hinders the binding of the following three molecules of DPNH. Although the authors have ascribed this behavior to relaxation effects, it is still possible for these results to be interpreted as being due to a modification of other subunit sites, as a consequence of the binding of the first mole of reduced DPN.

Another feature of dehydrogenases, which may be significant from an evolutionary point of view, is that the binding of the coenzyme helps to maintain the native structure against denaturating reagents [40]. The pyridine coenzymes appear to play an important role in promoting the folding and reassociation of subunits into the native enzymatically active form. An almost complete dependence on DPN has recently been reported for the association of yeast glyceraldehyde phosphate dehydrogenase subunits obtained after dissociation of the tetramer [41]. DPNH has been found to greatly accelerate the rate of conversion of chicken H_4 LDH subunits into a catalytic entity, similar to that of the native tetrameric structure [42]. It is of interest to note that in the absence of DPNH, an intermediate has been detected which has both catalytic and physical properties which are distinct from the native enzyme.

Most of the properties of dehydrogenase appear to be inherent in the subunit itself. That is the primary sequence of the subunit dictates the tertiary and quaternary structure of the enzyme, and that interaction with other subunits is a reflection of this sequence. For example, in LDH the proper folding and association into a tetramer is a result of the sequence. Subtle changes in primary structure may alter the capacity of some subunits to interact with polypeptide chains of a different sequence. This is indicated by the fact that in most higher vertebrates, the two types of LDH (H and M) hybridize readily [29]. However, in a number of lower vertebrates, this hybridization does not exist. For example, the frog H_4 does not hybridize with its M counterpart *in vivo* or *in vitro*, but it is of interest that the M enzyme will form hybrids with mammalian and avian H forms. This would imply that a change has occurred in the frog H polypeptide which is not compatible for interaction with the M subunit but still allows the H subunits to interact among themselves to form an active tetramer.

In hybrid enzymes, as a rule, the catalytic and physical properties are maintained for the subunit. Most of the data implies that little or no modification of the subunit properties occurs as the result of interaction with subunits of a different sequence. This may be taken as evidence to advance the hypothesis that evolutionary changes in the sequence of dehydrogenases are responsible for the modifications of properties of the enzyme; these changes must, however, not affect the capacity of the subunit to aggregate into an active polymeric protein.

5. Comparative structure of alcohol dehydrogenases

Total and partial sequence studies have been carried out on a few dehydrogenases which allows for some speculation regarding the evolution of a particular dehydrogenase. In the case of alcohol dehydrogenase, there are two enzymes which have been studied in some detail – namely the yeast and horse liver enzymes. Although the enzymes are of different molecular weights, they both contain zinc. Finding of zinc in the two alcohol dehydrogenases is of interest since other dehydrogenases have not definitely been shown to require this metal for activity. The horse liver and yeast enzymes differ considerably in their specifity to oxidize various alcohols.

A peptide containing an essential cysteine has been isolated from both the liver and yeast enzymes [9]. The sequence of this peptide is as follows:

	1	2	3	4	*5	6	7	8	9	10
Horse liver ADH	Ala	Thr	Gly	Ile	Cys	Arg	Ser	Asp	Asp	His
	1	2	3	4	*5	6	7	8	9	10
Yeast ADH	Tyr	Ser	Gly	Val	Cys	His	Thr	Asp	Leu	His

There are obviously certain similarities in the sequence of the two decapeptides. In a sense, it is surprising that two enzymes which are so different have such similarities. One might assume that positions 2 and 7 require a hydroxy amino acid or that in position 6 a basic amino acid is essential; position 10 appears to be conservative for histidine.

Theorell's group has almost completed the sequence of horse liver alcohol dehydrogenase. In the course of studies with the liver enzyme, it was noted that the enzyme possessed steroid dehydrogenase activity. Closer scrunity of this finding indicated that the steroid activity was associated only with certain isoenzymes of the liver ADH [11].

A separate entity has been isolated which has about two-thirds of the activity towards ethanol as does the ordinary alcohol dehydrogenase; but many times more active with steroids [43]. This entity is also a dimer containing zinc, and has been referred to by Theorell as LAD_S; the ordinary ADH, which has little or no steroid dehydrogenase activity, is referrred to as $LADH_E$ and is found in great excess over $LADH_S$. A hybrid containing both types of subunits has also been isolated. Analyses of the polypeptide chain of $LADH_S$ shows that there are five amino acid differences when compared to $LADH_E$ [44] (table 4).

The fact that five amino acid changes can induce such a great change in specificity is quite remarkable; furthermore, not only is the specificity with respect to the steroid changed, but also the $LADH_S$ has a much greater affinity for DPNH than does LAD_E. It seems likely that these two types, by gene duplication, bear relationship to the duplication of lactate dehydrogenase (see below). Like the M and H LDHs, the $LADH_S$ and $LADH_E$ can form a hybrid consisting of one unit of each type. It is of interest to note in the hybrid, the subunits are catalytically independent.

It seems reasonable to assume that the five displacements are at, or near, the binding and active sites of the enzyme. The changes are in a sense unusual, because immunologically there appears to be no cross reactivity with the two types of subunits [45]. We have observed cross reactivity in enzymes with enzymes that have a much larger number of replacements. The $LADH_E$ and $LADH_S$ systems would seem to be of great value to evolutionary studies dealing with the origin of new enzyme functions. In this connection, it is of

Table 4
Amino acid differences between ethanol and steroid dehydrogenase*.

Position from *N* terminal	Ethanol	Steroid
17	Glu	Gln
92	Thr	Ile
99	Arg	Ser
108	Phe	Leu
(362–370)	Glu	Lys

* H. Jörnvall, unpublished; quoted by H. Theorell [44].

interest to note that mutations in *Pseudomonas testeroni* result in dehydrogenases with altered steroid specificities [46].

6. Glyceraldehyde phosphate dehydrogenases

A good deal of chemical information is now available on the glyceraldehyde phosphate dehydrogenases. Harris and his associates have now achieved the complete sequence of both the lobster and pig muscle glyceraldehyde phosphate dehydrogenase [47,48]. Fig. 5 compares the sequences of the enzymes. There are approximately 90 amino acid differences between the two enzymes which is roughly one-fourth of the total chain length.

Cysteine 149 is the sole residue which is alkylated after treatment of the native enzyme with iodoacetate. This cysteine is also acetylated by acetyl phosphate and appears to be the residue which is critical for enzymatic activity. The sequence around the cysteine 149 has essentially been maintained as the same sequence which has been observed in all GPDs from all sources [48, 49]. This is illustrated in fig. 6.

It is of interest that the available information on the rabbit, ox, and pig enzymes indicate that the three mammalian GPDs are almost identical. Surprisingly, there is a 67% homology between the yeast and mammalian enzymes. A comparison of the pig, lobster, and yeast proteins show that 60% of the sequence is conserved. This degree of retention of homology appears to be considerably greater than that observed for other proteins such as hemoglobin or cytochrome c. It appears reasonable to assume that conservation within an enzyme such as GPD would be greater than a protein such as cytochrome c, because of the requirements for maintaining proper folding and subunit interaction for the retention of specificity and catalytic activity.

However, the amino acid changes which have been observed must be factors in promoting the kinetic and physical distinctions of the enzymes from different species. For example, the affinity of the mammalian GPDs for DPN is some 100-fold greater than that of the yeast enzyme.

7. Lactate dehydrogenases

Amino acid compositions of a number of vertebrate LDHs have shown the H type enzymes from the various species to be much more closely related to each other than are the H and M enzymes of one species; not only do they differ in chemical composition but also in physical and catalytic properties.

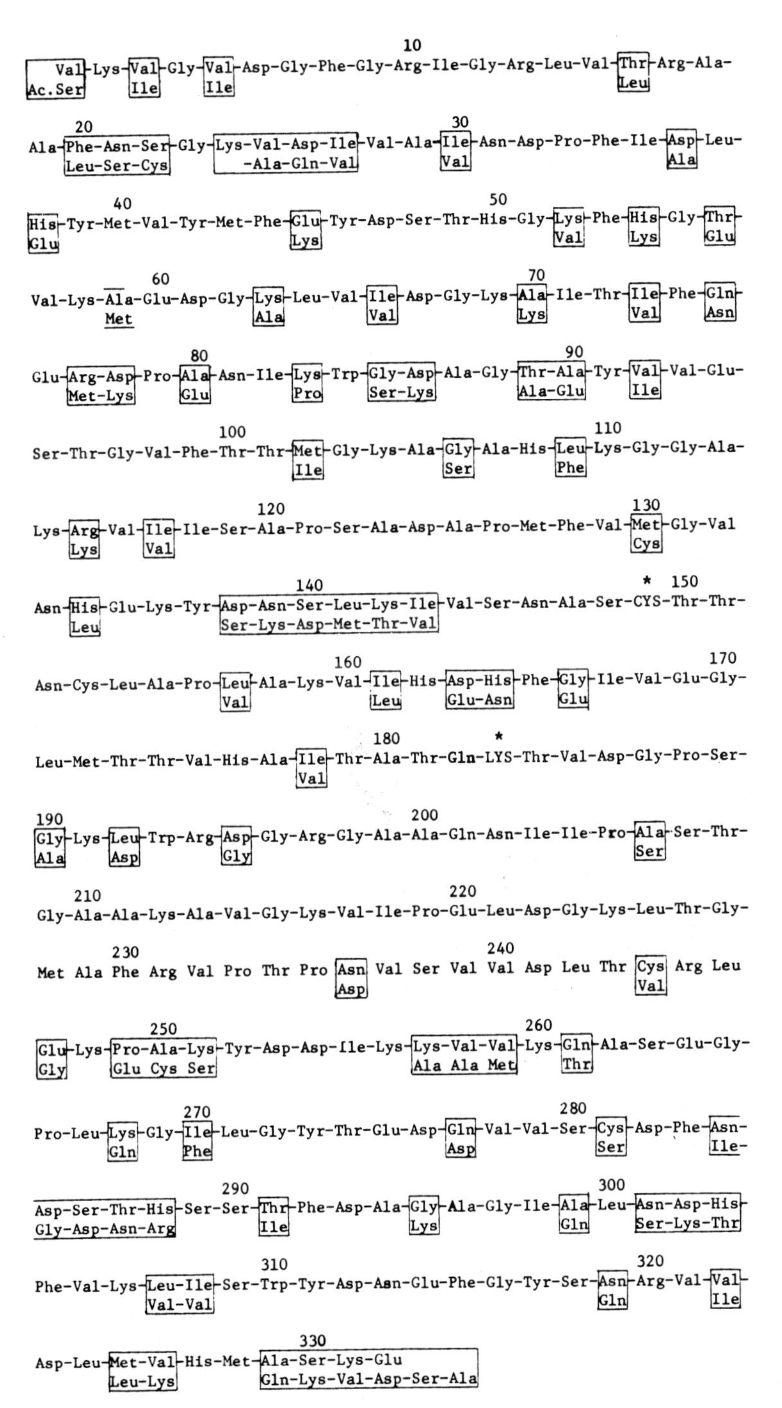

Fig. 5. Sequence of pig and lobster muscle GPDs [48].

					5			*		10					15			
Pig, Rabbit, Chicken, Ostrich, Sturgeon, Honey Bee, and Yeast	lys	ile	val	ser	asn	ala	ser	cys	thr	thr	asn	cys	leu	ala	pro	leu	ala	lys
Man	-	ile	ile	ser	asn	ala	ser	cys	thr	thr	asn	cys	leu	ala	pro	leu	ala	lys
Halibut	-	val	val	ser	asn	ala	ser	cys	thr	thr	asn	cys	leu	ala	pro	leu	ala	lys
Lobster	thr	val	val	ser	asn	ala	ser	cys	thr	thr	asn	cys	leu	ala	pro	val	ala	lys
Blue Crab	-	val	val	ser	asn	ala	ser	cys	thr	thr	asn	cys	leu	ala	pro	val	ala	lys
E. coli	gly	ile	val	ser	asn	ala	ser	cys	thr	thr	asn	cys	leu	ala	pro	leu	ala	lys

Fig. 6. Sequences of the active cysteine peptides of various glyceraldehyde phosphate dehydrogenases [48,49].

Several laboratories have reported the sequence of a peptide containing an SH group which is apparently essential for the normal binding of coenzyme. This sequence which has been obtained from the pig M_4, pig H_4, chicken M_4, beef H_4 and dogfish M_4 is as follows [50-53]:

1	2	3	4	5	6	7	8	9	10
Ile	Gly	Cer	Gly	Cys	Asn	Leu	Asp	Ser	Ala

It is of interest to note that chicken H_4 has a threonine substituted for a serine in the peptide. Although the peptide is essential for normal binding, the coenzyme still can bind when the cysteine is blocked. There is a possibility that the above sequence is not close to the binding site itself. There does not appear to be any significant homology between the LDH peptide and that described above for the alcohol dehydrogenases. Whether all DPN linked dehydrogenases have identical sequences for coenzyme binding is a matter of conjecture at the present time.

The existence of two types of LDH (H and M) is a good example of gene duplication and evolution into special functions. The two enzymes can carry out the same overall reaction although they are very different in their physical, chemical, and catalytic properties. It is not the aim of this paper to discuss in any detail the functional significance of the M and H LDHs, but to stress that specific changes in sequence lead to enzymes with survival advantages. These changes have led to the M type function as a pyruvate reductase and the H enzyme as a lactate dehydrogenase. Incomplete sequence data from Dr. Pfleiderer and his associates in Germany, as well as that of Dr. William Allison of our laboratory, indicate that there are areas of sequence homology in the two types, but also that there are considerable parts of the chains which are different. From these sequence studies, as well as from the amino acid composition and the trypsin peptide maps, I would guess that there will be a minimum difference of from 25 to 30 per cent between the H and M polypeptide chains. The establishment of the sequence of the two types should be of considerable importance in an understanding of the basic mechanisms involved in the evolution of a given enzyme.

A third type of LDH, which was first identified by its electrophoretic mobility, has been found in the sperm of birds and mammals; this enzyme has been termed LDH-X and may be the sole catalyst for pyruvate reduction in sperm [54]. The enzyme is much less specific for various keto acids and will reduce α-ketoglutarate with DPN at almost the same rate as it reduces pyruvate. The H and M dehydrogenase have little or no activity with α-ketoglutarate (see table 5). Oxidation of α-hydroxyglutarate, as well as lactate, is promoted by the LDH-X. The enzyme has recently been purified by Schatz and Segal [55] from rat testes. A molecular weight of 125,000 has been reported by these authors for LDH-X which is slightly lower than that found for the M and H LDHs of vertebrates. The turnover number of LDH-X also appears to be significantly lower. The enzyme not only differs from the H and M forms, with respect to substrate specificity, but also in its relative

Table 5
Reduction of α-keto acids by LDH-X (from testes) and other LDHs.

	Pyruvate	α-Ketoglutarate	Glyoxylate
Rat LDH-X*	1.0	0.62	0.48
Rat LDH M_4	1.0	0.0003	0.001
Rat LDH H_4	1.0	0.0005	——
Chicken LDH H_4	1.0	Not detectable	0.003

* Data of Schatz and Segal [55].

capacity to use coenzyme analogues. Although it is quite different from the H and M types, the LDH-X must be related to the other LDHs since there is some evidence that its subunits can hybridize with H or M units *in vitro* [56]. There is no information on its chemistry as yet. It will be of interest to learn whether the enzyme contains the same peptide essential for normal coenzyme binding which have been found in the M and H types. It may weel be that the LDH-X bears the same relationship to the other lactate dehydrogenase as does $LADH_S$ to $LADH_E$. A fourth type of LDH whose distribution has been found limited to nerve tissues of fish, has also been reported [57].

There are both D and L stereospecific DPN linked lactate dehydrogenases. In bacteria, the distribution of the D and L enzymes varies among different species. It had been generally thought that all animal enzymes were of the L specificity. Mr George Long in our laboratory has recently purified the LDH from the horseshoe crab (Limulus). To our surprise, he found that the enzyme was a D LDH. On examination of the LDH of other members of the subphylum Chelicerata, only D activity was found [58] (see table 6). Other arthropods were found to possess the L enzyme, except the barnacle which has a D specific LDH.* The mollusca all have the D form. In the annelids, the polychaetes have the D enzyme, whereas the oligochaetes have the L dehydrogenase (see table 6). The hydra LDH is specific for the L-lactate. In all species we have examined, only one type of enzyme is found, either the L or D, but never both activities.

It will be of interest to determine whether both the D and L enzymes arose from a common gene somewhere during the evolution of the invertebrates. Such a possibility does not seem to be too far fetched in view of the findings of the relationship of $LADH_E$ to $LADH_S$ by Theorell's group. To attack this problem, Mr. Long is now isolating the cysteine containing peptides of the Limulus D LDH and comparing their relationships to peptides from vertebrate LDHs as well as to the L LDHs of invertebrates.

All the invertebrate LDHs studied so far have been found to have molecular weights of approximately 70,000 and consist of two subunits. Hence, the evolution of the D LDH in animals seems to be related to the dimeric structure in contrast to the L enzyme in which a tetrameric structure has evolved. It is of interest to note that two different types of D LDH have been found in Limulus, one in heart and the second in muscle.

* There is still some controversary on the classification of the barnacle as to whether it is a true crustacean.

Table 6
Summary of invertebrate lactate dehydrogenase stereospecificities.

Phylum	Subphylum	Class	Common name	Lactate specificity	
				D-	L-
Arthropoda	Chelicerata	Arachnida	Spider A (legs, bodies)	+	
		Arachnida	Spider A (legs, bodies)	+	
			Spider B (bodies)	+	
			Tarantula (legs)	+	
			Wolf spider	+	
			Scorpion (legs, bodies)	+	
	Mandibulata	Crustacea	Shrimp (tail, muscle)		+
			Lobster (tail muscle)		+
			Crayfish (tail muscle)*		+
			Sand crab (leg muscle)		+
			Cancer crab (leg muscle)*		+
			Show bug		+
			Barnacle	+	
		Insecta	Gypsy moth (bodies)		+
			Grasshopper (legs)		+
		Chilopoda	Centipede		+
		Diplopoda	Millipede		+
Mollusca		Gastropoda	Moon snail (foot muscle)	+	
			Garden snail (foot muscle)	+	
			Abalone (foot muscle)	+	
		Cephalopoda	Octopus (leg muscle)	+	
Annelida		Polychaeta	Marine clam worm	+	
		Oligochaeta	Earthworm		+
			Common pond leech		+
Coelenterata		Hydrazoa	Hydra		+

* Assayed by Dr. F. Gleason in this laboratory; all other determinations by G. Long.

8. Stereospecificity and catalytic characteristics

Another characteristic which has been maintained in the evolution of a given dehydrogenase is the stereospecificity with respect to the 4 position of the reduced pyridine ring [59]. As an example, all LDHs, be they D or L enzymes, or whether they are from animal or microbial sources, are *A* specific enzymes. This is also true for the alcohol dehydrogenases. I would interpret this to indicate that the fundamental mechanism of a given dehydrogenase is not altered in the course of evolving new modified forms of the enzyme that are associated with the origin of new species.

There are certain properties and mechanisms which are common to a large number of dehydrogenases. In this connection, it may be of interest to mention the binary compounds of DPN. There are compounds in which α-keto acid adds to the position of the pyridine ring to form an adduct – which is equivalent to the redox state of DPNH. Mr. Everse and Dr. Zoll in our laboratory have studied these adducts and in fig. 7, the structure of the reduced

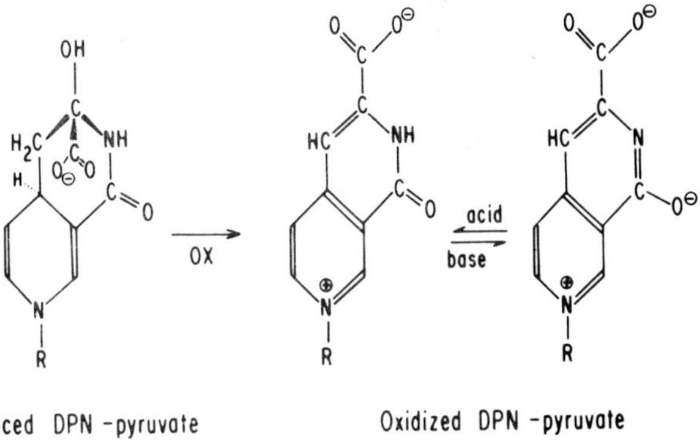

Reduced DPN -pyruvate Oxidized DPN -pyruvate

Fig. 7. Structure of reduced DPN-pyruvate and oxidized DPN-pyruvate adducts.

pyruvate adduct is given. This adduct will inhibit only lactate dehydrogenases and not other dehydrogenases (see table 7). In contrast the oxaloacetate adduct inhibits malic dehydrogenase, but not LDH. There is specificity with respect to the keto acid in the adduct.* These inhibition results indicate that the overall binding and mechanism of the various dehydrogenases are relative-

* Oxidized binary adducts have been prepared by Dr. Albert Winer (University of Kentucky) and his group. They show the same specific type of inhibition as observed with the reduced adducts.

Table 7
Inhibition of various dehydrogenases by reduced DPN^+ adducts.

Adduct	Chicken H_4 LDH	Mitochondrial MDH	Liver ADH	Yeast ADH	Liver GDH
	% Inhibition				
β-DPN^+-pyruvate	59	1	0	1	7
α-DPN^+-pyruvate	0	0	0	0	0
β-DPN^+-pyruvate ethyl ester	0	0	0	0	0
β-DPN^+-oxaloacetate	5	44	0	0	5
β-DPN^+-acetaldehyde	4	5	10	71	0
β-DPN^+-α-ketoglutarate	0	0	0	0	46
β-DPN^+-butyraldehyde	0	0	28	7	0
β-DPN^+-α-ketobutyrate	0	0	0	0	0

Concentration of adducts: 2×10^{-5} M.

ly similar, but the differences lie in the steric properties of the specific substrates.

There are many characteristics common to the various dehydrogenases which suggest evolutionary relationship. Preliminary X-ray diffraction studies show some overall similarities in LDH [60], GPD [34], and liver ADH [61]. The folding of subunits may be similar in most dehydrogenases, but it is likely that subunit interaction is an important factor in determining their dissimilarities. Elaboration of the detailed structure of several dehydrogenases will be of considerable help in the evaluation of the evolutionary relationships.

It seems likely that new dehydrogenases arose from gene duplication, and that the existence of duplicates allowed for types of modifications which eventually led to the occurence of forms with new functions. Specific amino acid changes at the "active site" may lead to the establishment of new specificities; changes away from the site can change the regulatory and physical properties of a dehydrogenase. Many of the dehydrogenases must have arisen early in life as we know it. For example, the GPDs show a great degree of similarity even when one compares the enzyme from *E. coli* or yeast, with that from mammals. On the other hand, enzymes such as the liver steroid dehydrogenase ($LADH_S$) and the LDH-X may have had a much later origin, and may have arisen during the evolution of the higher animals. A full knowledge of the sequence and structure of a number of different dehydrogenases as well as the same enzyme from different species, may be most rewarding in developing an understanding of the molecular changes which control evolution.

Acknowledgement

Some of the experimental work reported in this paper was supported by grants from the National Institutes of Health (CA–03611) and American Cancer Society (P–77J).

References

[1] K. Weber and M. Osborn, J. Biol. Chem. 244 (1969) 4406.
[2] W.F. Harrington and G.M. Kan, J. Mol. Biol. 13 (1965) 885.
[3] S.I. Harris and R.N. Pernham, J. Mol. Biol. 13 (1965) 876.
[4] R.Jaenicke, D. Schmid, and S. Knof, Biochemistry 7 (1968) 919.
[5] E. Appella and C.L. Markert, Biochem. Biophys. Res. Commun. 6 (1961) 171.
[6] R.D. Cahn, N.O. Kaplan, L. Levine, and E. Zwilling, Science 136 (1962) 962.
[7] A. Pesce, R.H. McKay, F.E. Stolzenbach, R.D. Cahn, and N.O. Kaplan, J. Biol. Chem. 239 (1964) 1753.
[8] A. Pesce, T.P. Fondy, F.E. Stolzenbach, F. Castillo, and N.O. Kaplan, J. Biol. Chem. 242 (1967) 2157.
[9] J.I. Harris, Nature 203 (1964) 30.
[10] M. Buhner and H. Sund, Europ. J. Biochem. 11 (1969) 73.
[11] R. Pietruszko, H.J. Ringold, T.K. Li, B.L. Vallee, A. Akeson, and H. Theorell, Nature 221 (1969) 440.
[12] H. Jornvall and J.I. Harris, Europ. J. Biochem. In press.
[13] H.A. White and N.O. Kaplan, J. Biol. Chem. 244 (1969) 6301.
[14] W.H. Murphey, C. Barnaby, F.L. Lin, and N.O. Kaplan, Biochemistry 6 (1967) 603.
[15] C. Wolfenstein, S. Englard, and I. Listowsky, J. Biol. Chem. 244 (1969) 6415.
[16] A. Yoshida, J. Biol. Chem. 240 (1965) 1113.
[17] W.H. Murphey, C. Barnaby, F.J. Lin, and N.O. Kaplan, J. Biol. Chem. 242 (1967) 1548.
[18] F.M. Huennekens, D.P. Mell, N.G.L. Harding, L.E. Gundersen, and J.H. Freisheim, 4th Intnl. Cong. on Pteridines (Toba, Japan 1970) (In Press).
[19] P. Dessen and D. Pantaloni, Europ. J. Biochem. 8 (1969) 292; H. Eisenbert and G.M. Tomkins, J. Mol. Biol. 31 (1968) 137.
[20] H. Eisenberg, Pyridine Nucleotide Dependent Dehydrogenases, ed. Horst Sund (Springer-Verlag, 1970) p. 293.
[21] J. Krause, K. Markau, M. Minssen and H. Sund, Pyridine Nucleotide Dependent Dehydrogenases, ed. Horst Sund (Springer-Verlag, 1970) p. 279.
[22] L. Corman, L.M. Prescott, and N.O. Kaplan, J. Biol. Chem. 242 (1967) 1383.
[23] Personal Communications from Dr. C. Frieden.
[24] R.C. Valentine, Abstracts Fourth European Regional Conference on Electron Microscopy Rome 2 (1968) 3.
[25] A. Yoshida and E. Freese, Biochem. Biophys. Acta 92 (1964) 33.
[26] M. Liss, S.B. Horwitz, and N.O. Kaplan, J. Biol. Chem. 237 (1962) 1343.
[27] H. Sund, In Biological Oxidations ed. T.P. Singer (Interscience Publishers, New York–London, 1968) p. 61.

[28] R. Jaenicke, Pyridine Nucleotide Dependent Dehydrogenases, ed. Horst Sund (Springer-Verlag, 1970) p. 71.
[29] N.O. Kaplan, Brookhaven Symposium 17 (1964) 131.
[30] C.L. Markert and E.J. Messaro, Science 162 (1968) 965.
[31] A. Conway and D.E. Koshland, Biochemistry 7 (1968) 4011.
[32] H.C. Watson and L.J. Banaszak, Nature 204 (1964) 918.
[33] I. Listowsky, G.S. Furfine, J.J. Betheil, and S. Englard, J. Biol. Chem. 240 (1965) 4253.
[34] H.C. Watson and L.J. Banaszak, Nature 204 (1964) 918.
[35] L. Kirschner, M. Eigen, R. Bittman and B. Voigt, Proc. Natl. Acad. Sci. 56 (1966) 1661; K. Kirschner and I. Schuster, Pyridine Nucleotide Dependent Dehydrogenases, ed. Horst Sund (Springer-Verlag, 1970) p. 217.
[36] D.E. Koshland, Jr., R.A. Cook and A. Cornish-Bowden, Pyridine Nucleotide Dependent Dehydrogenases, ed. Horst Sund (Springer-Verlag, 1970) p. 209.
[37] R. Jaenicke, Pyridine Nucleotide Dehydrogenases, ed. Horst Sund (Springer-Verlag, 1970) p. 209.
[38] H.D. Kaloustian and N.O. Kaplan, J. Biol. Chem. 244 (1969) 2902.
[39] G. Weber and S.R. Anderson, Biochemistry 4 (1965) 1942; S.R. Anderson and G. Weber, Biochemistry 4 (1965) 1948.
[40] G. DiSabato and N.O. Kaplan, J. Biol. Chem. 240 (1965) 1072.
[41] W.C. Deal, Biochemistry 8 (1969) 2795.
[42] O.P. Chilson, G.B. Kitto, J. Pudles, and N.O. Kaplan, J. Biol. Chem. 241 (1966) 2431; O.P. Chilson, G.B. Kitto, and N.O. Kaplan, Proc. Natl. Acad. Sci. 53 (1965) 1006.
[43] R. Pietruszko and H. Theorell, Arch. Biochem. Biophys. 131 (1969) 288.
[44] H. Jornvall, Quoted by H. Theorell, Pyridine Nucleotide Dependent Dehydrogenases, ed. Horst Sund (Springer-Verlag, 1970) p. 121.
[45] R. Pietruszko, H.J. Ringold, N.O. Kaplan, and J. Everse, Biochem. Biophys. Res. Commun. 33 (1968) 503.
[46] C.R. Roe and N.O. Kaplan, Biochemistry 8 (1969) 5093.
[47] J.I. Harris and R.N. Pernham, Nature 219 (1968) 1025; B.E. Davidson, M. Sajgo, H.F. Noller, and J.I. Harris, Nature 216 (1967) 1181.
[48] J.I. Harris, Pyridine Nucleotide Dependent Dehydrogenases, ed. Horst Sund (Springer-Verlag, 1970) p. 57.
[49] W.S. Allison, Ann. N.Y. Acad. Sci. 148 (1968) 180.
[50] T.P. Fondy, J. Everse, G.A. Driscoll, F. Castillo, F.E. Stolzenbach, and N.O. Kaplan, J. Biol. Chem. 240 (1965) 4219.
[51] J.J. Holbrook, G. Pfleiderer, J. Schnetzger, and S. Dumair, Biochem. 344 (1966) 1.
[52] W.S. Allison, J. Admiraal, and N.O. Kaplan, J. Biol. Chem. 244 (1969) 4743.
[53] G. Pfleiderer, Chr. Woenckhaus, D. Jeckel and K. Mella, Pyridine Nucleotide Dependent Dehydrogenases, ed. Horst Sund (Springer-Verlag, 1970) p. 145.
[54] W.H. Zinkham, Ann. N.Y. Acad. Sci. 151 (1968) 598; C. Hawtley and E. Goldberg, Ann. N.Y. Acad. Sci. 151 (1968) 611.
[55] L. Schatz and H.L Segal, J. Biol. Chem. 244 (1969) 4393.
[56] E. Goldberg, Arch. Biochem. Biophys. 109 (1965) 134; W.A. Zinkham, A. Blanco, and L. Kupchyk, Science 142 (1963) 1303.
[57] E. Goldberg, Science 151 (1966) 1091; C.L. Markert, and I. Faulhaber, J. Expt. Zool. 159 319.

[58] G.L. Long and N.O. Kaplan, Science 162 (1968) 685.
[59] N.O. Kaplan, In the Enzymes Vol III, eds. P.D. Boyer, H. Lardy, and K. Myrback, (Academic Press, Inc., New York, 1960) p. 105.
[60] M.J. Adams, D.J. Haas, B.A. Jeffey, A. McPherson, H.L. Mermall, M.G. Rossman, R.W. Schivitz, and A.J. Wonacott, J. Mol. Biol. 41 (1969) 159.
[61] C.I. Brănden, E. Zeppezauer, T. Boiwe, G. Söderlund, B.O. Söderberg and B. Nordström, Pyridine Nucleotide Dehydrogenases, ed. Horst Sund (Springer-Verlag, 1970) p. 129.

EVOLUTION OF CATABOLIC PATHWAYS IN BACTERIA

E.C.C. LIN

Department of Bacteriology and Immunology, Harvard Medical School, Boston, Massachusetts 02115, USA

Abstract: Lin, E.C.C. Evolution of Catabolic Pathways in Bacteria. *Miami Winter Symposia,* 1, pp. 89–102. North-Holland Publishing Company, Amsterdam, 1970.

A few mutational steps can enable a bacterial cell to grow rapidly on a novel carbon source. For example, wild-type *Aerobacter aerogenes* can not utilize xylitol, a five-carbon polyhydric alcohol. A succession of mutants was isolated by selection on xylitol as the sole source of carbon and energy. The first mutant, X1, acquired the ability to grow on the novel carbon source through a mutation in a regulatory gene, resulting in the constitutive synthesis of ribitol dehydrogenase. Since the enzyme has slight activity on xylitol, its derepression permits the organism to attack the compound which is thereby converted to D-xylulose, an intermediate in the D-arabitol pathway of *A. aerogenes*. Wild-type cells cannot grow on xylitol because their dehydrogenase is formed only when ribitol is present in their environment. From X1, a second-stage mutant, X2, was isolated which grows faster on xylitol than X1. Mutant X2 produces an altered ribitol dehydrogenase with apparent affinity for xylitol which is more than double that of the affinity of the wild-type enzyme. Although this alteration of the enzyme increased its efficacy in the catalysis as a xylitol dehydrogenase, the protein is less stable. A third-stage mutant, X3, in turn was isolated from strain X2. In a mineral medium with a xylitol concentration of 0.01 M, this tertiary mutant grows at a rate comparable to that on glucose, doubling every hour at 37°. The ribitol dehydrogenase in this mutant is enzymatically and physicochemically indistinguishable from the enzyme produced by strain X2. However, X3 produces constitutively an active transport system which accepts xylitol. The usual function of this system is apparently for the transport of D-arabitol, since the latter is not only a substrate but also acts as the inducer in parental strains. The evolutionary development of a novel pathway does not necessarily require sequential or alternating changes in different genes. In another study of a series of mutants of *Escherichia coli*, utilization of a novel carbon source, L-1,2-propanediol, was found to be associated with the appearance and increase in activity of a nicotinamide adenine dinucleotide-coupled dehydrogenase. In this case, repeated changes in one enzyme were apparently necessary for the achievement of faster growth rates on the nutrient. Nevertheless, the general emerging picture in this area of investigation confirms the hypothesis advanced by Horowitz that the acquisition of new biochemical reactions occurs in a sequential fashion; each new substrate is converted to a utilizable metabolite so that every addition bestows an advantage.

1. Introduction

To meet the challenges of both acute and long range changes in its chemical environment, a microbial species not only must be capable of rapid phenotypic adaptation but also must have a flexible genetic apparatus. It has long been known that novel genetic features in an organism might arise from mutational events as well as from the adoption of segments of foreign genomes. Although genetic exchanges play major roles at all levels of evolution, this kind of process serves mainly to shuffle various genetic units so that highly successful combinations might be established through natural selection. For the invention of new genes, mutations must play the primary role.

Some of the work in our laboratory has been aimed at learning how new metabolic functions in bacteria can arise without resorting to genetic importation. We decided to study the emergence of new catabolic pathways because it is relatively easy to apply strong selective pressures in favor of mutants which can grow on new compounds. Furthermore, we hoped that the changes which occur in such mutants would be amenable to analysis at the molecular level.

A. aerogenes has a relatively large genetic repertoire for the dissimilation of polyhydric alcohols; it is capable of growth on glycerol, ribitol, D-arabitol, D-mannitol, D-sorbitol, and several cyclitols [11,18–22,28,30,34]. The possession of a large number of genes carrying out one class of functions should make it statistically more likely that new variants playing similar roles might arise. We attempted to see if mutants could be selected from this species of bacteria that could utilize some of the polyhydric alcohols not attacked by wild-type cells.

2. The primary xylitol mutant

The first mutant we obtained which could utilize xylitol, designated as strain X1, was selected by plating about 10^7 cells of the parental strain on mineral agar with 0.2% (or 0.014M) xylitol. About one cell in half a million grew out as a colony. One such colony was purified and the cells were found to double once every four hours at 37° on xylitol (0.2%). Upon analysis of the cell-free extract, a feeble NAD-dependent xylitol dehydrogenase activity was detected. This activity was found in extracts of the mutant even when the cells were grown in the absence of xylitol. The relatively high frequency with which this kind of mutant occurs spontaneously in a bacterial population, together with the fact that the newly appearing enzyme is produced under most nutritional conditions, suggest that a random mutation had occurred in the regulator gene which controls the production of some normally occurring dehydrogenase with an unidentified function.

It was not difficult to uncover the customary function of this dehydrogenase because previously we had purified from the same species of *A. aerogenes* an NAD-linked ribitol dehydrogenase which exhibited slight activity on xylitol [11]. On examination, indeed, extracts of X1 cells exhibited a high ribitol dehydrogenase activity.

To test whether ribitol dehydrogenase was really responsible for this new metabolic activity, we isolated a mutant derivative from X1 which failed to grow on ribitol and in which the ribitol dehydrogenase activity was missing. This derivative strain concomitantly lost the ability to utilize the novel substrate xylitol. Ribitol dehydrogenase thus seems necessary for both the dissimilation of ribitol and xylitol in strain X1. To test whether the simple genetic derepression of the structural gene for this dehydrogenase was sufficient to permit growth on xylitol, a mutant was selected directly from the wild-type strain for the constitutive utilization of ribitol without any exposure to xylitol during the whole isolation procedure. When this ribitol constitutive mutant was tested on xylitol as the carbon source, it doubled every four hours on the novel carbon source exactly like strain X1 [16].

Ribitol dehydrogenase catalyzes the dehydrogenation of xylitol at carbon 2 to yield the pentose, D-xylulose. From what is known about the metabolism of *A. aerogenes* [1,4–8,11,18,24–26,29,35], it can be readily seen that once xylitol is converted to D-xylulose, further dissimilation of the metabolite presents no further problem to the organism because D-xylulose is an intermediate in the inducible pathway of another five carbon polyhydric alcohol, D-arabitol (fig.1).

It should be pointed out here that Mortlock, Fossitt and Wood, working on a different strain of *A. aerogenes*, PRL R3, of an independent origin, also found that a single mutation, causing constitutive expression of an NAD-linked ribitol dehydrogenase, allowed the cells to grow on xylitol [27]. This coincidence not only lends credence to bacterial taxonomy, but also shows that organisms with similar genetic make-up do find similar solutions for new environmental challenges.

Examination of the scheme presented in fig. 1 raises an obvious question: D-arabitol, when present as the sole source of carbon and energy, can support a doubling time of 40 to 50 min at 37°. This is true for all strains we have stud-

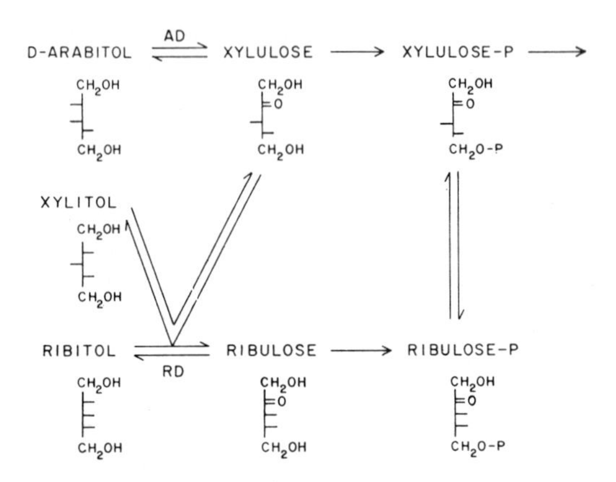

Fig. 1. Metabolic pathways for D-arabitol, ribitol, and xylitol. AD represents D-arabitol dehydrogenase and RD, ribitol dehydrogenase.

ied. Yet xylitol can only allow cells of X1 to double once in four hours. Since D-xylulose is the common intermediate in both pathways, this disparity of growth rate shows that there exists a constriction in the xylitol pathway at some point between the cellular uptake step and the dehydrogenation.

3. The secondary xylitol mutant

In an attempt to identify this rate-limiting step, we subjected cells of X1 to further selection in a liquid medium on xylitol [36]. A derivative with improved ability to utilize this compound did appear after numerous generations. This mutant, X2, doubled on xylitol every two hours (0.2%), in contrast to its parent which could grow only at half that rate. Table 1 compares the properties of ribitol dehydrogenase partially purified from X2 and X1. Whereas the X1 enzyme, as one would expect, is indistinguishable from the wild-type enzyme (not shown in table), the X2 enzyme exhibits a higher apparent affinity for the new substrate, xylitol. Most likely, a missense mutation had occurred in the structural gene of the dehydrogenase to make the protein more active on the novel substrate. On the other hand, this modification also reduced the thermal stability of the X2 enzyme.

Table 1
Properties of purified ribitol dehydrogenase.

Strain	Activity ratio (xylitol/ribitol)		K_m for xylitol[a]	K_m for ribitol[a]	Half-life[b] at 60°
	At 0.05 M substrate	At V_{max} (calculated)	(M)	(M)	(min)
X1	0.046	0.29	0.29	0.0031	2.8
X2	0.193	0.62	0.12	0.0025	0.7
X3	0.195	0.67	0.13	0.0020	0.7

a) The values of K_m were determined by the Lineweaver–Burk double reciprocal plot.
b) Heat inactivations were carried out with samples containing 0.1 mg of protein per ml in 0.01 M sodium phosphate buffer, pH 7.0. This table is quoted from [36].

4. The tertiary mutant

To pursue further the attempts to select a mutant which would grow on xylitol at a rate approaching that on D-arabitol, we serially transferred cells of X2, this time in media with the xylitol concentration reduced from 0.2% to 0.05%. A tertiary mutant, X3, capable of doubling on xylitol every 40 to 50 min was finally obtained.

However, when the dehydrogenase from this third mutant was purified and examined as described above, it was found to be indistinguishable from the X2 enzyme (table 1).

If the improved growth rate of the third stage mutant cannot be accounted for by a further structural modification of the dehydrogenase, it appears that the only remaining explanation would be that the power of the cells to extract the substrate from the growth medium has been improved. To test this possibility, we measured the growth rates of the three mutants and the wild-type strain as functions of the external concentration of xylitol. The following features are apparent in fig. 2: (1) The growth curves for both strains X1 and X2 are sigmoid. (2) Strain X3 grew readily in 0.05% xylitol, the condition under which this mutant was selected, while strains X1 and X2 grew hardly at all. (3) All three mutants grew at different rates in 0.2% xylitol. (4) Above 0.2% xylitol, the growth rates of the three mutants converged to approach a doubling time of about 50 min. (5) Wild-type cells (in which the ribitol dehydro-

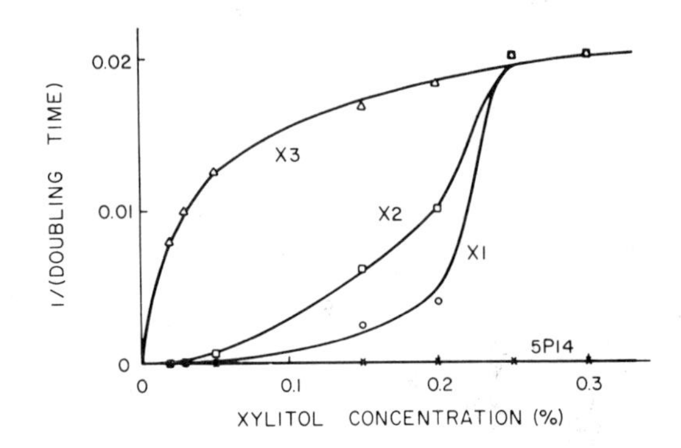

Fig. 2. The reciprocal of the doubling time (min) of the three mutants and the wild-type strain as a function of xylitol concentration in the medium. From ref. [36].

genase is inducible by ribitol but not by xylitol) did not grow even at the highest xylitol concentration tested.

An important fact that emerges from this set of growth experiments is that the concentration of xylitol affording half maximal growth rate for X3, is two orders of magnitude lower than the apparent K_m for xylitol of its dehydrogenase. This discrepancy, and the fact that all three mutants are capable of

Table 2
Uptake of ^{14}C-pentitols by suspended cells[a].

Strain	Growth medium	^{14}C- Xylitol uptake (counts/15 min)	^{14}C-D-Arabitol uptake (counts/2 min)
X1	Casein hydrolysate	60	440
X2	Casein hydrolysate	60	430
X3	Casein hydrolysate	775	6,350
A[c]	Casein hydrolysate	770	6,780
X1	D-Arabitol	210	1,720
X2	D-Arabitol	275	1,830
X3	D-Arabitol	760	6,550
A[c]	D-Arabitol	1,020	5,330

[a]Data are the values after subtraction of blanks (less than 50 counts/min) in which cells were heated at 60° for 30 min before exposure to substrate. This table is quoted from [36].

doubling about once in 50 minutes when the external xylitol concentration becomes sufficiently high, suggests that in X3 an active transport system is mobilized which permits more effective extraction of the nutrilite from the medium.

This interpretation was experimentally confirmed (table 2). Cells of X3 grown on an amino acid mixture and then incubated in a mineral medium with ^{14}C-xylitol at a low concentration (5×10^{-5}M), accumulated label rapidly. Cells of the predecessor strains, when grown under similar conditions, showed only slow uptake of xylitol. Apparently, an active transport system which normally serves to capture another substrate, had become constitutive in X3. The situation is analogous to the derepression of ribitol dehydrogenase in strain X1.

In order to identify the usual role of the transport system that became constitutive in X3, various straight chain polyhydric alcohols, known to be utilizable by the wild-type organism, were tested for competitive inhibition of xylitol uptake by cells of X3. The most potent inhibitor was D-arabitol. ^{14}C-D-arabitol itself was rapidly taken up by cells of X3, and moreover, the uptake process was constitutive, as in the case of xylitol uptake. In contrast, the predecessor strains X1 and X2 showed significant accumulation of labeled D-arabitol or xylitol only if this capability was previously induced.

A mutant, A^c, selected directly from wild-type cells for the constitutive utilization of D-arabitol without exposure to xylitol in the selection process, behaved very similarly to X3 in that both D-arabitol and xylitol were taken up without induction. This provides further evidence that the transport in question ordinarily functions for D-arabitol.

In another set of experiments (not reviewed here), it was shown with appropriately prepared mutants that labeled D-arabitol and xylitol, which accumulated at high concentrations in the cells could be recovered chemically unaltered. Thus, these uptake processes represent active transport without subsequent utilization [36].

5. Conclusions from the growth properties of the three xylitol mutants

With the realization that an active transport system for D-arabitol can be used to promote xylitol entry into the cell, a number of the growth features of the three xylitol mutants depicted in fig. 2 can now be interpreted.

Constitutivity of this transport system in X3 permits the cells to extract xylitol and grow in a manner reflecting the hyperbolic kinetics of the transport process.

The behaviors of X1 and X2 are more complicated because the D-arabitol transport system is inducible in these strains. The degree of induction of this transport system can be sensitively affected by the external concentration of xylitol. As xylitol is supplied in the medium at successively higher concentrations, its entry into the cell would be expected to increase via the basal D-arabitol transport system already present in the membrene and perhaps also through other less specific channels. But as more of the xylitol enters the cell, the rate of its dehydrogenation to D-xylulose would correspondingly be accelerated because the K_m values of the ribitol dehydrogenases for xylitol are rather high (table 1).

Finally it may be asked why the sharp rise in the growth rate occurs at different xylitol concentrations for mutants X1 and X2. Again, the K_m values of the dehydrogenases evidently are involved. The X2 enzyme has a greater affinity for xylitol than the X1 enzyme, so the production of D-xylulose increases faster in X2 cells than in X1 cells as the xylitol concentration rises. As the rate of D-xylulose production rises, the induction of the enzymes in the xylitol pathway becomes stronger. This chain reaction, behaving as a positive feedback loop, is most likely the basis for the sharp rise in the growth rates of X1 and X2 in the 0.2% region of xylitol concentration.

In summary, surprisingly few mutational steps can enable a bacterial cell to grow rapidly on a novel carbon source. In the case of the xylitol mutants

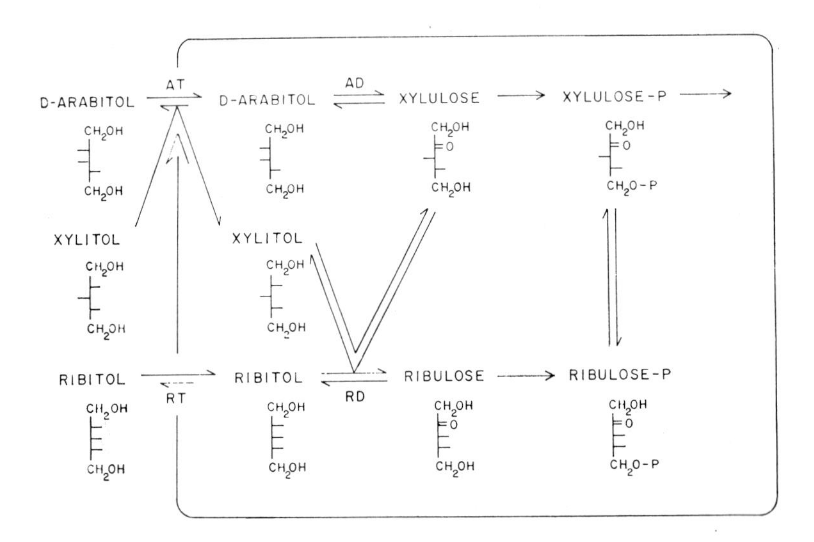

Fig. 3. The pathways for transport and catabolism of D-arabitol, ribitol, and xylitol. AT represents the D-arabitol transport system and RT, the ribitol transport system.

(fig. 3), we see that functional units belonging to different pathways can be readily recruited and adapted for new usage, presumably without gross alteration of the structure of any protein. The pathways normally employed for the dissimilation of D-arabitol and ribitol jointly contributed to the formation of a third pathway.

In general cells can be expected to have a much larger metabolic potential than is ordinarily expressed because enzymes rarely possess absolute specificity. The full metabolic potential of a cell is partially masked by control mechanisms so that most of the gene products are fabricated only when they can ordinarily be useful to the organism. The price for such economy is the curtailment of biochemical versatility.

It should, therefore, not be surprising that the first genetic change which enables a cell to attack a novel substrate is often a mutation which unleashes the expression of a structural gene coding for an enzyme which can act on substrates other than the usual physiological one. A strategy based on this principle had been devised by Joshua Lederberg in 1951 [15] when he showed that mutants that produce β-galactosidase constitutively can be selected with altrose-galactoside, a substrate of the hydrolase but not its inducer. Numerous other examples of this kind have since been reported and reviewed elsewhere [16, 23, 31, 33].

The case of the emergence of the xylitol pathway provides one of the first illustrations of how a sequence of mutations may be required to permit a novel compound to supply an organism with both building blocks and sources of metabolic energy sufficient to maintain growth (at a normal rate). A general theory of how metabolic pathways could emerge during evolution was advanced a quarter of a century ago by Norman Horowitz who proposed that the acquisition of a series of enzymes occurred in an order such that each new substrate could be converted to a utilizable metabolite. In this way each additional catalytic ability bestows an advantage [10]. The studies on the xylitol mutants essentially support this hypothesis: clearly the development of an active transport system for xylitol as the first step would be futile or even detrimental without a mode for its utilization.

6. Other examples

Continued search for examples of how new pathways can arise will reveal whether the large majority of the cases follow a general pattern. Additional

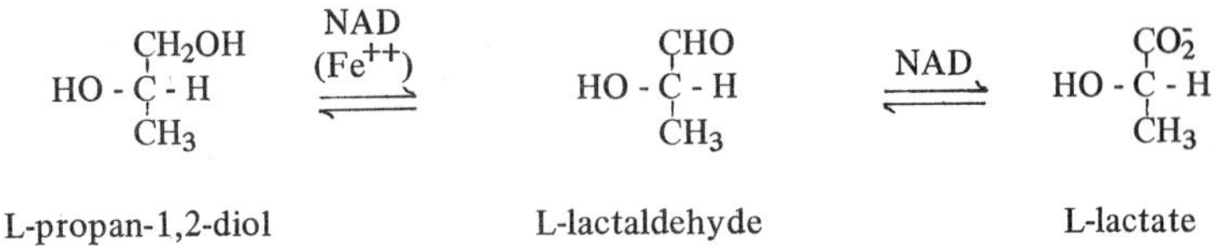

$$\underset{\text{L-propan-1,2-diol}}{\begin{array}{c}CH_2OH\\ HO-C-H\\ CH_3\end{array}} \underset{}{\overset{\text{NAD}\ (Fe^{++})}{\rightleftharpoons}} \underset{\text{L-lactaldehyde}}{\begin{array}{c}CHO\\ HO-C-H\\ CH_3\end{array}} \overset{\text{NAD}}{\rightleftharpoons} \underset{\text{L-lactate}}{\begin{array}{c}CO_2^-\\ HO-C-H\\ CH_3\end{array}}$$

Fig. 4. Metabolic pathway for propanediol of the mutant.

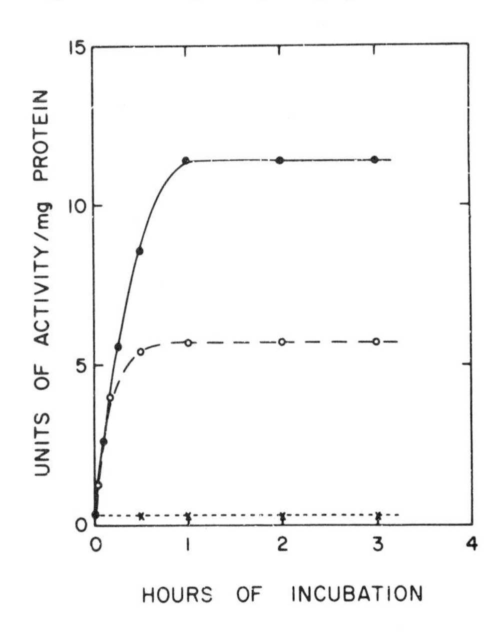

Fig. 5. Activation of a partially purified propan-1,2-diol dehydrogenase (X) by Fe^{++} (o) and Mn^{++} (•). From ref. [33].

patterns for the accretion of metabolic capacities have in fact been revealed *. These will be briefly reviewed below.

Recently a mutant of *E. coli* has been serially selected for growth on propan-1,2-diol in our laboratory [33]. In this case repeated changes in one enzyme might have occurred. The mutant produces a constitutive NAD-linked dehydrogenase which converts propan-1,2-diol to L-lactaldehyde which is then

* Mutants capable of growth on sugars not utilized by wild-type cells were obtained from *Pseudomonas putrefaciens* [14] and *Pseudomonas saccharophila* [3], but unfortunately the nature of these mutants was not fully established.

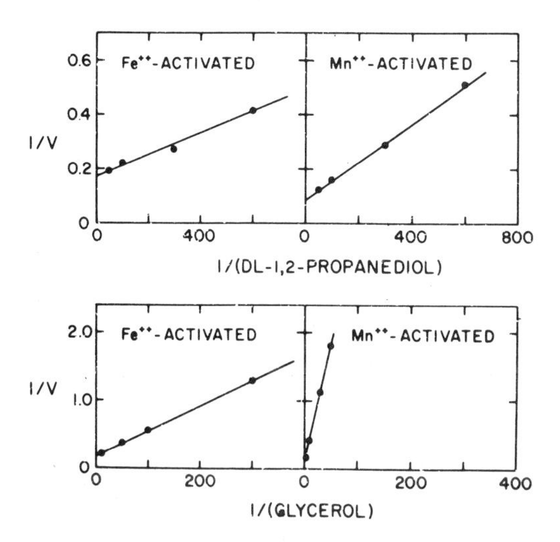

Fig. 6. Lineweaver-Burk plots of the Fe^{++}-activated and Mn^{++}-activated dehydrogenases with DL- propan-1,2-diol and glycerol as substrates. Velocities are expressed in terms of specific activity; substrate concentrations, in moles/liter. From ref. [33].

converted to L-lactate [32] (fig. 4). The propanediol enzyme is present only in the mutant, but the lactaldehyde enzyme is present both in the mutant and in the wild-type strain. The original function of the propanediol dehydrogenase gene remains a mystery, but the purified protein has an unusual feature: it requires either the divalent ferrous cation or the divalent manganous cation for activity. The activation process is slow (fig. 5). The metal ions not only are required for activity but impose an appreciable effect on substrate specificity as shown in fig. 6. Unfortunately, as in the case of the altered ribitol dehydrogenase of mutant X2 previously discussed, the propanediol dehydrogenase is also unstable.

Our suspicion that propanediol dehydrogenase is the object of repeated genetic modification is consonant with the finding that it is possible to transduce the capacity for rapid growth on the novel substrate from the mutant to the wild-type strain by phage Pl. Fig. 7 shows the position of this genetic locus on the *E. coli* chromosome [37].

More recently Mayo and Anderson [23] isolated from *A. aerogenes* a mutant which grows on L-mannose. In this case the enzymes, and perhaps the transport system too, are borrowed from the L-rhamnose pathway. Unexpectedly, L-mannose not only can be taken up and metabolized by the pre-exis-

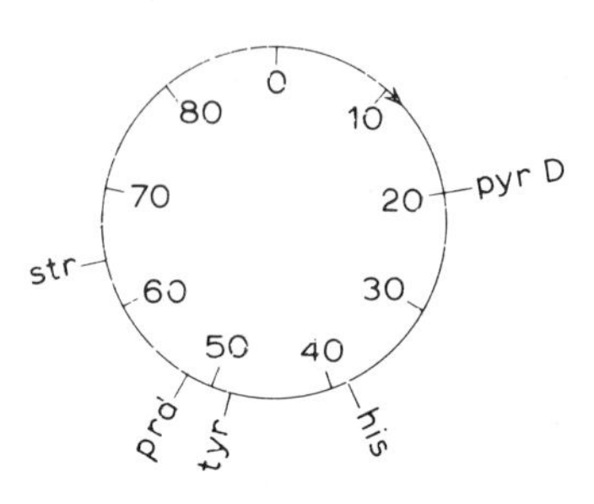

Fig. 7.

ting systems, but can cause their induction as well. The inability of wild-type cells to grow on L-mannose is apparently due to a metabolic imbalance following the uptake of the novel substrate. This imbalance is redressed genetically in the mutant.

7. On the accretion of genetic material necessary for the expansion of networks

Speculation on how the metabolic network in an organism can expand is incomplete without considering how its total genetic information can increase. To explain the growth of a genome before mechanisms of intercellular genetic exchanges became well developed, gene duplication in tandem [2] followed by gene divergence [17] has been invoked. This model has also been used to account for the frequent occurence, at least in bacteria, of multicistronic operons coding for proteins that catalyze a series of metabolic reactions [9]. Although the emergence of a series of proteins with homologous function, such as myoglobin and the subunits of hemoglobin, may very well have happened in this manner [12], a metabolic sequence involving dissimilar reaction mechanisms cannot be expected to arise this way. For example, to elaborate a xylitol pathway according to this theory would require that the gene for the NAD-linked "xylitol" dehydrogenase be descended from a copy of the gene for the ATP-dependent D-xylulose kinase and that the gene for the "xylitol" transport protein be in turn descended from the gene for the dehydrogenase. It is highly improbable from a biochemical viewpoint that the kind of reaction a protein catalyzes can be transformed by a few mutations.

On the other hand, it is more plausible that the role of a structural gene can first be widened by its constitutive expression. Only after such a gene has been allowed to be expressed freely would gene duplication be useful; one copy could retain its original function, while the other would be permitted to diverge and perfect the new function.

If the enzymes recruited into a new pathway are pre-empted from different pre-existent systems, again using the xylitol case as a model, it is not likely that the recruited genes are initially neighbors. Only if coordinate control is advantageous, would one expect the dispersedly located genes to be collected into a single operon by translocation [13]. Hence, the existence of a large operon does not necessarily imply that the cistrons occurring in it had a common ancestor.

The analysis given above is likely to be valid for the late stage evolution of new enzymes and for the establishment of genetic linkage in organisms already in possession of a sizable and sophisticated genome. The important questions of how new kinds of enzymes arise from very primitive organisms with only a few prototype genes, and of how the regulation of gene expression is gradually effected remain as challenging problems.

Acknowledgement

Most of the work conducted in our laboratory was carried out in collaboration with S.A. Lerner, Tai-te Wu, and S. Sridhara. The author is grateful to J.B. Alpers for editorial comments. Investigations of the problems in our laboratory have been made possible by Grant (GB-5854) from the National Science Foundation, and Grant (GM-11983) from the United States Public Health Service. The author was supported by a Research Career Development Award from the United States Public Health Service.

References

[1] B.K. Bhuyan and F.J. Simpson, Canad. J. Microbiol. 8 (1962) 737.
[2] C.B. Bridges, Science 83 (1936) 210.
[3] M. Doudoroff, N.J. Palleroni, J. MacGee and M. O'Hara, J. Bacteriol. 71 (1956) 196.
[4] D. Fossitt, R.P. Mortlock, R.L. Anderson and W.A. Wood, J. Biol. Chem. 239 (1964) 2110.
[5] D.D. Fossitt and W.A. Wood. In Methods in Enzymology, vol. 9 ed. W.A. Wood (Academic Press, Inc., New York, 1966) p. 180.
[6] H.J. Fromm and J.A. Bietz, Arch. Biochem. Biophys. 115 (1966) 510.
[7] H.J. Fromm and D.R. Nelson, J. Biol. Chem. 237 (1962) 215.

[8] H.J. Fromm, J. Biol. Chem. 233 (1958) 1049.
[9] N.H. Horowitz, in Evolving Genes and Proteins, ed. V. Bryson and H. Vogel (Academic Press, Inc., New York, 1965).
[10] N.H. Horowitz, Proc. Natl. Aca. Sci. 31 (1945) 153.
[11] S.B. Hully, S.B. Jorgensen and E.C.C. Lin, Biochim. Biophys. Acta. 67 (1963) 219.
[12] V.M. Ingram, The Hemoglobins in Genetics and Evolution. (Columbia University Press, New York, 1963).
[13] F. Jacob and E.L. Wollman, Sexuality and the Genetics of Bacteria (Academic Press, Inc., New York, 1961) p. 166.
[14] H.P. Klein and M. Doudoroff, J. Bacteriol. 59 (1950) 739.
[15] J. Lederberg, In Genetics in the Twentieth Century, ed. L.C. Dunn (Macmillan Co., New York, 1951) p. 263.
[16] S.A. Lerner, T.T. Wu and E.C.C. Lin, Science 146 (1964) 1313.
[17] E.B. Lewis, Cold Spring Harbor Symp. Quant. Biol. 16 (1951) 159.
[18] E.C.C. Lin, J. Biol. Chem. 236 (1961) 31.
[19] E.C.C. Lin, A.P. Levin and B. Magasanik, J. Biol. Chem. 235 (1960) 1824.
[20] E.C.C. Lin and B. Magasanik, J. Biol. Chem. 235 (1960) 1820.
[21] B. Magasanik, J. Biol. Chem. 205 (1953) 1007.
[22] B. Magasanik, M.S. Brooke, D. Karibian, J. Bacteriol. 66 (1953) 611.
[23] J.W. Mayo and R.L. Anderson, J. Bacteriol. 100 (1969) 948.
[24] R.P. Mortlock, D.D. Fossitt, D.H. Petering and W.A. Wood, J. Bacteriol. 89 (1965) 129.
[25] R.P. Mortlock and W.A. Wood, J. Bacteriol. 88 (1964) 838.
[26] R.P. Mortlock and W.A. Wood, J. Bacteriol. 88 (1964) 845.
[27] R.P. Mortlock, D.D. Fossitt and W.A. Wood, Proc. Natl. Acad. Sci. 54 (1965) 572.
[28] P. McPhedran, B. Sommer and E.C.C. Lin, J. Bacteriol. 81 (1961) 852.
[29] R.C. Nordlie and H.J. Fromm, J. Biol. Chem. 234 (1959) 2523.
[30] D. Rush, D. Karibian, M.L. Karnovsky and B. Magasanik, J. Biol. Chem. 226 (1957) 891.
[31] S. Schaefler, Bact. Proc. (1967) 54.
[32] S. Sridhara and T.T. Wu, J. Biol. Chem. 244 (1969) 5233.
[33] S. Sridhara, T.T. Wu, T.M. Chused and E.C.C. Lin, J. Bacteriol. 98 (1969) 87.
[34] S. Tanaka, S.A. Lerner and E.C.C. Lin, J. Bacteriol. 93 (1967) 642.
[35] W.A. Wood, M.J. McDonough and L.B. Jacobs, J. Biol. Chem. 236 (1961) 2190.
[36] T.T. Wu, E.C.C. Lin, and S. Tanaka, J. Bacteriol. 96 (1968) 447.
[37] T.T. Wu, T.M. Chused and E.C.C. Lin, Bact. Proc. (1967) p. 52.

MECHANISMS OF CARBOHYDRATE SYNTHESIS FROM C_2- AND C_3-ACIDS

H.L. KORNBERG

Department of Biochemistry, University of Leicester, Leicester, England

Abstract: Kornberg, H.L. Mechanisms of Carbohydrate Synthesis From C_2- and C_3-Acids. *Miami Winter Symposia* 1, pp. 103–121. North-Holland Publishing Co., Amsterdam, 1970.

Gluconeogenesis, the formation of carbohydrates from non-carbohydrate precursors, is a biosynthetic process of major importance. This is particularly evident when microorganisms grow on C_2- and C_3-compounds as sole carbon source since, under these conditions, many amino acids, cell wall components, the purine and pentose moieties of nucleic acids, and other cell constituents derived from carbohydrate precursors, must arise from phospho*enol*pyruvate (PEP). A variety of different reactions are used by microorganisms to effect this essential synthesis of PEP from C_2- and C_3-acids. The direct formation of PEP from pyruvate does not occur in higher organisms, but is achieved in some bacteria either via PEP-synthase, by which PEP, AMP and P_i are formed from pyruvate and ATP, or via pyruvate, phosphate dikinase, which is similar in mechanism but involves P_i as a reactant and PP_i as a product. Other microorganisms form PEP from pyruvate via the two-stage process also used by animal cells, in which pyruvate is first carboxylated to oxaloacetate which, in the presence of the appropriate nucleoside triphosphate, is then decarboxylated to PEP. Depending on the organism, the pyruvate carboxylase that catalyses the first part of this process is either unaffected by acetyl coenzyme A, stimulated by it, or completely dependent upon it; in thermophilic bacilli, a further point of control may regulate the conversion of biotin and the apo-enzyme to the active holo-enzyme. The direct formation of pyruvate from acetate and carbon dioxide occurs only in some microorganisms under strongly reducing conditions, with concomitant oxidation of reduced ferredoxin. The indirect formation of PEP from acetate, via the glyoxylate cycle and decarboxylation of the C_4-acids formed in that cycle, is widespread among microorganisms including protozoa; it does not occur in animal cells. In bacteria, regulation of this process is achieved through inhibition of the activity and of the synthesis of isocitrate lyase, by the C_3-end products of the overall process; in *Escherichia coli*, the synthesis of the key enzymes of the cycle is controlled as an operon. However, in organisms structurally more complex than bacteria additional controls must operate, since the enzymes of the cycle are compartmented in discrete particles when the cycle is required to play its gluconeogenic role.

1. Introduction

Gluconeogenesis, the formation of carbohydrates from non-carbohydrate precursors, is a biosynthetic process of major importance. Its quantitive significance may be relatively small, as in animals on a carbohydrate-rich diet; it may be considerable, as in animals fed a diet low in carbohydrate [1] ; or it may be of predominant importance, as in microorganisms growing on any compound that is not itself a carbohydrate. This predominance is apparent from fig. 1, in which the main biosynthetic routes are indicated as heavy arrows. If the carbon source is an intermediate of the tricarboxylic acid (TCA) cycle or can be directly converted to one, a variety of cell components – particularly the porphyrins, the aspartate and glutamate "families" of aminoacids and the pyrimidine rings of nucleic acids – can be derived from that cycle without involving gluconeogenic steps. However, the synthesis of carbohydrates, of both the purine and pentose moieties of nucleotides, of a variety

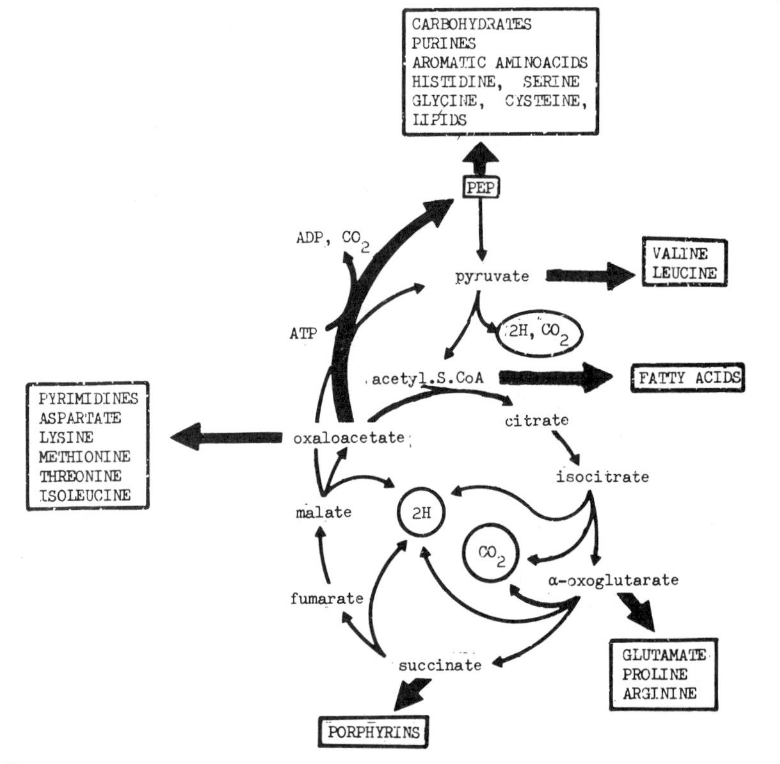

Fig. 1. The role of gluconeogenesis in the synthesis of cell constituents by *E. coli.*

of amino acids, and of a number of cell components not shown in this figure, involves a reversal (in effect, if not in mechanism) of the reactions that in other circumstances accomplish the fragmentation of carbohydrates to C_3-acids.

The starting material in this gluconeogenic pathway is phospho*enol*pyruvate (PEP) which is readily derived from C_4-acids via a nucleotide-linked decarboxylation of oxaloacetate (reaction i) catalysed by PEP-carboxykinase [2]:

$$HO_2C.CH_2.CO.CO_2H + NuTP \rightleftharpoons CH_2{:}C(OP).CO_2H + NuDP + CO_2 \qquad \text{(i)}$$

This enzyme appears to be ubiquitously distributed in animals, plants and microorganisms, although it differs in the identity of the nucleotide best fitted to participate in the reaction: the enzyme from animal cells prefers GTP, whereas that from many microorganisms prefers ATP as phosphate donor. Although the purified enzyme can effect the formation of oxaloacetate from PEP, CO_2 and the appropriate nucleoside diphosphate, its physiological role is to catalyse the decarboxylation of oxaloacetate. This view is supported by the observations that:

(a) animal cells in which lactate cannot give rise to the net formation of carbohydrate are virtually devoid of PEP-carboxykinase activity, whereas tissues capable of effecting gluconeogenesis (such as liver and kidney) are rich in this enzyme [1,3],

(b) the PEP-carboxykinase activity of yeasts [4,5] and bacteria [6] is repressed by growth on glucose but is induced during growth on C_4-acids,

(c) bacterial mutants devoid of PEP-carboxykinase activity fail to grow on C_4-acids, though they grow normally on carbohydrates [7].

The initial stages of gluconeogenesis from C_2- and C_3-acids must also be viewed in terms of the enzymatic reactions that effect the formation of PEP from these substances. It is in these reactions that living organisms manifest great variety, and it is their nature and regulation that forms the substance of this paper.

2. **Direct formation of PEP from pyruvate**

It has long been known (for review, see [8,9]) that the formation of pyruvate from PEP, catalysed by pyruvate kinase (reaction ii), is essentially irreversible under physiological conditions:

Table 1
Transformations of [U-^{14}C] AMP in the reversal of PEP-synthase.

Incubation system	Radioactivity (counts $\times 10^{-3}$/min/ml) AMP	ADP	ATP
Phosphate, complete	857	29	632
Arsenate, complete	1099	311	38.3
Phosphate, PEP omitted	1481	11.6	39.4
Arsenate, PEP omitted	1407	11.0	37.5

The complete incubation system contained, in 1 ml, 100 μmoles of sodium potassium buffer, pH 6.8; 5 μmoles of $MgCl_2$; 1 μmole of PEP; 1 μmole of [U-^{14}C] AMP (containing approximately 2 μCi of ^{14}C) and 180 μg. of purified PEP-synthase. When arsenate was used, 100 μmoles of sodium arsenate HCl buffer, pH 6.8, replaced the phosphate buffer. The phosphate-containing systems were incubated for 1 hr, the arsenate-containing ones for 2 hr, after which time the reactions were stopped by the addition of 0.1 ml of 20% (w/v) $HClO_4$. After neutralization with 0.2 ml of 2N-KOH, the precipitate was removed and the nucleotides were separated by paper chromatography in isobutyric acid: ammonia: water (66:1:33). The compounds were visualized under ultraviolet light the spots were cut out and their radioactivity was assayed with a Packard 4000 scintillation counter. (This table is modified from Cooper & Kornberg [11]).

$$CH_2{:}C(OP).CO_2H + ADP \rightleftharpoons CH_3.CO.CO_2H + ATP \qquad \text{(ii)}$$

An enzyme effecting the net formation of PEP from pyruvate and ATP was discovered in extracts of lactate-grown *Escherichia coli* by Cooper & Kornberg [10]. This enzyme (reaction iii), named PEP-synthase, differs from pyruvate kinase in that two of the energy-rich linkages of ATP are used in the formation of PEP, and that the products are AMP and P_i instead of ADP:

$$CH_3.CO.CO_2H + ATP \rightleftharpoons CH_2{:}C(OP).CO_2H + AMP + P_i \qquad \text{(iii)}$$

Unlike the pyruvate kinase reaction, the equilibrium of that catalysed by PEP-synthase is far over to the side of PEP formation. However, at pH values less than 7, the enzyme can effect the formation of ATP and pyruvate from PEP, AMP and P_i: if arsenate is added instead of inorganic phosphate, ADP is formed instead of ATP (table 1). This shows that it is inorganic phosphate and not PEP that contributes the terminal (γ) phosphate to ATP; clearly, the β-phosphate of ATP (or of ADP if arsenate is present) must be derived from PEP.

This conclusion was confirmed by analysis of the distribution of isotope from specifically-labelled [^{32}P] ATP, after its enzymic reaction with pyruvate

$$\mathrm{Enz + AdRPP\overset{\bullet *}{P} \rightleftharpoons [Enz - \overset{\bullet *}{P}P + AdRP] \rightleftharpoons Enz - \overset{\bullet}{P} + \overset{*}{P} + AdRP}$$

$$\mathrm{Enz - \overset{\bullet}{P} + CH_3.CO.CO_2H \rightleftharpoons CH_2 : C(O\overset{\bullet}{P})CO_2H + Enz}$$

$$\text{Sum:} \quad \mathrm{AdRPP\overset{\bullet *}{P} + CH_3.CO.CO_2H \rightleftharpoons AdRP + CH_2 : C(O\overset{\bullet}{P})CO_2H + \overset{*}{P}}$$

Fig. 2. Postulated mechanism for the reaction catalysed by PEP-synthase.

to form PEP, AMP and P_i, and by following the fate of O when this reaction was carried out in $H_2{}^{18}O$ [12,13]. As was expected from the results obtained with PEP-synthase working in reverse [11], terminally-labelled ATP gave rise to PEP and AMP devoid of ^{32}P, whereas β-labelled ATP produced [^{32}P] PEP. Moreover, isotope from $H_2{}^{18}O$ appeared exclusively in the inorganic phosphate produced in the course of the reaction. The formation of PEP from ATP and pyruvate must therefore involve the transfer of a pyrophosphoryl-group, and cannot involve the transfer of an adenosinepyrophosphoryl-group.

Studies of the exchange reactions catalysed by the highly purified enzyme [12] suggest a mechanism of the type shown in fig. 2. The evidence in support of this scheme may be summarized as follows:

(a) PEP-synthase catalyses a very rapid exchange of [^{14}C] pyruvate with

Table 2
Reactions catalysed by PEP-synthase.

	Rate (μmoles/mg protein/hr)
Overall reaction	
A. 1 μmole AMP + 1 μmole PEP	7.5
B. 0.1 μmole AMP + 1 μmole PEP + 1 μmole ATP	1.9
Exchange reaction	
C. [8 - ^{14}C] AMP with ATP:	
0.004 M-phosphate	0.082
0.10 M-phosphate	3.95
D. $^{32}P_i$ with ATP:	
minus AMP	0.129
plus AMP	3.25
E. [$^{14}C_3$] Pyruvate with PEP	122.5

(For experimental details, see Cooper and Kornberg [12]).

PEP, in the absence of inorganic phosphate or of AMP (table 2, reaction [E]);

(b) the enzyme catalyses a rapid exchange of $^{32}P_i$ with ATP in the presence of AMP but in the absence of pyruvate (table 2, reaction [D]).

(c) PEP-synthase catalyses a rapid exchange of [^{14}C] AMP with ATP, in the absence of pyruvate (table 2, reaction [C]). However, this latter reaction requires the presence of relatively high concentrations of inorganic phosphate. It may be that the hypothetical enzyme-PP complex is readily hydrolysed to enzyme-P and P_i, and that a concentration of enzyme-PP sufficient for exchange to occur can be maintained only if sufficiently high concentrations of P_i permit reversal of this hydrolysis. On the other hand, it may also be that AMP remains bound to the enzyme-PP and is liberated only when enzyme-P and P_i are formed; in this case, and as is suggested in fig.2, both AMP and P_i would have to bind to convert enzyme-P to enzyme-PP.

(d) When highly purified PEP-synthase was incubated with [$\beta\gamma$–^{32}P] ATP. the enzyme became labelled. This did not occur when [γ–^{32}P] ATP was used; it was thus enzyme-P and not enzyme -PP that was formed. This labelled protein readily lost its radioactivity when incubated with pyruvate; labelled PEP was formed in stoichiometric quantities. Similarly, with AMP and P_i. labelled ATP was formed, and with AMP and arsenate, ADP. Radioactivity was not lost rapidly from the enzyme when other possible phosphate acceptors were used [14]. Since phosphorylation of the enzyme was also readily demonstrated when PEP was the phosphate donor, the existence of enzyme-P and its participation in the PEP-synthase reaction must be regarded as established.

The rates of all the exchange reactions recorded in table 2 [C–E] are greater than the overall reaction (iii) from right to left (table 2, [B]). However, it will also be noted that the inclusion of 1 μmole of ATP, necessary for the measurement of the exchange reactions [C] and [D], reduces the rate of the overall reaction by about 75%. Similarly, assays of PEP-synthase activity in the direction of PEP formation show AMP to be a powerful inhibitor of this reaction. It is thus possible that the rate (and direction?) of PEP-synthase activity may be influenced by the relative proportions of AMP, ADP and ATP in the cell.

The physiological role of PEP-synthase is two-fold. Mutants devoid of this enzyme fail to grow on lactate, alanine or pyruvate as sole carbon source, although they grow readily on glucose, glycerol, acetate, or even on C_3-acids if these are supplemented with utilizable intermediates of the tricarboxylic acid cycle. This behaviour illustrates the anaplerotic [15] role of PEP-synthase: mutants devoid of the enzyme (pps$^-$) cannot form PEP from pyruvate and, although they possess PEP-carboxylase activity (iv):

Table 3
Formation of glycogen by pps^+ and pps^- *E. coli.*

Substrate added	Glycogen formed (mg/g dry wt of cells/hr) by	
	pps$^+$	*pps*$^-$
None	- 0.4	- 1.0
Glycerol	+3.6	+2.8
DL-lactate	+4.7	-1.6
Pyruvate	+4.3	-1.9

(For experimental details, see Cooper and Kornberg [11]).

$$CH_2 : C(OP).CO_2H + CO_2 \rightleftharpoons HO_2C.CH_2.CO.CO_2H + P_i \qquad \text{(iv)}$$

and although they can oxidize pyruvate, they cannot replenish the intermediates of the tricarboxylic acid cycle as they are removed in the course of biosynthesis (fig. 1). The inability of such pps^--mutants to form PEP from pyruvate results also in their inability to effect gluconeogenesis from pyruvate, although they are unimpaired in their ability to form glycogen from glycerol (table 3).

All *pps*$^-$-mutants of *E. coli* K12 that have been analysed so far map in one region of the genome [16] : the *pps*-allele is located at 32 min on the linkage map published by Taylor and Trotter [17] and is highly co-transducible with the *aro*D-marker that specifies the synthesis of dehydroquinase. However, the complexity of the enzymatic reaction (fig. 2) and the size of the enzyme (mol. wt. ca 180,000 [14]) make it unlikely that PEP-synthase is composed of only one polypeptide chain. It is thus suggestive that, although all pps^--mutants lack the ability to catalyse the overall reaction (iii), some are unimpaired in their ability to effect the rapid exchange of pyruvate with PEP (table 2, E) whereas others have virtually lost this component of the reaction system.

An enzyme analogous to PEP-synthase, termed pyruvate, phosphate dikinase, also effects the net formation of PEP from ATP and pyruvate, but in addition requires inorganic phosphate as a reactant; the products are those of the PEP-synthase reaction except that inorganic pyrophosphate is produced instead of phosphate. The enzyme catalysing this reaction (v)

$$CH_3.CO.CO_2H + ATP + P_i \rightleftharpoons CH_2 : C(OP).CO_2H + AMP + PP_i \qquad \text{(v)}$$

was first discovered in propionic acid bacteria [18] : the ability of these bacteria thus to effect the direct formation of PEP from pyruvate makes it likely that the physiological role of the PEP-carboxytransphosphorylase [19,20], previously discovered in these organisms, (reaction vi):

$$CH_2 : C(OP).CO_2H + P_i + CO_2 \rightleftharpoons HO_2C.CH_2.CO.CO_2H + PP_i \qquad \text{(vi)}$$

is in the direction of oxaloacetate formation (as written above) and not, as originally thought [19], in that of gluconeogenesis.

Studies with tropical grasses [21] and *Acetobacter xylinum* [22] suggested initially that these biological materials contained PEP-synthase; however, later work ([23]; M. Benziman, personal communication) revealed the enzyme to be pyruvate, phosphate dikinase. It is not clear which of the two enzymes effects the direct formation of PEP from pyruvate in the photosynthetic bacteria studied by Buchanan and Evans [24].

The mechanism of the pyruvate, phosphate dikinase reaction shows great similarities with that postulated for PEP-synthase. Both enzymes catalyse the exchange of PEP with pyruvate in the absence of nucleoside phosphates or inorganic phosphate (table 2, E), which suggests that a phospho-enzyme participates also in the pyruvate, phosphate dikinase reaction. However, some doubt exists on the manner in which this phospho-enzyme is formed. Evans and Wood [25] reported evidence, obtained with the enzyme from propionic acid bacteria, in favour of the mechanism shown in fig. 3A. They suggested that, like PEP-synthase, the pyruvate, phosphate dikinase enzyme reacted initially with ATP to form a pyrophosphoryl enzyme, which (unlike PEP-synthase) formed the phospho-enzyme by reacting with inorganic phosphate. As expected from this mechanism, the enzymic exchange of ATP with AMP pro-

A.

$$\begin{aligned}
\text{Enz} + \text{AdRPP}\overset{\bullet *}{\text{P}} &\rightleftharpoons \text{Enz - }\overset{\bullet *}{\text{PP}} + \text{AdRP}\\
\text{Enz - }\overset{\bullet *}{\text{PP}} + \overset{\triangle}{P_i} &\rightleftharpoons \text{Enz - }\overset{\bullet}{\text{P}} + \overset{*\triangle}{PP_i}\\
\text{Enz - }\overset{\bullet}{\text{P}} + CH_3.CO.CO_2H &\rightleftharpoons CH_2 : C(O\overset{\bullet}{\text{P}})CO_2H + \text{Enz}
\end{aligned}$$

B.

$$\begin{aligned}
\text{Enz} + \text{AdRPP}\overset{\bullet *}{\text{P}} &\rightleftharpoons [\text{Enz - }\overset{\bullet *}{\text{PP}} + \text{AdRP}]\\
[\text{Enz}\overset{\bullet *}{\text{PP}} + \text{AdRP}] + \overset{\triangle}{P_i} &\rightleftharpoons \text{Enz - }\overset{\bullet}{\text{P}} + \overset{*\triangle}{\text{PP}} + \text{AdRP}\\
\text{Enz - }\overset{\bullet}{\text{P}} + CH_3.CO.CO_2H &\rightleftharpoons CH_2 : C(O\overset{\bullet}{\text{P}})CO_2H + \text{Enz}
\end{aligned}$$

Fig. 3. Postulated mechanisms for the reaction catalysed by pyruvate, phosphate dikinase.

ceeded in the absence of added inorganic phosphate. This latter exchange was not found by Andrews and Hatch [26] to occur with the enzyme isolated from sugar cane leaves, if P_i and PP_i were omitted, but to proceed rapidly if they were added. This latter situation, reminiscent of that recorded for PEP-synthase in table 2, suggests the mechanism shown in fig. 3B.

It will be noted that both groups of workers agree that, of the PP_i formed, only one phosphate arises from ATP, the other being derived from P_i. Since most tissues are rich in pyrophosphatases, the role of P_i may be a purely catalytic one in the overall process: the virtually irreversible hydrolysis (vii) of PP_i would 'pull' the overall reaction (viii) in the direction of PEP synthesis, and would overcome the otherwise unfavourable energy barrier to the synthesis of PEP:

$$CH_3.CO.CO_2H + ATP + P_i \rightleftharpoons CH_2 : C(OP).CO_2H + AMP + PP_i \qquad \text{(v)}$$

$$H_2O + PP_i \rightleftharpoons P_i + P_i \qquad \text{(vii)}$$

$$\text{Sum:} \quad CH_3.CO.CO_2H + ATP \rightleftharpoons CH_2 : C(OP).CO_2H + AMP + P_i \qquad \text{(viii; cf iii)}$$

In view of the widespread distributions of pyrophosphatase and the consequent scarcity of inorganic pyrophosphate in biological material, it is the more remarkable that the pyruvate, phosphate dikinase reaction has been invoked as a means of forming pyruvate from PEP in organisms that appear to lack pyruvate kinase [27].

3. Indirect formation of PEP from pyruvate

The energy barriers associated with the formation of PEP from pyruvate are overcome in many organisms not by the direct processes, represented by reactions (iii) and (v), but by a two-stage mechanism in which pyruvate is first carboxylated to oxaloacetate, and this product is then transformed to PEP with loss of CO_2. The first of these stages is catalysed by pyruvate carboxylase (reaction ix), an enzyme discovered by Utter and Keech [28] in chicken liver mitochondria. The nature of the second stage, effected by PEP-carboxykinase, has already been discussed (reaction i). Although this two-stage process involves the cleavage of two pyrophosphate linkages of ATP as does the direct synthesis of PEP from pyruvate, it will be noted that these pyrophosphate bonds are now derived from two molecules of ATP (reaction x):

$$CH_3.CO.CO_2H + ATP + CO_2 \rightleftharpoons HO_2C.CH_2.CO.CO_2H + ADP + P_i \quad (ix)$$

$$HO_2C.CH_2.CO.CO_2H + ATP \rightleftharpoons CH_2 : C(OP).CO_2H + ADP + CO_2 \quad (i)$$

$$\text{Sum:} \quad CH_3.CO.CO_2H + 2ATP \rightleftharpoons CH_2 : C(OP).CO_2H + 2ADP + P_i \quad (x)$$

The pyruvate carboxylase of avian liver is virtually inactive unless catalytic quantities of acetyl-coenzyme A are also present [28]. This is true also for the enzyme as isolated from all vertebrate sources so far examined (M.F. Utter, personal communication) and for a variety of microorganisms, such as *Arthrobacter globiformis* [29], *Rhodopseudomonas spheroides* [30] and both mesophilic and thermophilic bacilli [31]. Activation of pyruvate carboxylase by acetyl-coenzyme A represents a mode of control of gluconeogenesis of obvious utility [9]; however, it is not universally encountered. Thus, the pyruvate carboxylase of yeast is quite active in the absence of acetyl-coenzyme A, though the activity of the enzyme is roughly doubled by its presence [32]. The pyruvate carboxylases of *Pseudomonas citronellolis* [33] and of *Aspergillus niger* [34], on the other hand, appear to be unaffected by acetyl-coenzyme A.

Just as a stimulation of oxaloacetate formation from pyruvate by acetyl-coenzyme A represents a useful mode of control of gluconeogenesis, so would it be advantageous to an organism to be able to "switch off" the action of pyruvate carboxylase if oxaloacetate is already available to a cell. Such a situation is observed with the enzyme isolated from baker's yeast [35], from *Arthrobacter* (K.M. Jones, personal communication) and from thermophilic bacilli ([36]; fig. 4): aspartate, the immediate precursor of oxaloacetate, inhibits the enzyme, probably by competing with the activator acetyl-coenzyme A.

The formation of pyruvate carboxylase, unlike that of PEP-carboxykinase, does not appear to be repressible by glucose [5]. However, recent studies with thermophilic bacilli [36] suggested that the formation of the active enzyme from its enzymically inactive apo-protein and biotin may also be subject to metabolic control. It was observed that the specific activity of pyruvate carboxylase in a strain of *Bacillus stearothermophilus* growing on succinate at 55°, was much lower than when the cells grew in media similar but containing glucose as carbon source. This difference was not indicative of enzymic induction, since the addition of biotin to the succinate-grown cells resulted in a rapid formation of the active enzyme, even when chloramphenicol prevented the *de novo* synthesis of protein (table 4); it was probable that the

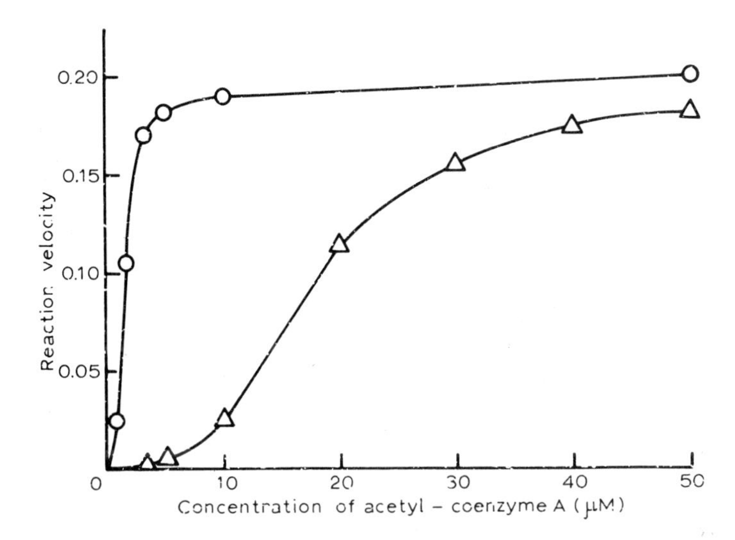

Fig. 4. Activity of pyruvate carboxylase purified from a thermophilic bacillus in the absence (○) and presence (△) of 10 mM L-asparate. (From [36]).

cells had formed the apo-enzyme and the synthetase that normally effects the combination of that apo-enzyme with biotin, but were unable to form the active holoenzyme because of a partial biotin deficiency. This interpretation was confirmed with cell-free extracts, incubated with biotin, ATP and a divalent metal ion. As indicated by the open triangles of fig. 5, the specific activity of the pyruvate carboxylase in the extract rose nearly twenty-fold in 45 min, from 0.5 to 9.3. However, even this specific activity is only about one-tenth of that found in extracts of cells grown on biotin-adequate media: although the holoenzyme had indeed been formed under these conditions (and no such synthesis was observed if ATP or biotin were omitted), the efficiency of this process was low and of the same order as that described for other carboxylating enzymes [37,38]. But when acetyl-coenzyme A was also added to the incubation system, the specific activity of the enzyme rose more rapidly and to a level close to that observed with freshly prepared extracts. As shown by the open circles of fig. 5, it reached a value of 100 in 30 min, which represented a virtually total reconstruction of the holoenzyme.

Inclusion of L-aspartate in the incubation mixture to a large extent overcame this stimulation by acetyl-coenzyme A of holoenzyme formation (fig. 5, closed circles). This effect of aspartate is much greater than its effect on the activity of the enzyme (fig. 4): whereas the inhibitory effect of even 10 mM-

Table 4
Formation of pyruvate carboxylase from biotin and the apo-enzyme by succinate-grown *B. stearothermophilus.*

Culture	Time after biotin addition (min)	Specific activity of pyruvate carboxylase (μmoles of $^{14}CO_2$ fixed/min/mg of protein)	E_{680nm} of culture	[^{14}C]Leucine incorporated (c/min/ml of culture)
Succinate	0	4.7	0.53	0
(control)	30	128	0.71	7460
	60	140	0.99	14980
Succinate	0	1.6	0.53	0
plus	30	162	0.57	136
chloramphenicol	60	152	0.55	190

Two cultures were grown at 55°C with succinate as carbon source in the presence of L-leucine (40 μg/ml). When the absorbance of the cultures at 680 nm reached about 0.5 (at a cell density of about 0.18 mg dry weight/ml), D-biotin (1 μg/ml) and [^{14}C]-leucine (27 nCi/ml) were added to one culture (control); the other culture (test) received at this time chloramphenicol (500 μg/ml) and, 60 min later, biotin and [^{14}C]leucine. Samples were withdrawn at the times indicated in table 4 and monitored for extinction at 680 nm as an index of growth, for pyruvate carboxylase activity and for protein synthesis.

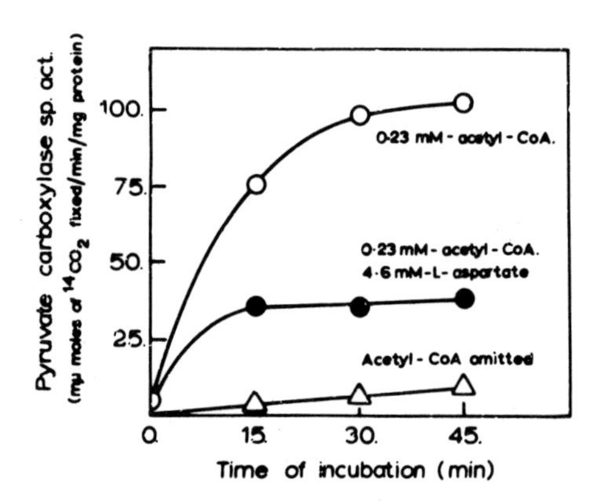

Fig. 5. Action of effectors on the reconstitution of pyruvate carboxylase from the apo-enzyme, biotin and ATP in cell-free extracts of a thermophilic bacillus (Data from [36])

aspartate on pyruvate carboxylase activity was overcome by as little as 50 μM-acetyl-coenzyme A, the stimulation of holoenzyme synthesis by 230 μM-acetyl-coenzyme A was markedly reduced by less than 5 mM-aspartate.

These stimulatory and inhibitory effects on holoenzyme formation of the same compounds that affect its activity is probably a further manifestation of the allosteric nature of the enzyme, and reflect the conformational changes needed before the apo-protein can react with biotin; they may, however, also reveal a novel metabolic control point in the formation of an active enzyme.

4. Formation of PEP from acetate

Two means of synthesizing PEP from acetate have been described. One, confined to strict anaerobes that can effect the reductive carboxylation of acetyl-coenzyme A, with the mediation of ferredoxin (reaction xi):

$$CH_3.CO.S.CoA + CO_2 + Fd_{red} + 2H^+ \rightleftharpoons CH_3.CO.CO_2H + Fd_{ox} + CoASH \quad \text{(xi)}$$

leads directly to the formation of pyruvate [39,40]. The C_3-acid thus formed can then be either carboxylated further to oxaloacetate (reaction ix), which is then decarboxylated to PEP (reaction i), or can be directly transformed to PEP, either via PEP-synthase (reaction iii) or via pyruvate, phosphate dikinase (reaction v). This more direct route is thought to occur in some photosynthetic bacteria [41].

Aerobic organisms cannot effect the reductive carboxylation of acetate and have to employ indirect pathways for PEP synthesis. The most commonly encountered of these is the glyoxylate cycle, a biosynthetic variant of the TCA cycle [42], shown as heavy arrows in fig. 6. The reactions of the two key enzymes of that cycle, isocitrate lyase and malate synthase, (reactions xii and xiii), together with other reactions of the TCA cycle and with PEP-carboxykinase, effect the net formation of one PEP unit from two units of acetate.

$$\underset{\text{isocitric acid}}{HO_2C.CH_2.CH.(CO_2H).CH(OH).CO_2H} \rightleftharpoons \underset{\text{succinic acid}}{HO_2C.CH_2.CH_2.CO_2H} + \underset{\text{glyoxylic acid}}{O{:}CH.CO_2H} \quad \text{(xii)}$$

$$CH_3.CO.S.CoA + O{:}CH.CO_2H + H_2O \rightleftharpoons \underset{\text{malic acid}}{HO_2C.CH_2.CH(OH)CO_2H} + CoASH \quad \text{(xiii)}$$

The operation of this cycle is regulated in a variety of ways, the complexity of which reflect also the complexity of the organism under study.

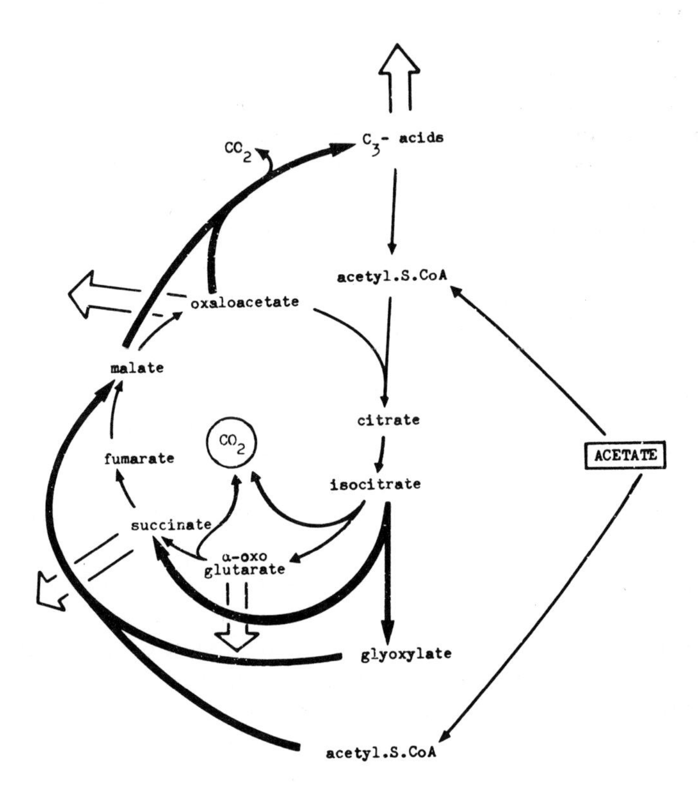

Fig. 6. The tricarboxylic acid cycle (light arrows) and glyoxylate cycle (heavy arrows) as a route for the synthesis of PEP from acetate.

(a) "*Fine control*" The first key enzyme of the cycle, isocitrate lyase, is strongly inhibited by C_3-acids such as PEP [43] or pyruvate [44]. It will be apparent from fig. 6 that these compounds may be regarded as end-products of the cycle, in so far as they form the major carbon compounds produced by the cycle and converted to other cell components; their inhibitory action on the first enzyme of the sequence that leads to their formation is thus an example of a 'feedback' or 'end-product' inhibition [45]. Since isocitrate lyase is considerably more sensitive to inhibition by PEP than by pyruvate at $pH < 8.5$, but more sensitive to pyruvate than to PEP at higher pH values [46], the possibility of regulating the activity of this enzyme by either C_3-compound may be of advantage to the numerous organisms that use this cycle and that grow on acetate under a variety of conditions.

(b) *"Coarse control"* The regulatory reactions so far discussed in this paper act on enzymes already present in the cell and, as implied by their description as "fine control" processes, are immediate in their effect. In addition to them, microorganisms possess the striking ability to "select" the metabolic routes most appropriate to the utilization of their carbon sources, by regulating the rate at which key enzymes of these pathways are synthesized. This type of control through "enzymic adaptation" may be viewed as a "coarse control" process, inevitably less immediately responsive to environmental changes than is a "fine control"; it is achieved by regulating the activity not of enzymes but of genes.

Studies with *E. coli* [44] showed that the preferential rate of synthesis of both isocitrate lyase and malate synthase increased dramatically as cells were transferred from media containing lactate to media containing acetate, and that growth on this latter medium occurred only after the specific activity of both enzymes had risen by a factor of about 10. This suggested that both enzymes might be controlled by one regulator gene, a conclusion supported by the finding [47] that mutants constitutive for isocitrate lyase are constitutive also for malate synthase. The gene specifying this constitutivity (*iclR*) has been located on the *E. coli* genome [48] : it is highly co-transducible with the structural genes for isocitrate lyase (*icl*) and the *metA* marker. As has been elegantly shown by Vanderwinkel and de Vlieghere [49] , this region of the genome carries also the gene specifying the malate synthase of the glyoxylate cycle (*masA*), presumably between the *metA* and the *icl* markers (see lower portion of fig. 7). This close proximity of the structural and regulator genes raises the possibility that their action is controlled as an 'operon'.

Experiments with partial diploids lend support to this view [50] . The recipient in the cross pictured in fig. 7 was a F^- strain of *E. coli* that required

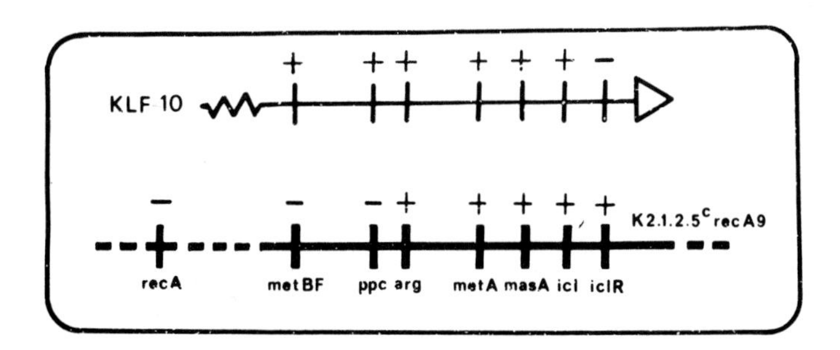

Fig. 7. Construction of a merodiploid from an episome carrying the 'inducibility' character for *icl*, and a *recA*-recipient constitutive for *icl*.

Table 5
Test of dominance of *iclR*$^-$ (inducible) over *iclR*$^+$ (constitutive).

	Organism	Phenotype	Genotype	Carbon source for growth	Specific activity of isocitrate lyase
(A)	K2.1.2.5^c.recA9	glucose$^-$ met$^-$	*metBF, pps, ppc, iclR, recA*	succinate	42 - 51
(B)	KLF10/K2.1.2.5^c.recA9	glucose$^+$ met$^+$	*pps*, $ppc^+/_{ppc^-}$, $met^+/_{met^-}$, $iclR^-/_{iclR^+}$, *recA*	glucose	0.3 - 0.9
				succinate	2 - 6
(C)	Segregrants from (B)	glucose$^-$ met$^-$	as (A)	succinate	27 - 40
(D)	(B) after acridine orange treatment	glucose$^-$ met$^-$	as (A)	succinate	38 - 59

methionine for growth (*metBF*), lacked PEP-carboxylase activity (*ppc*) but formed isocitrate lyase constitutively (*iclR*); it also carried a gene preventing the integration into its genome of DNA introduced by genetic donors (*recA*, [51]). Such "foreign" DNA was introduced from an F′ strain that transferred a short piece of genetic material to the recipient: since, in consequence of the *recA*-lesion, virtually no recombination occurred between this episome and the genome of the recipient, this organism had become a partial diploid with the characteristics ppc^+/ppc^-; met^+/met^-, $iclR^-/iclR^+$. Diploidy in the first two characters resulted in the restoration of PEP-carboxylase activity and the loss of the methionine requirement; more importantly in the present context, the $iclR^-/iclR^+$ diploidy led also to the loss of constitutivity of isocitrate lyase synthesis (table 5). After treatment of the merodiploid with acridine orange, the episome was eliminated: the cells recovered from this treatment were again ppc^-, met^- and produced isocitrate lyase constitutively. Segregrants obtained from the merodiploid without acridine orange treatment had the same phenotype.

These results thus indicate that *iclR*-mutants have lost the ability to form a repressor for the anaplerotic enzymes of the glyoxylate cycle, which can be restored by introduction of an episome carrying the wild-type allele. This *trans*-dominance of $iclR^-$ (inducible) over $iclR^+$ (constitutive) is analogous to the *trans*-dominance of the inducibility character in the lac operon [52].

(c) *"Compartmental control"* Although there is no convincing evidence for the net formation of PEP from acetate in animals, gluconeogenesis from fatty acids (and hence the glyoxylate cycle) is not confined to bacteria [53]. Indeed, one of the earliest pieces of evidence for the operation of this cycle was the demonstration [54] that germinating castor bean seeds contained high levels of isocitrate lyase and malate synthase when converting their stores of fat to carbohydrate; these enzymes cease to be formed when this gluconeogenesis is completed [55].

This type of "coarse control" process is, however, more complex than occurs in bacteria. Studies with *Tetrahymena pyriformis* [56] first suggested that the enzymes of the glyoxylate cycle were bound in a particulate structure during gluconeogenesis from fatty acids; these particles have since been identified as peroxisomes [57]. A similar situation was revealed by studies of a strain of *Chlorella* that formed isocitrate lyase constitutively [58]. After autotrophic growth or growth on glucose, this enzyme could not be sedimented by the high-speed centrifugation of *Chlorella* extracts; however, after growth on acetate (during which the enzymes of the cycle were, of course, required to function as components of the cycle) both isocitrate lyase and malate synthase appeared in a particulate form, together with other enzymes of the glyoxylate cycle.

Similar glyoxylate cycle particles, quaintly termed 'glyoxysomes' by Breidenbach, Kahn and Beevers [59], were recognised to harbor the enzymes of the cycle in castor beans; such particles have subsequently been found in all plant tissues capable of effecting carbohydrate formation from fatty acids via the glyoxylate cycle [60].

The factors that control not only the preferential synthesis of the glyoxylate cycle enzymes but also their incorporation into such particulate structures remain to be elucidated.

References

[1] H.A. Krebs, Proc. Roy. Soc. B 159 (1964) 545.
[2] M.F. Utter and K. Kurahashi, J. Biol. Chem. 207 (1954) 787.
[3] M.C. Scrutton and M.F. Utter, Ann. Rev. Biochem. 37 (1969) 249.
[4] G. de Torrontegui, E. Palacian and M. Losada, Biochem. Biophys. Res. Commun. 22 (1966) 227.
[5] J.J. Cazzulo, L.M. Claisse and A.O.M. Stoppani, J. Bact. 96 (1968) 623.
[6] E. Shrago and A.L. Shug, Arch. Biochem. Biophys. 130 (1969) 393.
[7] A.W. Hsie and H.V. Rickenberg, Biochem. Biophys. Res. Commun. 25 (1966) 676.
[8] H.A. Krebs and H.L. Kornberg, Energy Transformations in Living Matter, (Springer Verlag, Heidelberg 1957).
[9] M.F. Utter, Iowa State J. Sci. 38 (1963) 97.
[10] R.A. Cooper and H.L. Kornberg, Biochim. Biophys. Acta 104 (1965) 618.
[11] R.A. Cooper and H.L. Kornberg, Proc. Roy. Soc. B 168 (1967) 263.
[12] R.A. Cooper and H.L. Kornberg, Biochim. Biophys. Acta 141 (1967) 211.
[13] K. Berman, N. Itada and M. Cohn, Biochim. Biophys. Acta 141 (1967) 214.
[14] R.A. Cooper and H.L. Kornberg, Biochem. J. 105 (1967) 49C.
[15] H.L. Kornberg, Essays in Biochemistry 2 (1966) 1.
[16] C.B. Brice and H.L. Kornberg, Proc. Roy. Soc. B 168 (1967) 281.
[17] A.L. Taylor and C.D. Trotter, Bact. Rev. 31 (1967) 332.
[18] H.J. Evans and H.G. Wood, Fed. Proc. 27 (1968) 588.
[19] P.M. Siu, H.G. Wood and R.L. Stjernholm, J. Biol. Chem. 236 (1961) PC21.
[20] H. Lochmuller, H.G. Wood and J.J. Davis, J. Biol. Chem. 241 (1966) 5678.
[21] M.D. Hatch and C.R. Slack, Arch. Biochem. Biophys. 120 (1967) 224.
[22] M. Benziman, Biochem. Biophys. Res. Commun. 24 (1966) 391.
[23] M.D. Hatch and C.R. Slack, Biochem. J. 106 (1968) 141.
[24] B.B. Buchanan and M.C.W. Evans, Biochem. Biophys. Res. Commun. 22 (1966) 484.
[25] H.J. Evans and H.G. Wood, Proc. Natl. Acad. Sci. 61 (1968) 1448.
[26] T.J. Andrews and M.D. Hatch, Biochem. J. 114 (1969) 117.
[27] R.E. Reeves, J. Biol. Chem. 243 (1968) 3202; R.E. Reeves, R.A. Menzies and D.S. Hsu, J. Biol. Chem. 243 (1968) 5486.
[28] M.F. Utter and D.B. Keech, J. Biol. Chem. 235 (1960) PC17.
[29] E.S. Bridgeland and K.M. Jones, Biochem. J. 104 (1967) 9P.
[30] J. Payne and J.G. Morris, J. Gen. Microbiol. 58 (1969) 222.
[31] T.K. Sundaram, J.J. Cazzulo and H.L. Kornberg, Biochim. Biophys. Acta 192 (1969)

[31] T.K. Sundaram, J.J. Cazzulo and H.L. Kornberg, Biochim. Biophys. Acta 192 (1969) 355.
[32] M. Ruiz-Amil, G. de Torrontegui, E. Palacian, L. Catalina and M. Losada, J. Biol. Chem. 240 (1965) 3485.
[33] W. Seubert and U. Remberger, Biochem. Z. 334 (1961) 401.
[34] S.J. Bloom and M.J. Johnson, J. Biol. Chem. 237 (1962) 2718.
[35] E. Palacian, G. de Torrontegui and M. Losada, Biochem. Biophys. Res. Commun. 24 (1966) 644.
[36] J.J. Cazzulo, T.K. Sundaram and H.L. Kornberg, Nature 223 (1969) 1137.
[37] D.P. Kosow and M.D. Lane, Biochem. Biophys. Res. Commun. 4 (1961) 92.
[38] M.D. Lane, D.L. Young and F. Lynen, J. Biol. Chem. 239 (1964) 2858.
[39] R. Bachofen, B.B. Buchanan and D.I. Arnon, Proc. Natl. Acad. Sci. 51 (1969) 690.
[40] I.G. Andrew and J.G. Morris, Biochim. Biophys. Acta 97 (1965) 176.
[41] M.C.W. Evans, B.B. Buchanan and D.I. Arnon, Proc. Natl. Acad. Sci. 53 (1966) 928.
[42] H.L. Kornberg and H.A. Krebs, Nature 179 (1957) 988.
[43] J.M. Ashworth and H.L. Kornberg, Biochim. Biophys. Acta 73 (1963) 519.
[44] H.L. Kornberg, Biochem. J. 99 (1966) 1.
[45] H.E. Umbarger, Science 123 (1956) 848.
[46] P.J. Syrett and P.C.L. John, Biochim. Biophys. Acta 151 (1968) 295.
[47] E. Vanderwinkel, P. Liard, F. Ramos and J.M. Wiame, Biochem. Biophys. Res. Commun. 12 (1963) 157.
[48] C.B. Brice and H.L. Kornberg, J. Bact. 96 (1968) 2185.
[49] E. Vanderwinkel and M. de Vlieghere, Eur. J. Biochem. 5 (1968) 81.
[50] H.L. Kornberg, in: Metabolic Regulation and Enźyme Action, FEBS Symp., Madrid, Vol. 19 (Academic Press, 1968) p. 5.
[51] B. Low, Proc. Natl. Acad. Sci. 60 (1968) 160.
[52] F. Jacob and J. Monod, J. Mol. Biol. 3 (1961) 318.
[53] H.L. Kornberg and S.R. Elsden, Advanc. Enzymol. 23 (1961) 410.
[54] H.L. Kornberg and H. Beevers,Biochim. Biophys. Acta 26 (1957) 531.
[55] W.D. Carpenter and H. Beevers, Plant. Physiol. 34 (1959) 403.
[56] J.F. Hogg and H.L. Kornberg, Biochem. J. 86 (1963) 462.
[57] M. Müller, J.F. Hogg and C. de Duve, J. Biol. Chem. 243 (1968) 5385.
[58] L.C. Harrop and H.L. Kornberg, Proc. R. Soc. B 166 (1967) 11.
[59] R.W. Breidenbach, A. Kahn and H. Beevers, Plant Physiol. 43 (1968) 705.
[60] T.G. Cooper and H. Beevers, J. Biol. Chem. 244 (1969) 3507.

REGULATORY MECHANISMS IN THE BIOSYNTHESIS OF α-1,4-GLUCANS IN BACTERIA AND PLANTS

J. PREISS, S. GOVONS, L. EIDELS, C. LAMMEL,
E. GREENBERG, P. EDELMANN and A. SABRAW
Department of Biochemistry and Biophysics, University of California, Davis, California 95616, USA

Abstract: Preiss, J., Govons, S., Eidels, L., Lammel, C., Greenberg, E., Edelmann, P. and Sabraw, A. Regulatory Mechanisms in the Biosynthesis of α-1, 4-Glucans in Bacteria and Plants. *Miami Winter Symposia*, 1, pp. 122–138. North-Holland Publishing Company, Amsterdam, 1970.

The biosynthesis of glycogen in bacteria and of starch in plant leaves and green algae occurs via the following reactions: (a) ATP + G-1-P ⇌ ADP-glucose + PP_i, (b) ADP-glucose + α-1,4-glucan → glucosyl α-1,4-glucan + ADP.
synthesis. It has been shown that glycolytic intermediates activate and that AMP, ADP or P_i inhibit the synthesis of ADP-glucose by affecting the activity of the enzyme (ADP-glucose pyrophosphorylase) catalyzing the first reaction. On the basis of activator specificity, one can classify the ADP-glucose pyrophosphorylases isolated from various sources into four groups. Leaves of higher plants and green algae that fix CO_2 via the Calvin-Benson cycle or the Hatch-Slack cycle contain an ADP-glucose pyrophosphorylase that is activated by 3-phosphoglycerate and is inhibited by orthophosphate. The *Enterobacteriaceae* that catabolize glucose via glycolysis contain an ADP-glucose pyrophosphorylase activated by either fructose diP, TPNH or pyridoxal-P and inhibited by 5′-adenylate. Those microorganisms using the Entner-Doudoroff pathway as the main route of carbohydrate catabolism have ADP-glucose pyrophosphorylases activated by either fructose 6-P or pyruvate and inhibited by 5′-AMP, ADP or P_i. *Rhodospirillum rubrum*, an organism unable to utilize glucose for growth but which grows well on TCA intermediates, has an ADP-glucose-pyrophosphorylase activated by pyruvate. It thus appears that there is a correlation between the carbon utilization pathway found in an organism and the nature of activator specificity observed for its ADP-glucose-pyrophosphorylase. Furthermore, for many of the ADP-glucose pyrophosphorylases, the ATP saturation curve appears to be sigmoidal. In view of the inhibition of the enzyme by either 5′AMP, ADP or P_i, ADP-glucose synthesis and, therefore, glycogen synthesis is considered to be under control of the energy charge of the cell. High energy charge (i.e., high ATP levels relative to the level of the total adenine nucleotides in the cell) allows glycogen synthesis to proceed. In a low energy charge situation ADP-glucose synthesis and, therefore, glycogen synthesis would be inhibited. Another aspect in the regulation of glycogen synthesis has been the finding that in exponentially growing cells of *Escherichia coli* the enzymes cata-

lyzing reactions (a) and (b) are repressed and are coordinately derepressed when the cells enter the stationary phase of growth. The magnitude of repression is greater in cells growing in enriched media and with high growth rates than in cells growing in basal media and with slower growth rates.

Several mutants have been isolated from a culture treated with N-methyl-N′-nitro-N-nitrosoguanidine that are affected in their ability to accumulate glycogen. A number of these mutants are deficient in the storage of glycogen and also are deficient in the enzyme catalyzing reaction (b) (ADP-glucose: α-1,4-glucan 4-glucosyl transferase). One "glycogen deficient" mutant contains an ADP-glucose pyrophosphorylase that has less affinity for its substrates and activators. A number of mutants accumulate higher levels of glycogen than the parent strain. One contains higher levels of both ADP-glucose pyrophosphorylase and transferase than the parent strain. Both enzymes are found at derepressed levels in the exponential phase of growth. Another "glycogen excess" mutant contains normal levels of ADP-glucose pyrophosphorylase and transferase. However, kinetic studies reveal that the ADP-glucose pyrophosphorylase has been altered so that its affinitv for its activators, TPNH, pyridoxal-P, and fructose diphosphate is greater than the parent strain enzyme and that its sensitivity to 5′AMP, ADP and orthophosphate inhibition is less than what is observed for the parent strain enzyme. These studies indicate that the regulatory effects seen for the *E. coli* ADP-glucose pyrophosphorylase *in vitro* are physiologicallv pertinent to the *in vivo* control of glycogen synthesis.

1. The function of α-1,4-glucans in cells

Most cells, whether they are derived from plant, bacterial or from mammalian sources, contain an α-1,4-glucan. In algae and higher plants the glucan is usually starch and in bacteria and in animals it is known as glycogen. The accumulation of these α-1,4-glucans occurs as a result of the energy produced from the metabolic activities carried out by the individual cells and therefore the α-glucans may be considered as end products of energy metabolism. These storage products may be utilized either to maintain the energy level of the cell or to supply carbon or energy under conditions where no other carbon source is available. For example, glycogen accumulation in bacteria occurs as a result of limiting growth conditions in the presence of excess carbon source. It is thought that under these conditions the rate of ATP production is in excess of its utilization since the biosynthesis of various cell constituents associated with cell growth and division is not required. The accumulated glycogen is then utilized in the absence of other available carbon sources to provide energy necessary to preserve cell integrity during the starvation or nongrowing period. This "energy of maintenance", as defined by Dawes and Ribbons [1], is the energy required under conditions of starvation, to allow for the continuation of the various cellular activities needed for preserving cell integrity and cell viability. Under conditions of starvation, energy would be required for processes such as turnover of components like RNA and protein, for the maintenance of motility, as well as for other chemical and mechanical activities needed for cell survival.

2. **Regulation of glycogen synthesis by metabolities**

The importance of glycogen accumulation to the cell is suggested by the evidence that synthesis of the α-1,4-glucans in most organisms is regulated by multiple mechanisms. The type of regulation observed in each cell is usually consistent with the metabolic activity going on in that particular cell. For instance, in mammalian muscle the biosynthesis of the α-1,4-glucosidic linkages occurs by transfer of glucose from uridine diphosphate glucose to a pre-existing α-1,4-glucan primer (Reaction 1) [2-7].

1. UDP-glucose + α-1,4-glucan → α-1,4-glucosyl-glucan + UDP

Regulation of the biosynthesis of glycogen in muscle occurs by 1) activation of the transferase by glucose 6-P, an important metabolite in the muscle, and 2) by interconversion of inactive and active forms of the UDP-glucose transferase [2-7]. This latter process is under hormonal control and is unique to the mammalian α-1,4-glucan system [5,7].

In leaves of higher plants, green algae and bacteria, the pertinent reactions for synthesis of α-1,4-glucosidic linkages in starch and glycogen occur by synthesis of ADP-glucose (reaction 2) and then transfer of the glucose to the starch or glycogen primer (reaction 3).

2. ATP + α-glucose 1-P ⇌ ADP-glucose + PPi
3. ADP-glucose + α-1,4-glucan → α-1,4-glucosyl-glucan + ADP

Regulation in these systems occurs at the level of ADP-glucose synthesis. Usually the ADP-glucose pyrophosphorylases from these systems are stimulated by glycolytic intermediates and inhibited by 5′-AMP and/or ADP and/or orthophosphate. These observations are consistent with the concept that under conditions of "ATP excess" (or high energy charge as defined by Atkinson et al. [8,9]) α-1,4-glucan synthesis will occur in bacteria and plants. The activation of ADP-glucose synthesis by a glycolytic intermediate provides a mechanism whereby maximum ADP-glucose synthesis and therefore maximum α-1,4-glucan synthesis occurs only in the presence of excess carbon. Since ATP is a substrate of the enzyme catalyzing ADP-glucose synthesis and either AMP, ADP or Pi may be inhibitors of the enzyme, ADP-glucose synthesis as well as

Table 1
Activators and inhibitors of ADP-glucose pyrophosphorylases from various sources.

Source	Primary activators	Secondary activators	Inhibitor	Possible mode of carbon metabolism
Leaves of higher plants green algae	3-Phosphoglycerate	Fructose-6-P, fructose-di-P, phosphoenolpyruvate	Pi	Calvin cycle or Hatch-Slack cycle
Escherichia coli, Aerobacter aerogenes, Aerobacter cloacae, Salmonella typhimurium, Citrobacter freundii, Escherichia aurescens	Fructose-di-P, TPNH, Pyridoxal-P,	2-Phosphoglycerate, 3-phosphoglyceraldehyde, phosphoenolpyruvate	AMP	Glycolysis
Arthrobacter viscosus *Agrobacterium tumefaciens* *Rhodopseudomonas capsulata*	Fructose-6-P	Pyruvate Ribose-5-P Deoxyribose-5-P	Pi,AMP,ADP	Entner-Doudoroff
Rhodospirillum rubrum	Pyruvate	None	None	Does not grow on glucose; grows well on TCA intermediates
Serratia marcescens	None	—	5′AMP	Glycolysis

A primary activator is defined as a metabolite causing the greatest increase in V_{max}. Secondary activators are those compounds causing activation but to a lesser degree than the primary activators.

α-1,4-glucan synthesis may be considered to be regulated by the energy charge of the cell.

The ADP-glucose pyrophosphorylases of green plants, algae and bacteria may be classified into five groups that are distinguished with respect to their activator specificity and they are listed in table 1. The ADP-glucose pyrophosphorylases from the leaves of higher plants [10-13] or green algae [14] are activated by 3-phosphoglycerate and to lesser extents by phosphoenolpyruvate, fructose 6-phosphate and fructose 1,6-diphosphate. The *Enterobacteriaceae* [15,16] have fructose 1,6-diphosphate, TPNH and pyridoxal phosphate (PLP) as their primary activators while other bacteria, such as *Arthobacter* [17,18], *Agrobacterium tumefaciens* [15], or *Rhodopseudomonas capsulata* [19], have fructose 6-P as the primary activator for their ADP-glucose pyrophosphorylases. *Rhodospirillum rubrum* ADP-glucose pyrophosphorylase has as its sole activator pyruvate [20], while *Serratia marcescens* ADP-glucose pyrophosphorylase does not appear to be activated by any glycolytic intermediate [16].

One can correlate the nature of the primary activator of the ADP-glucose pyrophosphorylase with the general mode of carbon metabolism that each group exhibits. For example, the leaves of many higher plants and green algae usually fix CO_2 during photosynthetic activity to form 3-phosphoglycerate. Photosynthetic activity also generates ATP and reducing power to enable the newly formed 3-phosphoglycerate to be converted to hexose phosphate and ultimately to starch. In this case, therefore, starch is considered as an end product of photosynthesis. The first glycolytic intermediate in this pathway is 3-PGA, and thus the activation of the leaf and algal ADP-glucose pyrophosphorylase by 3-PGA may be considered as a "feed forward" activation.

Table 1 also shows that those bacteria that catabolize glucose via glycolytic action contain ADP-glucose pyrophosphorylases activated by FDP, TPNH and PLP; those utilizing the Entner-Doudoroff pathway have ADP-glucose pyrophosphorylases activated by fructose 6-P. *R. rubrum*, which cannot utilize glucose as a carbon source for growth but grows well on tricarboxylic acid cycle intermediates, has an ADP-glucose pyrophosphorylase activated solely by pyruvate. There is insufficient time here to attempt to rationalize a physiological function for the different activators found for the ADP-glucose pyrophosphorylases in the various species with the particular type of carbon metabolism observed. However, this has been done in a recently published review [21]. Suffice to say here that the activator specificity of an ADP-glucose pyrophosphorylase of an organism appears to correlate with the nature of the carbon utilization pathway found in that organism. It is interesting to speculate whether originally there was only one type of genetic segment responsible for the activator site of the ADP-glucose pyrophosphorylase which by a number of small mutational modifications evolved into the four classes described here. Alternatively a number of independent genes could have arisen for the four different classes. The similarity of the various classes of sites with respect to activators makes the first speculation more feasible. The driving force for the particular selection of the regulatory site would be the compatibility of the site with the carbon metabolism that the organism is carrying out. It would also appear that the genetic information for the ADP-glucose pyrophosphorylase activator site has been lost in *Serratia marcescens.*

Table 1 also shows the inhibitors for the various classes of ADP-glucose pyrophosphorylases. Pi appears to be the most effective inhibitor for the plant enzymes while AMP is the most effective inhibitor for the ADP-glucose pyrophosphorylases of the *Enterobacteriacea.* The organisms metabolizing glucose via the Entner-Doudoroff pathway have ADP-glucose pyrophosphorylases that may be inhibited by either Pi, AMP or ADP.

In summary, the regulatory effects by metabolities observed *in vitro* on either partially purified or homogeneous ADP-glucose pyrophosphorylase suggest that regulation of ADP-glucose synthesis and therefore glycogen synthesis is under control of the "energy charge state" of the cell and that the rate of synthesis of ADP-glucose proceeds maximally only when glycolytic intermediates are present in sufficient amounts to cause activation of the ADP-glucose pyrophosphorylase.

3. Genetic regulation of glycogen synthesis in bacteria

3.1. Appearance of glycogen-synthesizing enzymes during the growth cycle of *Escherichia coli* B

It has been shown by several groups [1,22-25] that accumulation of glycogen occurs in the stationary part of the growth curve. Fig. 1 shows that in enriched media (containing 1% glucose and 0.6% yeast extract), the levels of the glycogen biosynthetic enzymes in *E. coli* B (ADP-glucose pyrophosphorylase and α-1,4-glucan synthetase) as well as glycogen accumulation begin to increase dramatically at the end of late logarithmic growth and reach their maximum in the stationary phase. In the enriched media, there is an 11- to 12-fold elevation in the specific activities of the synthetase and pyrophosphorylase in the stationary phase of growth. In limiting nitrogen, minimal media containing glucose, the synthetase and pyrophosphorylase levels in the stationary phase of the growth curve are about the same as those found in the enriched media. However, in minimal media, the levels of these enzymes are less repressed in the exponential phase of growth. Thus as observed in fig. 1b, only a 2- to 3-fold increase in the levels of these enzymes is seen during the transition from logarithmic phase to stationary phase. It therefore appears that the syntheses of the α-1,4-glucan synthetase and ADP-glucose pyrophosphorylase are coordinately derepressed as soon as the microorganism ceases to grow. That this mechanism of derepression involves protein synthesis was suggested by the studies of Cattaneo et al. [28] which showed that glycogen accumulation in *E. coli* in the beginning of stationary phase was inhibited by the addition of chloramphenicol to the media.

The levels of the ADP-glucose pyrophosphorylase and α-1,4-glucan synthetase present in the cell during the maximal rate of glycogen synthesis are about 9- to 15-fold in excess of that required for the observed *in vivo* rate of glycogen accumulation. This would suggest that other factors besides genetic regulation control the activities of the glycogen biosynthetic enzymes. It is quite possible then that the activity of the ADP-glucose pyrophosphorylase, and therefore the rate of α-1,4-glucan synthesis, may be modulated by the concentrations of its allosteric effectors.

The increase of the glycogen synthetic enzymes in *E. coli* in the post exponential phase of growth is independent of the nature of the carbon source in the media (e.g. succinate, acetate, pyruvate, glycerol, nutrient broth media) the organism is grown on.

Other strains of *E. coli* (K12 3000, CP79, KS) show the same type of derepression of the activity of glycogen biosynthetic enzymes upon going from logarithmic phase to stationary phase. Fig. 2 shows that in another bacterium, *Agrobacterium tumefaciens,* derepression of the glycogen biosynthetic enzymes also occurs when the growth of the organism becomes limited. Greater repression is seen when the organism is grown on enriched media (fig. 2a). However, this phenomenon does not occur in the photosynthetic organism *Rhodopseudomonas capsulata* (table 2). Glycogen accumulation, α-1,4-glucan synthetase activity and ADP-glucose pyrophosphorylase activity appear to be at their maximum during the logarithmic phase of growth when the cells are grown on malate as a carbon source either in the light or in the dark.

Recently we have been able to isolate a mutant of *E. coli B*, SG-3, that is partially derepressed with respect to the activities of the α-1,4-glucan synthetase and ADP-glucose pyrophosphorylase [27]. This mutant was isolated from a culture of *E. coli* B mutagenized with N-methyl-N′-nitro-N-nitrosoguanidine by the technique of Adelberg et al. [30]. Colonies of this mutant give a dark brown stain with I_2-solution that is more intense than that given by the parent colony when grown on glucose-yeast extract agar media [27]. Table 3 shows the levels of glycogen and the glycogen synthetic enzymes during the

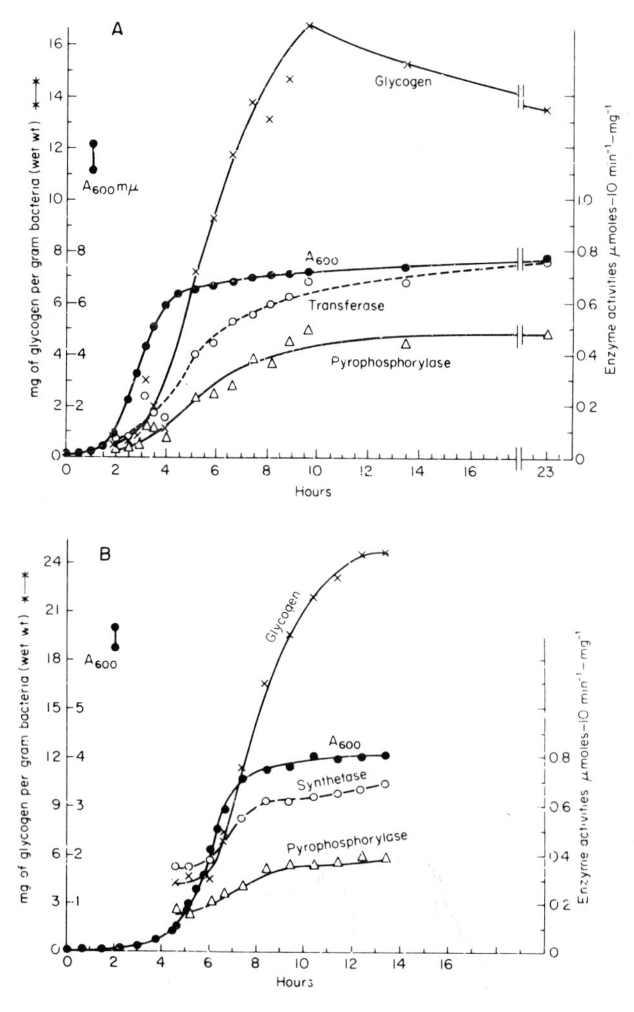

Fig. 1. Changes in glycogen, ADP-glucose pyrophosphorylase and α-1,4-glucan synthetase (transferase) levels during the growth of *E. coli* B. A. Cells were grown in a New Brunswick 14 liter fermentor on media containing 1% glucose, 1.1% K_2HPO_4, 0.85% KH_2PO_4 and 0.6% yeast extract at 37°. Aliquots of the culture were taken out at indicated times, and the cells were harvested by centrifugation for 10 minutes at 13,000 × g. In B the media contained 0.75% glucose, 0.12% $(NH_4)_2SO_4$ and basal salts solution P, pH 7.0 [26]. Enzyme and glycogen assays were done as described [27]. ADP-glucose pyrophosphorylase was measured in the direction of ATP formation.

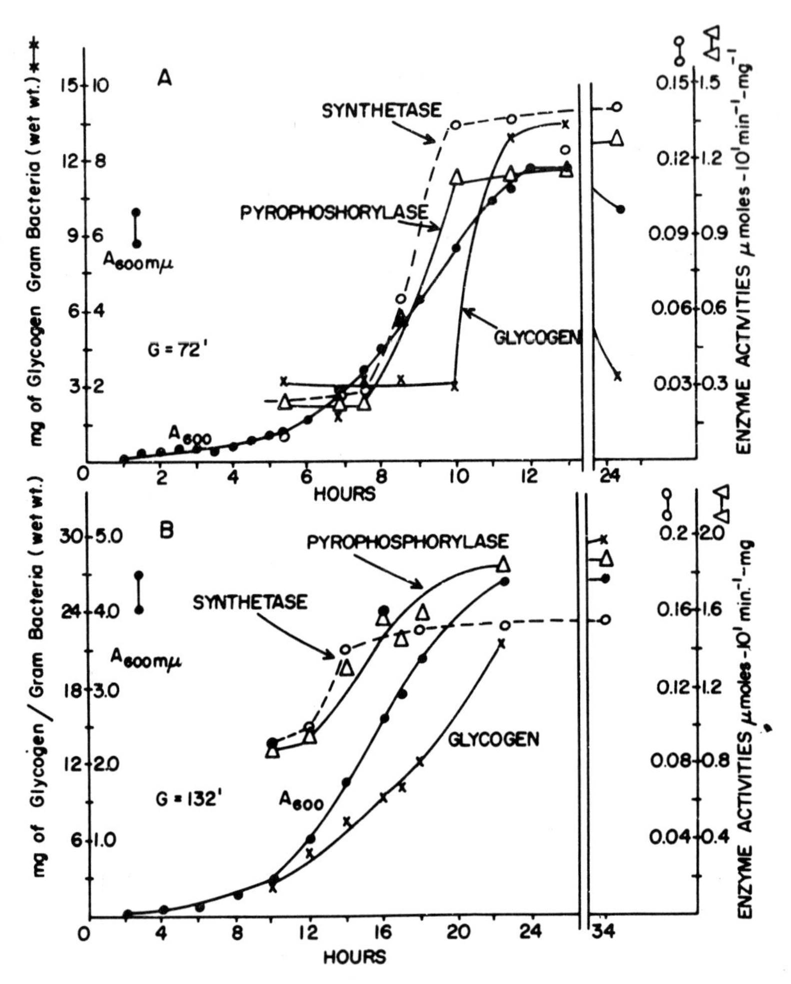

Fig. 2. Changes in glycogen, ADP-glucose pyrophosphorylase and α-1,4-glucan synthetase levels during the growth of *Agrobacterium tumefaciens* B-6. A. Cells were grown in a New Brunswick 14 liter fermentor on media containing 1% glucose, 1.1% K_2HPO_4, 0.85% KH_2PO_4 and 0.6% yeast extract at 37°. B. Cells were grown in media containing 0.6% glucose, 0.12% $(NH_4)_2SO_4$ and basal salts solution P, pH 7.0 [26]. Enzyme [18] and glycogen assays [27] were done as previously described. ADP-glucose pyrophosphorylase was measured in the direction of ATP formation. The generation times (G) of the organisms in the various media are indicated in the figure.

Table 2
Polyglucose polymer, ADP-glucose pyrophosphorylase, and α-1,4-glucan synthetase levels in *Rhodopseudomonas capsulata.*

Conditions	Polymer (mg/g dry wt.)	ADP-glucose pyrophosphorylase (μmoles/mg/10 min)	α-1,4-glucan synthetase (μmoles/mg/10 min)
Malate			
Aerobic, dark, middle log	47	0.70	0.22
late log	41	0.76	0.24
stationary	10	0.82	0.24
Malate			
Anaerobic, light, middle log	69	0.40	0.19
late log	60	0.39	0.19
stationary	21	0.41	0.21

Polyglucose polymer was determined by the method of Sigal et al. [25]. Conditions for preparation of extracts were essentially done as described previously [27]. The assay of α-1,4-glucan synthetase has been described [29] and the assay for ADP-glucose pyrophosphorylase was done in the direction of ATP synthesis [18].

Table 3
Glycogen, ADP-glucose pyrophosphorylase and α-1,4-glucan synthetase levels in *E. coli* B and mutant SG-3. The growth of the organisms, composition of the media, glycogen and enzyme assays are the same as those indicated in fig. 1.

Growth conditions	Glycogen (mg/g wet wt.)		ADPG pyrophosphorylase (μmoles/mg/10 min)		α-1,4-glucan synthetase (μmoles/mg/10 min)	
1% Glucose yeast extract	B	SG-3	B	SG-3	B	SG-3
exponential phase cells	1.3	2.9	0.04	0.39	0.10	0.41
stationary phase cells	19	26	0.53	0.94	0.85	1.23
0.6% Glucose, minimal media						
exponential phase cells	4.2	7.2	0.21	0.79	0.43	0.85
stationary phase cells	20	37	0.43	1.29	0.93	1.38

Table 4
Enzyme levels in *E. coli* B and mutant SG-3. Enzyme activity is expressed in μmoles product formed /mg/10 min. Enzyme activity was measured in extracts of cells grown in 0.6% glucose-minimal media [26] prepared as previously described [27]. Either the 30,000 × *g* or the 105,000 × *g* supernatant fraction of the crude extract was used for assay.

Enzyme	*E. coli* B		*E. coli* SG-3	
	Log phase	Stationary phase	Log phase	Stationary phase
Glucose 6-P dehydrogenase	0.94	1.40	0.84	1.40
Phosphoglucomutase	0.24	2.0	0.36	1.50
Hexokinase	0.12	0.50	0.10	0.56
Glycogen phosphorylase	0.001	0.0028	0.0014	0.0022
Phosphofructokinase	0.57	0.63	0.53	0.63
Pyruvate kinase	0.25	0.32	0.37	0.35
Isocitrate dehydrogenase	6.0	4.98	8.35	6.58
Glutamate-aspartate amino-transferase	4.07	4.02	4.76	3.87
Phosphohexoseisomerase	7.34	7.24	6.78	7.50
*PEP carboxykinase	0.39	0.55	0.25	0.64

* Enzyme was tested in cells grown on succinate-containing yeast extract media.

growth of the mutant on glucose-yeast extract media and on minimal media. In the early logarithmic phase SG-3 has 2- to 4-times more glycogen per gram of wet weight cells, contains 8- to 11-times more ADP-glucose pyrophosphorylase activity and 3-times more α-1,4-glucan synthetase activity than the parent strain. Thus both enzymes seem to be elevated to appreciable extents in the exponential phase of growth. There is further elevation of 2.7- and 4- to 5-fold respectively, of the pyrophosphorylase and synthetase activities when SG-3 enters the stationary phase. Table 3 shows also that in minimal media SG-3 contains derepressed levels of the glycogen biosynthetic enzymes and about 2 times more glycogen than the wild type.

Other enzymes, particularly the glycolytic enzymes, were assayed for in extracts of mutant SG-3 and the levels compared to those found in *E. coli* B (table 4). All enzymes listed in table 4 appeared to be at the same level in the parent and mutant strain. Noteworthy are the levels of glycogen phosphorylase in *E. coli* B and SG-3. The level of phosphorylase activity is only $\frac{1}{400}$ of the α-1,4-glucan synthetase activity in crude extracts. Although phosphorylase activity is increased in stationary phase cells of both *E. coli* B and SG-3 its level is not derepressed as the biosynthetic enzymes are in the mutant strain. SG-3 also behaves normally with respect to induction of β-galactosidase by the inducers, isopropyl thiogalactoside and methyl thiogalactoside. The induced β-galactosidase activity in SG-3 is sensitive to catabolite repression [31,32] when glucose is present in the media in the same way as the parent strain.

Recent kinetic studies on partially purified preparations of SG-3 ADP-glucose pyrophosphorylase and α-1,4-glucan synthetase indicate that they are identical to the parent strain *E. coli* B enzymes with respect to inhibitor and activator specificity, substrate requirements and affinity and inhibition characteristics. Thus more of the normal enzyme activities seem to be synthesized in SG-3 rather than more reactive enzyme molecules.

The derepression phenomenon of the glycogen synthetic enzymes in the

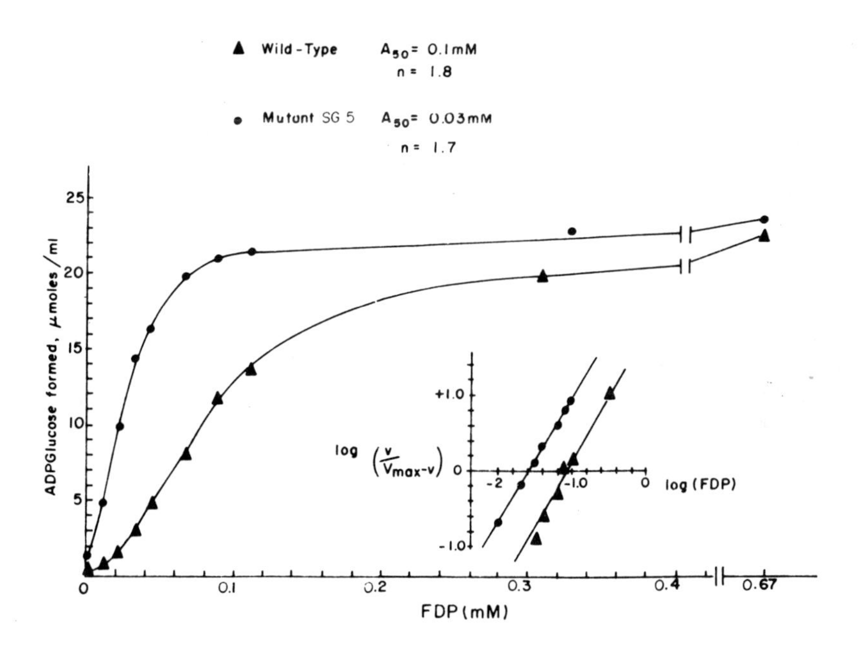

Fig. 3. The effect of fructose diphosphate concentration on ADP-glucose pyrophosphorylase activity of *E. coli B* and mutant SG-5. The reaction mixtures contained 20 μmoles of Tris pH 8.5, 0.3 μmole of ATP, 0.1 μmole of glucose-^{14}C-1-P (specific activity 8 × × 10^5 cpm/μmole), 1 μmole of $MgCl_2$, 100 μg of bovine plasma albumin enzyme and FDP in varying amounts as indicated in the figure in a volume of 0.2 ml. The assay was carried out as previously described [15]. A_{50} corresponds to the concentration of activator required for 50% of maximal activation; *n* is the Hill Coefficient obtained from the data plotted according to the Hill equation (inset).

stationary phase is reminiscent of the derepression of the synthesis of various proteins required for spore formation in the *Bacillus* genus (see review by Kornberg et al. [33]). Both processes, glycogen synthesis and spore formation, may be regarded as survival mechanisms. The controlling or initiating factors for both processes may possibly have some basic similarities.

3.2. *Isolation of a glycogen excess mutant, SG-5, containing an ADP-glucose pyrophosphorylase modified at its regulatory sites*

The ability to detect mutants by staining colonies of *E. coli* with an I_2-KI solution has enabled us to isolate a number of mutants from mutagenized cultures: some that are glycogen-deficient and others accumulating glycogen in excess of the parent strain. One "glycogen excess" mutant contains normal amounts of both α-1,4-glucan synthetase and ADP-glucose pyrophosphorylase. However, subsequent studies revealed that the concentration of the activators, PLP, FDP and TPNH, needed for 50% maximal activation were lower than the concentrations needed for half-maximal activation of the parent enzyme. Fig. 3. shows the FDP activation curve for both the *E. coli* B and mutant SG-5 ADP-glucose pyrophosphorylases. About 0.1 mM FDP is needed to give 50% of the maximum activation for *E. coli* B enzyme while only 0.03 mM FDP is required to give 50% of the maximal activation for the SG-5 ADP-glucose pyrophosphorylase. Table 5 summarizes the data for other activators for the mutant and parent strain ADP-glucose pyrophosphorylases. In addition to having a higher apparent affinity for the activators, the mutant ADP-glucose pyrophosphorylase also is less sensitive to inhibition by AMP. Fig. 4 compares the inhibition by AMP of the parent strain ADP-glucose pyrophosphorylase

Table 5
Kinetic parameters of activators of *E. coli* B and SG-5 ADP-glucose pyrophosphorylases.

Activator	*E. coli* B		SG-5	
	A_{50} (mM)	$\bar{n}$	A_{50} (mM)	$\bar{n}$
Fructose diphosphate	0.14	1.8	0.044	1.6
TPNH	0.16	2.2	0.054	2.0
Pyridoxal-5-P	0.0079	3.3-3.9	0.0046	2.6

The reaction mixture for assaying ADP-glucose synthesis is described in the legend of fig. 3. A_{50} corresponds to the concentration of activator required for 50% of maximal activation; $\bar{n}$ is the Hill coefficient obtained from the data plotted according to the Hill equation.

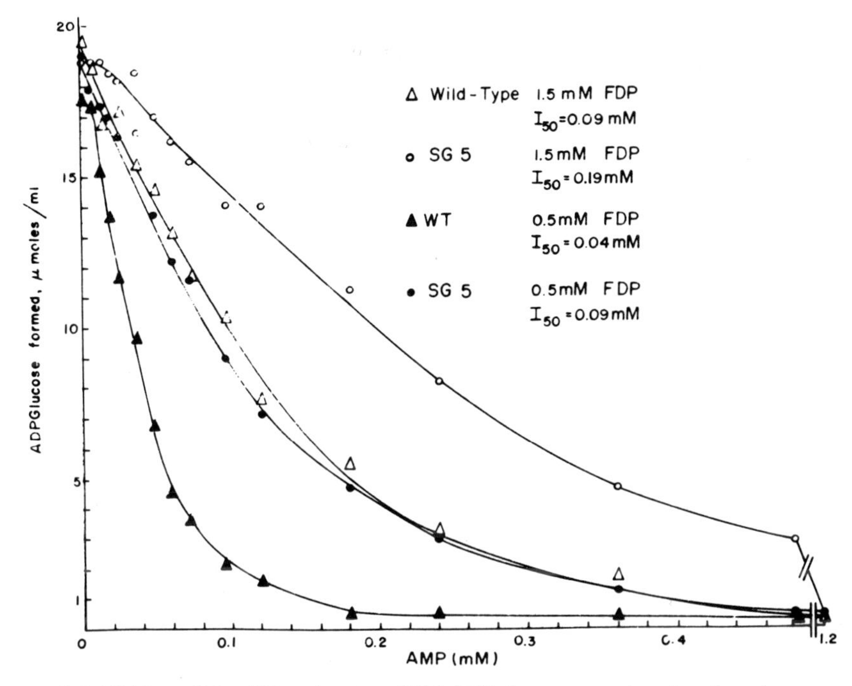

Fig. 4. Inhibition of *E. coli* B and mutant SG-5 ADP-glucose pyrophosphorylases by AMP. The reaction mixture and assay is the same as described for fig. 3 except that FDP and AMP were varied as indicated in the figure. I_{50} is the concentration of inhibitor giving 50% inhibition under the conditions of the assay.

with the mutant SG-5 enzyme. When the concentration of activator, FDP, is 1.5 mM, 0.09 mM AMP is required to inhibit the wild type enzyme 50% while 0.19 mM AMP is needed for 50% inhibition of the mutant enzyme. Previous reports [34,35] have indicated that the activator concentration modulates the sensitivity of the *E. coli* B ADP-glucose pyrophosphorylase to AMP inhibition. Thus as seen in fig. 4, a decrease of activator concentration from 1.5 mM to 0.5 mM decreases the concentration of AMP needed for 50% inhibition for the wild type enzyme from 0.09 to 0.04 mM. Likewise with 0.5 mM FDP, the concentration of AMP needed for 50% inhibition of the mutant enzyme is decreased from 0.19 mM to 0.09 mM. The mutant enzyme remains less sensitive to AMP even at the lower activator concentration. Table 6 shows data for the interaction of AMP with the other activators of the *E. coli* enzyme. In each instance, the activators modulate the sensitivity of the parent and mutant ADP-glucose pyrophosphorylases to AMP inhibition. In all cases the mutant

Table 6
Inhibition constants of *E. coli* B and SG-5 ADP-glucose pyrophosphorylases.

Activator	Concn. (mM)	*E. coli* B AMP I_{50} (mM)	*E. coli* B $\bar{n}$	SG-5 AMP I_{50} (mM)	SG-5 $\bar{n}$
FDP	1.5	0.09	1.7	0.19	1.8
	0.5	0.04	1.7	0.09	1.5
	0.11	0.007	1.6	-----	---
	0.03	-----	---	0.007	1.4
TPNH	1.5	0.032	2.1	0.066	2.5
	0.5	0.014	2.1	0.026	1.9
	0.15	0.0042	1.5	0.0076	1.6
	0.05	------	---	0.0033	1.1
PLP	0.05	0.077	2.8	0.35	2.9
	0.015	0.0074	1.7	0.059	2.8
	0.0075	0.0027	1.2	0.017	2.1

The reaction mixtures for assaying ADP-glucose synthesis are described in fig. 3. The concentration of activators used in the experiment are listed in the table. I_{50} corresponds to the concentration of inhibitor giving 50% inhibition and $\bar{n}$ is the Hill coefficient.

enzyme is more resistant to AMP inhibition. SG-5 accumulates about 4 times as much glycogen in the logarithmic growth phase and about 2 to 3 times as much in the stationary phase as the wild-type *E. coli* B. The above studies suggest that the increased glycogen synthesis in SG-5 is due to a modification in the ADP-glucose pyrophosphorylase causing a greater affinity for its activators and a lesser affinity for the inhibitor. These data strongly suggest that the regulatory effects observed *in vitro* are important for the *in vivo* regulation of glycogen synthesis in *E. coli* B. The relative insensitivity of the SG-5 ADP-glucose pyrophosphorylase to AMP inhibition and the increased rate of accumulation of glycogen in this mutant certainly suggests that glycogen synthesis is controlled by the energy charge of the cell.

The isolation of "glycogen mutants" has enabled us to determine the physiological significance of the various regulatory phenomena seen *in vitro*. Both regulation of enzyme synthesis as well as regulation of enzyme activity appear to play important roles in the accumulation of glycogen in the cell. The two mechanisms of regulation observed, the multiple number of activators for ADP-glucose pyrophosphorylase, underline the importance of glycogen accumulation for the bacterial cell. However, at present the known physiological function of glycogen in bacteria is at best speculative. Isolation of the proper mutants may perhaps give some insight of this function. Although only two mutants have been discussed here, a number of other mutants have been isolated. "Glycogen deficient" mutants defective in α-1,4-glucan synthetase have been isolated. Some mutants seem to be deficient in branching enzyme since their colonies stain blue with the I_2 solution. Hopefully, some of the deficient mutants may be very useful in studying the physiological function of the polysaccharide and the consequences of being deficient with respect to this polymer.

References

[1] E.A. Dawes and D.W. Ribbons, Bacteriol. Rev. 28 (1964) 126.
[2] L.F. Leloir, J.M. Olavarria, S.H. Goldemberg and H. Carminatti, Arch. Biochem. Biophys. 81 (1959) 508.
[3] E.D. Algranati and E. Cabib, J. Biol. Chem. 237 (1962) 1007.
[4] P.W. Robbins, R.R. Traut and F. Lipmann, Proc. Natl. Acad. Sci. U.S. 45 (1959) 6.
[5] C. Villar-Palasi and J. Larner, Biochim. Biophys. Acta 30 (1958) 440.
[6] R. Kornfeld and D.H. Brown, J. Biol. Chem. 237 (1962) 1772.
[7] M. Rosell-Perez and J. Larner, Biochem. 3 (1964) 773.
[8] D.E. Atkinson and G.M. Walton, J. Biol. Chem. 242 (1967) 3239.
[9] D.E. Atkinson, Biochem. 7 (1968) 4030.
[10] H.P. Ghosh and J. Preiss, J. Biol. Chem. 240 (1965) 960.
[11] H.P. Ghosh and J. Preiss, J. Biol. Chem. 241 (1966) 4491.
[12] J. Preiss, H.P. Ghosh and J. Wittkop in: Biochemistry of Chloroplasts, vol. II, ed. T.W. Goodwin (Academic Press, New York, 1966) p. 131.
[13] G.G. Sanwal, E. Greenberg, J. Hardie, E.C. Cameron and J. Preiss, Plant Physiol. 43 (1968) 417.
[14] G.G. Sanwal and J. Preiss, Arch. Biochem. Biophys. 119 (1967) 454.
[15] J. Preiss, L. Shen, E. Greenberg and N. Gentner, Biochem. 5 (1965) 1833.
[16] J. Preiss, G.R. Gayon, A. Sabraw and C. Lammel - unpublished results.
[17] L. Shen and J. Preiss, Biochem. Biophys. Res. Commun. 17 (1964) 424.
[18] L. Shen and J. Preiss, Arch. Biochem. Biophys. 116 (1966) 374.
[19] L. Eidels and J. Preiss - Arch. Biochem. Biophys. (1970) in press.
[20] C.E. Furlong and J. Preiss, J. Biol. Chem. 244 (1969) 2539.
[21] J. Preiss in: Current Topics of Cellular Regulation, Vol. I, eds. B.L. Horecker and E.R. Stadtman (Academic Press, New York, 1969) p. 125.
[22] T. Holme and H. Palmstierna, Acta Chem. Scand. 10 (1956) 578.
[23] E.G. Mulder, M.H. Deinema, W.L. Van Ween and L.P.T.M. Zevenhuizen, Rec. Trav. Chim. 81 (1962) 797.
[24] I.H. Segel, J. Catteneo and N. Sigal, Proc. Intern. C.N.R.S. Symp. Mech. Regulation Cellular Activities Microorganisms 1963 (1965) 337.
[25] N. Sigal, J. Cattaneo and I.H. Segel, Arch. Biochem. Biophys. 108 (1964) 440.

[26] D.G. Fraenkel and F. Neidhart, Biochim. Biophys. Acta 53 (1961) 96.
[27] S. Govons, R. Vinopal, J. Ingraham and J. Preiss, J. Bact. 97 (1969) 970.
[28] J. Cattaneo, N. Sigal, A.Favard and I.H. Segel, Bull. Soc. Chim. Biol. 48 (1966) 441.
[29] E. Greenberg and J. Preiss, J. Biol. Chem. 239 (1964) 4314.
[30] E. Adelberg, M. Mandel and G. Chen, Biochem. Biophys. Res. Commun. 18 (1965) 788.
[31] F.C. Neidhart and B. Magasanik, Nature 178 (1956) 801.
[32] B. Magasanik, Cold Spring Harbor Symp. Quant. Biol. 26 (1961) 249.
[33] A. Kornberg, J.A. Spudich, D.L. Nelson and M.P. Deutscher, Ann. Rev. Biochem. 37 (1968) 51.
[34] N. Gentner and J. Preiss, Biochem. Biophys. Res. Commun. 27 (1967) 417.
[35] N. Gentner and J. Preiss, J. Biol. Chem. 243 (1968) 5882.

HOMOLOGIES IN GLYCOGEN AND STARCH CATABOLISM

E.Y.C. LEE, E.E. SMITH and W.J. WHELAN
Department of Biochemistry, University of Miami School of Medicine, Miami, Florida 33152, USA

Abstract: Lee, E.Y.C., Smith, E.E. and Whelan, W.J. Homologies in Glycogen and Starch Catabolism. *Miami Winter Symposia* 1, pp. 139–150. North-Holland Publishing Company, Amsterdam, 1970.

The mechanisms by which glycogen and starch are catabolized seem to differ markedly at the stage at which the molecules are debranched. It is argued that this difference is more apparent than real, and that a plant glycosyl transferase, D-enzyme, whose function has hitherto been obscure, plays an important role in the overall starch catabolic process. Both glycogen and amylopectin require the concerted actions of phosphorylase, a transferase and a debranching enzyme for their conversion into monosaccharides. In the case of yeast and rabbit muscle, the transferase and debranching enzyme are associated into a multi-enzyme system. In potato the corresponding enzymes are not so associated, but may be the ancestors of the more efficient multi-enzyme complex.

The specificities of amylopectin and glycogen phosphorylases are compared in regard to their actions on linear and branched molecules. Glycogen phosphorylase acts much more rapidly on a branched polysaccharide, and the debranching system cooperates by never providing linear substrate. Amylopectin phosphorylase acts equally well on linear and branched substrates and is thus adapted to the specificity of its associated debranching enzyme, which does provide linear substrate, and to the presence in starch of the essentially linear amylose component.

1. Mechanisms of glycogen and starch debranching*

The close similarity in chemical structure between glycogen and the branched component of starch, amylopectin, has been paralleled by the

* The principal enzymes mentioned in this article and their abbreviated names, in parenthesis, are:
α-1,4-glucan:orthophosphate glucosyltransferase (phosphorylase) EC 2.4.1.1;
dextrin 6-glucanohydrolase (amylo-1,6-glucosidase) EC 3.2.1.33;
α-1,4-glucan: α-1,4-glucan 4-glycosyltransferase (referred to as transferase if associated with amylo-1,6-glucosidase, as D-enzyme if from potatoes) EC 2.4.1.25;
amylopectin 6-glucanohydrolase (R-enzyme, pullulanase) EC 3.2.1.9.

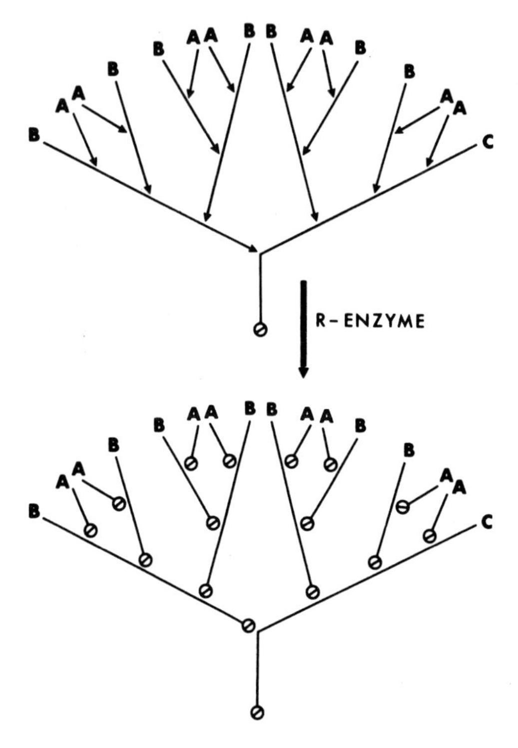

Fig. 1. The action of R-enzyme in hydrolysing the 1→6-branch linkages of amylopectin. Key: ——— = chain of 1→4-bonded α-glucose units; → = 1→6-bond; ϕ = reducing end glucose unit. A, B and C denote the three chain types in amylopectin and glycogen [3].

finding of equally close similarities between the enzymic mechanisms by which the molecules are synthesized and degraded. There are, of course, regulatory processes that control the activities of the enzymes of glycogen metabolism in higher animals that have no counterpart in starch metabolism, but the basic similarity between the ways in which the molecules are formed and dismembered is very real. There appears to be one exception, namely the way in which the branch linkages are removed in catabolism.

The enzymic mechanisms for the removal of the amylopectin and glycogen branch linkages were discovered almost simultaneously. Hobson, Whelan and Peat [1] found an enzyme in broad beans, which they named R-enzyme, that acted on amylopectin to hydrolyse the 1→6-branch points directly, so that the molecule fell apart into its unit chains (fig. 1). Cori and Larner [2] reported

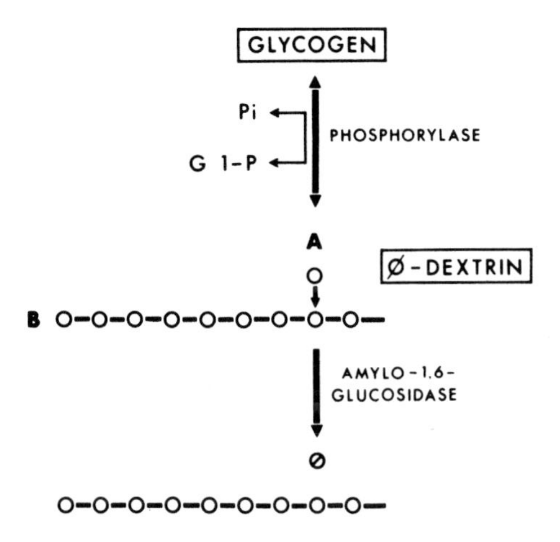

Fig. 2. Enzymic debranching of glycogen (Cori and Larner [2]). Key: 0 = α-glucose unit; ϕ = free glucose; —— = 1→4-bond; → = 1→6-bond. See text and fig. 1 for designation of chains as A and B.

that rabbit muscle contained an enzyme that hydrolysed the glycogen branch linkages, but only after the glycogen has first been degraded by the exo action of phosphorylase to the macromolecular phosphorylase limit dextrin (ϕ-dextrin). The debranching enzyme liberated glucose from the ϕ-dextrin and was termed amylo-1,6-glucosidase. In the terminology of Peat et al. [3] , the glucose so liberated must have arisen from an A chain (fig. 1) that phosphorylase had attenuated to a single glucose unit. Since ϕ-dextrin was known to be capable of still further degradation by β-amylase [2] , also an exo enzyme, it followed that the B chains (fig. 1) had not been attenuated by phosphorylase to the same degree as had the A chains. It could be calculated that if there are assumed to be equal numbers of A and B chains, and if the A chain was only one glucose unit long in the ϕ-dextrin, the outer portion of the B chain was seven units in length (fig. 2).

Walker and Whelan [4] investigated the structure of glycogen and amylopectin ϕ-dextrins with R-enzyme and by other means and found no evidence of single-unit A chains. On the contrary, R-enzyme liberated maltotetraose from amylopectin ϕ-dextrin. R-Enzyme does not attack glycogen or its ϕ-dextrin but a similar enzyme, pullulanase, discovered later, does do so and

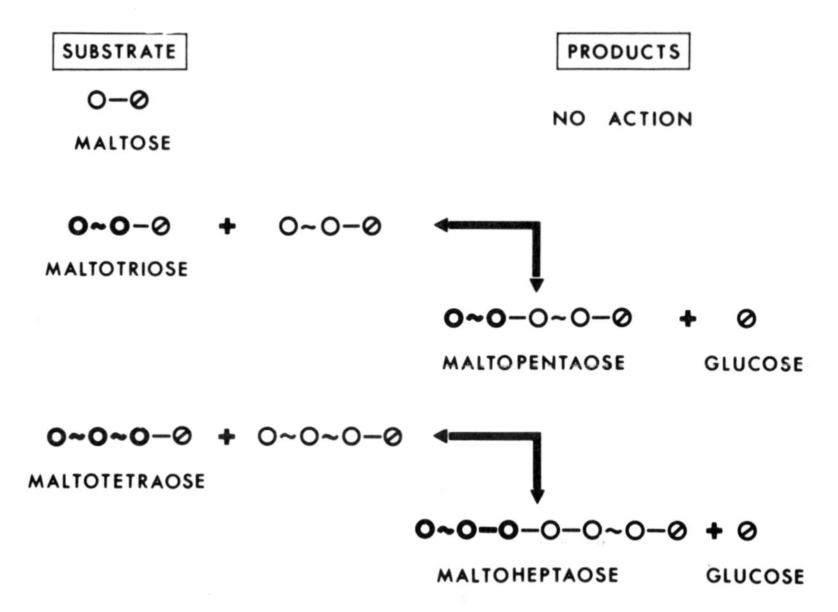

Fig. 3. Substrate specificity of D-enzyme [6,7]. Key: 0 = α-glucose unit; ϕ = free glucose; —— = transferable 1→4-bond; ∿ = non-tranferable 1→4-bond. Bold circles indicate glucosyl residues undergoing transfer.

Abdullah, Taylor and Whelan [5] used pullulanase to demonstrate maltotetraose formation also from glycogen ϕ-dextrin.

Walker and Whelan [4] concluded that in the ϕ-dextrin the A chains were not one but four glucose units in length. This being so, the attack of β-amylase on ϕ-dextrin, and the extent of the attack, could be accommodated by proposing a symmetrical 4/4 arrangement of the lengths of the A and outer B chains of ϕ-dextrin, rather than the 1/7 arrangement that had been concluded from the work of Cori and Larner [2]. This revision of the ϕ-dextrin structure, however, immediately raised the problem of the origin of the glucose formed when the 1→6-bonds were split by amylo-1,6-glucosidase. Walker and Whelan [4] offered an explanation based on a finding arising from studies on starch metabolism. This was the discovery by Peat, Whelan and Rees [6] of D-enzyme, a transglycosylase occurring in the potato, which has the capacity to redistribute groups of α-glucosyl units between starch-type oligo- or polysaccharides. Thus a maltosyl residue is removed from maltotriose, and, in a reversible manner, is transferred to an acceptor maltotriose molecule to give maltopentaose (fig. 3). The enzyme is unable to act on certain 1→4-bonds

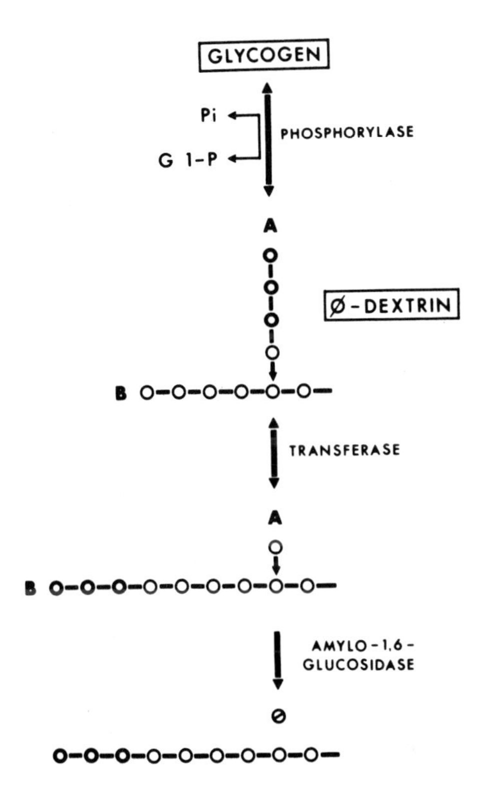

Fig. 4. Enzymic debranching of glycogen (Walker and Whelan [4]). Symbols as in figs. 2 and 3.

in such a polymer, specifically that at the reducing chain end and that penultimate to the non-reducing end [7]. This is explained in fig. 3, and it will be seen, for example, that these constraints prevent the enzyme from using maltose either as a donor or acceptor substrate. A maltosyl residue is the only group that can be transferred from maltotriose, and maltotriosyl from maltotetraose.

Walker and Whelan [4] speculated, and offered some experimental evidence, that associated with the rabbit muscle amylo-1,6-glucosidase was a transferase of the D-enzyme type. This, by transfer, could reduce the A chain in the ϕ-dextrin to a single glucose unit, permitting the subsequent action of amylo-1,6-glucosidase. Thus the removal of the glycogen branch point required phosphorolysis, glycosyl transfer and finally hydrolysis by amylo-1,

6-glucosidase (fig. 4). The existence of the transferase was confirmed in the Cori [8] and Whelan [9] laboratories and Brown and Illingworth [8] made the important finding that the specificity of the muscle transferase seemed very like that of D-enzyme in that it had a preference for the transfer of a maltotriosyl residue. This meant that when the transferase acted on ϕ-dextrin, the conversion of the 4-unit A chain into a glucose "stub" for glucosidase action would occur in the most efficient manner possible, that is, in a single step, by removal of maltotriose (fig. 4). Brown and Brown [10] also showed that the rabbit muscle glucosidase and transferase activities could not be separated from each other by conventional enzyme purification methods. Taylor and Whelan [11] suggested that the two enzymes are linked to form a multienzyme complex. Lee, Carter, Nielsen and Fischer [12] have purified the yeast transferase/glucosidase system to homogeneity without separation of the two activities and Lee and Carter [13] the rabbit muscle system also to homogeneity but without separation.

A picture therefore emerges of two very different types of debranching process as between amylopectin and glycogen. It is the purpose from this point on to argue that the differences are more apparent than real. Similarities between the processes begin to emerge if the question is posed of the ultimate fate of the catabolized polysaccharides. For glycogen this is easy to see. Phosphorylase acting with glucosidase/transferase will convert glycogen into n-1 moles of α-glucose 1-phosphate and 1 mole of glucose, where n is the average length of the glycogen unit chains. The further metabolism of the two products needs no elaboration.

If potato phosphorylase is imagined to act on amylopectin with R-enzyme, the products will be n-4 moles of glucose 1-phosphate and 1 mole of maltotetraose. Potato phosphorylase, like animal phosphorylase, will not reduce a maltosaccharide chain to fewer than four units in length [14]. What is the fate of the maltotetraose? Plants are, of course, equipped with amylases and α-glucosidases that would convert the tetraose into glucose. This is a possible answer to the fate of the tetrasaccharide. A more plausible answer comes from asking why potato contains D-enzyme [15]. Would the ability of D-enzyme to act on maltotetraose assist in the breakdown of amylopectin? If phosphorylase is present, the answer is positive. By converting maltotetraose into maltoheptaose (fig. 3) D-enzyme now renders three additional glucose units susceptible to phosphorolysis; viz:

$$\text{2 maltotetraose} \xleftrightarrow{\text{D-enzyme}} \text{maltoheptaose + glucose}$$

$$\text{maltoheptaose} \xleftrightarrow[\text{+ P}_i]{\text{phosphorylase}} \text{3 } \alpha\text{-glucose 1-phosphate + maltotetraose}$$

$$\text{Sum: Maltotetraose} \xleftrightarrow[\text{phosphorylase + P}_i]{\text{D-enzyme +}} \text{3 } \alpha\text{-glucose 1-phosphate + glucose}$$

Therefore, if D-enzyme acted in concert with potato phosphorylase and R-enzyme on amylopectin, the end result would be exactly the same as when phosphorylase and glucosidase/transferase act on glycogen.

We can, therefore, picture the overall processes of glycogen and amylopectin breakdown as in fig. 5. Glycogen is broken down by the enzyme sequence phosphorylase → transferase → glucosidase. Amylopectin must be considered to break down by the sequence phosphorylase → R-enzyme → D-enzyme. The reason for placing transferase action last in the latter case is that if transferase action were to precede debranching there would be no debranching. This is because of the interesting contrast in specificity between R-enzyme and amylo-1,6-glucosidase. The latter enzyme can only split the 1→6-bond when a single glucose unit is involved in the A chain. R-Enzyme by contrast will not act on such a structure. The smallest A chain that it will remove is maltose [16]. The same behaviour was found for the bacterial analogue, pullulanase [17]. Therefore if D-enzyme were to act on the ϕ-dextrin, as does muscle transferase, the still-branched product would be immune to debranching. It is not known whether D-enzyme can carry out this unproductive transfer, but since it would be reversible it is of no consequence to the argument.

We may therefore speculate that during evolution the demands for more rapid production of metabolic energy have resulted in the two separate enzymes of starch debranching, R-enzyme and D-enzyme, becoming locked into the obviously more efficient type of debranching system represented by muscle glucosidase/transferase. A parallel is to be seen in the enzymes of fatty acid synthesis, described elsewhere in this volume by Lynen [18]. In *Escherichia coli* they are separate enzymes. In yeast and mammalian liver they are found as multi-enzyme complexes. We might speculate that *E. coli*, which stores glycogen, will have separate glucosidase and transferase activities. The evolutionary concept is not, however, quite so simple since in the case of glycogen and starch debranching we must envisage also the change in specificity from R-enzyme, which is a maltosaccharylase, to amylo-1,6-glucosidase, which has no ability to remove maltosaccharides. Such a change in specificity might however demand little change in protein structure. It may be of note

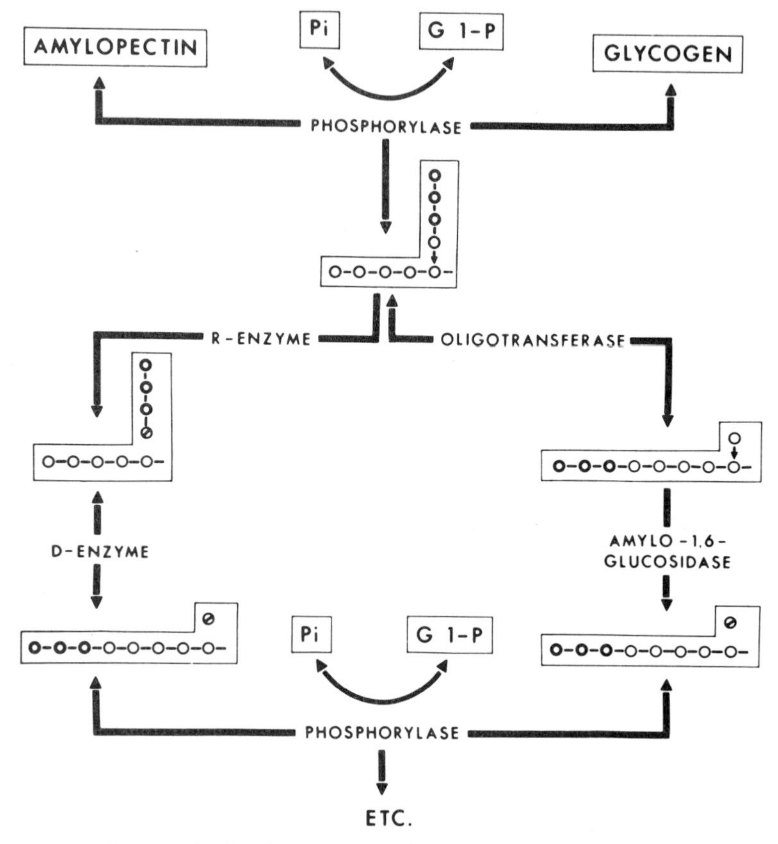

Fig. 5. Mechanism of phosphorolysis and debranching of glycogen and amylopectin. Symbols as in figs. 2 and 3.

that Walker and Builder [19] have reported the presence in *Streptococcus mitis* of a separate α-1,6-glucosidase and transglycosylase which in combination would debranch certain branched oligosaccharides, and have suggested that these two enzymes might play a role in the debranching of glycogen in the presence of the proper supporting enzymes. The α-1,6-glucosidase, however, has a specificity different from that of mammalian and yeast amylo-1, 6-glucosidase.

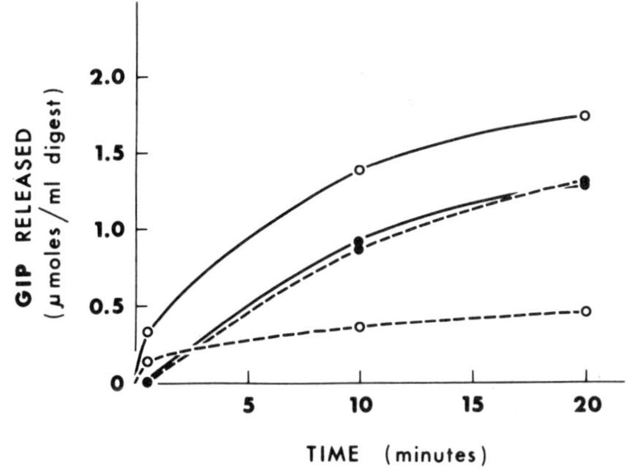

Fig. 6. Phosphorolysis by potato (•) and rabbit muscle (○) phosphorylases of amylopectin (—) and its debranched products (- - -). Amylopectin (2.5 mg/ml) was treated at 30° with pullulanase (0.5 units/ml) in a digest containing 50 mM phosphate buffer pH 6.0. The reaction was stopped when the iodine-staining and reducing powers of the digest reached maximum levels (3 hr) indicating that the debranching reaction was at an end. The digest was heated at 100° for 10 min to inactivate the pullulanase. Debranched amylopectin (1 mg/ml) was incubated at 25° in digests containing 0.2 M phosphate buffer pH 6.5, 1 mM AMP and either potato phosphorylase or rabbit muscle glycogen phosphorylase *b* (0.5 units/ml). Similar digests were prepared, containing untreated amylopectin (1 mg/ml). Samples were removed at intervals, the enzyme inactivated by heating at 100° for 3 min, and the α-glucose 1-phosphate released determined by a slight modification of a method [26] that employs a coupled phosphoglucomutase-glucose 6-phosphate dehydrogenase-NADP system. The reduced nucleotide was measured by the increase in absorbancy at 340 nm.

2. Specificities of starch and glycogen phosphorylases

Related to the possible evolutionary connexion between the amylopectin and glycogen debranching systems is the possibility of a similar connexion between the main agent, in each case, of polymer degradation, namely phosphorylase. Both the plant and the animal enzymes catalyse exactly the same reaction and share in common pyridoxal phosphate covalently bound to the protein [20]. Among the differences between the enzymes is their specificity towards linear and branched substrates.

Potato phosphorylase attacks linear and branched molecules with equal avidity. Fig. 6. shows that the enzyme attacks amylopectin at the same rate, regardless of whether this is the intact molecule or whether the 1→6-branch linkages have first been broken by pullulanase. The situation with rabbit muscle phosphorylase is very different (fig. 6). Intact amylopectin is attacked smoothly, but the debranched material, after an initially rapid but slight hydrolysis, is attacked at a small fraction of the rate of the branched polymer. The initial attack is probably attributable to a small amount of branched material that remains after pullulanolysis [21]. (Amylopectin was chosen for the experiment since pullulanase has only a slight debranching action on glycogen [17].) Debranching makes no difference to the content and concentration of non-reducing chain ends available to the enzyme. The effect on glycogen phosphorylase of debranching must have its origin in other reasons which are not difficult to enumerate. A glycogen or amylopectin molecule can be likened to a sundew flower and the enzyme to a fly. Once the enzyme encounters its substrate, a chain end, its escape from the environment of the molecular surface may well be extremely difficult. As soon as it breaks with one chain end it is captured by another. The same will not be true when the unit chains of the molecule are dispersed uniformly through the solution, when set free by debranching. The contrast is one of high (undebranched) and low (debranched) local micro-environmental concentrations of end groups, and has been dealt with fully by Smith [22]. The logic of the more rapid attack on the branched polymer seems obvious. It is clearly of considerable relevance that the glycogen debranching system acts in such a way that the unit chains are not set free. The only low molecular weight products of the combined actions of phosphorylase and the debranching system are α-glucose 1-phosphate and glucose. Therefore the undegraded portion of the glycogen is always optimally susceptible to phosphorolysis. Glycogen phosphorylase would clearly be a very inefficient enzyme if it had to work in cooperation with an R-enzyme-like system.

The above arguments, though plausible, offer no explanation of the behaviour of amylopectin phosphorylase. Indeed the behaviour of this enzyme in acting equally well on branched and linear substrates (fig. 6) seems to argue against our explanation for glycogen phosphorylase. An explanation of the behaviour of both types of enzyme is, however, found in their different orders of enzyme-substrate binding. Glycogen phosphorylase binds only weakly to the chain ends of oligosaccharides (K_m maltopentaose = 1.5×10^{-2} M [23]), but the weak binding is compensated for when attacking glycogen by the high local concentration of substrate chain ends. Potato phosphorylase binds much more firmly (K_m maltopentaose = 7×10^{-5} M [24]), permitting multiple attack, once it has encountered its substrate. In this regard Parrish et al. [24] have already proposed that potato phosphorylase binds so firmly to an oligosaccharide substrate that it only relinquishes its grasp in exchange for another substrate molecule. It might be possible to test the hypothesis of different binding affinities by measuring the extents of multiple attack by the two enzymes using the technique of Bailey and French [25].

That amylopectin phosphorylase should act efficiently on linear polysaccharides is required not only by the nature of the amylopectin debranching system but also by the fact that starches commonly contain around 25% of the essentially linear, high molecular weight amylose, having only one chain end per thousand or more glucose units. One would think this a poor substrate for an exo enzyme, but perhaps this is compensated for by multiple attack and strong binding affinity.

Acknowledgements

This work was supported by the National Institutes of Health (Grant no. AM-12532). E.Y.C. Lee is an Investigator of the Howard Hughes Medical Institute.

References

[1] P.N. Hobson, W.J. Whelan and S. Peat, J. Chem. Soc. (1951) 1451.
[2] C.F. Cori and J. Larner, J. Biol. Chem. 188 (1951) 17.
[3] S. Peat, W.J. Whelan and G.J. Thomas, J. Chem. Soc. (1956) 3025.
[4] G.J. Walker and W.J. Whelan, Biochem. J. 76 (1960) 264.
[5] M. Abdullah, P.M. Taylor and W.J. Whelan, in: Control of Glycogen Metabolism, eds. W.J. Whelan and M.P. Cameron (Churchill, London, 1964) p. 123.
[6] S. Peat, W.J. Whelan and W.R. Rees, J. Chem. Soc. (1956) 44.
[7] G. Jones and W.J. Whelan, Carbohyd. Res. 9 (1969) 483.
[8] D.H. Brown and B. Illingworth, Proc. Natl. Acad. Sci. U.S. 48 (1962) 1784; D.H. Brown, B. Illingworth and C.F. Cori, Nature 197 (1963) 980.
[9] M. Abdullah and W.J. Whelan, Nature 197 (1963) 979.
[10] D.H. Brown and B.I. Brown, in: Methods in Enzymology, Vol. 8, eds. E.F. Neufeld and V. Ginsburg (Academic Press, New York, 1966) p. 515.
[11] P.M. Taylor and W.J. Whelan, in: Control of Glycogen Metabolism, ed. W.J. Whelan (Academic Press, London and New York, 1968) p. 101.
[12] E.Y.C. Lee, J.H. Carter, L.D. Nielsen and E.H. Fischer, Biochemistry 9 (1970) 2347.
[13] E.Y.C. Lee and J.H. Carter, unpublished results.
[14] W.J. Whelan and J.M. Bailey, Biochem. J. 58 (1954) 560.
[15] W.J. Whelan, in: Encyclopaedia of Plant Physiology, Vol. 6, ed. W. Ruhland (Springer, Berlin, Heidelberg, New York, 1958) p. 154.
[16] W.J. Whelan, Biochemical Society Symposium No. 11 (1953) p. 17.
[17] M. Abdullah, B.J. Catley, E.Y.C. Lee, J. Robyt, K. Wallenfels and W.J. Whelan, Cereal Chem. 43 (1966) 111.
[18] F. Lynen, this volume, p. 151.
[19] G.J. Walker and J.E. Builder, Biochem. J. 105 (1967) 937.
[20] D.H. Brown and C.F. Cori, in: The Enzymes, Vol. 5, eds. P.D. Boyer, H. Lardy and K. Myrback (Academic Press, New York, 1961) p. 207.

[21] C. Mercier, unpublished results.
[22] E.E. Smith, in: Control of Glycogen Metabolism, ed. W.J. Whelan (Academic Press, London and New York, 1968) p. 203.
[23] E.E. Smith, unpublished results.
[24] F.W. Parrish, E.E. Smith and W.J. Whelan, Arch. Biochem. Biophys. 137 (1970) 185.
[25] J.M. Bailey and D. French, J. Biol. Chem. 226 (1957) 1.
[26] H.-U. Bergmeyer and H. Klotzch, in: Methods of Enzymatic Analysis, ed. H.-U. Bergmeyer (Verlag Chemie, Weinheim and Academic Press, New York, London, 1963) p. 131.

COMPARATIVE ASPECTS OF FATTY ACID SYNTHESIS

F. LYNEN
With the experimental collaboration of D. Oesterhelt, W. Pirson, M. Sumper, W. Winnewisser and J. Ziegenhorn
Max-Planck-Institut für Zellchemie, Karlstrasse 23, München 2, Germany

Abstract: Lynen, F. Comparative Aspects of Fatty Acid Synthesis. *Miami Winter Symposia* 1, pp. 151–175. North-Holland Publishing Company, Amsterdam, 1970.

Enzyme systems capable of fatty acid synthesis from malonyl-CoA, called fatty acid synthetase, are widely distributed in nature and can be isolated from various unicellular organisms, plants and animal tissues. All enzyme systems require, in addition to malonyl-CoA, TPNH as reducing agent and acetyl-CoA as a primer of the synthetic process.

Fatty acid synthetases may be arranged into two groups: (1) stable multienzyme complexes, having molecular weights between 5×10^5 and 2.3×10^6, in which all the individual enzymes of the pathway are assembled into a compact unit, e.g. yeast and various animal tissues, etiolated *Euglenia gracilis* and *Mycobacterium phlei*; and (2) fatty acid synthetases in which the constituent enzymes readily separate on conventional protein fractionation, e.g. *E. coli, Clostridium, Pseudomonas* and photoauxotrophic *Euglena gracilis.*

Both groups of fatty acid synthetases contain protein bound 4′-phosphopantetheine (acyl carrier protein, ACP) which serves as a carrier of acyl groups during chain elongation. In yeast, adipose tissue and liver, ACP is integrated into the multienzyme complex. The swinging arm of about 20 Å length carries the intermediates bound covalently to its SH-group to and from each of the enzymes which have little freedom of motion in the rather rigid structure of the multienzyme complex. In contrast, bacterial and plant ACPs are not firmly attached to the individual enzymes involved in fatty acid synthesis. Consequently, the fatty acyl ACP derivatives themselves can migrate and serve as substrates in subsequent reactions. In the stable multienzyme complexes, the fatty acids must be released from the complex associated ACP component either by transfer to CoA (yeast) or by hydrolysis (animal tissues).

All synthetases produce fatty acids of different chain-length. In the case of the yeast enzyme system, the distribution of fatty acids can be described with a mathematical model, based on the assumption that the fatty acyltransferase involved experiences hydrophobic interaction with the growing carbon chain, starting at the level of the C_{13}-acid. The intensity of this interaction increases by an energy increment of -0.9 kcal per additional CH_2-group.

Fatty acid synthetases of group I may be visualized as compartments in which all the intermediates are covalently bound and in which the contact with the outside is mediated by the various acyl transferases. According to our experiments, the active sites of acetyltransferase and malonyl transferase of the yeast complex contain serine residues which have carrier functions in acyltransfer between CoA and ACP. A stable ^{14}C-malonyl pentapeptide containing Ser, His, Gly, Ala and Leu was isolated from the peptic hydrolysate of ^{14}C-malonyl enzyme. Tryptic digestion of ^{14}C-acetyl enzyme at pH 6.9 led to the isolation and characterization of a stable ^{14}C-acetyl octapeptide:

(^{14}C-Ac)-Ser-Gln-Gly-Leu-Thr-Val-Ala-Val.

The absence of basic amino acids in this peptide indicates that it is derived from the *C*-terminal end of the transferase protein.

The stability of group I fatty acid synthetases can be influenced by ionic strength. The *Mycobacterium phlei* complex and the pigeon liver synthetase dissociate in dilute buffer solutions. In order to dissociate the yeast synthetase several freezings and thawings in 1 M LiCl or NaCl at -70° are required. The molecular weight of the proteins in the subunit fraction are between 200-250,000, which may indicate that the dissociation did not result in single enzyme components. After dissociation, both the total synthetase activity and the β-ketoacid reductase activity (one of the partial reactions assayed) disappeared. Reactivation occurs on decrease of ionic strength by dilution or dialysis, suggesting that the catalytically active conformation of the reductase enzyme requires some protein-protein interactions occurring only in more complex structures.

1. Introduction

In a Symposium directed towards the discussion of "Homologies in Enzymes and Metabolic Pathways" it seems to be very appropriate also to include a discussion of enzyme systems involved in fatty acid biosynthesis. These enzyme systems are playing a most important role in cellular metabolism in view of the fact that fatty acids are the principal constituents of various lipids and as such are involved in the building up of the various membrane structures inside the cells and surrounding them.

The biosynthesis of fatty acids is achieved in two steps. In the first step acetyl-CoA, the common building block for numerous natural products is carboxylated to form malonyl-CoA. In this process protein-bound biotin acts as a CO_2 carrier with the intermediary formation of CO_2-biotin.

$$\text{ATP} + \text{HCO}_3^- + \text{biotin enzyme} \rightleftharpoons \text{CO}_2\text{-biotin enzyme} + \text{ADP} + \text{P}_i$$

$$\text{CO}_2\text{-biotin enzyme} + \text{acetyl-CoA} \rightleftharpoons \text{biotin enzyme} + \text{malonyl-CoA}.$$

The energy required to bind CO_2 to biotin is delivered by the cleavage of ATP into ADP and inorganic phosphate. CO_2-biotin once formed then transfers its carboxylate group to the active methyl group of acetyl-CoA. Acetyl-CoA carboxylases have been isolated from a variety of cells and it would be possible to compare the various individual enzymes. I would, however, like to restrict my discussion to enzyme systems involved in the second step of fatty synthesis, that is the formation of the long carbon chains of fatty acid.

$$RCO\text{-}SCoA + n\,HOOCCH_2CO\text{-}SCoA + 2n\,TPNH + 2n\,H^+ \rightarrow$$

$$R(CH_2.CH_2)_n\text{-}CO\text{-}SCoA + n\,CO_2 + n\,H_2O + n\,HSCoA + 2n\,TPN^+$$

In this process, two carbon units derived from malonyl-CoA are successively added to the "priming" acyl-CoA derivative which principally is represented by acetyl-CoA. The carbon dioxide previously fixed in the carboxylation reaction is ejected again during the formation of fatty acids and this is quite important from the thermodynamic point of view.

2. Groups of fatty acid synthetases

The ability to synthesize fatty acids from malonyl-CoA is widespread in nature and enzyme systems required have been isolated from various unicellular organisms [1,2] including *E. coli* [1,3,4] and yeast [5,6] as well as from many plants [7–9] and animal tissues [10–18]. All enzyme systems require, in addition to malonyl-CoA, TPNH as a reducing agent and an acyl-CoA as a "priming" substrate of the synthetic process. A deeper insight into the detailed mechanism of fatty acid synthesis was first obtained in our studies on the purified enzyme system from yeast [5,19] which led to the following scheme (fig. 1). An essential feature of the biosynthetic process can be seen in the fact that all the intermediates are covalently bound by sulfur bridges to protein. It was recognized that two different types of sulfhydryl groups participate in the synthesis and we named them "central" and "peripheral" sulfhydryl groups before their chemical identification could be achieved. In the scheme the two sulfhydryl groups are distinguished by the prefix *c* for the central and *p* for the peripheral sulfhydryl groups. Fatty acid synthesis starts with a transfer of an acyl residue from the "priming" acyl-CoA to the "central" sulfhydryl group and from there to the "peripheral" one which is part of a cysteine residue [20] and represents the active site of the condensing enzyme component [19,21,22]. The process is followed by the transfer of a malonyl residue from malonyl-CoA to the "central" sulfhydryl group. In a condensation accompanied by decarboxylation, the acylmalonyl enzyme is transformed into a β-ketoacyl enzyme. The stepwise conversion of the β-keto-

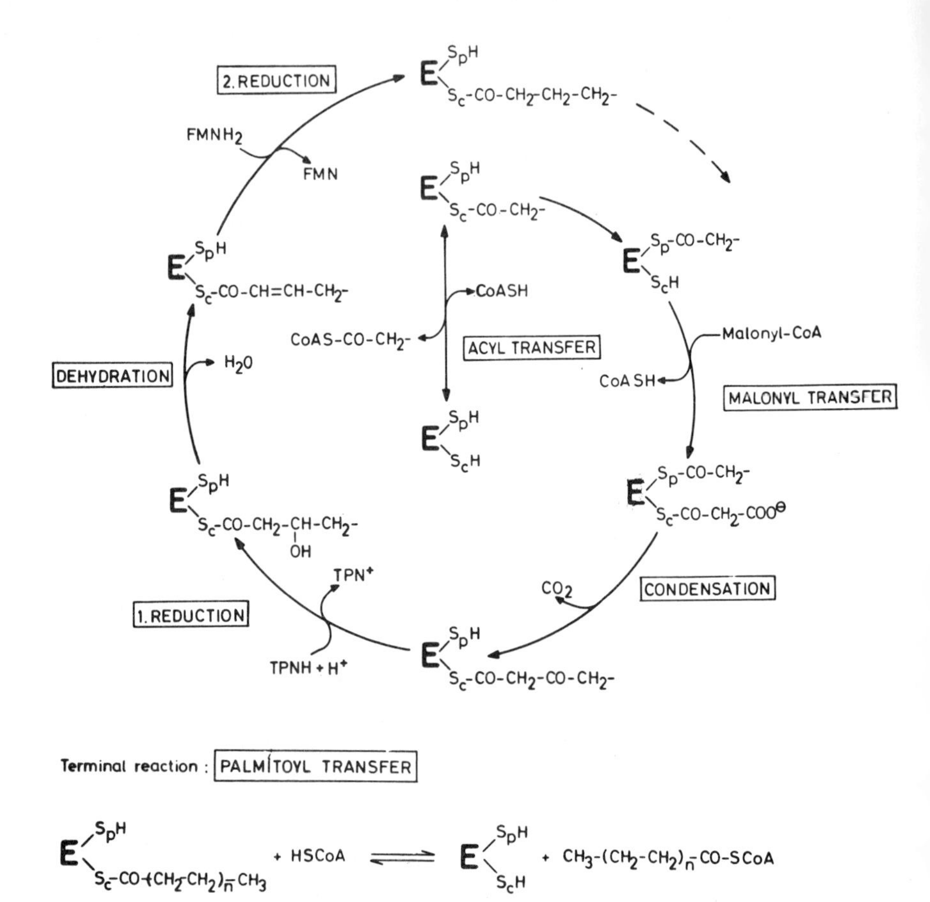

Fig. 1. Mechanism of fatty acid synthesis.

acid into the saturated acid is accomplished by reduction with TPNH to the β-hydroxylacyl enzyme, followed by dehydration to the α,β-dehydroacyl enzyme and another TPNH linked reduction to form the acyl enzyme which is by two carbon atoms longer than the acyl group used as a primer. In the second reduction step flavin mononucleotide sometimes serves as hydrogen carrier between TPNH and the double bond. This is the case with the yeast enzyme system as shown in studies of our laboratory [5]. After transferring the saturated carboxylic acid from the "central" sulfhydryl group to the "peripheral" one, the cycle of reactions, is repeated and repeated until the

Table 1
Fatty acid synthetases existing as multienzyme complexes.

Source	M.W.	Products
Pigeon liver	4.5×10^5	C_{16}; free
Rat liver	5.4×10^5	C_{16}; free
Adipose tissue	?	C_{14}-C_{16}; free
Mammary gland	($S_{20,w}$ = 12.9)	C_8 -C_{18}; free
Yeast	2.3×10^6	C_{16},C_{18}; CoA
Penicillium patulum	$\sim 2.5 \times 10^6$	C_{16},C_{18}; CoA
Euglena gracilis (etiolated)	$\sim 1 \times 10^6$	C_{16}; CoA
Mycobacterium phlei	1.7×10^6	C_{14}-C_{26}, esterified

Table 2
Fatty acid synthetases existing as individual enzymes.

Source	Products
E. coli	C_{16},C_{18},$C_{16:1}$,$C_{18:1}$, β-OH-C_{10}-C_{14}; ACP
Clostridium species	”
Pseudomonas species	”
Bacillus subtilis	C_{15} and C_{17} iso- and anteiso
Avocado mesocarp	C_{16}, C_{18}
Lettuce chloroplasts	C_{16}, C_{18}
Spinach chloroplasts	C_{16}, C_{18}
Euglena gracilis (photoauxotrophic)	C_{18}; ACP

chain is grown to the stage of fatty acids with 16 or sometimes even more carbon atoms. In the yeast system, the process is terminated by tranfer of palmitic or stearic acid from the “central” sulfhydryl group to free CoA. The acids are thus released from the enzyme system. In the mammalian fatty acid synthetases, the terminal reaction involves hydrolysis and in this case free fatty acids are produced instead of the CoA derivatives [10,23].

It is possible to arrange the fatty acid synthetases into two groups. One group includes stable multienzyme complexes having molecular weights between 5×10^5 and 2.3×10^6 in which all the individual enzymes of the pathway are assembled into a compact unit. In this group belong the fatty acid synthetases listed in table 1. They are classified as multienzyme complexes because functionally and operationally they behave as single units. The fatty acid synthetases listed in table 2 belong to a second category. They are systems only in the sense that they jointly catalyze a multistep metabolic pathway. At the cell free level, however, the constituent enzymes fail to show any signs of physical interaction. The best studied example in this category is the enzyme system of *E. coli*. As the experiments in the laboratories of Vagelos [25] and Wakil [26] have shown, all the individual enzymes of the pathway exist as monofunctional units and they readily separate on conventional protein fractionation.

3. Acyl carrier protein

Both groups of fatty acid synthetases contain protein-bound 4′-phosphopantetheine (acyl carrier protein = ACP) which represents the "central" SH-group and serves as carrier of acyl groups during chain elongation. In the first group of fatty acid synthetases, the acyl carrier protein component is integrated into the multienzyme complexes [17,24,27-29]. In contrast, the bacterial and plant acyl carrier proteins (table 2) are not firmly attached to the individual enzymes involved in fatty acid synthesis [30-32,7-9]. Consequently, the acyl carrier protein derivatives themselves can migrate and serve as substrates in subsequent reactions [33,34]. In the stable multienzyme complexes, the fatty acids must be released from the complex-associated acyl carrier protein component either by transfer to coenzyme A or by hydrolysis.

Due to the work of Hill in Wakil's laboratory [35], the amino acid sequence of the acyl carrier protein of *E. coli* is now known (fig. 2). The polypeptide contains 77 amino acids and three serine residues. Serine in the position 36 carries the functional 4′-phosphopantetheine in phosphodiester linkage. Similar acyl carrier proteins were isolated from *M. phlei* [36] and from plant material [9]. The various proteins are heat-stable and contain an abundance of acidic amino acids. Their total amino acid compositions are remarkably similar [9]. Recent observations indicate that portions of the amino acids around the 4′-phosphopantetheine may be similar among *E. coli, Arthrobacter viscosus*, and spinach [37]. The amino acid sequences await to be elucidated and when this has been achieved, the question of the very probable homology between these proteins may be discussed. Using a ^{14}C-pantothenate labelled fatty acid synthetase from yeast, Willecke et al. [28] could demonstrate that 4′-phosphopantetheine belongs to an individual structural element also in the stable multienzyme complexes as listed in table 1. It turned out to be smaller than the other proteins aggregated in the complex and to be rather firmly at-

1 10
NH_2-Ser-Thr-Ile-Glu-Glu-Arg-Val-Lys-Lys-Ile-Ile-Gly-Glu-

20
Gln-Leu-Gly-Val-Lys-Gln-Glu-Glu-Val-Thr-Asp-Asn-Ala-Ser-

30 * 40
Phe-Val-Glu-Asp-Leu-Gly-Ala-Asp-Ser-Leu-Asp-Thr-Val-Glu-

50
Leu-Val-Met-Ala-Leu-Glu-Glu-Glu-Phe-Asp-Thr-Glu-Ile-Pro-

60
Asp-Glu-Glu-Ala-Glu-Lys-Ile-Thr-Thr-Val-Gln-Ala-Ala-Ile-

70 77
Asp-Tyr-Ile-Asn-Gly-His-Gln-Ala-COOH

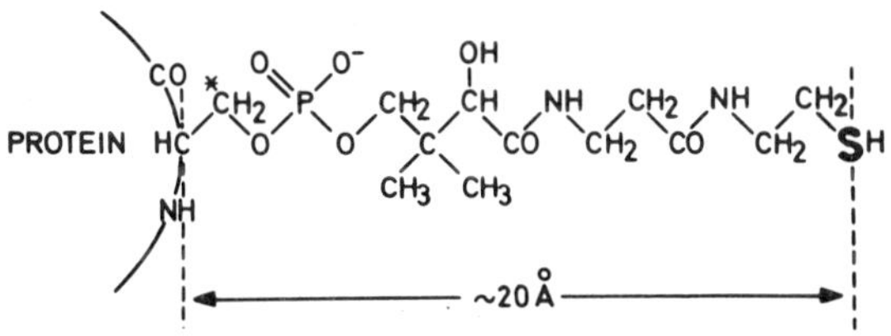

Fig. 2. Chemical structure of the *E. coli* acyl carrier protein.

tached to the complex. Dissociation of the complex with 3.5 M guanidine hydrochloride was required in order to release the ^{14}C-labelled ACP component.

The attachment of the carboxylic acid intermediates through 4′-phosphopantetheine seems to be very important for the functioning of the stable multienzyme complexes. This way the system gains a flexible arm of about 20 Å length when fully extended (fig. 2) but which is shorter when partially folded. In this manner, it is possible to bring the carboxylic acid intermediates bound covalently to the sulfhydryl group in close contact with the active site of each component enzyme which might have only limited freedom of motion in the rather rigid structure of a multienzyme complex [19].

At this point an important advantage should be mentioned which results from catalyzing a multistep reaction sequence in a structurally organized multienzyme complex. In order to build the carbon chain of stearic acid from malonyl-CoA and the primer acetyl-CoA, at least 32 intermediates have to be passed. What seems to be most important, is that all these intermediates are covalently bound to the enzyme system. If we assume, for simplicity, that the acyl carrier protein is present only once in the complex, we realize that at any given moment only one of the 32 intermediates can be present (per multienzyme complex). From the known volume of the fatty acid synthetase, found by X-ray small angle scattering technique to be 3.5×10^6 Å^3 [38], we may calculate the concentration of any intermediate to be 5×10^{-4} M. This locally high concentration of each intermediate facilitates the formation of the Michaelis-complexes and is reflected in the kinetics of the overall process. We may assume therefore, that the rate of fatty acid synthesis is principally only determined by the concentration of the diffusible substrates, such as acetyl-CoA, malonyl-CoA and TPNH.

The multienzyme complex may be visualized as strict compartmentation of a metabolic multistep reaction sequence in the smallest possible space. As a result the covalently bound intermediates within the complex are not accessible to competing processes, as for example to the enzymes of fatty acid degradation. According to our concept, the multienzyme complexes represent intracellular factories in which the building blocks are assembled piece by piece and only the finished product is released from the complex. With this picture in mind it is tempting to draw an analogy between fatty acid synthetase of type 1 and the mitochondrion. The function of the mitochondrial membrane, serving as a selective barrier, which guarantees the availability of certain metabolites and the exclusion of others, in the fatty acid synthetase is taken over by the acyltransferase enzyme components (see fig. 1), which deliver acetyl and malonyl fragments to the complex and release the long-chain fatty acids from it. In fatty acid synthesis the transferases play the critical role in compartmentalization.

4. The "priming" reaction

This leads us immediately to the specificity of the "priming" reaction in fatty acid synthesis and the reaction terminating the growth of the carbon chain. The initial reaction effects a transfer of the "primer" acyl group from the SH-group of acyl-CoA to the SH-group of the acyl carrier protein ("central" SH-group). In most of the fatty acid synthetases, the enzyme which catalyzes this acyltransfer has a fairly narrow specificity, the activity being high with acetyl-CoA and decreasing with increasing chain-length of the acid. In all these cases the products of fatty acid synthesis are mostly saturated straight-chain fatty acids (see table 1 and 2). Some microbial fatty acid synthetases produce also unsaturated straight-chain fatty acids (see table 2). This is related to the presence of a specific β-hydroxydecanoyl thiolester dehydrase

[39], which transforms β-hydroxyldecanoyl-ACP into Δ^3-decenoyl-ACP. On the other hand, *B. subtilis* synthesizes branched-chain fatty acids predominantly [40] and uses branched-chain acyl-CoA esters such as isobutyryl-CoA, isovaleryl-CoA or 3-methylbutyryl-CoA as "priming" substrates [2]. Products of synthesis are therefore C_{15} and C_{19} *iso-* or *anteiso* acids. The *B. subtilis* cell-free system was unable to utilize acetyl-CoA. However, if acetyl-ACP was used as the "priming" substrate, the initial acyl-CoA: ACP acyltransferase was by-passed and acetate was rapidly incorporated into straight-chain fatty acids [2]. These experiments of Butterworth and Bloch clearly demonstrated the important role the initial acyltransferase of the various synthetases plays in dictating the nature of the fatty acids formed.

The specificity of yeast fatty acid synthetase in regard to the "priming" substrate was extensively studied by Pirson [41] in our laboratory. Fig. 3 summarizes his results. As can be seen from this figure, acetyl-CoA is the far superior "priming" substrate. The capacity to prime fatty acid synthesis falls abruptly with the next higher homologues of acetyl-CoA but then increases again and reaches a new maximum at decanoyl-CoA indicating the presence of at least two acyltransferase components with different chain-length specificity. When studying the concentration dependence of the "priming" reaction with long-chain acyl-CoA derivatives an inhibitory effect at higher substrate concentrations became apparent. By kinetic analysis it could be shown that this inhibition was competitive with respect to malonyl-CoA [42] indicating a competition between acyl and malonyl residues for the SH-carrier group of the acyl carrier protein component.

At the level of C_{10} straight-chain acids Pirson [41] also studied the influence of functional groups in the hydrocarbon chain on the capacity to serve as "priming" substrates. As can be seen from table 3, in which the relative rates at optimal substrate concentrations are summarized, introduction of a hydroxy group or a double bond vicinal to the carbonyl group strongly depresses the "priming" activity. Substitutions at positions more distant from the carboxyl group however result in much smaller effects. The reaction products obtained with various "priming" substrates are listed in the last column of table 3. It was found that functional groups further away from the thiolester group were mostly retained during chain elongation whereas the conjugated double bond or the β-hydroxy group such as in Δ^2-decenoic acid or 3-hydroxydecanoic acid become eliminated. This was to be expected because the latter acids when bound to the acyl carrier protein-component of the complex are known to be normal intermediates of fatty acid synthesis. The low "priming" activity of the corresponding CoA-derivatives could be explained by the slow transfer of the acyl residues from CoA to the acyl carrier

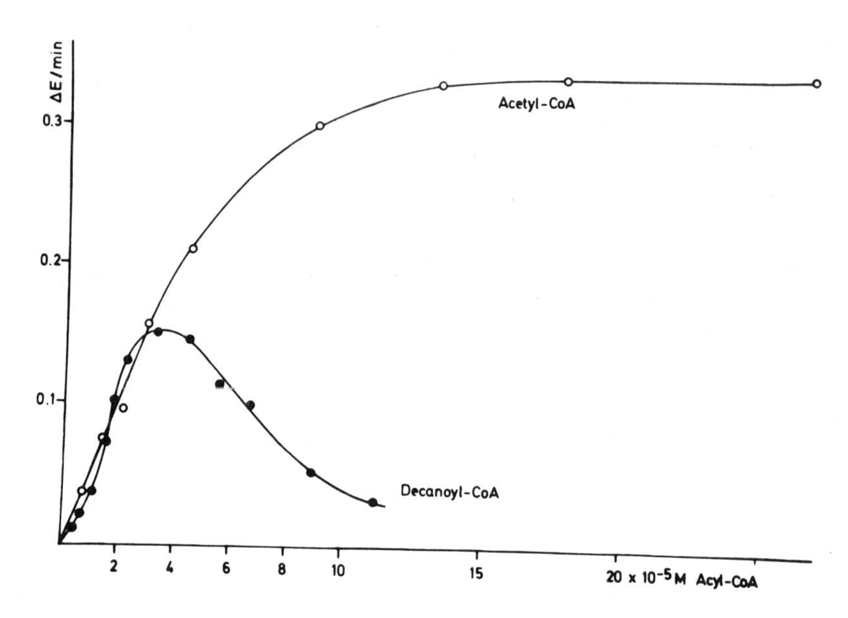

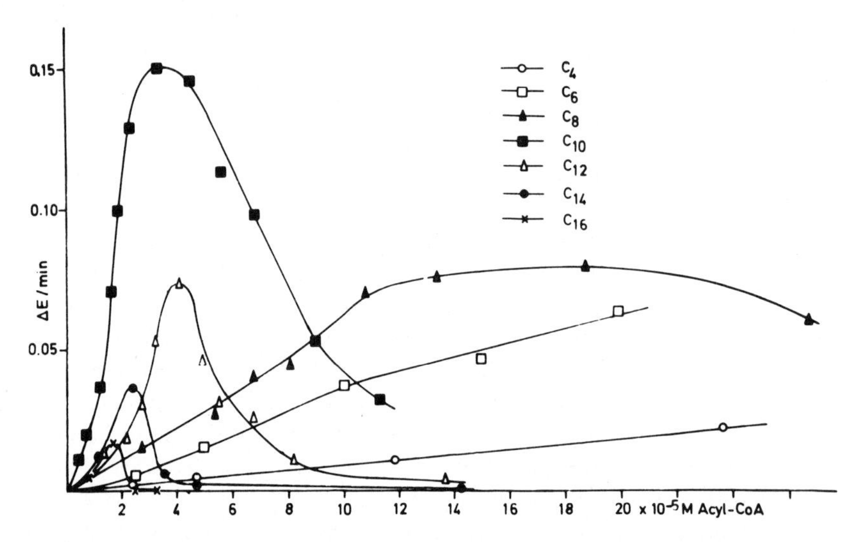

Fig. 3. Activity of various saturated acyl-CoA derivatives as "priming" substrates [41]. The cuvette (d = 1 cm) contained in a total volume of 2 ml: 200 μmoles phosphate buffer, pH 6.5; 20 μmoles cysteine; 0.6 mg serum albumin; 0.3 μmoles TPNH; 24.8 μg fatty acid synthetase (2.2 enzyme units/mg) and the amount of acyl-CoA (acetyl-CoA, butyryl-CoA, etc.) as shown on the abscissa. The reaction was started by the addition of 0.13 μmoles of malonyl-CoA. The rate of TPNH-consumption (Δ E/min) was measured at 25° and at a wavelength of 334 nm.

Table 3
Relative capacity of decanoyl-CoA and its derivatives to serve as "priming" substrates.

	Relative V_{max}	Products
Decanoyl-CoA	100	C_{12}-C_{18}; CoA
4-Oxodecanoyl-CoA	40	Oxo-C_{12}-C_{18}; CoA (Hydroxy-C_{12}-C_{18})
7-Oxodecanoyl-CoA	24	–
Δ^3-Decenoyl-CoA	34	$C_{12:1}$-$C_{18:1}$; CoA (C_{12}-C_{18})
Δ^9-Decenoyl-CoA	127	–
10-Hydroxydecanoyl-CoA	8	–
D,L-3-Hydroxydecanoyl-CoA	2	C_{12}-C_{18}; CoA
Δ^2-Decenoyl-CoA	4	C_{12}-C_{18}; CoA

protein catalyzed by the acyltransferase component. In passing it may be remarked that yeast fatty acid synthetase can also reduce the double bond in Δ^3-decenoyl-CoA or the carbonyl group in 4-oxodecanoyl-CoA at very slow rates. These slow side reactions therefore became only apparent when the "priming" substrate was preincubated with the enzyme complex in presence of TPNH before chain elongation was initiated by the addition of malonyl-CoA. The enzyme components of yeast fatty acid synthetase responsible for these side reactions have not yet been elucidated.

5. Termination of chain growth

Let us turn now to the rather puzzling problem as to why in experiments, in which acetyl-CoA was used as "priming" substrate, all even numbered fatty acids from 4 to 18 C-atoms exist as enzyme-bound intermediates, but that palmitic and stearic acid are mainly released from the complex. When propionyl-CoA is substituted for acetyl-CoA, odd numbered fatty acids are produced. In this case, the derivative with 17 C-atoms is the principal product amounting to about 75% of the fatty acids formed (fig. 4). Apparently the information to stop the synthesis specifically at the C_{17}-level must in some way be incorporated in the multienzyme complex. There are several explanations of this phenomenon available. In one of them it was assumed that the long-chain fatty acids were expelled from the enzyme complex as a result of lipophobic repulsion [19]. Sumper and Oesterhelt [43] in our laboratory have transformed this rather vague hypothesis into a model which considers some of the thermodynamic problems involved. This model considers the saturated

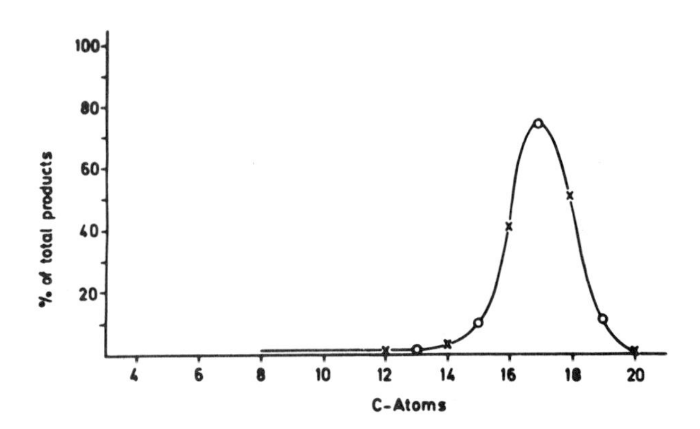

Fig. 4. Percentage distribution of products formed by fatty acid synthetase [43]. The reaction mixture contained in a volume of 2 ml: 200 μmoles of phosphate buffer, pH 6.5; 0.8 mg serum albumin; 1.2 μmoles TPNH; 200 μg fatty acid synthetase (1.5 – 2.5 enzyme units/mg) and either 0.25 μmoles ^{14}C-acetyl-CoA or 0.25 μmoles propionyl-CoA. $T = 25^{\circ}$. The reaction was started by the addition of 0.5 μmoles of ^{12}C- or ^{14}C-malonyl-CoA, and the rate of reaction was followed optically at 366 nm. The reaction was stopped in the linear rate region by the addition of 1 ml of 10% KOH in methanol. Fatty acid methyl esters were separated by gas chromatography. x——x acetyl-CoA as "primer"; o——o, propionyl-CoA as "primer".

acyl group of the growing fatty acid to be attached to the "central" SH-group of the complex. In this situation the acyl group has two possibilities for further reaction.

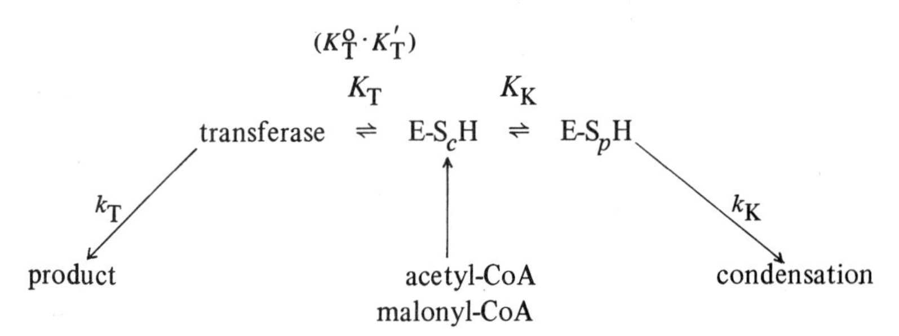

It can be transferred reversibly to the "peripheral" SH-group and can there be elongated by a C_2-unit derived from malonyl-CoA, or it can be transferred via a transferase site from the "central" SH-group to CoA and in this way leaves the multienzyme complex.

The model is based on two assumptions that are supported by experimental evidence:

a) the probability of any covalently enzyme-bound saturated acyl residue forming a product by transfer to CoA is determined by the relative velocities of the condensing and transferring reactions.

b) The growing carbon chain interacts with the enzyme protein only after a chain-length of 13 C-atoms has been attained; this interaction changes the relative velocities in favour of product formation by an energy increment of -0.9 kcal for each additional methylene group beyond the thirteenth.

To calculate the probability of product release at a particular chain-length (P_n) the following equation was derived from the model describing quantitatively the observed product distribution.

$$P_n = \frac{1}{1 + \frac{k_K}{k_T} \times \frac{K_K}{K_T^o} \times \frac{[\text{Mal-CoA}]}{[\text{Mal-CoA + Ac-CoA}]} \times e^{(m \times \Delta F/RT)}}$$

where k_K and k_T are rate constants of condensation and transfer respectively, K_K and K_T^o are equilibrium constants of acyltransfer between "central" SH-group and the active sites of condensing enzyme or transferase respectively. ΔF represents the energy increment of -0.9 kcal, that is the energy of interaction per methylene group. The variable m is the number of C-atoms in the growing acyl chain minus 13. The expression

$$\frac{[\text{Mal-CoA}]}{[\text{Mal-CoA + Ac-CoA}]}$$

takes into consideration the fact that both, malonyl and acetyl residues compete for the "central" SH-group but condensation can only occur if malonic acid is present there.

The formula suggests conditions under which either short-chain acyl-CoA derivatives or mainly stearyl-CoA can be produced. Synthesis under these conditions was examined experimentally and results indicated that the formula can be applied to a wide range of experimental conditions [43]. As an example in fig. 5 two experiments are represented. In these experiments adequate TPNH-concentrations were present but the ratio of acetyl-CoA to malonyl-CoA was varied. In one experiment this ratio was 1:2, whereas in the other it was close to infinity. The latter condition could be experimentally achieved by incubating fatty acid synthetase with acetyl-CoA and TPNH as the sole sub-

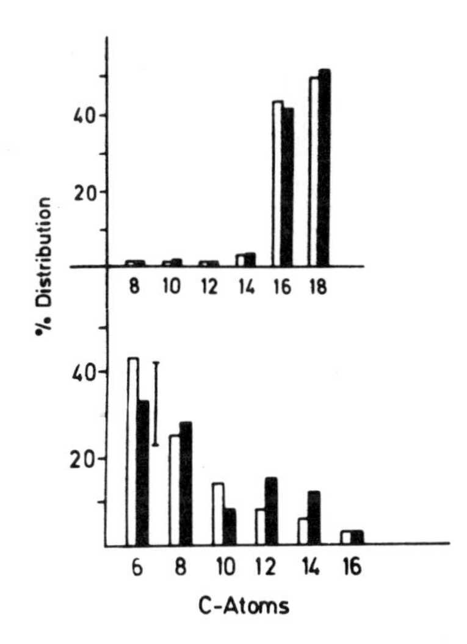

Fig. 5. Influence of the malonyl-CoA/acetyl-CoA ratio on the fatty acid products formed by fatty acid synthetase [43]. The reaction mixture in the upper part of the figure contained in a total volume of 2 ml: 200 μmoles phosphate buffer, pH 6.5; 0.8 mg serum albumin; 1.2 μmoles TPNH; 0.25 μmoles 1-^{14}C-acetyl-CoA and 200 μg fatty acid synthetase (1.5–2.5 enzyme units/mg). The reaction was started by the addition of 0.5 μmoles malonyl-CoA. The reaction mixture in the experiment on the lower part of the figure contained in a total volume of 2 ml: 100 μmoles tris-buffer, pH 7.5; 20 μmoles $MgCl_2$; 20 μmoles potassium citrate; 7 μmoles glutathione; 7 μmoles KOH; 50 μmoles $KHCO_3$; 1.5 mg serum albumin; 1.2 μmoles TPNH; 0.4 μmoles 1-^{14}C-acetyl-CoA; 0.6 enzyme unit of fatty acid synthetase and 0.01 enzyme unit of acetyl-CoA carboxylase (chicken liver). The reaction was started with the addition of 7 μmoles of ATP. $T = 25^{\circ}$. Further details see fig. 4. The open bars represent the values calculated with the question given in the text, the black bars represent the experimental values.

strates together with small amounts of acetyl-CoA carboxylase. This permitted the continuous generation of malonyl-CoA at a very slow rate in the reaction mixture. Under these conditions chain elongation was slowed down so extensively that short-chain fatty acids obtained the chance to be released from the complex. The good agreement between theoretical predictions (open bars) and experimental values (black bars) in both experiments should be noticed.

In order to explain that the growing hydrocarbon chain interacts with the enzyme protein only after a chain-length of 13 C-atoms has been attained,

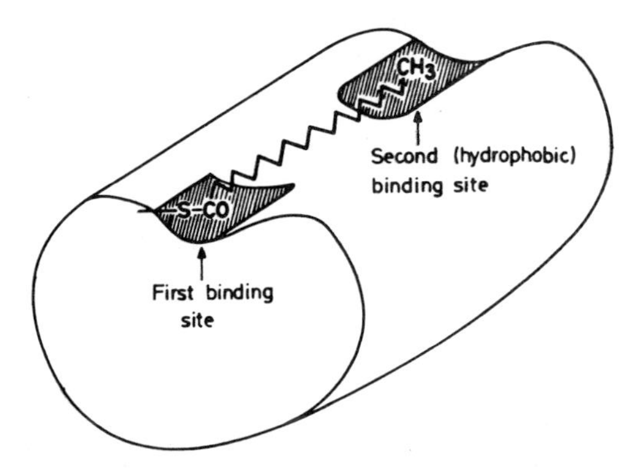

Fig. 6. Hypothetical structure of the acyltransferase responsible for the release of the fatty acids from the multienzyme complex. (For explanation see text).

the following hypothesis was made (fig. 6). It was assumed that the transferase possesses two active sites, one of which binds to the thiolester group and must have the same affinity to all the saturated acyl intermediates. The other one, presumably an area of apolar amino acid side chains, is so far distant from the first site that it comes into play only after a critical length of the fatty acyl chain has been exceeded. When this stage has been reached, each additional methylene group in the fatty acid would add an energy increment to the hydrophobic interaction and thereby increase the affinity of the transferase for the acyl residue. A ΔF-value of -0.9 kcal which permits maximal agreement between theory and experiment lies in the range of hydrophobic interactions as calculated by Nemethy and Scheraga [44].

Finally it should be pointed out that we have developed a model assuming that the increasing chain-length of the acyl group causes an increased affinity at the transferase site. We may just as well have assumed that the variation took place in K_K and that with increasing chain-length the affinity for the condensing site ("peripheral" SH-group) decreased. In this case, the ΔF value oer methylene group would have a positive sign so that the mathematical expression developed would still be applicable. Recent experiments of Ayling [45] in our laboratory actually support the latter assumption.

Most of the fatty acid synthetases behave like the yeast enzyme and with sufficient supply of malonyl-CoA and TPNH produce preferentially C_{16} and C_{18}-acids (see tables 1 and 2). In the *E. coli*-system unsaturated and β-hydroxy acids are also formed as acyl carrier protein-derivatives, and this enzyme system contains the specific β-hydroxydecanoyl thiolester dehydrase of Bloch [39] and does not exist as a stable multienzyme complex. In the group of stable multienzyme complexes the enzyme system of *M. phlei* is exceptional in so far as it affords products with a biphasic chain-length distribution, maxima occurring at C_{18} and C_{24} [32]. It is probable therefore that in this case two chain-length specific enzymes are involved in termination of the elongation process.

Considering the deciding role the specific acyltransferases occupy in fatty acid synthesis by stable multienzyme complexes, it was tempting to obtain further insight into the chemical make up of their active sites. In this respect some progress was made recently by Ziegenhorn [46,47] in our laboratory. He fractionated tryptic and peptic digests of ^{14}C-acetyl enzyme, prepared by incubating yeast fatty acid synthetase with ^{14}C-acetyl-CoA, and could identify one serine residue bound in the following *C*-terminal amino acid sequence as being the carrier of acetic acid.

```
        14C-Ac
          |
          O
          |
- Lys-Ser-Gln-Gly-Leu-Thr-Val-Ala-Val.
```

Previous experiments of Schweizer [27] on ^{14}C-malonyl enzyme had shown that the carrier group of the malonyltransferase component is also a serine residue. Experiments are now in progress which will elucidate the active site of the acyltransferases involved in termination of the elongating process. We have already some indication that in this case also polypeptide-bound serine residues are involved [45].

6. Subunit structure of fatty acid synthetases

The two extremes of fatty acid synthetase, those that represent tightly-aggregated multienzyme complexes and those that lack any higher order of structural organization might represent two stages in the evolution of the enzyme system. According to this concept the synthesis of fatty acids from malonyl-CoA and TPNH and acetyl-CoA may originally have been carried out by separate enzymes with an easily dissociable acyl carrier protein as in the case of the *E. coli* and related systems (table 2). In the course of evolution the various component enzymes and the acyl carrier protein came to be associated into a form of fatty acid synthetase in which the components were firmly bound to each other as in the multienzyme complexes (table 1). The forces which stabilize the multienzyme complexes seem to be of the same kind as those stabilizing quarternary structures of single proteins. This is indicated by the fact that treatment of the enzyme complexes with urea, guanidine hydrochloride, or detergents like sodium deoxycholate, dissociates the complexes. Therefore, the various protein components in the complexes must possess both:

1) catalytic sites responsible for the enzyme activity, and
2) binding or recognition sites responsible for the interaction with other proteins to form a specific spatial structure.

The stability of the various synthetase complexes varies in this group. As an example, successful isolation of the complex from *M. phlei* requires a high ionic strength environment. When the buffer concentration is lowered from 0.1 M to 0.01 M phosphate enzyme activity is rapidly lost and, as sucrose density centrifugation indicates, probably with fragmentation of the complex in-

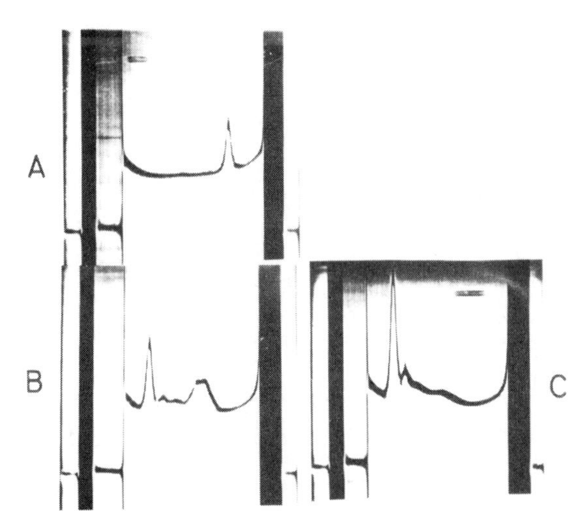

Fig. 7. Changes in sedimentation behaviour after freezing of enzyme solutions in 1 M LiCl [49].

A) Fatty acid synthetase (3 mg/ml) in 1 M LiCl
B) Fatty acid synthetase (10 mg/ml) in 1 M LiCl after a single freezing and thawing
C) Same preparation as in (B) after a second freezing and thawing.

All photographs were taken 14 min after reaching the maximum speed of 59,780 rpm in the analytical ultracentrifuge at 20°. Sedimentation from left to right.

to units of smaller sizes [32]. The pigeon liver synthetase also loses activity and dissociates into subunits in dilute buffer solutions [11,48]. The yeast multienzyme complex appears to be much more stable and requires more drastic treatment for its dissociation [49].

It has been shown that freezing in salt solution causes dissociation or denaturation of a number of enzymes [50]. Although the nature of this process is not clearly understood, a "concentration effect" is assumed [51]. The liquid regions of the frozen system contain a high concentration of salt which changes such critical parameters as ionic strength, pH and water structure and may lead to disturbances of electrostatic and hydrophobic bonds between subunits. Native yeast fatty acid synthetase can be dissolved in 1 M NaCl or LiCl without loss of its enzymatic activity or change in its sedimentation and electrophoretic behavior. The freezing of such solutions two or more times for 15 min at -70° and thawing at room temperature results in a drastic alteration of enzymatic and physical properties. Fig. 7 shows the sedimentation behavior in an analytical ultracentrifuge of native and treated enzyme. After a

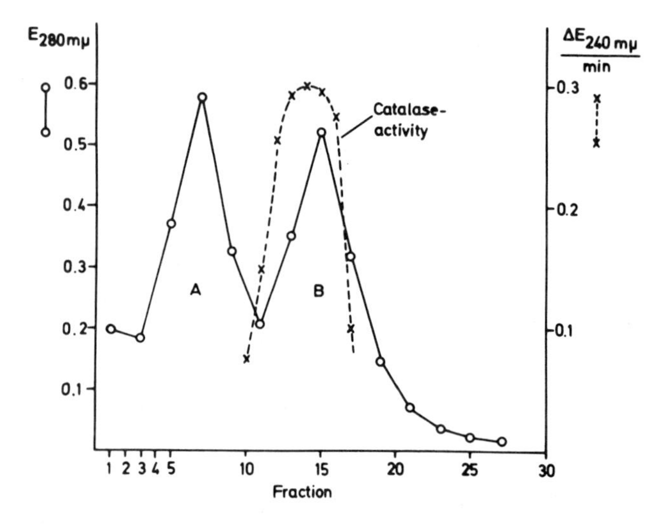

Fig. 8. Separation of subunit fractions by sucrose gradient centrifugation [52]. Linear sucrose gradient (5–20% sucrose) containing 0.1 M tris buffer, pH 7.5 and 0.5 M NaCl; volume: 30 ml. 0.7 ml solution, containing 7 mg of dissociated fatty acid synthetase in 1 M NaCl (frozen twice) and catalase as a marker, was layered on top of the gradient and the whole was centrifuged for 35–50 hr at 24,000 rpm and about 0°. After centrifugation, 1 ml fractions were taken from the bottom of the tube. Protein concentration was measured optically at 280 nm. Catalase activity was measured at 240 nm.

single freezing and thawing the appearance of more slowly sedimenting species can be noted. After two freezings native fatty acid synthetase is no longer present.

In order to obtain some information about the molecular weights of the subunits formed, the solution was spun in a sucrose density gradient together with added catalase as a marker (fig. 8). After one freezing-cycle in 1 M NaCl two fractions could be isolated. After two freezing-cycles the heavier, faster sedimentating protein peak had disappeared and all the proteins sedimented at a rate similar to the sedimentation of catalase [52]. The results obtained in the analytical ultracentrifuge and by sucrose density gradient centrifugation can be explained by assuming that the dissociation of fatty acid synthetase into subunits occurs stepwise, first to species having a molecular weight of about 5×10^5 and then to species of molecular weight 2×10^5 similar to that of catalase.

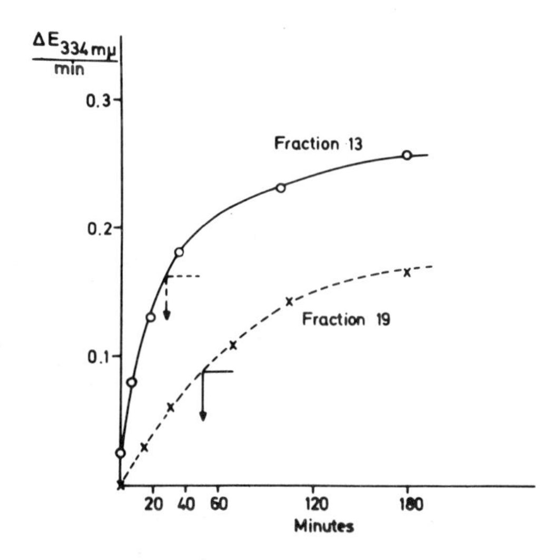

Fig. 9. Reconstitution of fatty acid synthetase activity [52]. The subunit fractions used in this experiment are derived from a sucrose gradient centrifugation analogous to the one presented in fig. 8. In this experiment the maxima of peak A and B (see fig. 8) were present in fractions 13 (larger subunits) and 19 (smaller subunits) respectively. For reconstitution the fractions taken from the sucrose gradient were diluted 5-times by the addition of 0.1 M phosphate buffer (containing 0.04 M cysteine) and incubated at 20°. At various times, given on the abscissa, aliquots were taken and fatty acid synthetase activity was measured. Rates of TPNH oxidation, as measured at 334 nm, are listed on the ordinate.

$$\begin{array}{ccc} \text{fatty acid synthetase} & \rightleftharpoons & n \text{ larger subunits} \\ (\text{M.W. } 2.3 \times 10^6) & & (\text{M.W. } 5 \times 10^5) \\ & & \upharpoonleft\downharpoonright \\ & & m \text{ smaller subunits} \\ & & (\text{M.W. } 2.5 \times 10^5) \end{array}$$

When native fatty acid synthetase is no longer present enzymatic activity has vanished. Reactivation occurs on decrease of ionic strength by dilution or dialysis [49,52]. The kinetics of reactivation are shown in fig. 9. As can be seen from this figure, activity reappears faster with the heavy fraction. In this experiment in which the reactivation was measured at 20° half the altogether attainable fatty acid synthetase activity was obtained within 30 min with the heavy fraction whereas it took over 50 min to reach the same value with the lighter fraction. From this and other experiments we come to the conclusion that the reactivation of the light subunits occurs stepwise. These subunits become first associated to form the subunits of molecular weight 5×10^5 which then reaggregate to form the complete fatty acid synthetase.

The reactivation process is strongly temperature dependent and has an optimum at 20°. At 0° scarcely any reactivation occurs while at 37° an initial

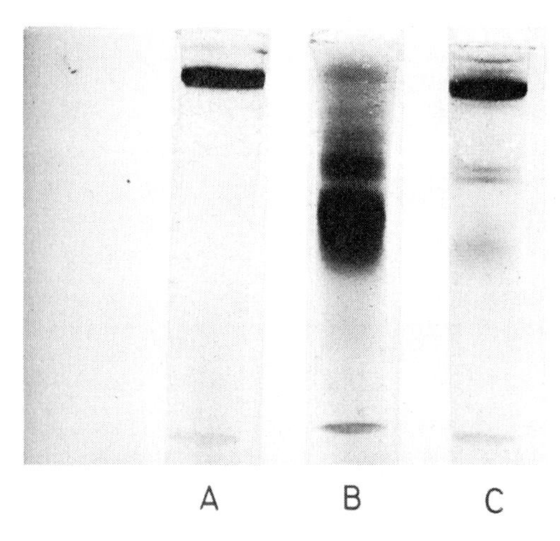

Fig. 10. Disc-gel electrophoresis of native and pretreated fatty acid synthetase at pH 9.5 [49] 3.4% Polyacrylamide-gel; 100 μg protein applied.

A) Fatty acid synthetase in 1 M LiCl

B) Same preparation after two freezings and thawings.

C) Same as (B) after reactivation by dilution (see fig. 9).

fast reactivation turns after 90 min again in the direction of decreased activity. As expected, the protein concentration also is influential. When the reconstituted enzyme was compared with native fatty acid synthetase no difference in the sedimentation behavior could be observed [49].

Dissociation and reconstitution of fatty acid synthetase can also be followed by analytically disc-gel electrophoresis [49]. On the left part of fig. 10 the electrophoretic pattern of native fatty acid synthetase dissolved in 1 M LiCl is presented. The middle part shows the same sample after freezing and thawing twice. The patterns indicate that the multienzyme complex has been split into subunits which migrate at different rates during electrophoresis. The diagram on the right side shows the electrophoretic behavior of the dissociated sample after reactivation by dilution.

The construction of yeast fatty acid synthetase requires the proper orientation of at least seven different enzymes: three acyltransferases, two reductases, one condensing enzyme and one dehydrase, together with the acyl carrier protein which is present about five-fold in the native complex of molecular weight 2.3×10^6 according to recent measurement of the pantothenate content of yeast fatty acid synthetase [53]. The question whether the different enzymes are present in the complex only once or in duplicates or triplicates cannot yet be answered. The observation that native yeast fatty acid synthetase also contains about 5 molecules of flavin mononucleotide [54] may indicate a structure of higher order. On the basis of our analytical data we must also conclude that in the light subunit fraction of molecular weight about 250,000 two different molecular species must be present, one of which contains the acyl carrier protein component whereas the other is free of it. Our next goal must be to prove this conclusion by separating the two subunit species by chromatographic procedures.

The close association between the acyl carrier protein and some enzyme protein in the subunit fractions is also indicated by an experiment in which ^{14}C-pantothenate labelled fatty acid synthetase was dissociated with the freezing technique and then subjected to disc-gel electrophoresis. Radioactivity measurements of the various protein bands revealed that most of them carried radioactive pantothenic acid. Only the band which had travelled the greatest distance appears to be free of radioactivity (fig. 11). These results prove anew that the binding forces between the acyl carrier protein and some enzyme proteins are rather strong. Thus it seems possible that the acyl carrier protein may have an important structural function in the architecture of the multienzyme complex besides its role as an acyl carrier.

Since fatty acid synthesis depends on the proper functioning of all component enzymes, loss of total synthetic activity does not imply loss of all partial

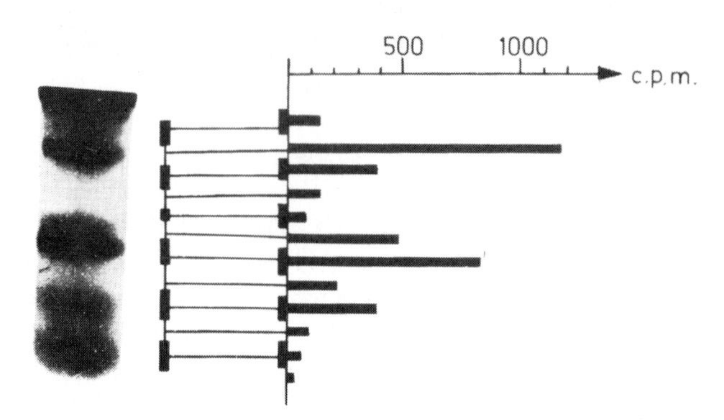

Fig. 11. Distribution of ^{14}C-pantothenate in the various protein bands separated by disc-gel electrophoresis [52]. ^{14}C-Pantothenate labelled fatty acid synthetase [24] was dissociated with the freezing technique and then subjected to disc-gel electrophoresis (see fig. 10) After staining, the gel was cut in slices of 1–3 mm thickness. For radioactivity measurements the material was eluted from the gel by boiling with 6 N HCl for 12–16 hr. A photograph of the stained electropherogram is shown on the left side. The vertical black bars in the middle indicate the location and size of the gel slices. The horizontal bars on the right side reproduce the radioactivity measured.

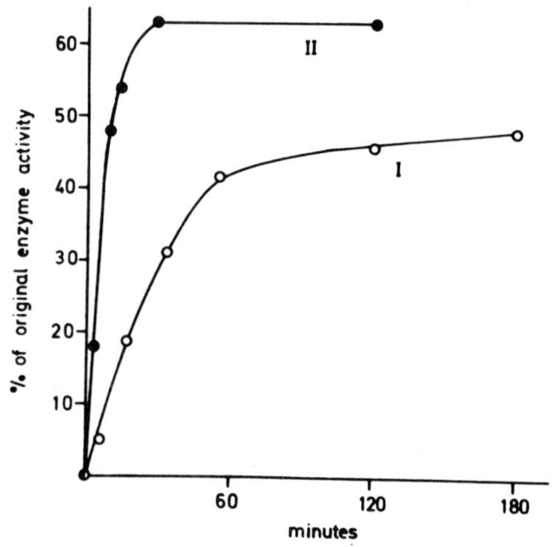

Fig. 12. Kinetics of reconstitution of fatty acid synthetase and reductase activities [49]. Fatty acid synthetase, dissolved in 1 M LiCl, was dissociated by two freezings and thawings. The inactive enzyme solution was diluted five-fold by the addition of 0.01 M phosphate, pH 7.5, containing 0.04 M cysteine, and incubated at 20°. Samples were taken at several time intervals, as listed on the abscissa, and their enzymic activities were measured.

I) Fatty acid synthetase activity

II) Reductase activity.

reaction. The partial reactions can be measured independently with model substrates [5,55].

To date, we have confined ourselves to study of the reduction step leading from β-ketoacid to β-hydroxyacid, the step which we have called the first reduction. This step can be measured by using *S*-acetoacetyl *N*-acetyl cysteamine as a model substrate [55]. After two freezings and thawings of the fatty acid synthetase solution in 1 M NaCl or LiCl the reductase activity has also completely disappeared. Reactivation occurs on decrease of ionic strength by dilution or dialysis. In fig. 12 the kinetics of reconstitution of both fatty acid synthetase and reductase activities are shown. As can be seen the partial activity reappears more quickly than total synthetase activity. Two explanations are conceivable for the fact that after dissociation of the complex the reductase activity had sunk to zero. Either the native structure of the partial enzyme was altered by the freezing procedure and must first reactivate itself before association with neighboring enzymes becomes possible. Or possibly the enzymatically active conformation of the reductase depends on specific protein-protein interactions with several neighboring enzymes which become physically separated during dissociation.

In summary, the pathways of fatty acid synthesis have been discussed and theories have been proposed which would explain the controls of the length of the carbon chain in fatty acid synthesis. Properties and stabilities of the enzyme systems involved are discussed.

References

[1] P. Goldman, A.W. Alberts and P.R. Vagelos, J. Biol. Chem. 238 (1963) 1255.
[2] P.H.W. Butterworth and K. Bloch, European J. Biochem. 12 (1970) 496.
[3] W.J. Lennarz, R.J. Light and K. Bloch, Proc. Natl. Acad. Sci. U.S. 48 (1962) 840.
[4] E.L. Pugh, F. Sauer, M. Waite, E.R. Toomey and S.J. Wakil, J. Biol. Chem. 241 (1966) 2635.
[5] F. Lynen, Federation Proc. 20 (1961) 941.
[6] H.P. Klein, J. Bacteriol. 92 (1966) 130.
[7] P. Overath and P.K. Stumpf, J. Biol. Chem. 239 (1964) 4103.
[8] J.L. Brooks and P.K. Stumpf, Arch. Biochem. Biophys. 116 (1966) 108.
[9] R.D. Simoni, R.S. Criddle and P.K. Stumpf, J. Biol. Chem. 242 (1967) 573.
[10] R. Bressler and S.J. Wakil, J. Biol. Chem. 236 (1961) 1643.
[11] R.Y. Hsu, G. Wasson and J.W. Porter, J. Biol. Chem. 240 (1965) 3736.
[12] P.C. Yang, P.H.W. Butterworth, R.M. Bock and J.W. Porter, J. Biol. Chem. 242 (1967) 3501.
[13] R.O. Brady, R.M. Bradley and E.G. Trams, J. Biol. Chem. 235 (1960) 3093.

[14] D.N. Burton, A.G. Haavik and J.W. Porter, Arch. Biochem. Biophys. 126 (1968) 141.
[15] R.O. Brady, J. Biol. Chem. 235 (1960) 3099.
[16] D.M. Martin, M. G. Horning and P.R. Vagelos, J. Biol. Chem. 236 (1961) 663.
[17] A.R. Larrabee, E.G. McDaniel, H.A. Backerman and P.R. Vagelos, Proc. Natl. Acad. Sci. U.S. 54 (1965) 267.
[18] S. Smith and S. Abraham, Federation Proc. 28 (1969) 537.
[19] F. Lynen, Biochem. J. 102 (1967) 381.
[20] A. Hagen, Ph.D. Thesis, University of Munich (1963).
[21] M.D. Greenspan, A.W. Alberts and P.R. Vagelos, J. Biol. Chem. 244 (1969) 6477.
[22] R.E. Toomey and S.J. Wakil, J. Biol. Chem. 241 (1966) 1159.
[23] J.W. Porter and A. Tietz, Biochim. Biophys. Acta 25 (1957) 41.
[24] W.W. Wells, J. Schultz and F. Lynen, Biochem. Z. 346 (1967) 474.
[25] P.R. Vagelos, Ann. Rev. Biochem. 33 (1964) 139.
[26] R.E. Toomey, M. Waite, I.P. Williamson and S.J. Wakil, Federation Proc. 24 (1965) 290.
[27] F. Lynen, D. Oesterhelt, E. Schweizer and K. Willecke, in: Cellular Compartmentalization and Control of Fatty Acid Metabolism (Oslo Universitetsforlaget, Oslo, 1968), p. 1.
[28] K. Willecke, E. Ritter and F. Lynen, European J. Biochem. 8 (1969) 503.
[29] C.J. Chesterton, P.H.W. Butterworth, A.S. Abramovitz, E.J. Jacob and J.W. Porter, Arch. Biochem. Biophys. 124 (1968) 386.
[30] P.W. Majerus, A.W. Alberts and P.R. Vagelos, Proc. Natl. Acad. Sci. U.S. 51 (1964) 1231.
[31] S.J. Wakil, E.L. Pugh and F. Sauer, Proc. Natl. Acad. Sci. U.S. 52 (1964) 106.
[32] D.N. Brindley, S. Matsumara and K. Bloch, Nature 224 (1969) 666.
[33] J. Nagai and K. Bloch, J. Biol. Chem. 243 (1968) 4626.
[34] H. Goldfine, G.P. Ailhaud and P.R. Vagelos, J. Biol. Chem. 242 (1967) 4466.
[35] T.C. Vanaman, S.J. Wakil and R.L. Hill, J. Biol. Chem. 243 (1968) 6420.
[36] S. Matsumara, D.N. Brindley and K. Bloch, Biochem. Biophys. Res. Commun. 38 (1969) 369.
[37] S. Matsumara and P.K. Stumpf, Arch. Biochem. Biophys. 125 (1968) 932.
[38] I. Pilz, M. Herbst, O. Kratky, D. Oesterhelt and F. Lynen, European J. Biochem. 13 (1970) 55.
[39] D.J.H. Brock, L.R. Kass and K. Bloch, J. Biol. Chem. 242 (1967) 4432.
[40] K. Saito, J. Biochem. (Tokyo) 47 (1960) 710.
[41] W. Pirson, Ph.D. Thesis, University of Munich (1970).
[42] G. Lust and F. Lynen, European J. Biochem. 7 (1968) 68.
[43] M. Sumper, D. Oesterhelt, C. Riepertinger and F. Lynen, European J. Biochem. 10 (1969) 377.
[44] G. Nemethy and H.A. Scheraga, J. Chem. Phys. 36 (1962) 3401.
[45] J. Ayling, unpublished observations.
[46] J. Ziegenhorn, Ph.D. Thesis, University of Munich, (1970).
[47] J. Ziegenhorn, R. Niedermeier and F. Lynen, Z. Physiol. Chem. 351 (1970) 137.
[48] P.C. Yang, R.M. Bock, R.Y. Hsu and J.W. Porter, Biochim. Biophys. Acta 110 (1965) 608.
[49] M. Sumper, C. Riepertinger and F. Lynen, FEBS Letters 5 (1969) 45.
[50] O.P. Chilson, L.A. Costello and N.O. Kaplan, Federation Proc. 24 (1965) 55.

[51] S.P. Leibo and R.F. Jones, Arch. Biochem. Biophys. 106 (1964) 78.
[52] M. Sumper, Ph.D. Thesis, University of Munich (1970).
[53] W. Winnewisser, unpublished observations.
[54] D. Oesterhelt, H. Bauer and F. Lynen, Proc. Natl. Acad. Sci. U.S. 63 (1969) 1377.
[55] F. Lynen, in: Methods in Enzymology, Vol. 14, eds. S.P. Colowick and N.O. Kaplan (Academic Press, New York, 1969) p. 17.

COMPARATIVE ASPECTS OF STEROID HORMONE-FORMING CELLS*

Kenneth SAVARD, D.Sc.

Endocrine Laboratory, Departments of Biochemistry and Medicine, University of Miami School of Medicine, Miami, Florida, USA

Abstract: Savard, K. Comparative Aspects of Steroid-forming Cells. *Miami Winter Symposia* 1, pp. 176–190. North-Holland Publishing Company, Amsterdam, 1970.

Steroid hormones are formed from cholesterol in the cells of highly specialized tissues of normal animals: the adrenal cortex, the testis, the follicle, corpus luteum and interstitium of the ovary, and in pregnancy, the placenta. The enzymic systems responsible for the transformation of cholesterol into pregnenolone (the initial step in steroidogenesis) are unique to these tissues and are located in the mitochondrial elements of all tissues studied so far. This enzyme complex, the "cholesterol side-chain cleavage" system, is composed of "mixed function" oxidases, an NADPH-generating system, and an oxygen-transport system involving cytochrome *P*-450.

Pregnenolone, derived from cholesterol by the mitochondrial process described above, is the key steroidal precursor to all of the various hormones formed by the different steroidogenic cells. The enzymes affecting this biosynthetic process are with only 2 exceptions found in the microsomal or endoplasmic reticular elements of the steroid-producing cells; 11-β-hydroxylase and 18-hydroxylase of adrenal tissue are mitochondrial. Thus the biochemical individuality of the steroid-producing cells, that of synthesizing a given steroid by means of comparable pathways, is for the most part, due to parameters (enzymes, cofactors, etc.) which are localized in the smaller subcellular particles and the soluble components of the endoplasmic reticulum.

Because these tissues have common, or proximal, embryologic origins, it seems likely that in the course of normal development, highly specialized genetic information is transcribed or translated. At least 2 separate enzyme systems are common to all steroid-producing cells. These are the mitochondrial "cholesterol side-chain cleavage" enzyme system. and the 3-β-hydroxysteroid dehydrogenase(s)-isomerase system. The other enzymes which confer specific chemical structures do not seem to be subject to the same "coordinated control". They are selectively found in one or the other, but not all, of the steroid-producing tissues within a single species. Some selectively occurs in the same tissues in different animals; viz. the corpus luteum of the cow and human; the testes of the fish compared to that of the human and other mammals.

* Supported in part by grant No. HD 03248 National Institutes of Health, U.S. Public Health Service.

Of the enzymes found in the corpus luteum of the human ovary, 17-hydroxylase and the 19-hydroxylase-aromatase are absent in the corpus luteum of the cow. The enzyme 11-β-hydroxylase essential to the adrenal cortex, is found in the normas testis of the traut and the salmon. In these fish the normal androgen is 11-oxytestosterone. Some transcriptional or translational selectivity occurs in the adrenal cortex of humans; absence of 21-hydroxylase, 11-β-hydroxylase and others, results in a group of genetically-linked diseases known as congenital adrenal hyperplasia (or the adrenogenital syndrome). The process of tumorogenesis in some tissues or cells evokes the occurrence of enzymes not found in the normal tissue of the same species. Examples of these are 11-β-hydroxylase in interstitial cell tumors of the human testis; 21-hydroxylase in induced tumors of the testis in the BALB/c strain of mice.

1. Introduction

The chemical nature of the steroid hormones was established over a 10 year period during the early 1930's. At that time, Butenandt suggested that in view of their common carbon-skeleton, all the steroid hormones were very probably derived biologically from cholesterol [1]. It has taken the ensuing 30 or so years, with the development of microanalytical techniques, the availability of isotopically-labelled steroids and steroid precursors, and a voluminous literature, to prove that he was correct in his prediction.

Today, the enzymes of the biosynthetic pathway are well defined. They have been identified in virtually all the steroid-forming tissues from at least one species. This literature is well documented in the comprehensive work of Dorfman and Ungar [2].

The biochemical knowledge of the enzymes of the pathway is very limited. The enzymes have not received the careful attention and study that this field deserves. In consequence, and with very few exceptions, nothing is known of their kenetic properties, none has been isolated from glandular sources in pure form, none has been crystallized. There are probably several good reasons for this, the main one being the difficulty and extreme limitations in acquiring fresh glandular tissues. The second and most important is that the subject of these biologically critical hormones is given little attention in the traditional teaching of biochemistry. The subject has been mainly in the hands of specialists whose major talent has been in the microanalytical methods for steroid analysis, and not in the area of modern chemistry. In the few instances where work of high caliber has been done, the enzyme has been isolated from non-endocrine tissues, or from micro-organisms. The enzymes of the steroidogenic pathway have been characterized and their existence demonstrated, principally by incubation or perfusion studies where the technique used was the biosynthetic conversion of isotopically-labelled steroids or steroidal precursors into various transformation products.

It will be apparent to the knowledgeable reader that the seemingly broad picture of enzyme and tissue homologies drawn here, is really a sketchy one. Demonstration of specific enzymes and their subcellular distributions has not been made in every steroidogenic tissue. Nor have all the tissues of any single species been systematically studied. However, if the data from one tissue are overlain on the data from another, a reasonably clear overall picture can be drawn. For more detailed reviews of the particular tissues, the reader is referred to several authoritative and more comprehensive papers on the adrenal cortex [3,4], the testis [4,5], the ovary [6] and the placenta [7]. For a large series of reviews, both broad and restricted, the reader is referred to the Proceedings of the Third International Congress of Endocrinology [8] and to Sih and Witlock [9].

2. Biosynthesis of steroid hormones

The cholesterol-side-chain cleavage complex (I abc)

The initial steps in steroid biosynthesis involve a mitochondrial system of enzymes which are unique to steroid-forming tissues and whose effect is to transform cholesterol (C_{27}) into isocaproate (C_6) and pregnenolone (3-β-hydroxy-Δ^5-pregnen-20-one (C_{21}). These reactions and the enzymes involved are depicted in fig. 1 (Ia,b,c), and consist of a) 20-α-hydroxylation, b) 22-hydroxylation, and c) 20, 22-dihydroxycholesterol lyase, or side-chain cleavage. The first two steps involve two highly specific mixed-function oxidases for cholesterol, require atmospheric oxygen, and NADPH, and together with the lyase, appear to constitute the rate-limiting step in the entire steroidal process. Con-

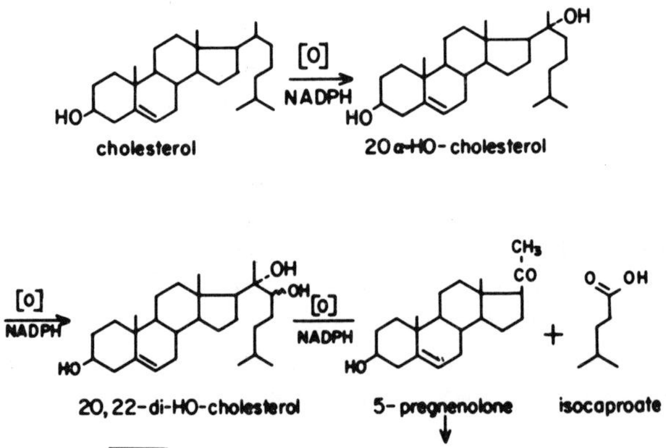

Fig. 1. Cholesterol "side-chain cleavage" enzyme system.

trol of the rate of steroidogenesis by tropic hormone LH (or ICSH) in gonadal tissues (10, 11), ACTH in the adrenal cortex [12], and by exogenous 3′,5′-AMP in all tissues, appears to be exerted at this step. The subject has been reviewed for adrenal tissue by Garren [3,13] and by Koritz and Kumar [14]. The system has been solubilized from mitochondria of adrenal cortex, and the activating effect of exogenous 3′,5′-AMP has been reported by Roberts and Creange [15]. It has also been studied in the mitochondria of testes [16], ovarian elements of the rat [17] and cow [18]. G.S. Boyd [19] has indicated, that the intermediates in the biosynthesis are probably protein-bound and the substrate molecule is released only after completion of the reaction sequence. The system from adrenal tissue has recently been explored by dynamic studies in a cell-free system [20]; these authors present data which suggest that the sequence of reactions shown in fig. 1 may not be correct.

The cholesterol involved in this process is a small fraction of the total tissue cholesterol. It appears to be highly sequestered and compartmentalized; its exact morphologic and/or chemical form is most elusive and at present entirely unknown [21]. This "steroidogenic cholesterol" pool is derived principally from tissue and plasma stores [22,23]. However, the biosynthetic labelling of squalene, sterols and cholesterol from mevalonate-^{14}C and from acetate-^{14}C has been reported in isolated tissue studies with bovine corpus luteum slices [24,25] and adrenal tissues [26]. Influences of tropic hormone treatment on the size of this "pool of steroidogenic cholesterol" have been studied in adrenal tissue [27].

3-β-Hydroxysteroid dehydrogenase-isomerase (II a and b)

The prevalence of this enzyme in all steroid producing tissues (table 1) is apparent, and has been reviewed in the comprehensive work of Dorfman and Ungar [2] and more recently by Sih and Whitlock [9]. Although the dehydrogenase belongs to a broad class of alcohol dehydrogenases, the enzyme(s) associated with the conversion of pregnenolone into progesterone is(are) highly specific for the 3-β-hydroxy-Δ^5-steroidal structure. It is not known if a family of enzymes is responsible for the several comparable enzymatic steps shown in fig. 2. In one of the few instances in this area, the enzyme system has been isolated from sheep adrenal microsomes, purified, and its kinetic properties studied [28].

The dehydrogenase-isomerase system is generally considered to be located in the microsomes of the tissues studied. It has however not been possible to separate it entirely from the mitochondria of adrenal preparations [19]. This suggests that the system may in part be localized on the outer side of the mitochondrial membrane, an important position for the enzyme when mitochondrial-microsomal transport mechanisms are considered.

Table 1
Enzymes of steroidogenic pathway of specialized endocrine tissues.

Steroidogenic tissues		Mito-chondria 20,22-Hydroxylases + lyase	Microsomal - soluble (endoplasmic reticulum) 3-β-ol-Dehydrogenase + Isomerase		21- Hydroxylase	17- Hydroxylase	17,20- Lyase	19-Hydroxylase + Aromatase		17-Keto reductase C_{18}	C_{19}	Mito-chondrial 11-β Hydroxylase	18- Hydroxylase	Steroidal product
		Iabc	IIa	IIb	III	IVa	IVb	Va	Vb	VIa	VIb	VII	VIII	
Adrenal (human)	z. glomerul.	Iabc	IIa	IIb	II	-	-	-	-	-	-	VII	VIII	Aldosterone
	z. fascicul. (rat)	Iabc	IIa	IIb	III	-	-	-	-	-	-	VII	-	Corticosterone
	”	Iabc	IIa	IIb	III	IVa	-	-	-	-	-	VII	-	Cortisol
	z. reticular.	Iabc	IIa	IIb	?	IVa	IVb	-	-	-	-	VII	-	11-β-HO-and.dione
		Iabc	-	-	?	IVa	IVb	-	-	-	?	-	-	DHA
Placenta	Human	Iabc	IIa	IIb	-	-	-	Va	Vb	VIa	-	-	-	Progest.+estrogen
Ovary	Follicle	Iabc	IIa	IIb	-	IVa	IVb	Va	Vb	VIa	-	-	-	Estrogen
	Corp.lut (cow)	Iabc	IIa	IIb	-	-	-	-	-	-	?	-	-	Progesterone
	Corp.lut (human)	Iabc	IIa	IIb	-	IVa	IVb	Va	Vb	VIa	-	-	-	Prog.+estrogen
	Interstitium	Iabc	IIa	IIb	-	IVa	IVb	-	-	-	VIb	-	-	And.dione+testost.
Testis	Interstit. (human)	Iabc	IIa	IIb	-	IVa	IVb	-	-	-	VIb	-	-	Testost.+and.dione
	” (fish)	Iabc	IIa	IIb	-	IVa	IVb	-	-	-	VIb	VII	-	11-β-HO-Testosterone
	Tumor (human)	Iabc	IIa	IIb	-	IVa	IVb	-	-	-	VIb	VII	-	11-β-HO-Testosterone
	” (mice)	Iabc	IIa	IIb	III	IVa	IVb	-	-	-	-	-	-	21-HO-Prog.+androg.

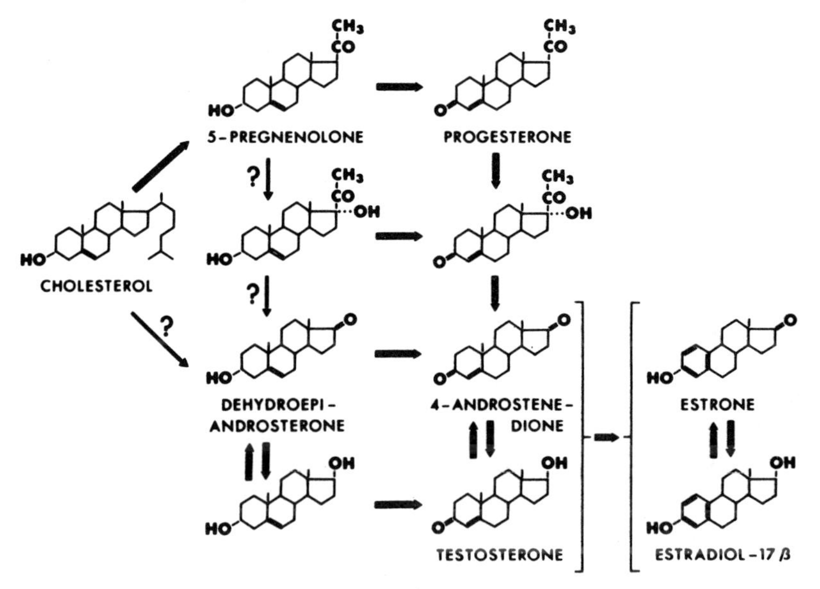

Fig. 2. Enzymatic steps in biosynthesis of gonadal hormones.

2.1. *The gonadal hormones*

The enzymes involved in the formation of androgens and estrogens are associated principally with the stepwise removal of carbon atoms from the carbon skeleton of pregnenolone and progesterone. These transformations involve the modification of a pregnane ring (C_{21}) → androstane ring (C_{19}) → estrogen ring (C_{18}) (fig. 2). Most literature serves to document this as the principal, if not the only, pathway in mammalian tissues. The direct formation of androgen (C_{19}) from cholesterol (C_{27}) has been proposed on the basis of *in vitro* studies with labelled cholesterol [29]. In studies with adrenal tissues, the mechanism has been reported to occur, giving rise to labelled methylheptanone as the side-chain fragment [30]. This pathway has not been confirmed in other laboratories and has not received general acceptance.

Formation of androgens – 17-hydroxylase and 17, 20-lyase (IVa and IVb)

The C_{19}-steroids are referred to generally as the androgenic steroids as they are all androgenic in conventional bioassay systems; testosterone, however, is by far the most potent one, at least in some species like the human and bovine. From a biosynthetic point of view, the formation of androgen from progesterone or pregnenolone involves: a) a 17-hydroxylase (a microsomal mixed-function oxidase) (IVa) and b) a carbon-17,20-lyase (microsomal) (IVb). The transformation was first described in 1956 [31,32] and since then has been extensively studied in testicular tissue in rats, mice and certain fish by Tamaoki and his associates [5].

The dual system for the side-chain cleavage of pregnenolone and progesterone has been found in most tissues where androgens and estrogens are formed (table 1). A notable exception is the human placenta, which is incapable of forming C_{19} steroids from progesterone, or pregnenolone; this is due to the apparent absence of the 17-hydroxylase enzyme from this tissue [7]. Estrogens are formed in the placenta, from C_{19} steroids originating in the adrenal tissue of the fetus [reviewed in 33].

The adrenal cortex of the human is an important source of Δ^4-androstenedione and its 11-β-hydroxylated derivative. The enzymes involved are presumed to be in the zona fasciculata since this is the principal component of the adult tissue; the fetal zone (z. reticularis), which produces substantial DHA (3-β-hydroxy-Δ^5- androstene-17-one) during pregnancy [34] regresses shortly after birth.

Formation of estrogen – 19-hydroxylase and aromatase (Va and Vb)

The formation of estrogen from androgen was first observed in mammalian tissues in 1956 [35]. This was the first indication of the biosynthetic intermediary role of one hormonal substance serving as the precursor of another. This particular obligatory role of an androgen molecule serving as precursor for estrogen seems to be a very ancient concept. Its first mention is in the creation of Eve from the rib of Adam [36]. The enzymes involved in this transformation consist of: a) a 19-hydroxylase (Va) [37], a mixed-function oxidase; and b) an aromatase complex (Vb), which causes the simultaneous removal of the angular carbon-19 between rings A and B and the aromatization of ring A. The enzyme system is found in all estrogen-producing tissues (placenta, ovary and testis [38]), and has been recorded as being microsomal in human placenta [39] and human corpus luteum [40]. A review of the distribution of this system in placental tissues of various animal species has appeared [7].

2.2. *The adrenal hormones*

Unlike the previous transformations leading to the biosynthesis of androgen and estrogen, the formation of the corticosteroids involves a series of specific hydroxylases. These enzymes, in their respective tissue and cell types, cause the insertion of structurally specific as well as stereochemically specific hydroxyl groups on the skeleton of progesterone, without modification of the carbon skeleton itself.

The steroid hydroxylases associated with the synthesis of the corticosteroids, together with those associated with the synthesis of estrogens, androgens and the cleavage of the side-chain of cholesterol, belong to the class of mixed-function oxidases of Mason [41] or the mono-oxygenase classification of Hayaishi [42]. The reader is referred to several reviews of these complex systems [9,43,44].

The synthesis of cortisol involves the sequential enzymatic steps of 17-hydroxylation (IVa), 21-hydroxylation (III) and 11-β-hydroxylation (VII) (presumably in that sequence). Steric considerations preclude 17-hydroxylation if 21-hydroxylation has already occurred (fig. 3).

The rat adrenal biosynthesizes corticosterone exclusively [45] because of the absence of 17-hydroxylase in adrenal tissue of the rat. Most mammalian species biosynthesize mixtures of cortisol and corticosterone in a reasonably consistent ratio for each species [45]. The human and guinea-pig appear to form cortisol exclusively, suggesting a relatively active 17-hydroxylase system in the zona fasciculata of the adrenal in these species.

The inter-renal tissues of certain fish biosynthesize 1-α-hydroxycorticosterone as the principal product [46]. Thus an adrenal enzyme, 1-α-hydroxylase, appears uniquely in the elasmobranchs and does not occur in the tissues of warm-blooded animals [47]. Its absence in the latter may be due to its thermal (37°) instability or its presence may be due to unique genetic "read-out" during the cellular differentiation in the particular species; it is not found in all poikilothermal creatures.

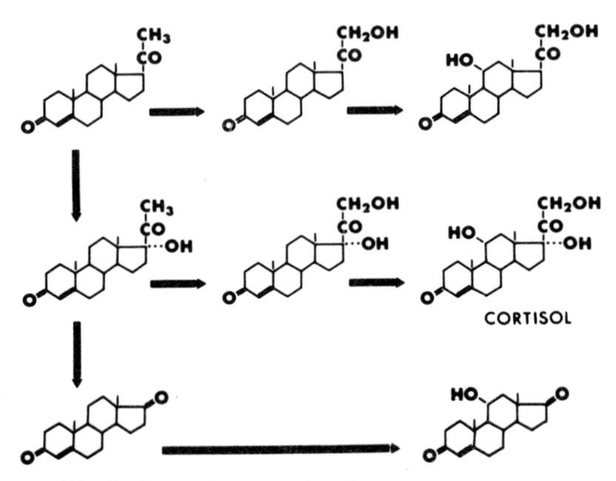

Fig. 3. Steroidogenesis in adrenalcortical tissue.

Formation of aldosterone (18-hydroxylase)

Aldosterone is formed in the zona glomerulosa [48] of the adrenal of most species including man. Its biosynthesis from progesterone via corticosterone involves 21-hydroxylase (III), 11β-hydroxylase (VII) and a further enzyme, 18-hydroxylase (VIII), which is not found in any other steroid-producing tissue.

3. Distribution of enzymes in steroidogenic tissues

The enzymes of the steroidogenic process, are distributed among the specialized tissues of a given species in a highly systematized way. When the common or proximal embryologic origins of these tissues (reviewed recently by Jirasek [49]), are considered, it is apparent that highly specific genetic information has been translated or transcribed during this fetal differentiation and development. Studies of fetal endocrine structures are quite sparse but those carried out in the human reveal definitive steroidogenic function at various stages [34,49].

The enzymes which are common to all the steroid-forming tissues would appear to be under some coordinated control at unknown, but early stages in fetal development. Other enzymes, like the specific steroid hydroxylases, must be selectively translated or transcribed during this process since they are found in only certain tissues but not others. The distribution of these enzymes in human tissues is shown in table 1: (drawn in part from Christensen and Gillim [50]).

Enzymes common to all or to most steroidogenic tissues

The cholesterol side-chain cleavage enzyme system (Iabc) (fig. 1), is found in all tissues where steroid hormones are formed. It has not been isolated from all tissues of any species, but has been studied in tissues of several domestic and experimental animals.

The 3-β-hydroxysteroid dehydrogenase-isomerase (IIab) enzymes system is found in all tissues except the fetal human adrenal, where the enzyme is absent or inhibited [51]. This enzyme pair or family is essential for the formation of all steroid hormones with the Δ^4-3-ketosteroid structure of ring A. Its occurrence in all tissue (except the fetal adrenal) is understandable.

Not included, but deserving of mention, are the many enzymes associated indirectly with steroidogenesis and found in all steroidogenic tissues studied so far. These include the many NADP-linked dehydrogenases of these tissues [52,53]. The adenyl cyclase of the membranes of adrenal [54], ovary [55], and testis [56], respond to tropic hormone, and causes increased concentrations of cyclic 3′,5′-AMP in each of these tissues. This cyclic nucleotide is presumed to mediate the tropic hormone's stimulatory effect on the mitochondrial side-chain cleavage enzyme system [reviewed in references 3,6,13, 43].

Selective distribution of certain enzymes involved in sex hormone biosynthesis

Formation of estrogens involves the system 19-hydroxylase-aromatase (Vab), which occurs in all estrogen-forming tissues, the ovarian follicle, the human corpus luteum, the testis and the placenta. Because of the multi-cellular nature of the sources of steroid in the male and female gonads, some consideration must be given to the cell types rather than the whole gland. It is well recognized that the trace amounts of estrogen formed in the adult male testis are not likely to be formed in the Leydig cells [57]. The testicular source of testicular estrogens is presumed to be the Sertoli cells because of excessive female hormone production in cases of Sertoli cell tumors. In the ovary, it is the follicle, and in particular, the theca cell of the follicle, that is presumed to be the source of ovarian estrogen in all species. It would thus appear that the 19-hydroxylase-aromatase enzyme system occurs in the follicular theca cells and the placenta of all mammalian species. It is not certain whether this enzyme system can be associated with the Sertoli cells of the testes; it seems to be so for the dog and the human.

In developmental terms, only in certain selected gonadal cell types and the placenta, is this enzyme present. The enzyme system appears to be absent from normal adrenal tissues, the testicular Leydig cells, and, as will be seen below, from the ovarian corpus luteum of most species.

The enzymes associated with the formation of androgen from C_{21} precursors are a 17-hydroxylase (IVa) and a carbon 17-20-lyase (IVb). This pair of enzymes can be presumed to be, and is in fact present in all tissues or cell types where androgens and estrogens are formed. This will include the above mentioned estrogen-forming tissues and cells, plus the androgen-forming testicular Leydig cells of all species, the fetal adrenal (human-fetal zone) and the human adult adrenal cortex. A notable exception is the human placenta, which, being devoid of 17-hydroxylase [7,33], cannot form androgen from progesterone, but must utilize C_{19}-steroid of fetal adrenal origin for its biosynthesis of estrogen. It is not known if the enzyme "17-hydroxyprogesterone lyase" (IVb) is also absent from human placental tissue.

The enzymes of corticosteroidogenesis

The highest selectivity in distribution involves the enzymes peculiar to the biosynthesis of corticosterone, cortisol and aldosterone. These enzymes, 11-β-hydroxylase (VII), 21-hydroxylase (III) and 18-hydroxylase (VIII), are almost never found in normal gonadal tissues (with one notable exception, the gonads of the fish, described below). Three hydroxylases, 11-β, 21- and 17-hydroxylase are associated with the formation of cortisol; 11-β-hydroxylase and 21-hydroxylase are associated with the formation of corticosterone. These enzymes are presumed to be confined to the zona fasciculata cells where these two hormones are elaborated. As already noted, the adrenal of the rat is devoid of 17-hydroxylase.

The inter-renal tissues of certain fish, the elasmobranchs, are devoid of 17-hydroxylase. Instead they possess a relatively unusual enzyme, 1-α-hydroxylase, not encountered in any mammalian adrenal tissue [57] ; the consequence of this is the formation of 1-α-hydroxycorticosterone as principal the steroidal product of these tissues.

The zona glomerulosa of most species studied elaborates aldosterone. This adrenal zone therefore possesses 11-β-hydroxylase (VII), 21-hydroxylase (III) and 18-hydroxylase, but contains no 17-hydroxylase.

Species differences in enzyme distribution

It has already been noted that the adrenal of the rat elaborates corticosterone while the adrenals of the human and the guinea-pig elaborate cortisol; most other mammals elaborate both steroids [45] . It would appear that the rat and certain fish [57] do not possess the enzyme 17-hydroxylase (IVa) in their adrenal tissue. The rat testis and ovary, however, elaborate androgen and estrogen respectively, and consequently possess 17-hydroxylase. A certain discrete selectivity therefore occurs with respect to this enzyme in the developmental stages of the adrenal and gonads of the rat, which does not seem to occur generally in other species (the little studied fish tissues may be similar).

Some discrete species differences exist in the corpora lutea of several animals. The bovine corpus luteum elaborates progesterone only and is devoid of 17-hydroxylase (IVa) and the 19-hydroxylase-aromatase system (VIab) for estrogen formation [58] . The human corpus luteum, in contrast, elaborates progesterone, androgen (C_{19} steroids) and estrogen and consequently possesses all the enzymes of the biosynthetic pathway [59] shown in fig. 2. Corpora lutea of other species, rabbit [60] , sheep [61] and sows [62] do not seem to possess the 19-hydroxylase-aromatase system for estrogen synthesis. Genetic programming of the enzymes in this ovarian tissue appears to be highly selective among species; it also differs from other steroidogenic tissue since this ovarian element develops anew with each estrus cycle, from "dormant" granulosa cells of the follicles.

Perhaps the most intriguing aspect of the distribution of these enzymes in normal steroidogenic tissues, is the seemingly "inappropriate" occurrence of selected enzymes in some tissues. Because of the regular occurrence in the adrenals of all species, and the absence from the gonads of all mammals (table 1), the two enzymes 11-β-hydroxylase (VII) and 21-hydroxylase (III) are considered to be enzymes of the adrenal cortex. It is interesting therefore and somewhat surprising that the testes of several species of fish [5,57] should elaborate 11-oxygenated testosterone as principal androgen and that the testes tissue in this species of fish should contain 11-β-hydroxylase (VII) which normally is confined to the adrenal in other species.

Very likely, as more comparative studies of the distribution of the steroidogenic tissues are carried out in domestic and wild animals and fish, an evolutionary pattern of selectivity will emerge. To date, all too few thorough studies exist even of domestic animals. It is unusual, and indeed fortuitous that the most thoroughly studied species (table 1) is man.

Anomalous enzymic distribution in genetic and malignant disease

Studies of humans have revealed that a family of inheritable diseases of the adrenal exist in which one or another of the enzymes essential for the biosynthesis of the corticosteroids is absent [63]. These are referred to as the adrenogenital syndrome or congenital adrenal virilism, in which the young patient is virilized irrespective of the sex. The enzyme defects, in order of their incidence are: 21-hydroxylase (III), 11-β-hydroxylase (VII) and others.

Tumors of the endocrine gland have historically contributed much to the understanding of the functioning of these specialized tissues. For the most part, most tumors result in the overproduction of the steroid hormones normally formed in the tissue; the changes in steroid formation are generally of a quantitative nature.

Morphologic disturbances in the anatomical positioning of adrenal tissue during fetal development have resulted in the localization of anomalous adrenalcortical tissue in the gonads of humans. Tumors of these "adrenal rests" have the expected microscopic appearance of adrenal tissue, and they normally elaborate the expected adrenal corticosteroids. It is possible that these tumors develop spontaneously, and their morphologic as well as enzymic resemblance to adrenal tissue results from a process of "de-differentiation" at both the morphologic and molecular level.

There have however been reported testicular tumors of the Leydig-cells [64], with usual microscopic attributes of testicular tissue, which elaborated 11-oxygenated testosterone; the presence of 11-β-hydroxylase in this tumor was clearly demonstrated. Induced testicular tumors in rats [65] and in mice

[66] have been shown to acquire 21-hydroxylase (III), an enzyme normally found only in the adrenal. It would appear therefore that the process of tumorigenesis in these testicular tissues has caused a "de-differentiation" event to occur at a translational or transcriptional level, which results in the formation of an enzyme which is usually found only in adrenal tissue, and which is "inappropriate" to the gonadal tissue.

It is obvious that much valuable and exciting work lies ahead in the field of the steroid-producing tissues, where the genetic aspects of the enzymes of the steroidogenic process are entirely unexplored.

References

[1] A. Butenandt, Nature 130 (1932) 238.
[2] R.I. Dorfman and F. Ungar, Metabolism of Steroid Hormones (Academic Press, New York, 1965).
[3] L.D. Garren, Vitamins Hormones 26 (1968) 119.
[4] L.T. Samuels and K.B. Eik-Nes, in: Metabolic Pathways, Vol. II, 3rd ed. (Academic Press, New York, 1968) p. 169.
[5] B.I. Tamaoki, H. Inano and H. Nakano, in: The Gonads, ed. K.W. McKerns (Appleton-Century-Crofts, New York, 1969) p. 547.
[6] K. Savard, J.M. Marsh and B.F. Rice, Recent Progr. Hormone Res. 21 (1965) 285.
[7] L. Ainsworth and K.J. Ryan, in: Progress in Endocrinology, ed. C. Gual (Excerpta Media ICS-184, Amsterdam, 1969) p. 755.
[8] Progress in Endocrinology, ed. C. Gual (Excerpta Media, ICS-184, Amsterdam,1969).
[9] C.J. Sih and H.W. Whitlock Jr., Ann. Rev. of Biochem. 37 (1968) 661.
[10] N.R. Mason, J.M. Marsh and K. Savard, J. Biol. Chem. 237 (1962) 1801.
[11] P.F. Hall and K.B. Eik-Nes, Biochim. Biophys. Acta 90 (1962) 411.
[12] D. Stone and O. Hechter, Arch. Biochem. Biophys. 51 (1954) 457.
[13] L.D. Garren, W.W. Davis, G.N. Gill, H.L. Moses, R.L. Ney and R.M. Crocco, in: Progress in Endocrinology, ed. C. Gual (Excerpta Medica ICS-184, Amsterdam, 1969) p. 120.
[14] S.B. Koritz and A.M. Kumar, J. Biol. Chem. 245 (1970) 152.
[15] S. Roberts and J.E. Creange, in: Functions of the Adrenal Cortex, Vol. I, ed. K.W. McKerns (Appleton-Century-Crofts, New York, 1969) p. 339.
[16] D. Toren, K.M.J. Menon, E. Forchielli and R.I. Dorfman, Steroids 3 (1964) 381.
[17] S. Sulimovici and G.S. Boyd, Biochem. J. 103 (1967) 16.
[18] P.F. Hall and S. Koritz, Biochemistry 3 (1964) 129.
[19] G.S. Boyd and E.R. Simpson, in: Funtions of the Adrenal Cortex, Vol. I, ed. K. McKerns (Appleton-Century-Crofts, New York, 1969) p. 49.
[20] S. Bustein and M. Gut, Steroids 14 (1969) 207.
[21] K. Savard, W. LeMaire and L. Kumari, in: The Gonad, ed. K. McKerns (Appleton-Century-Crofts, New York, 1969) 119.
[22] E.A. Solod, D.T. Armstrong and R.O. Greep, Steroids 7 (1966) 607.
[23] H. Werbin and G.V. LeRoy, J. Am. Chem. Soc. 76 (1954) 5260.

[24] H.R. Hellig and K. Savard, J. Biol. Chem. 240 (1965) 1957.
[25] H.R. Hellig and K. Savard, Biochemistry 5 (1966) 2944.
[26] R.B. Billiar and K.B. Eik-Nes, Arch. Biochem. Biophys. 115 (1966) 318.
[27] M. Matsuba, S. Ichii and S. Kobayaski, in: Steroid Dynamics, eds. G. Pincus, T. Nato and J.F. Tait (Academic Press, New York, 1966) p. 357.
[28] M.G. Ward and L.L. Engel, J. Biol. Chem. 234 (1964) PC 3604; 241 (1966) 3154.
[29] R.I. Dorfman, Acta Endocrinol. 40 (1962) 188.
[30] R.A. Jungmann, Biochim. Biophys. Acta 164 (1968) 110.
[31] W.R. Slaunwhite Jr. and L.T. Samuels, J. Biol. Chem. 220 (1956) 341.
[32] K. Savard, R.I. Dorfman, B. Baggett and L.L. Engel, J. Clin. Endocrinol. Metab. 16 (1956) 1629.
[33] S. Solomon, C.E. Bird, W. Ling, M. Iwamiya and P.C.M. Young, Recent Progr. Hormone Res. 23 (1967) 297.
[34] E. Bloch, in: Functions of the Adrenal Cortex, Vol. II, ed. K. McKerns (Appleton-Century-Crofts, New York, 1968) p. 721.
[35] B. Baggett, L.L. Engel, K. Savard and R.I. Dorfman, J. Biol. Chem. 221 (1956) 931.
[36] Genesis, Chapter 2; verses 21-23.
[37] A. Meyer, Biochim. Biophys. Acta 17 (1955) 441.
[38] B. Baggett, L.L. Engel, L. Balderas, G. Lamman, K. Savard and R.I. Dorfman, Endocrinology 64 (1959) 600.
[39] K.J. Ryan, J. Biol. Chem. 234 (1959) 268.
[40] R.B. Arceo and K.J. Ryan, Acta Endocrinol. 56 (1967) 225.
[41] H.S. Mason, Ann. Rev. Biochem. 34 (1965) 595.
[42] O. Hayaishi, Proc. 6th Internat. Congress Biochemistry, Vol. 33 (New York, 1964) p. 31.
[43] Functions of the Adrenal Cortex, Vols. I and II, ed. K. McKerns (Appleton-Century-Crofts, New York, 1968).
[44] O. Hayaishi, Ann. Rev. Biochem. 38 (1969) 21.
[45] I.E. Bush, J. Endocrinol. 9 (1953) 95.
[46] B. Truscott and D.R. Idler, J. Endocrinol. 40 (1968) 515.
[47] D.R. Idler, B. Truscott and H.C. Stewart, in: Progress in Endocrinology, ed. C. Gual (Excerpta Medica ICS-184, Amsterdam, 1969) p. 724.
[48] P.J. Ayres, J. Eichhorn, O. Hechter, N. Saba, J.F. Tait and S.A.S. Tiet, Acta Endocrinol. 33 (1960) 27.
[49] J.E. Jirasek, in: Progress in Endocrinology, ed. C. Gual (Excerpta Medica ICS-184, Amsterdam, 1969) p. 1100.
[50] A.K. Christensen and S.W. Gillim, in The Gonads, ed. K. McKerns (Appleton-Century-Crofts, New York, 1969) p. 415.
[51] E. Bloch, in: Functions of the Adrenal Cortex, Vols. II., ed. K. McKerns (Appleton-Century-Crofts, New York, 1968) p. 721.
[52] G.E. Glock and P. McLean, Biochem. J. 56 (1954) 171.
[53] K. Savard, J.M. Marsh and D.S. Howell, Endocrinology 73 (1963) 554.
[54] D.G. Graham-Smith, R.W. Butcher, R.L. Ney and E.W. Sutherland, J. Biol. Chem. 242 (1967) 5535.
[55] J.M. Marsh, R.W. Butcher, K. Savard and E.W. Sutherland, J. Biol. Chem. 241 (1966) 5436.
[56] F. Murod, B.S. Strauch and M. Vaughn, Biochim. Biophys. Acta 177 (1969) 591.

[57] L.M. Fishman, G.A. Sarfaty, H. Wilson and M.B. Lipsett, in: Endocrinology of the Testis, Colloquia on Endocrinology Vol. 16, Ciba Foundation, eds. G.E.W. Wolstenholme and M. O'Connor (Churchill, London, 1966) p. 156.
[58] K. Savard and G. Telegdy, Steroids 5 (Supplement II, 1975) 205.
[59] W.Y. Huang and W.H. Pearlman, J. Biol. Chem. 238 (1963) 1308.
[60] G. Telegdy and K. Savard, Steroids 8 (1966) 685.
[61] C.C. Kaltenbach, B. Cook, G.D. Niswender and A.V. Nalbandov, Endocrinology 81 (1967) 1407.
[62] B. Cook, C.C. Kaltenbach, H.W. Norton and A.V. Nalbandov, Endocrinology 81 (1967) 573.
[63] A.M. Bongiovanni, W.R. Eberlein, A.S. Golman and M. New, Recent Prog. Hormone Res. 23 (1967) 375.
[64] K. Savard et al., J. Clin. Invest. 39 (1960) 534.
[65] O.V. Dominquez, H.F. Acevedo, R.A. Huseby and L.T. Samuels, J. Biol. Chem. 235 (1960) 2608.
[66] H. Inano and B.I. Tamaoki, Endocrinology 83 (1968) 1074.

SERINE PROTEINASES

B.S. HARTLEY
Medical Research Council Laboratory of Molecular Biology, Cambridge, England

Abstract: Hartley, B.S. Serine Proteinases. *Miami Winter Symposia* 1, pp. 191–209. North-Holland Publishing Co., Amsterdam, 1970. Determination of the tertiary structures of α-chymotrypsin and of various enzyme-inhibitor complexes have provided plausible explanations for the major features of the catalytic activity, specificity and activation of the zymogen. Homologies in sequence between chymotrypsin and other pancreatic serine proteinases now appear in a new light. Internal residues in chymotrypsin are largely hydrophobic and corresponding positions in other pancreatic proteinases show a high degree of identity or chemical similarity. Hypothetical models of trypsin and elastase can, therefore, be built assuming identity with chymotrypsin in the conformation of the polypeptide chain and no serious chemical or steric problems are thereby encountered. Identical conformations of the catalytic groups, Asp-102, His-57 and Ser-195 are observed, although 60% of the surface residues are different. In the trypsin model, the substrate-binding cavity is identical to that in chymotrypsin except that Asp-189 replaces Ser-189. This explains the specificity of trypsin for arginine and lysine substrates. The hypothetical elastase model has been essentially validated by a 3.5 Å Fourier map of the crystals. Here the entrance to the substrate-binding cavity is very constricted by replacement of Val-216 and Thr-226 for Gly-216 and Gly-226 in chymotrypsin and trypsin, which explains why benzoyl alanine methyl ester is among the small best synthetic substrates. Thrombin, a plasma serine proteinase, exhibits the same high characteristic sequence homologies as the pancreatic proteinases having, for example, Ile-16, His-57, Asp-102, Asp-189, Asp-194, Ser-195, Gly-216 and Gly-226 like trypsin. However, substantial insertions in the thrombin sequence and the presence of a glycopeptide moiety prevents us from building a definitive model. The important questions of the specificity of thrombin for fibrinogen and the complex activation process of prothrombin remain to be explained, although the chymotrypsinogen-chymotrypsin model must be the basis for the structure, activation and activity of prothrombin and thrombin.

1. Introduction

The story of homologies in amino acid sequences of the serine proteinases has been told many times, but I have no hesitation in telling it again in the context of this Symposium. The story exemplifies George Wald's dictum that "You can't work long in Biology without running into evolution", and though I cannot pretend to offer any conclusive answers to the big biological questions, I do think that we can put these questions in a new framework. This framework is provided by a reassessment of the catalytic activity and homologies of sequence of these enzymes in the light of recent determinations of the tertiary structure of chymotrypsin, elastase and subtilisin. These structures were presented at a Discussion Meeting of the Royal Society in December, 1968, and I recommend interested readers to a series of papers in this publication [1-6].

2. Catalytic activity of chymotrypsin

Since bovine α-chymotrypsin-A has been the most extensively studied serine proteinase, it would be well to remind ourselves of some of the outstanding properties of this enzyme which we would like the tertiary structure to explain. Its specificity towards phenylalanyl, tryptophanyl, tyrosyl or leucyl bonds was first elucidated by studies with small peptide analogues, and we now know that it will hydrolyse esters, amides, hydrazides, anilides, etc., of such residues at rates parallel to the efficiency of the leaving group. It is therefore not selective for the leaving group, and the main elements of specificity appear to be a hydrophobic planar group on the side chain of the amino acid and an acyl substituent on the amino group. Some "locked" substrates [7,8] define the geometry of the bond split with respect to the aromatic ring of the side-chain.

The hydroxyl group of a single serine residue is uniquely nucleophilic. It is uniquely phosphorylated by organophosphorus compounds such as di-isopropyl phosphorofluoridate (DFP) – which property was proposed as the definition of a serine proteinase [9] – and also uniquely sulphonylated [10,11] or acylated [12]. The "pseudo-substrate", *p*-nitrophenyl acetate, seems to be the model for "true" substrates with good leaving groups in that an acyl Ser-195 is an intermediate in the catalysis [13].

The pH-activity curve [14] suggests that a group in the enzyme with pKa approximately 7 is essential in the unprotonated form for both acylation and deacylation [15] whereas substrate-binding is inhibited by deprotonation of a group above pH 8.

3. The primary structure of chymotrypsin

Determination of the complete amino acid sequence of bovine chymotrypsinogen-A (fig. 1) did not help significantly to explain the catalytic activity but allowed us to define the positions of interesting chemical modifications in the peptide chain. Thus the reactive serine was identified as Ser-195, and His-57 and Met-192 had to be close to the substrate-binding site since they reacted specifically with reagents which were analogues of substrates [16,18], but alkylation of Met-192 with iodoacetate did not destroy the catalytic activity [19]. Tyrosine-146 iodinates preferentially in solution without inactivating the enzyme [20] but Tyr-171 iodinates preferentially in the crystal [21]. All of the amino groups of chymotrypsinogen can be acetylated or guanidinated and the product activates with trypsin to yield acetylated δ-chymotrypsin identical in activity to the unmodified enzyme [22].

One amino group in chymotrypsin which could be important, namely that of Ile-16 released by the tryptic activation, is unreactive towards fluorodinitrobenzene in the native enzyme [25]. Oppenheimer, Labouesse and Hess [24] showed that acetylated δ-chymotrypsin resembles the unmodified enzyme in its pH-activity curve and in the dependence of its specific optical rotation on an ionising group with pK_a 8-9, whereas acetylated chymotrypsinogen or acetylated DIP-δ-chymotrypsin showed no such dependence. They concluded that a protonated amino group of Ile-16 is essential for the active conformation of the enzyme.

4. The tertiary structure of α-chymotrypsin

The amino acid sequence showed its true value when combined with the results of X-ray crystallography of bovine α-chymotrypsin-A by Blow and his colleagues [25,26]. The model which was constructed from these data showed several structural features which help to explain the properties of the enzyme mentioned above.

Looking at the model from the "front", with Ser-195 front centre and the "methionine loop" at the bottom, we find that Cys-1 is at the centre of the back surface and the A-chain (residues 1-13) winds round the back and right hand surface. The core of the roughly spherical molecule appears to be composed of two halves formed by the B-chain (residues 16-146) which convolutes on itself to form the "N-W" hemisphere and the C-chain (residues 149--245) which convolutes to form the "S-E" hemisphere. The internal junction between these two hemispheres is largely formed by two roughly parallel

	16	17	18	19	20	21	22	23	24	25	26	27	28	29	30	31
CA:	ILE	- VAL	- Asn	- GLY	- GLU	- GLU	- ALA	- Val	- Pro	- GLY	- SER	- TRP	- PRO	- TRP	- GLN	- VAL
CB:	”	”	”	”	”	- ASP	”	”	”	”	”	”	”	”	”	”
T :	ILE	- VAL	- GLY	- GLY	- Tyr	- Thr	- Cys	- Gly	- Ala	- ASN	- THR	- Val	- PRO	- TYR	- GLN	- VAL
E :	VAL	- VAL	- GLY	- GLY	- Thr	- GLU	- ALA	- GLN	- ARG	- ASN	- SER	- TRP	- PRO	- Ser	- GLN	- ILE
Th:	ILE	- VAL	- Glu	- GLY	- GLN	- ASP	- ALA	- GLU	- Val	- GLY	- Leu	- Ser	- PRO	- TRP	- GLN	- VAL

	32	33	34	35	36				37	38	39	40	41	42	43	44
CA:	SER	- LEU	- GLN	- Asp	- LYS	——	——	——	THR	- GLY	- PHE	- HIS	- PHE	- CYS	- GLY	- GLY
CB:	”	”	”	”	Ser	——	——	——	”	”	”	”	”	”	”	”
T :	SER	- LEU	- ASN	——	——	——	——	——	SER	- GLY	- TYR	- HIS	- PHE	- CYS	- GLY	- GLY
E :	SER	- LEU	- GLN	- Tyr	- ARG	- Ser	- Gly	- SER	- SER	- TRP	- Ala	- HIS	- Thr	- CYS	- GLY	- GLY
Th:	Met	- LEU	- Phe	- Arg	- LYS	——	——	SER	- Pro	- Gln	- Glu	- Leu	- Leu	- CYS	- GLY	- Ala

	45	46	47	48	49	50	51	52	53	54	55	56	57	58	59	60
CA:	SER	- LEU	- ILE	- ASN	- GLU	- ASN	- TRP	- VAL	- VAL	- THR	- ALA	- ALA	- HIS	- CYS	- Gly	- Val
CB:	”	”	”	SER	”	ASP	”	”	”	”	”	”	”	”	”	”
T :	SER	- LEU	- ILE	- ASN	- Ser	- GLN	- TRP	- VAL	- VAL	- SER	- ALA	- ALA	- HIS	- CYS	- Tyr	- Lys
E :	THR	- LEU	- ILE	- Arg	- GLN	- ASN	- TRP	- VAL	- Met	- THR	- ALA	- ALA	- HIS	- CYS	- Val	- Asp
Th:	SER	- LEU	- ILE	- SER	- ASP	- Arg	- TRP	- VAL	- Leu	- THR	- ALA	- ALA	- HIS	- CYS	- Leu	- Leu

	61	62	63	64	65												
CA:	THR	- Thr	- Ser	- Asp	- Val	——	——	——	——	——	——	——	——	——	——	——	——
CB:	”	”	”	”	”	——	——	——	——	——	——	——	——	——	——	——	——
T :	SER	- Gly	- ILE	- Gln	——	——	——	——	——	——	——	——	——	——	——	——	——
E :	Arg	- Glu	- LEU	- Thr	- Phe	- ARG	——	——	——	——	——	——	——	——	——	——	——
Th:	Tyr	- Pro	- (Trp,	Pro,	Asx,	LYS,	Asx,	Phe)	Thr	- Val	- Val	- Asx	- Asx	- Leu	- Leu	- (Ser,	- Val,
							CBH										

			66	67	68	69	70	71	72	73	74	75	76	77	78	79
CA:	——	——	VAL	VAL	Ala	GLY	GLU	Phe	ASP	Gln	Gly	Ser	Ser	SER	Glu	LYS
CB:	——	——	”	”	”	”	”	”	”	”	”	Leu	Glu	THR	”	ASP
T :	——	——	VAL	ARG	LEU	GLY	GLN	Asp	ASN	ILE	ASN	Val	Val	GLU	GLY	ASN
E :	——	——	VAL	VAL	VAL	GLY	GLU	HIS	ASN	LEU	ASN	Gln	Asn	Asn	GLY	Thr
Th:	Val,	Glx,	Trp)	ARG	ILE	GLY	Lys,	HIS	Ser	Arg	Thr	Arg	Tyr	GLU	Arg	LYS

	80	81	82	83	84		85	86	87	88	89	90	91	92	93	94
CA:	ILE	GLN	LYS	LEU	Lys	——	ILE	Ala	LYS	VAL	PHE	Lys	Asn	Ser	LYS	TYR
CB:	Thr	”	Val	”	”	——	”	Gly	”	”	”	”	”	PRO	”	PHE
T :	GLN	GLN	PHE	ILE	SER	——	Ala	Ser	LYS	Ser	ILE	VAL	HIS	PRO	Ser	TYR
E :	GLU	GLN	TYR	VAL	Gly	——	VAL	Gln	LYS	ILE	VAL	VAL	HIS	PRO	Tyr	TRP
Th:	VAL	GLU	LYS	ILE	SER	Met	LEU	Asp	LYS	ILE	TYR	ILE	HIS	PRO	ARG	TYR

	95	96	97	98	99			100	101	*102*	103	104	105	106	107	108
CA:	ASN	SER	Leu	THR	ILE	——	——	ASN	ASN	ASP	ILE	Thr	LEU	LEU	LYS	LEU
CB:	Ser	ILE	”	”	VAL	——	——	Arg	”	”	”	”	”	”	”	”
T :	ASN	SER	ASN	THR	LEU	——	——	ASN	ASN	ASP	ILE	Met	LEU	ILE	LYS	LEU
E :	ASN	THR	ASP	ASP	VAL	Ala	Ala	Gly	Tyr	ASP	ILE	ALA	LEU	LEU	ARG	LEU
Th:	ASN	Trp	Lys	GLU	Asn	Leu	——	ASP	Arg	ASP	ILE	ALA	LEU	LEU	LYS	LEU

	109	110	111	112	113	114	115	116	117	118	119	120	121	122	123	124
CA:	Ser	THR	ALA	ALA	SER	Phe	SER	GLN	Thr	VAL	Ser	Ala	VAL	CYS	LEU	PRO
CB:	ALA	”	PRO	”	GLN	”	”	GLU	”	”	”	”	”	”	”	”
T :	LYS	SER	ALA	ALA	SER	LEU	ASN	SER	Arg	VAL	Ala	Ser	ILE	Ser	LEU	PRO
E :	ALA	Gln	Ser	VAL	THR	LEU	ASN	SER	TYR	VAL	Gln	Leu	Gly	Val	LEU	PRO
Th:	LYS	Arg	PRO	ILE	GLU	LEU	SER	ASP	TYR	ILE	His	Pro	VAL	CYS	LEU	PRO

	125	126	127	128		129	130	131	132	133	134	135	136	137	138	139
CA:	SER	ALA	SER	ASP	—	Asp	Phe	ALA	ALA	GLY	THR	Thr	CYS	Val	Thr	THR
CB:	”	”	Asp	GLU	—	”	”	Pro	”	”	Met	Leu	”	Ala	”	”
T :	THR	—	SER	Cys	—	Ala	Ser	—	ALA	GLY	THR	Gln	CYS	Leu	ILE	SER
E :	ARG	ALA	Gly	THR	—	ILE	LEU	ALA	Asn	Asn	SER	Pro	CYS	Tyr	ILE	THR
Th:	LYS,	ALA	SER	THR	Arg,	LEU	LEU	His	ALA	GLY	Phe	Lys	Gly	Arg	VAL	THR

	140	141	142	143	144	145	146	147	148								
CA:	GLY	TRP	GLY	LEU	THR	ARG	Tyr	THR	Asn -								
CB:	”	”	”	Lys	”	LYS	”	ASN	Ala								
T :	GLY	TRP	GLY	ASN	THR	LYS	SER	SER	GLY -								
E :	GLY	TRP	GLY	LEU	THR	ARG	THR	ASN									
Th:	GLY	TRP	GLY	ASN	Arg,	Thr	THR	SER	Val	Ala	Glu	Val	Gln	Pro	Ser	Val	

	149	150	151	152	153	154	155	156	157	158	159	160	161	162	163	164
CA:	Ala	ASN	THR	PRO	ASP	ARG	LEU	GLN	GLN	ALA	SER	LEU	PRO	LEU	LEU	SER
CB:	LEU	Lys	”	”	”	LYS	”	”	”	”	THR	”	”	ILE	VAL	”
T :	Thr	Ser	Tyr	PRO	ASP	Val	LEU	Lys	Cys	Leu	Lys	Ala	PRO	ILE	LEU	SER
E :	Gly	GLN	Leu	ALA	Gln	Thr	LEU	GLN	GLN	ALA	Tyr	LEU	PRO	Thr	VAL	ASP
Th:	LEU	GLN	THR	ALA	Ala	LYS	LEU,	GLN	Val	Val	Asn	LEU	PRO	LEU	VAL	GLU

	165	166	167	168	169	170			171	172	173	174	175	176	177	178
CA:	ASN	THR	ASN	CYS	LYS	Lys	—	—	TYR	TRP	GLY	THR	LYS	ILE	LYS	ASP
CB:	”	”	ASP	”	ARG	”	—	—	”	”	”	SER	ARG	VAL	THR	”
T :	ASN	SER	Ser	CYS	LYS	SER	—	—	Ala	TYR	Pro	Gly	Gln	ILE	THR	Ser
E :	Tyr	Ala	ILE	CYS	Ser	SER	Ser	Ser	TYR	TRP	GLY	SER	Thr	VAL	LYS	ASN
Th:	Arg	Pro	VAL	CYS	LYS						Arg	Ile	ARG	ILE	THR	ASX

	179	180	181	182	183	184		185	186	187	188					*189*
CA:	Ala	MET	ILE	CYS	ALA	GLY	—	Ala	Ser	GLY	VAL	—	—	—	—	SER
CB:	Val	”	”	”	”	”	—	”	”	”	”					”
T :	ASN	MET	PHE	CYS	ALA	GLY	TYR	Leu	Glu	GLY	Gly	LYS	—	—	—	ASP
E :	Ser	MET	VAL	CYS	ALA	GLY	—	Gly	Asn	GLY	VAL	ARG	—	—	—	SER
Th:	ASX	MET	PHE	CYS	ALA	GLY	TYR	Lys	Pro	GLY	Glu	Gly	Lys	Arg	Gly	ASP

	190	*191*	192	193	*194*	*195*	196	197	198	199	200	201	202	203	204	
CA:	SER	CYS	Met	GLY	ASP	SER	GLY	GLY	PRO	LEU	VAL	CYS	LYS	Lys	ASN	—
CB:	”	”	”	”	”	”	”	”	”	”	”	”	Gln	”	”	—
T :	SER	CYS	GLN	GLY	ASP	SER	GLY	GLY	PRO	Val	VAL	CYS	Ser	Gly	Lys	—
E :	Gly	CYS	GLN	GLY	ASP	SER	GLY	GLY	PRO	LEU	His	CYS	Leu	Val	ASN	—
Th:	Ala	CYS	GLU	GLY	ASP	SER	GLY	GLY	PRO	Phe	VAL	Met	LYS	Ser	Pro	Tyr

		205	206	207	208	209	210	211	212	*213*	214	215	*216*	217		218
CA:	—	GLY	Ala	TRP	Thr	LEU	Val	GLY	ILE	VAL	SER	TRP	GLY	SER	—	Ser
CB:	—	”	”	”	”	”	Ala	”	”	”	”	”	”	”	—	”
T :	—	—	—	—	—	LEU	Gln	GLY	ILE	VAL	SER	TRP	GLY	SER	—	—
E :	—	GLY	Gln	TYR	Ala	Val	His	GLY	VAL	Thr	SER	PHE	Val	SER	Arg	Leu
Th:	Asn	Asn	Arg	TRP	Tyr	Gln	Met	GLY	ILE	VAL	SER	TRP	GLY	Glu	—	—

	219	*220*	221		222	223	224	225	*226*	*227*	*228*	229	230	231	232	233
CA:	Thr	CYS	Ser	—	THR	Ser	Thr	PRO	GLY	VAL	TYR	Ala	ARG	VAL	THR	ALA
CB:	”	”	”	—	”	”	”	”	Ala	”	”	”	”	”	”	”
T :	GLY	CYS	Ala	Gln	Lys	Asn	LYS	PRO	GLY	VAL	TYR	THR	LYS	VAL	Cys	Asn
E :	GLY	CYS	ASN	Val	THR	Arg	LYS	PRO	Thr	VAL	PHE	THR	ARG	VAL	SER	ALA
Th:	GLY	CYS	ASP	Arg	Asn	Gly	LYS	Tyr	GLY	Phe	TYR	THR	His	VAL	Phe	Arg

	234	235	236	237	238	239	240	241	242	243	244	245		
CA:	Leu	- VAL	- Asn	- TRP	- VAL	- GLN	- GLN	- THR	- LEU	- ALA	- Ala	- ASN		
CB:	”	Met	- Pro	”	”	”	GLU	”	”	”	”	”		
T :	TYR	- VAL	- SER	- TRP	- ILE	- Lys	- GLN	- THR	- ILE	- ALA	- SER	- ASN		
E :	TYR	- ILE	- SER	- TRP	- ILE	- ASN	- ASN	- VAL	- ILE	- ALA	- SER	- ASN		
Th:	Lys	- LEU	- Lys	- TRP	- ILE	- GLN	- Lys	- VAL	- ILE	- Asp	- Arg	- Leu	- Gly	- Ser

Fig. 1. Sequence homologies in the “B-chains” of serine proteinases. CA, bovine chymotrypsin-A [27,48,49]; CB, bovine chymotrypsin-B [50]; T, bovine trypsin [2,51,52]; E, porcine elastase [42] and Th, bovine thrombin [53]. Numbering is that of chymotrypsinogen-A. Only differences between CA and CB are shown. Residues underlined are identical in any pair of enzymes, and those in capitals are chemically similar. Numbers underlined show side chains internal in the tertiary structure of α-chymotrypsin A and numbers in “italic” type show side chains forming the catalytic site or substrate binding site of chymotrypsin.

disulphide-bridged loops: the "histidine loop" (residues 42-58) of the B-chain and the serine" loop" (residues 191-220) of the C-chain. At the front surface of this junction, Nϵ2 of the imidazole ring of His-57 is hydrogen-bonded to the hydroxyl group of Ser-195, but Nδ 1 of this ring was surprisingly found to be hydrogen-bonded to the carboxyl group of Asp-102 in a buried environment shielded from water by the aromatic ring of Tyr-94. Fig. 2 shows the conformation of these residues.

Another significant feature, also shown in fig. 2, is that the amino group of Ile-16 is buried in another hydrophobic pocket to form an internal ion-pair with the carboxyl group of Asp-194. Sigler et al. [26] have pointed out that the amino group of Ile-16 must have been at the surface in the zymogen, and that the movement into this cleft on activation might pull Asp-194 into the pocket, thereby swinging Ser-195, as by a hinge, to hydrogen bond with His-57. If this internal ion pair were essential for the conformation of the active structure, then the conformational change and inactivation at high pH would be explained by deprotonation of Ile-16 so as to release Asp-194 to a more hydrophobic site, with consequent movement of Ser-195.

The structure also suggests an explanation for the unusual nucleophilicity of the hydroxyl group of Ser-195 [27]. The "buried" charge on Asp-102 could be 'relayed' to the surface through tautomeric forms of His-57 and the

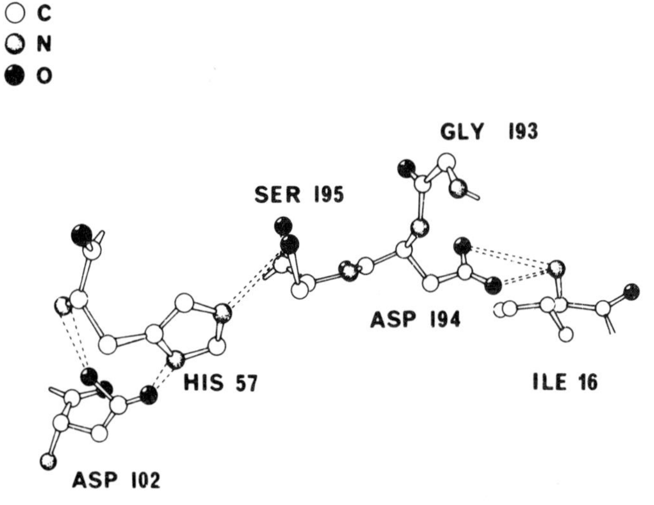

Fig. 2. The conformation of some residues close to Ser-195, viewed from the solvent [28]. All the side-chains except Ser-195 and the upper half of the imidazole ring are buried beneath other groups in the surface.

His-Ser hydrogen bond. It would thereby finish up on the serine oxygen and enhance its nucleophilicity.

Difference Fourier studies of α-chymotrypsin crystals with "virtual" substrates have elucidated the nature of the substrate-binding site. It is, of course, impossible to study most enzyme-substrate complexes crystallographically because of their short life. However, chymotrypsin catalyses O^{18} exchange between water and formyl L-tryptophan or formyl L-phenylalanine, where product and substrate are chemically identical, so we may hope to see such a complex. Steitz, Henderson and Blow [28] have shown that crystals of α-chymotrypsin at pH 5.7 bind these substrates in a hydrophobic pocket im-

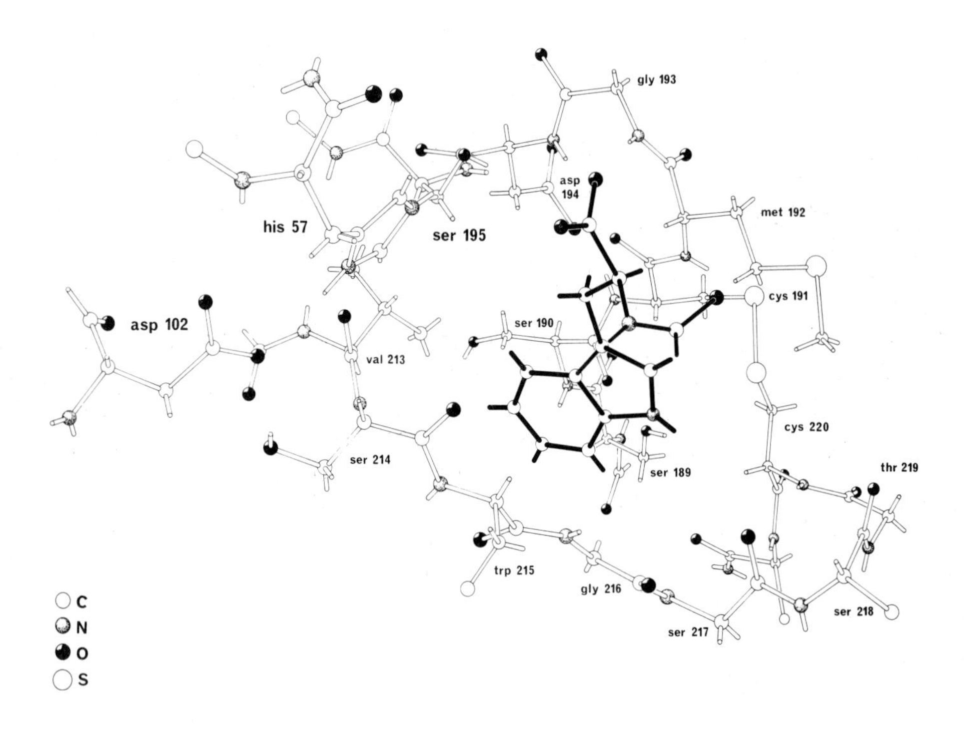

Fig. 3. The conformation of formyl L-tryptophan bound to α-chymotrypsin [28]. The indole ring lies in a hydrophobic pocket whose lower lip is formed by the peptide chain of residues 213-218.

mediately adjacent to Ser-195, as shown in fig. 3. The conformation is essentially identical to the 'equatorial' conformation of locked substrates discussed by Cohen and Schulz [29] rather than the 'axial' conformation suggested by Awad, Neurath and Hartley [30]. A productive orientation of the substrate appears to be governed by a hydrogen bond between the acylamido group of the substrate and the α-carbonyl of Ser-214. As postulated by Blow et al. [27] the carboxyl group is found to lie close to the oxygen of Ser-195. In the native enzyme, water molecules are found within this hydrophobic cleft, and the displacement of these by the hydrophobic side chain probably provides most of the binding energy.

5. Other mammalian serine proteinases

Five zymogens of endopeptidases have been found in porcine pancreatic juice: chymotrypsinogen-A, -B and -C, trypsinogen [31] and pro-elastase [32]. Bovine juice is similar except that chymotrypsinogen-C seems to be part of a 'procarboxypeptidase' complex [33] and proelastase has not been observed (it may have been missed because of its tendency to form a euglobulin precipitate). The chymotrypsins are very similar in specificity but trypsin attacks lysyl or arginyl bonds and elastase attacks non-aromatic hydrophobic side chains and is uniquely reactive against elastin. Despite these differences in specificity, all the enzymes seem to have a common catalytic mechanism, being inhibited by DFP and having a pH dependence for catalytic activity implying a role for unionised imidazole. In trypsin, a histidine residue homologous with His-57 of chymotrypsin reacts with a substrate analogue reagent [34,35].

Another group of serine proteinases are enzymes concerned with the complex processes of blood clotting and fibrinolysis, of which the best categorised is thrombin. Thrombin is secreted by liver (an organ with a common embryological origin to pancreas) into the plasma as its zymogen, prothrombin. This is a single chain of molecular weight 68,000 (chymotrypsinogen is 25,800) which yields thrombin (molecular weight 35,000) in a complex activation process which is the control step of blood clotting. Like most plasma proteins, but unlike the pancreatic proteinases, thrombin is a glycoprotein containing glucosamine, mannose, galactose and sialic acid [36]. Like trypsin, it splits esters of lysine or arginine, but has negligible activity on most lysyl or arginyl bonds in denatured peptides. It does however split an arginine residue in secretin, nine residues away from an aromatic residue [37] – a sequence which is common to most fibrinopeptides-A but not fibrinopeptides--B. The physiological specificity of thrombin must therefore involve multiple binding sites in fibrinogen, and may require the orientation of these sites in an ordered tertiary structure of the substrate. Thrombin seems to be a typical serine proteinase, being inhibited by DFP or by reaction with tosyl lysyl chloromethylketone [34] (which reacts with His-57* in trypsin).

6. Sequence homologies in mammalian serine proteinases

To what extent can we explain the similarities in mechanism or differences in specificity between these enzymes knowing their amino acid sequences? Without the tertiary structure of one member of the class, this would have been a foolish question, but the sequence comparisons become much more illuminating when viewed in the light of the tertiary structure of chymotrypsin.

The *N*-terminal sequences of the zymogens show no similarity. Activation of trypsinogen liberates the *N*-terminal peptide, Val-Asp_4-Lys, but 13 of the 15 *N*-terminal residues of chymotrypsinogen remain attached to the enzyme because of the Cys-1 to Cys-122 disulphide bridge. A similar bridge links position 22 in the 49 residue A-chain of bovine thrombin to Cys-122 in the B--chain [2]. We suspect that the B-chain of thrombin (260 residues) is C-terminal in prothrombin (ca. 520 residues) since it has *C*-terminal serine, but cannot yet say that the A- and B-chains were directly connected in the zymogen. The *N*-terminal sequence of proelastase is still unknown, but amino acid analyses suggest that the zymogen is about 10 residues longer than the enzyme [32].

Beyond Ile-16, however, the sequences of each protein are notably homologous (fig. 1). The homology is greatest in those side chains which form the internal hydrophobic core of chymotrypsin. Side chains which appear in the chymotrypsin model to be in contact with solvent are much more variable. This pattern of homology is repeated throughout the sequences, and it appeared likely that the internal sections of the polypeptide chains in each of these enzymes might have almost identical conformations.

This conclusion was fortified by comparison of the disulphide-bridge structures in these proteins (fig. 4). At identical positions in each protein a disulphide bridge forms A, the 'histidine loop' (42-58), B, the 'methionine loop' (168-182) and C, the 'serine loop' (191-220). Bridge D (136-201), which connects the B-chain of α-chymotrypsin to the C-chain, is present in all pan-

* All residues are numbered according to their homology with bovine chymotrypsinogen, as in fig. 1.

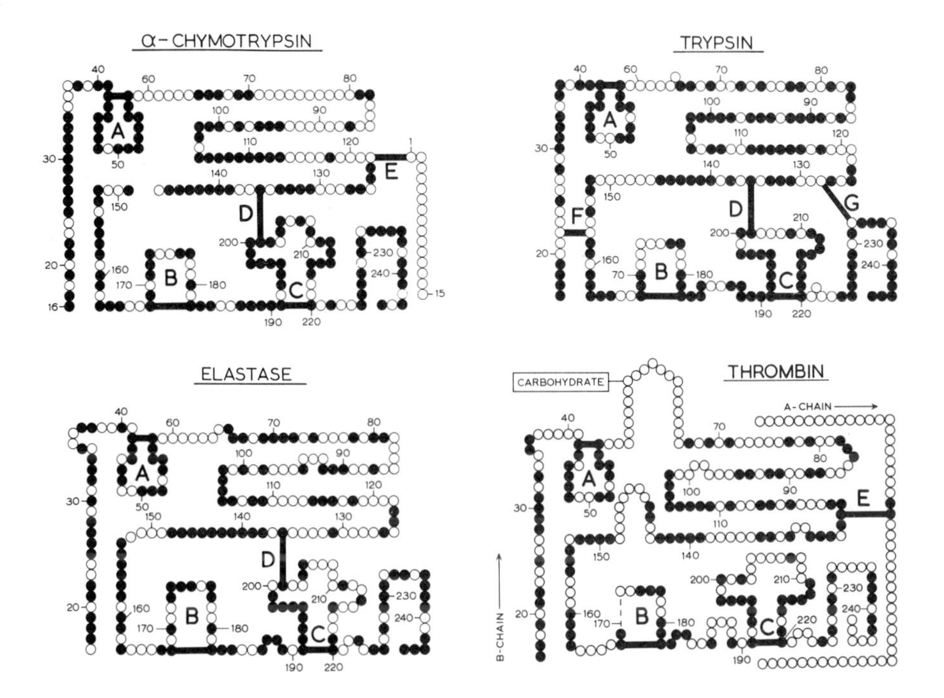

Fig. 4. Disulphide bridges in serine proteinases. Residues identical in any pair are shown as solid circles. Numbering is that of chymotrypsinogen. 'Insertions' in one sequence relative to another are shown as kinks or small loops. The thrombin pattern is still tentative, being based on homology.

creatic proteinases but absent in thrombin. As mentioned above, both thrombin and chymotrypsin have a bridge at position 122 which connects the A-chain, and trypsin has two unique bridges at positions 22-157 and 128-132. The discovery of great similarity in tertiary structure in all these enzymes was foreshadowed by this homologous pattern of bridges, and by the observation that the 'extra' bridges of trypsin could be accommodated in the chymotrypsin model without disturbing the polypeptide chain [26].

7. Models of trypsin and elastase

The sequence homology was sufficiently striking to encourage us to build models of trypsin and elastase on the assumption that the conformation of the internal parts of the polypeptide chains were identical to those of chymotrypsin. The act of building these models supported this assumption since we found that even where there were changes in the internal side-chains, these accommodated neatly to a close-packed hydrophobic structure. Those places where 'deletions' had been arbitrarily introduced into one or other sequence to maximise homology were all found to be external loops of the chain, and these residues could be easily removed without disturbing the rest of the chain. All of the changes from hydrophobic to hydrophilic side chains occurred at the surface of our models.

The vital residues of the 'charge relay system', Asp-102, His-57 and Ser-195, occupied identical positions in each model, and the environment of each of these groups was essentially identical. Some regions of the internal structures were different, for example, a cluster comprising residues 29, 45, 53, 200, 209, 210 and 212 were almost identical in chymotrypsin and trypsin but differed in their packing in the elastase model. This cluster occurs towards the 'back' of the molecule and did not seem to be significant in explaining specificity differences. Looking at the side chains which form and line the substrate binding pocket of chymotrypsin, we found to our surprise that they were identical in the trypsin model except that Asn-189 replaced Ser-189. This prompted a reinvestigation of the trypsin sequence, which led us to conclude that residue 189 is aspartic acid [2]. When *N*-formyl L-arginine was built into this cavity in the same conformation as the formyl L-tryptophan (fig. 3), we found that an excellent internal ion pair was indicated between the positively charged guanidine group of the substrate and the negatively charged carboxyl of Asp-189. There seems little doubt that the specificity differences between trypsin and chymotrypsin are due solely to this single change, and that no major conformational changes are involved.

We were less happy about the elastase model, for we found that our model obliged us to close up the entrance to the substrate pocket, since Val-216 and Thr-226 replace Gly-216 and Gly-226 of trypsin and chymotrypsin. This provided no obvious explanation for the substrate specificity and indeed raised a problem about why Ser-195 in elastase should react so specifically with DFP or tosyl fluoride, which are found to occupy the substrate cavity in the chymotrypsin model. We were forced to conclude that the diisopropyl phosphoryl group or the tosyl group in elastase must protrude into solution.

8. The tertiary structure of elastase

Convincing as these hypothetical models appeared to be, it was obviously necessary that the structural predictions should be verified. This became possible when it was found independently in Cambridge and Toronto [39], that elastase formed beautiful crystals which were very favourable for X-ray crystallography*. Moreover, the strategy which had been used by Sigler et al. [25] for producing heavy atom derivatives of chymotrypsin proved readily applicable to the elastase crystals [41]. Tosyl elastase was crystallised from 10 mM sodium acetate, pH 5.0, with 20-200 mM sodium sulphate. Crystals of *p*-chloromercuribenzene sulphonyl-(PCMBS-) elastase, similarly prepared, proved extremely isomorphous and gave phase information which was used to calculate difference Fouriers for other heavy atom derivatives. Uranyl derivatives of both tosyl and PCMBS-elastase were used to collect data for a 3.5 Å electron density map. These extremely favourable circumstances allowed model building to begin 18 months after the enzyme was first crystallised.

With the aid of the amino acid sequence, which had just been completed [42], a satisfactory interpretation of the electron density map was possible. The model confirmed the 'hypothetical' structure in all major particulars. The conformation of the interior of the molecule was almost identical to that in the hypothetical model, except in the region at the 'back' of the molecule mentioned above where the side chains are different to those of chymotrypsin. Here an alternative packing of the side chains was found, but the conformation of the peptide chain itself was unaffected. Other errors in the hypothetical model occurred in regions of the elastase sequence where 'insertions' occur relative to chymotrypsin, e.g. positions 36-37, where the hypothetical structure was necessarily a mere guess. Another expected difference between the two models lay in the conformations of Asp-102, His-57 and Ser-195. The hypothetical model was built according to the structure of native chymotrypsin, whereas the crystals were of tosyl-elastase. As predicted, the tosyl group was found to be excluded from the 'tosyl hole' (see below) and instead interacted with the imidazole ring of His-57, pulling it away from and out of the plane of the 'native' structure. The α-carbon positions of the components of the 'charge-relay' system were, however, identical to those in chymotrypsin. With these allowable exceptions, the model of tosyl-elastase was almost identical to the hypothetical model.

The correspondence was particularly striking in the substrate-binding pocket. The conformation of Val-216 and Thr-226 was exactly as predicted, so that the entrance to the pocket was almost entirely blocked. In the early

* These crystals of the pure enzyme are quite distinct from previously described 'crystalline elastase, which is a mixture containing about 50% elastase with other acidic components [40].

stages of the model building we had noted that the residual pocket could accept only the side chains of glycine or alanine, so it was pleasing to learn subsequently that benzoyl L-alanine methyl ester was the best small synthetic substrate for chymotrypsin [43]. It is likely, however, that the substrate specificity of elastase for peptides, and particularly for elastin which it binds strongly, depends on interaction of the enzyme with groups in the substrate other than the side chain immediately *N*-terminal to the bond split.

9. The structure of thrombin

The sequence of bovine thrombin is not yet quite complete, so it is premature to attempt to build models such as those of trypsin and elastase. Several conclusions can, however, already safely be made. Homology predicts that the enzyme will have an identical conformation of residues at the catalytic site to that shown in fig. 3. Moreover the side chains forming and lining the substrate-binding pocket in the trypsin model are identical in the thrombin sequence, so we can guess that, as in trypsin, Asp-189 is responsible for the specificity of thrombin towards lysyl and arginyl esters. Some features of the primary structure of thrombin may help to explain its special properties, such as inactivity towards lysyl or arginyl bonds in most peptides but unique specificity towards fibrinogen. The A-chain of thrombin is disulphide-bridged at position 22 to Cys-122 of the B-chain, as in chymotrypsin. Though this bridge is at the 'back' of the molecule the thrombin A-chain of 49 residues is much bigger than the chymotrypsin A-chain of 13 residues, so it is possible that parts of this chain could wind to the 'front' close to the catalytic site. Even more significant is a large 'insertion' of 12 residues in thrombin between positions 65 and 66 which carries the carbohydrate residues. Portions of this large additional loop could certainly interact with substrates. Another 'insertion' in thrombin of 7 residues at positions 148-149 occurs at the same place as the cleavage separating the B- and C-chains of chymotrypsin, and this loop could also interact with protein substrates. Hence it is unlikely that we will be able to build hypothetical models of thrombin with the same confidence as those of trypsin and elastase, but we will try nevertheless.

10. Serine proteinases of microorganisms

The satisfying similarities which we see in the mammalian serine proteinases vanish when we look at the extracellular serine proteinases of *B. subtilis* or *B. amyloliquefaciens.* Although these enzymes have a nucleophilic serine residue which can be phosphorylated, sulphonylated or acetylated, and a unique histidine residue essential for the catalytic activity [44], it has long been known that the active serine sequence, Thr-Ser-Met-Ala, differs from the Gly-Asp-Ser-Gly found in mammals [45]. The complete amino acid sequences of the two subtilisins show 67% identity, but no significant homology was detected between either of these and any mammalian serine enzyme [46]. Kraut and his colleagues [6,47] determined the 2.5 Å electron density map of phenylmethanesulphonyl-subtilisin and were able to build an atomic model with the aid of the sequence of Smith et al. [46]. The conformation of the polypeptide chain showed no resemblance to that of chymotrypsin, being composed of eight segments of α-helices. Yet at the catalytic site of the enzyme was found a grouping of a 'buried' Asp-32 hydrogen bonded to one nitrogen of the imidazole ring of His-64, with a hydrogen bond between the other nitrogen and the nucleophilic hydroxyl of Ser-214! This structure is essentially similar to the catalytic grouping in the mammalian enzymes, but appears to be grafted on to a quite different protein backbone. One cannot avoid the conclusion that an identical catalytic mechanism has evolved convergently from two quite different protein ancestors.

A few years ago I believed that all serine proteinases of microorganisms would prove to be like subtilisin, since a proteinase from another bacterium, *Proteus vulgaris* and from a mould, *Aspergillus oryzae*, had already been shown to have the characteristic Thr-Ser-Met-Ala sequence [47]. These bacterial enzymes also differed from the mammalian type in having no disulphide bridges, and one thought of the mammalian enzymes with their zymogen origin and stabilising disulphide bridges as a later evolutionary development from a different protein ancestor. Yet, as Dr. Whitaker will tell us, an extracellular proteinase from *B. sorangium* has disulphide bridges and a Gly-Asp-Ser-Gly sequence homologous with chymotrypsin, and a proteinase from the mould *Streptomyces griseus* bears an even more astonishing resemblance to bovine trypsin. I really am very surprised, for I would have thought that bacteria or moulds would have specialised on either the Thr-Ser-Met-Ala or the Gly-Asp-Ser-Gly type of serine proteinase.

11. Conclusions

I therefore confess that I can no longer make a convincing argument about the evolution of the serine proteinases, but the above facts must be taken into account by those who ask where new enzymes come from. In mammals, the pancreatic zymogens are secreted as a synergistic group, with an activating 'trigger' (trypsinogen) and a 'safety catch' (pancreatic trypsin inhibitor). In the intestine, the interlocking specificities of these enzymes combine with carboxypeptidases and aminopeptidase to provide amino acids from the peptic digest which arrives from the stomach. One can see that this interdependence might encourage specialisation by gene doubling of a common pancreatic ancestor followed by divergence of specificity. The changes responsible for different specificities seem to be remarkably minor, yet over half the surface residues change, apparently aimlessly. One would have thought that *some* divergence of the folding of the chain would have been allowed in selecting for a new enzyme specificity?

Between subtilisin and chymotrypsin, we see a case of convergent evolution of an identical constellation of catalytic groups on different protein 'carriers', Two thoughts arise: first, this catalytic mechanism must be a very efficient way of hydrolysing peptide bonds if it has twice arisen and twice been refined by evolution to the same point. Secondly, I am impressed by the strictness by which the folding of the peptide chain is conserved in classes of proteins, e.g. the serine proteinases, α- or β-haemoglobins and myoglobin, or (probably) lysozyme and α-lactalbumin. It suggests that unique protein conformations may be fewer than many of us would have dared to guess, and that Nature has done a very clever thing in evolving a stable, soluble protein.

References

[1] J.J. Birktoft, D.M. Blow, R. Henderson and T.A. Steitz, Phil. Trans. Roy. Soc. Lond. 257 (1970) 67.
[2] B.S. Hartley, ibid. p. 77.
[3] G.P. Hess, ibid. p. 89.
[4] S.A. Bernhard and H. Gutfreund, ibid. p. 105.
[5] D.M. Shotton and H.C. Watson, ibid. p. 111.
[6] R.A. Alden, C.S. Wright and J. Kraut, ibid. p. 119.
[7] G.E. Hein, R.B. McGriff and C. Niemann, J. Amer. Chem. Soc. 82 (1960) 1830.
[8] B. Belleau and R. Chevalier, J. Amer. Chem. Soc. 90 (1968) 6864.
[9] B.S. Hartley, Ann. Rev. Biochem. 29 (1960) 45.
[10] B.S. Hartley and V. Massey, Biochim. Biophys. Acta, 21 (1956) 58.
[11] D.E. Fahrney and A.M. Gold, J. Amer. Chem. Soc. 85 (1963) 997.
[12] R.A. Oosterbaan and M.E. Van Adrichem, Biochim. Biophys. Acta 27 (1958) 423.
[13] B.S. Hartley and B.A. Kilby, Biochem. J. 56 (1954) 288.
[14] B.R. Hammond and H. Gutfreund, Biochem. J. 61 (1955) 187.
[15] H. Gutfreund and J.M. Sturtevant, Proc. Natl. Acad. Sci. 42 (1956) 719.
[16] E.B. Ong, E. Shaw and G. Schoellman, J. Amer. Chem. Soc. 86 (1964).

[17] L.B. Smillie and B.S. Hartley, Abstr. 1st Meet. Fed. Europ. Biochem. Soc. Lond. (Academic Press, London, 1964) p. 26.
[18] J.R. Brown and B.S. Hartley, ibid. p. 25.
[19] D.E. Koshland, D.H. Strumeyer and W.R. Ray, Brookhaven Symp. Biol. 15 (1962) 101.
[20] A.N. Glazer and F. Sanger, Biochem. J. 90 (1963) 92.
[21] P.B. Sigler, Ph. D. Thesis, University of Cambridge (1967).
[22] C.H. Chervenka and P.E. Wilcox, J. Biol. Chem. 222 (1956) 635.
[23] V. Massey and B.S. Hartley, Biochim. Biophys. Acta 21 (1956) 361.
[24] H.L. Oppenheimer, B. Labouesse and G.P. Hess, J. Biol. Chem. 241 (1966) 2720.
[25] B.W. Matthews, P.B. Sigler, R. Henderson and D.M. Blow, Nature 214 (1967) 652.
[26] P.B. Sigler, D.M. Blow, B.W. Matthews and R. Henderson, J. Mol. Biol. 35 (1968) 143.
[27] D.M. Blow, J.J. Birktoft and B.S. Hartley, Nature 221 (1969) 337.
[28] T.A. Steitz, R. Henderson and D.M. Blow, J. Mol. Biol. 46 (1969) 337.
[29] S.G. Cohen and R.M. Schulz, J. Biol. Chem. 243 (1968) 2607.
[30] E.S. Awad, H. Neurath and B.S. Hartley, J. Biol. Chem. 235 (1960) PC35.
[31] D. Gratecos, O. Guy, M. Rovery and P. Desnuelle, Biochim. Biophys. Acta. 175 (1969) 82.
[32] A. Gertler and Y. Birk, Europ. J. Biochem. 12 (1970) 170.
[33] J.R. Brown, R.N. Greenshields, M. Yamasaki and H. Neurath, Biochemistry 2 (1963) 867.
[34] E. Shaw, M. Mares-Guia and W. Cohen, Biochemistry 4 (1965) 2219.
[35] E. Shaw and S. Springhorn, Biochem. Biophys. Res. Commun. 27 (1967) 391.
[36] S. Magnusson, unpublished data.
[37] V. Mutt, S. Magnusson, J.E. Jorpes and E. Dahl, Biochemistry 4 (1965) 2358.
[38] M.O. Dayhoff, Atlas of Protein Sequence and Structure 4 (1969) D70.
[39] D.M. Shotton, B.S. Hartley, N. Camerman, T. Hofmann, S.C. Nyburg and L. Rao, J. Mol. Biol. 32 (1968) 155.
[40] U.J. Lewis, D.E. Williams and N.G. Brink, J. Biol. Chem. 222 (1956) 705.
[41] H.C. Watson, D.M. Shotton, H. Muirhead and J.M. Cox, Nature (1970) in press.
[42] D.M. Shotton and B.S. Hartley, Nature, in press.
[43] H. Kaplan and H. Dugas, Biochem. Biophys. Res. Commun. 34 (1969) 681.
[44] F.S. Markland, E. Shaw and E.L. Smith, Proc. Natl. Acad. Sci. 61 (1968) 1440.
[45] F. Sanger and D.C. Shaw, Nature 187 (1960) 872.
[46] E.L. Smith, R.J. DeLange, W.H. Evans, M. Landon and F.S. Markland, J. Biol. Chem. 243 (1968) 2184.
[47] D.C. Shaw, Ph. D. Thesis, University of Cambridge (1962).
[48] B.S. Hartley, Nature 201 (1964) 1284.
[49] B.S. Hartley and D.L. Kauffman, Biochem. J. 101 (1966) 229.
[50] L.B. Smillie, A. Furka, N. Nagabhushan, K.J. Stevenson and C.O. Parkes, Nature 218 (1968) 343.
[51] K.A. Walsh, D.L. Kauffman, K.S.V.S. Kumar and H. Neurath, Proc. Natl. Acad. Sci. 51 (1964) 301.
[52] B. Meloun, I. Kluh, V. Kostka, L. Moravek, Z. Prusik, J. Vanecek, B. Keil and F. Sorm, Biochim. Biophys. Acta 130 (1966) 543.
[53] S.M. Magnusson and B.S. Hartley, unpublished evidence.

A BACTERIAL HOMOLOGUE OF ELASTASE, THE α-LYTIC PROTEASE OF A MYXOBACTERIUM

D.R. WHITAKER

Biochemistry Laboratory,
National Research Council of Canada, Ottawa, Canada

Abstract: Whitaker, D.R. A Bacterial Homologue of Elastase, The α-Lytic Protease of a Myxobacterium. *Miami Winter Symposia* 1, pp. 210–222. North-Holland Publishing Company, Amsterdam, 1970.

The catalytic properties of this enzyme are indistinguishable from those of the pancreatic serine proteases. Its substrate-binding properties are similar to those of porcine elastase and the two enzymes have other common properties which are not shared by chymotrypsin and trypsin. The α-enzyme matches the pancreatic enzymes in the sequences around its catalytically functional serine and histidine residues and in major features of construction, but most of its sequence shows few direct matches with that of elastase. It appears to be an instance where homology is confined to the essentials.

1. Introduction

The bacterium which produces the α-enzyme is a Myxobacterium ("Myxobacter 495") which was isolated from local soil at the Institute of Microbiology of the Canada Dept. of Agriculture in Ottawa [1,2]. Its taxonomy is still under investigation [3] but it has been reported to be a species of *Sorangium* [2]. It attracted attention because its culture filtrates were intensely bacteriolytic towards several other soil bacteria and were capable of lysing soil nematodes [1]. Two basic proteases proved to be responsible for the lytic activity [4] but the organism produces a number of other proteases as well. The following have been isolated:

1. *α-Lytic protease* – the most basic protease of the set and one of the two bacteriolytic enzymes. It is a serine protease with the same sequence around its reactive serine residue as the pancreatic enzymes, i.e. Gly.Asp.Ser*.Gly.Gly. As discussed later, its specificity resembles that of pancreatic elastase.

2. *β-Lytic protease* – the other bacteriolytic enzyme and the enzyme re-

sponsible for lytic activity towards nematodes. It is not a serine protease and it is an endopeptidase [5], not an exopeptidase, but it has properties which suggest a relationship with carboxypeptidase [6].

3. δ-*Protease* – a serine protease with a trypsin-like specificity but with a different sequence, Ser.Ser*Gly., around its reactive serine residue [7]. This sequence is known to occur in the serine protease of *Arthrobacter sp.*, another soil bacterium [8].

4. ε-*Protease* – a serine protease with a specificity which might be described as chymotrypsin-like in that the enzyme has a strong affinity for aromatic amino acids. However it cleaves the linkage at the amino end, not at the carboxyl end, of an aromatic residue. It has the same active-serine sequence as the δ-protease [9].

Taken individually, only one of these enzymes provides an unequivocal match with a pancreatic enzyme but, taken as a set, they show a pattern which is not unreminiscent of the pattern of enzymes secreted by the pancreas. In some respects then, the homology to be discussed – the homology between the α-enzyme and pancreatic elastase – is a homology within a broader homology.

2. Isolation and general properties of the α-enzyme

The organism is grown in a glucose-amino acid medium in a fermenter [10] or in shake-culture [11]. After about 40 hr, the culture solution is centrifuged to remove the cells and freed of heavy metals by mixed-bed ion-exchange. The enzymes are then adsorbed on Amberlite CG50 at pH 5. The less basic enzymes are stripped from the resin by a titration with alkali. The β- and the α-enzyme are displaced selectively by a gradient [4] or a stepwise inflow [11] of citrate buffers of increasing ionic strength. The α-enzyme is refractionated on a column of the same resin, dialysed and freeze-dried.

The preparation of crystals suitable for X-ray diffraction studies is described by James and Smillie [12].

Table 1 gives the amino acid composition [13,14]. In agreement with it, the α-enzyme is a basic protein (iso-electric point > pH 10) with a molecular weight of approximately 20,000 [15]. The dimensions of the unit cell in the trigonal crystals prepared by James and Smillie are $a = b = 66.5$ Å, $c = 80.2$ Å [12]. The optical rotatory dispersion spectrum suggests that the enzyme has virtually no α-helices in its native conformation [16].

The α-enzyme is an active and very stable protease. Its bacteriolytic activity stems from its ability to cleave the peptides which cross-link chains of amino

Table 1
Amino acid composition of α-lytic protease.

Amino acid	No. of residues	Amino acid	No. of residues
Histidine	1	Glycine	32
Lysine	2	Alanine	24
Arginine	12	Valine	19
Aspartic acid	2	Isoleucine	8
Asparagine	13	Leucine	10
Glutamic acid	4	Phenylalanine	6
Glutamine	9	Tyrosine	4
Serine	20	Tryptophan	2
Threonine	18	Methionine	2
Proline	4	Half-cystine	6

Total number of residues: 198

sugars in the cell-walls of susceptible bacteria such as *Arthrobacter globiformis* [17]. It is inhibited extremely rapidly by diisopropyl phosphorofluoridate (DFP) and isopropyl phosphonofluoridate (Sarin); for example, treatment of 2.4×10^{-4} M enzyme at pH 7.7 with 6×10^{-4} M DFP at 25° gives a 95% inhibition within 10 min. The inhibition is accompanied by an irreversible esterification of one serine residue [18]; in other words, the α-enzyme is a serine protease.

A relationship with porcine elastase was first suggested by the action of the α-enzyme on the A and B chains of performate-oxidized insulin [5]. The cleavage points were at linkages which involved the carbonyl groups of neutral, aliphatic amino acids and, almost without exception, they were cleavage points on Naughton and Sanger's list for porcine elastase [19]. This suggestion was strengthened by subsequent findings that assay methods for one enzyme were also suitable for the other. For example, the α-enzyme has elastase activity as measured by the rate of dye-release from orcein-impregnated elastin [20] (its activity is roughly half that of porcine elastase); porcine elastase has bacteriolytic activity towards *Arthrobacter globiformis* cells [20] (its activity is roughly half that of the α-enzyme) and *N*-benzoyl-L-alanine methyl ester, an excellent substrate for assaying the esterase activity of the α-enzyme [34], is an equally good substrate for porcine elastase [20].

The original comparison of cleavage points on the insulin chains suggested that the α-enzyme was the more selective enzyme: its initial attack on the B chain was confined to 3 linkages – the Ala_{14}-Leu_{15} linkage, the Val_{12}-Glu_{13}

linkage and the Val_{18}-Cys_{19} [SO_3H] linkage. However a recent study [21] suggests that the reverse is true: thoroughly purified porcine elastase is still more selective – its initial attack is confined to the Ala_{14}-Leu_{15} linkage. As shown later, the esterase activities of the two enzymes support this finding.

3. The homology in structure with the pancreatic enzymes

The sequence around the active serine residue was the first feature of the α-enzyme to be examined [18]. It was determined on ^{32}P-sarin labelled enzyme by the methods used by Naughton et al. [22] for porcine elastase. The sequence Thr.Ser.Met. was anticipated as the evidence at the time suggested that it was the sequence characteristic of a microbial serine protease but, as mentioned earlier, it proved to be Gly.Asp.Ser.Gly.Gly., the sequence in the pancreatic enzymes.

The sequence around the single histidine residue of the α-enzyme was the second target. It was of particular interest for this reason. The pancreatic enzymes have two histidine residues, His_{40} and His_{57}†, in a cystine peptide whose disulfide bridge might be considered to have the function of bringing the two imidazole, close enough to function cooperatively as catalysts. This consideration had led to several proposals of reaction mechanisms which required the pancreatic enzymes to have two catalytically functional histidine residues. However there was no compelling kinetic evidence to support this mechanism and only one histidine residue, His_{57}, was alkylated by the chloromethyl ketone inhibitors of chymotrypsin and trypsin [23,24]. The α-enzyme, of course, could not possibly function in this way.

The histidine sequence [25], determined in Dr. L.B. Smillie's laboratory by Brown and Hartley's performic acid diagonal procedure [26], proved to be homologous with the sequences around His_{57} of the pancreatic enzymes. The following comparison indicates the degree of homology:

Enzyme	*Residue*: 52 53 54 55 56 57 58 59 60 61
α-Lytic protease	PHE.VAL.THR.ALA.GLY.HIS.CYS.GLY.THR.VAL.
Chymotrypsin	VAL.VAL.THR.ALA.ALA.HIS.CYS.GLY.VAL.THR.
Trypsin	VAL.VAL.SER.ALA.ALA.HIS.CYS.TYR.LYS.SER.
Elastase	VAL.MET.THR.ALA.ALA.HIS.CYS.VAL.ASP.ARG.

This finding, in conjunction with the kinetic data discussed later, suggested that the pancreatic enzymes, like the α-enzyme, had only one catalytically

† These residues are numbered with chymotrypsinogen as the reference enzyme. The residues in question are residues 28 and 45 in table 3.

Table 2
Amino acid sequence of α-lytic protease [14].

	1	2	3	4	5	6	7	8	9	10	11	12	13	14	15	16	17	18	19	20
1	ALA.	ASN.	ILE.	VAL.	GLY.	GLY.	ILE.	GLU.	TYR.	SER.	ILE.	ASN.	ASN.	ALA.	SER.	LEU.	CYS.	SER.	VAL.	GLY.
21	PHE.	SER.	VAL.	THR.	ARG.	GLY.	ALA.	THR.	LYS.	GLY.	PHE.	VAL.	THR.	ALA.	GLY.	HIS.	CYS.	GLY.	THR.	VAL.
41	ASN.	ALA.	THR.	ALA.	ARG.	ILE.	GLY.	GLY.	ALA.	VAL.	VAL.	GLY.	THR.	PHE.	ALA.	ALA.	ARG.	VAL.	PHE.	PRO.
61	GLY.	ASN.	ASP.	ARG.	ALA.	TRP.	VAL.	SER.	LEU.	THR.	SER.	ALA.	GLN.	THR.	LEU.	LEU.	PRO.	ARG.	VAL.	ALA.
81	ASN.	GLY.	SER.	SER.	PHE.	VAL.	THR.	VAL.	ARG.	GLY.	SER.	THR.	GLU.	ALA.	ALA.	VAL.	GLY.	ALA.	ALA.	VAL.
101	CYS.	ARG.	SER.	GLY.	ARG.	THR.	THR.	GLY.	TYR.	GLN.	CYS.	GLY.	THR.	ILE.	THR.	ALA.	LYS.	ASN.	VAL.	THR.
121	ALA.	ASN.	TYR.	ALA.	GLU.	GLY.	ALA.	VAL.	ARG.	GLY.	LEU.	THR.	GLN.	GLY.	ASN.	ALA.	CYS.	MET.	GLY.	ARG.
141	GLY.	ASP.	SER.	GLY.	GLY.	SER.	TRP.	ILE.	THR.	SER.	ALA.	GLY.	GLN.	ALA.	GLN.	GLY.	VAL.	MET.	SER.	GLY.
161	GLY.	ASN.	VAL.	GLN.	SER.	ASN.	GLY.	ASN.	ASN.	CYS.	GLY.	ILE.	PRO.	ALA.	SER.	GLN.	ARG.	SER.	SER.	LEU.
181	PHE.	GLU.	ARG.	LEU.	GLN.	PRO.	ILE.	LEU.	SER.	GLN.	TYR.	GLY.	LEU.	SER.	LEU.	VAL.	THR.	GLY.		

Disulfide bridges: 17-37, 101-111, 137-170.

Table 3
Amino acid sequence of porcine elastase [29].

	1	2	3	4	5	6	7	8	9	10	11	12	13	14	15	16	17	18	19	20
1	VAL.	VAL.	GLY.	GLY.	THR.	GLU.	ALA.	GLN.	ARG.	ASN.	SER.	TRP.	PRO.	SER.	GLN.	ILE.	SER.	LEU.	GLN.	THR.
21	ARG.	SER.	GLY.	SER.	SER.	TRP.	ALA.	HIS.	THR.	CYS.	GLY.	GLY.	THR.	LEU.	ILE.	ARG.	GLN.	ASN.	TRP.	VAL.
41	MET.	THR.	ALA.	ALA.	HIS.	CYS.	VAL.	ASP.	ARG.	GLU.	LEU.	THR.	PHE.	ARG.	VAL.	VAL.	VAL.	GLY.	GLU.	HIS.
61	ASN.	LEU.	ASN.	GLN.	ASN.	ASN.	GLY.	THR.	GLU.	GLN.	TYR.	VAL.	GLY.	VAL.	GLN.	LYS.	ILE.	VAL.	VAL.	HIS.
81	PRO.	TYR.	TRP.	ASN.	THR.	ASP.	ASP.	VAL.	ALA.	ALA.	GLY.	TYR.	ASP.	ILE.	ALA.	LEU.	LEU.	ARG.	LEU.	ALA.
101	GLN.	SER.	VAL.	THR.	LEU.	ASN.	SER.	TYR.	VAL.	GLN.	LEU.	GLY.	VAL.	LEU.	PRO.	ARG.	ALA.	GLY.	THR.	ILE.
121	LEU.	ALA.	ASN.	ASN.	SER.	PRO.	CYS.	TYR.	ILE.	THR.	GLY.	TRP.	GLY.	LEU.	THR.	ARG.	THR.	ASN.	GLY.	GLN.
141	LEU.	ALA.	GLN.	THR.	LEU.	GLN.	GLN.	ALA.	TYR.	LEU.	PRO.	THR.	VAL.	ASP.	TYR.	ALA.	ILE.	CYS.	SER.	SER.
161	SER.	SER.	TYR.	TRP.	GLY.	SER.	THR.	VAL.	LYS.	ASN.	SER.	MET.	VAL.	CYS.	ALA.	GLY.	GLY.	ASN.	GLY.	VAL.
181	ARG.	SER.	GLY.	CYS.	GLN.	GLY.	ASP.	SER.	GLY.	GLY.	PRO.	LEU.	HIS.	CYS.	LEU.	VAL.	ASN.	GLY.	GLN.	TYR.
201	ALA.	VAL.	HIS.	GLY.	VAL.	THR.	SER.	PHE.	VAL.	SER.	ARG.	LEU.	GLY.	CYS.	ASN.	VAL.	THR.	ARG.	LYS.	PRO.
221	THR.	VAL.	PHE.	THR.	ARG.	VAL.	SER.	ALA.	TYR.	ILE.	SER.	TRP.	ILE.	ASN.	ASN.	VAL.	ILE.	ALA.	SER.	ASN.

Disulfide bridges: 30-46, 127-194, 158-174, 184-214.

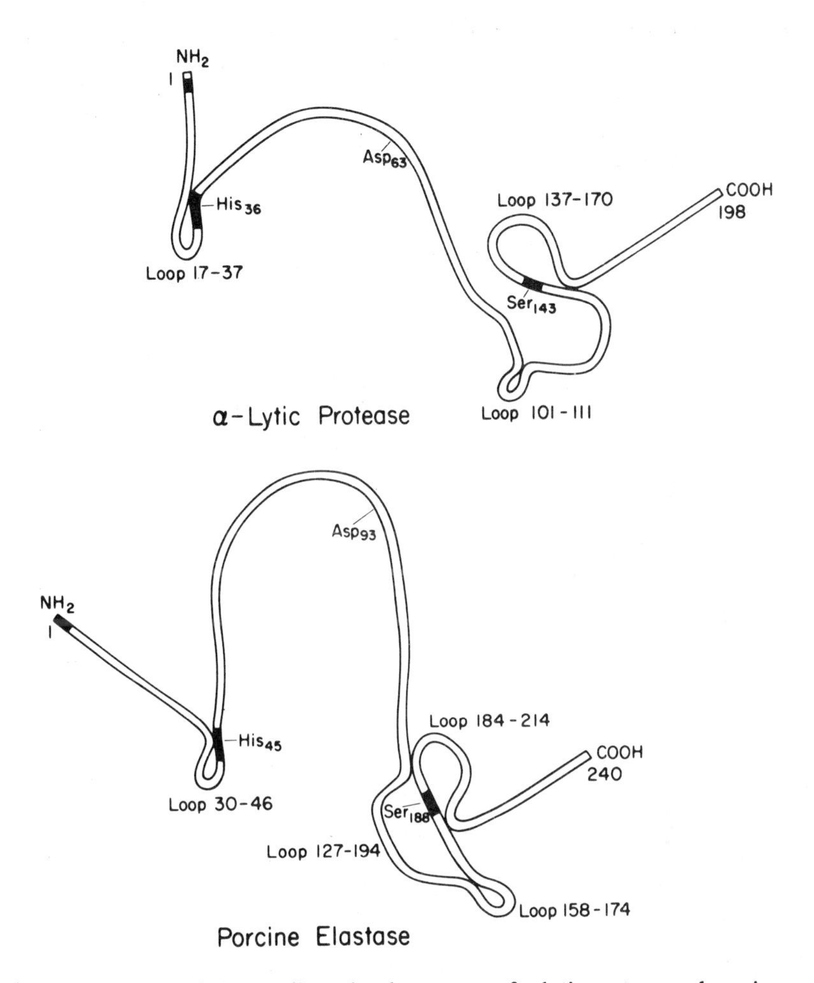

Fig. 1. A comparison of the two-dimensional structures of α-lytic protease and porcine elastase. The shaded regions are regions with homology in sequence. The functional residues of the active site and the disulfide bridges at the loops are numbered relative to the *N*-terminal residue of each enzyme.

functional histidine residue. The determination by Sigler et al. [27,28] of the three dimensional structure of chymotrypsin provided the conclusive evidence.

The complete amino acid sequence was determined in collaboration with Dr. L.B. Smillie and his co-workers [14]. Tables 2 and 3 give its sequence and that of porcine elastase as determined by Shotton and Hartley [29]. The complete sequence added very little in the way of direct matches with the sequence of the pancreatic enzymes; the only extended homology is at residues 3-6. These are identical with residues 1-4 of trypsin and homologous with residues 1-4 of elastase. However it is evident from fig. 1 that the α-enzyme and elastase are constructed in much the same way with the functional histidine and serine residues in similar loops at similar locations. According to the "charge-relay" mechanism of Blow et al. [30] a buried aspartyl residue is also an essential component of the active site and Shotton and Watson [31] have demonstrated that it is present in the active site of porcine elastase. The three-dimensional structure of the α-enzyme is still to be determined but there is only one aspartyl residue which could function in a relay (the only other aspartyl residue is adjacent to the active serine residue) and, as shown in fig. 1, its location is not unlike that of the buried aspartyl residue of elastase.

One other common feature of the α-enzyme and elastase is evident from the following comparison:

Enzyme	*Residues per mole of enzyme*	
	Lysine	*Arginine*
α-Enzyme	2	12
Porcine elastase	3	11
Bovine trypsin [32]	14	2
Bovine chymotrypsin [33]	14	3

The α-enzyme and elastase, unlike trypsin and chymotrypsin, are "low-lysine-high arginine" enzymes and hence are subject to much smaller changes in net charge when the pH is raised from 8 to 10.5. The optical rotations of the two enzymes are essentially constant between pH 5 and pH 10 [16,20] and, as discussed later, their esterase activities are not depressed by mild alkali.

4. The correspondence in enzymatic properties

Fig. 2 illustrates the main features of catalysis by the α-enzyme as determined by measurements of esterase activity in a pH-stat [20,34]. The rate depends on an ionization with a pK_a of 6.7 in H_2O and 7.3 in D_2O and, as can be shown by estimates of k_c and K_m as a function of pH, the pH-dependence of k_c/K_m is determined by the pH-dependence of k_c. The values of pK_a are consistent with catalysis by an unprotonated imidazole group; the value of 1.0 for $\Delta \log (k_c/K_m)/\Delta pH$ is consistent with catalysis by a single imidazole group [35] and the 50% reduction in k_c when H_2O is replaced by D_2O is consistent

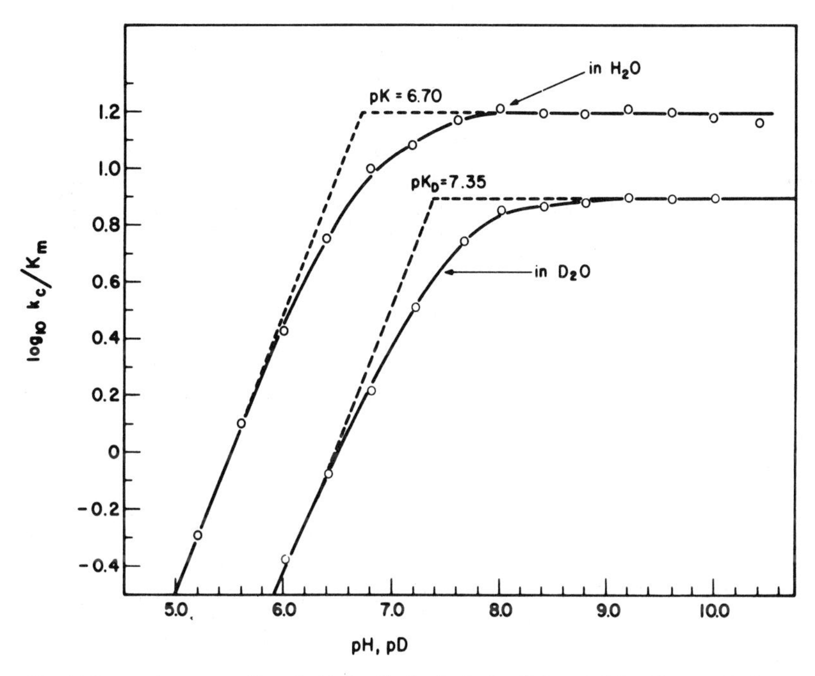

Fig. 2. Dependence on pH and pD for the hydrolysis of *N*-acetyl-L-valine methyl ester by α-lytic protease. The ordinate is the log of the second-order velocity constant which, in the Michaelis-Menten formulation, is equal to k_c/K_m, where k_c is the catalytic rate coefficient ($V_{max}/[E]_o$) and K_m is the Michaelis-Menten constant.

with general basic catalysis by that group [36]. These catalytic properties are essentially identical with those of chymotrypsin [37,38] and porcine elastase [20].

The difference from chymotrypsin is that K_m remains constant at pH's in excess of pH 8 and thus shows no evidence of any dependence on the state of ionization of α- or ϵ- amino groups. These groups can, in fact, be fully acetylated [34] without loss of esterase activity. Porcine elastase behaves in exactly the same way [20,39].

Tables 4, 5 and 6 compare the specificities of the two enzymes with respect to the side chain, the *N*-acyl substituent, and the alcoholic group of amino acid esters [20]. Both enzymes are very sensitive to changes in the side chain and the *N*-acyl substituent and respond differently to some of the changes. Elastase, for example, is much more sensitive than the α-enzyme to a replacement of L-alanine's methyl side chain by L-valine's isopropyl side chain

Table 4
Esterase activity of α-lytic protease and porcine elastase towards *N*-benzoyl amino acid methyl esters at pH 8.0 in 0.10 N KCl at 25°C.

Substrate	α-Lytic protease k_c/K_m (M^{-1} sec^{-1})	Porcine elastase k_c/K_m (M^{-1} sec^{-1})
Bz-Gly-OMe	8.1	2.9
Bz-Ala-OMe	720	640
Bz-Val-OMe	230	19
Bz-Leu-OMe	28	21
Bz-Ile-OMe	7.6	16
Bz-D-Ala-OMc	0.0	2.8

(table 4) while the α-enzyme is more sensitive to changes in the alkyl component of *N*-alkyloxycarbonyl-L-alanine methyl ester (table 5). The alcoholic group is not so critical a determinant of reactivity but the series in table 6 (chosen to mimic the side chains of glycine, alanine, valine, leucine and phenylalanine) shows a rough trend for reactivity to increase with the size of the alcoholic group. Another feature of the data is discussed below.

Hydrolyses by these and related enzymes are usually considered to proceed in the following way:

$$\mathrm{RCOR'} + \mathrm{E} \underset{k_{-1}}{\overset{k_1}{\rightleftharpoons}} (\mathrm{RCOR'})\cdot\mathrm{E} \xrightarrow[\searrow \mathrm{R'OH}]{k_2} \mathrm{RCO{\cdot}E} \xrightarrow[\searrow \mathrm{RCOOH}]{k_3} \mathrm{E}$$

Table 5
Esterase activity of α-lytic protease and porcine elastase towards *N*-acyl-L-alanine methyl esters at pH 8.0 and 25°C.

Acyl group	α-Lytic protease K_m (mM)	α-Lytic protease k_c (sec^{-1})	α-Lytic protease k_c/K_m ($M^{-1}sec^{-1}$)	Porcine elastase K_m (mM)	Porcine elastase k_c (sec^{-1})	Porcine elastase k_c/K_m ($M^{-1}sec^{-1}$)
Methyloxycarbonyl	32	1.4	43	13	0.44	33
Ethyloxycarbonyl	26	10	390	22	0.99	46
n-Propyloxycarbonyl	20	13	650	21	3.5	160
n-Butyloxycarbonyl	23	9.4	410	21	5.5	260
Isobutyloxycarbonyl	26	11	430	36	7.8	220
t-Butyloxycarbonyl	8.8	11	1200	41	4.7	120
Benzyloxycarbonyl	38	6.4	170	13	3.9	300

Table 6
Esterase activity of α-lytic protease and porcine elastase towards *N*-n-propyloxycarbonyl-L-alanine esters at pH 8.0 in 0.10 N KCl at 25°C.

Ester of *N*-n-propyloxy-carbonyl-L-alanine	α-Lytic protease			Porcine elastase		
	K_m (mM)	k_c (sec^{-1})	k_c/K_m ($M^{-1} sec^{-1}$)	K_m (mM)	k_c (sec^{-1})	k_c/K_m ($M^{-1} sec^{-1}$)
Methyl ester	20	13	650	21	3.5	160
Ethyl ester	30	4.9	160	19	3.6	190
Isobutyl ester	11	9.5	840	6.3	2.1	340
Isopentyl ester	~10	~10	830	~9	~5	570
Phenylethyl ester			1300			580

where (RCOR′)·E is a Michaelis-Menten complex and RCO·E is an acyl enzyme, esterified at the active serine residue. The pre-steady state kinetics for the hydrolysis of *p*-nitrophenyl trimethylacetate by elastase [40] and the α-enzyme [34] are very similar and they are consistent with the formation of an acyl enzyme as an intermediate. Acyl-enzyme intermediates of chymotrypsin have been isolated [41,42] but the only conclusive evidence of their existence has been obtained with "activated" substrates, i.e. substrates such as *p*-nitrophenyl acetate in which group R′ is such a good leaving group that the substrate is a powerful acylating agent. A point of current interest is whether acyl enzymes have an equally real existence with "specific" substrates.

The data on the hydrolysis of *p*-nitrophenyl trimethylacetate by elastase and the α-enzyme are consistent with the rate of acylation being much faster than the rate of deacylation (i.e. $k_2 \gg k_3$). If this were a general condition, deacylation would always be the rate-limiting step and the catalytic rate coefficient, k_c, for the hydrolysis of amino acid esters would be independent of the nature of the alcoholic group. Data supporting this conclusion have been reported for chymotrypsin [43] and trypsin [44]. However the values of k_c in table 6 are not constant. As a more extreme example, a value of 110 sec^{-1} has been reported for the elastase-catalysed hydrolysis of the *p*-nitrophenyl ester of *N*-benzyloxycarbonyl-L-alanine [45], a value nearly 30-fold greater than the value in table 5 for the methyl ester.

One possible explanation is that only the relative magnitudes of k_2 and k_3 are at issue – with "activated" substrates, k_3 is rate-limiting, with specific substrates, k_2 may be partly or completely rate-limiting. A second possibility,

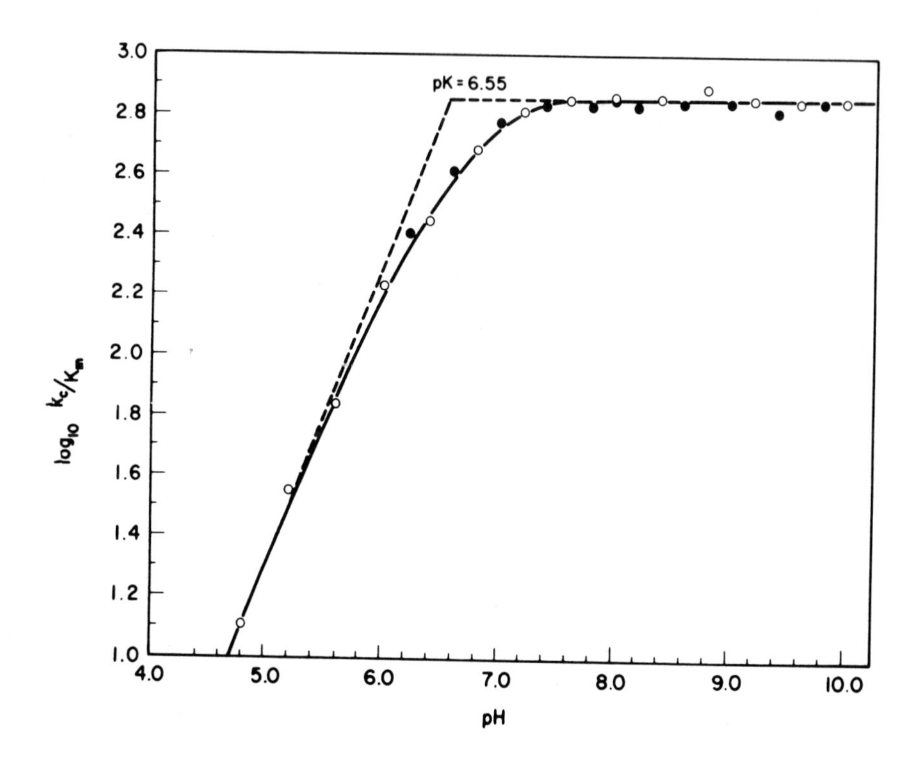

Fig. 3. Hydrolysis of *N*-benzoyl-L-alanine methyl ester by α-lytic protease (○) and acetylated α-lytic protease (●).

suggested by Kaplan et al. [20], is that the course of the reaction depends on the nature of the leaving group. According to this suggestion, there are two limiting pathways:

(i) in that for substrates with a good leaving group (*p*-nitrophenyl esters for example), the leaving group departs prior to the attack by water and thereby provides the condition for formation of an acyl-enzyme intermediate;

(ii) in that for substrates with a poor leaving group, the leaving group departs as a result of the attack by water; there is no acyl-enzyme intermediate. This explanation assumes a distinction in pathways much like the distinction between an S_N1 and an S_N2 reaction mechanism.

5. Significance of the homology

The homology between the α-enzyme and porcine elastase appears to be confined to essentials. The main matches in sequence are at residues with a catalytic function; the homology at the *N*-terminal end may merely reflect a common need for hydrophobic side chains at that position. The homology may well be but another instance of convergent evolution. However the possibility that the two enzymes are expressions of a common evolutionary pathway has been strengthened recently by the findings of Jurášek et al. [46] with regard to the trypsin-like component of Pronase, an enzyme preparation from a Streptomocyte. This enzyme matches the α-enzyme in several respects for example, it also has only one histidine residue, but the direct matches with trypsin are much more extensive and range over all three disulfide bridges: An evolutionary pathway has been traced from the mammals to the spiny dogfish [47] to the sea-anemone [48]; a continuation to the streptomycetes and bacteria would raise the question as to where the connecting pathway lies.

Acknowledgement

The writer wishes to thank Dr. D.M. Shotton of the University of Bristol for the opportunity to read manuscripts on elastase prior to their publication. Figs. 2 and 3 reproduced by permission of the National Research Council of Canada from the Canadian Journal of Biochemistry, 47 (1969) 308.

References

[1] H. Katznelson, D.C. Gillespie and F.D. Cook, Canad. J. Microbiol. 10 (1964) 699.
[2] D.C. Gillespie and F.D. Cook, Canad. J. Microbiol. 11 (1965) 109.
[3] F.D. Cook, personal communication.
[4] D.R. Whitaker, Canad. J. Biochem. 43 (1965) 1935.
[5] D.R. Whitaker, C. Roy, C.S. Tsai and L. Jurášek, Canad. J. Biochem. 43 (1965) 1961.
[6] N.B. Oza and D.R. Whitaker, Manuscript in preparation.
[7] G.M. Paterson and D.R. Whitaker, Manuscript in preparation.
[8] S. Wählby, Biochim. Biophys. Acta 151 (1968) 409.
[9] W.H. Newsome and D.R. Whitaker, Manuscript in preparation.
[10] D.R. Whitaker, F.D. Cook and D.C. Gillespie, Canad. J. Biochem. 43 (1965) 1927.
[11] D.R. Whitaker, Canad. J. Biochem. 45 (1967) 991.
[12] M.N.G. James and L.B. Smillie, Nature 224 (1969) 694.
[13] L. Jurášek and D.R. Whitaker, Canad. J. Biochem. 45 (1967) 917.

[14] M.O.J. Olson, N. Nagabushnan, M. Dzwiniel, L.B. Smillie and D.R. Whitaker. Manuscript in preparation.
[15] L. Jurášek and D.R. Whitaker, Canad. J. Biochem. 43 (1965) 1955.
[16] G.M. Paterson and D.R. Whitaker, Canad. J. Biochem. 47 (1969) 317.
[17] G.S. Tsai, D.R. Whitaker, L. Jurášek and D.R. Whitaker, Canad. J. Biochem. 43 (1965) 1971.
[18] D.R. Whitaker and C. Roy, Canad. J. Biochem. 43 (1965) 1935.
[19] M.A. Naughton and F. Sanger, Biochem. J. 78 (1961) 156.
[20] H. Kaplan, V.B. Symonds, H. Dugas and D.R. Whitaker, Canad. J. Biochem. 48 (1970) 649.
[21] A.S. Narayanan and R.A. Anwar, Biochem. J. 114 (1969) 11.
[22] M.A. Naughton, F. Sanger, B.S. Hartley and D.C. Shaw, Biochem. J. 77 (1960) 149.
[23] E.B. Ong, E. Shaw and G. Schoellman, J. Biol. Chem. 240 (1965) 694.
[24] E. Shaw, M. Mares-Guia and W. Cohen, Biochemistry 4 (1965) 2219.
[25] L.B. Smillie and D.R. Whitaker, J. Am. Chem. Soc. 89 (1967) 3350.
[26] J.R. Brown and B.S. Hartley, Biochem. J. 101 (1966) 214.
[27] P.B. Sigler, D.M. Blow, B.W. Mathews and R. Henderson, J. Mol. Biol. 35 (1968) 143.
[28] B.W. Matthews, P.B. Sigler, R. Henderson and D.M. Blow, Nature 214 (1967) 652.
[29] D.M. Shotton and B.S. Hartley, in: Atlas of Protein Sequence and Structure, Vol. 4, ed. M.O. Dayhoff (National Biochemical Research Foundation, Silver Spring) p. 120.
[30] D.M. Blow, J.J. Birktoft and B.S. Hartley, Nature 221 (1969) 337.
[31] D.M. Shotton and H.C. Watson, Phil. Trans. Roy. Soc. Lond., 257 (1970) 111.
[32] K.A. Walsh, D.L. Kauffman, K.S.V. Sampath Kumar and H. Neurath, Proc. Natl. Acad. Sci. U.S. 51 (1964) 301.
[33] B.S. Hartley, Nature 201 (1964) 1284.
[34] H. Kaplan and D.R. Whitaker, Canad. J. Biochem. 47 (1969) 305; J. Am. Chem. Soc. 89 (1967) 3352.
[35] M. Dixon, Biochem. J. 55 (1953) 161.
[36] M.L. Bender, E.J. Pollock and M.C. Neveu, J. Am. Chem. Soc. 84 (1962) 595.
[37] M.L. Bender, G.E. Clement, F.J.Kézdy and H. d'A. Heck, J. Am. Chem. Soc. 86 (1964) 3680.
[38] F.J. Kézdy, G.E. Clement and M.L. Bender, J. Am. Chem. Soc. 86 (1964) 3690.
[39] H. Kaplan and H. Dugas, Biochem. Biophys. Res. Commun. 34 (1969) 681.
[40] M.L. Bender and T.H. Marshall, J. Am. Chem. Soc. 90 (1967) 201.
[41] R.A. Oosterbaan and M.E. van Andorchem, Biochim. Biophys. Acta 27 (1958) 423.
[42] Y. Shalitin and J.R. Brown, Biochem. Biophys. Res. Commun. 24 (1966) 817.
[43] B. Zerner, R.P.M. Bond and M.L. Bender, J. Am. Chem. Soc. 86 (1964) 3674.
[44] N.J. Barnes, J.B. Baird and D.T. Elmore, Biochem. J. 90 (1964) 470.
[45] D.M. Shotton, Methods in Enzymology. In press.
[46] L. Jurásek, D. Fackre and L.B. Smillie, Biochem. Biophys. Res. Comm. 37 (1969) 99.
[47] H. Neurath, R.A. Bradshaw and R. Arron, Abstr. Intern. Union Biochem. Symp. Copenhagen (1969). In press.
[48] D. Gibbon and G.H. Dixon, Nature 222 (1969) 753.

PHOSPHORYLASE KINASE: COMPARISON BETWEEN NORMAL AND DEFICIENT MICE AND MEN

F. HUIJING

Department of Biochemistry, University of Miami School of Medicine, Miami, Florida 33136, U.S.A.

Abstract: Huijing, F. Phosphorylase Kinase: Comparison Between Normal and Deficient Mice and Men. *Miami Winter Symposia* 1, pp. 223–232. North-Holland Publishing Company, Amsterdam, 1970.
Glycogen phosphorylase exists in an active and an inactive form. The inactive may or may not be activated by AMP depending on the tissue investigated. The difference between the active and the inactive form is one phosphate group on a hydroxyl group of one specific serine residue per phosphorylase monomer. In muscle a phosphorylase monomer has a molecular weight of approximately 90,000. The enzyme responsible for the phosphorylation of the serine hydroxyl and thus for the activation is phosphorylase kinase. It requires ATP and magnesium; the latter has to be present in excess of ATP. The kinase apparently does not distinguish its high molecular weight substrate, phosphorylase, from that of other tissues or species. Phosphorylase kinase deficiency is an inborn error of control of metabolism. In humans it was designated as glycogen storage disease type VIa after it became clear that some of the patients in which previously a partial liver phosphorylase deficiency had been shown (type VI), in fact have a partial phosphorylase kinase deficiency. The study of the disease in humans is greatly facilitated by the fact that the low phosphorylase activity and the kinase deficiency are not limited to liver but extend to blood cells. Thus biopsies are not necessary.

Comparison of phosphorylase kinase deficiency in man and mouse reveals many similarities. In both species the defect is inherited as a sex-linked trait indicating that the affected gene is located on the X-chromosome. Adult humans with the kinase deficiency and adult *I*-strain mice do not show symptoms or signs of the deficiency. Children with type VIa glycogen disease show liver enlargement, muscular weakness and mild asymptomatic fasting hypoglycemia as well as some other mild disturbances. Deficient mice show a high incidence of perinatal death from unknown causes. Litter sizes are small. Crosses between *I*-strain mice and C57 black strain mice produce larger litters. In the *I*-strain mice phosphorylase kinase is completely absent from the striated muscle. On the contrary, in humans the deficiency is only partial and extends to liver, muscle and blood cells. Kinetic studies show that the phosphorylase kinase of blood cells of deficient male patients has a normal maximal velocity but that the affinity for its high molecular weight substrate is markedly impaired, as indicated by an approximately 10-fold increase of the K_m for phosphorylase over the normal enzyme. The K_m for ATP is normal. This might be explained by a mutation in the structural gene for the kinase, but the possibility remains that the structural gene is normal and there is a defect in the activating system of the kinase since the kinase also occurs in an active and an inactive form. The inactive form of rabbit muscle phosphorylase kinase has an extremely high K_m for phosphorylase. Activation of phosphorylase kinase in leukocytes of patients with type VIa glycogen disease is presently under investigation.

1. Introduction

Phosphorylase kinase (ATP: phosphorylase phosphotransferase, EC 2.7.1. 38), the subject of this paper, is the enzyme responsible for the activation of phosphorylase (α-1,4-glucan: orthophosphate glucosyl-transferase, EC 2.4.2.1) (fig. 1) through phosphorylation of a specific serine hydroxyl group in the phosphorylase [1]. Phosphorylase kinase of rabbit muscle was extensively studied by Krebs et al. [2,3]. Although phosphorylase itself shows a clear tissue specificity (e.g., even the phosphorylases of striated and smooth muscle are genetically and biochemically different [4]), phosphorylase kinase lacks tissue specificity, e.g., phosphorylase kinase of liver will activate heart phosphorylase [5]. There is also no clearcut species specificity, e.g., human leucocyte kinase activates both rabbit muscle phosphorylase *b* and rabbit liver dephosphophosphorylase, while Hug et al. [6] reported that rabbit muscle phosphorylase kinase activates human liver dephosphophosphorylase.

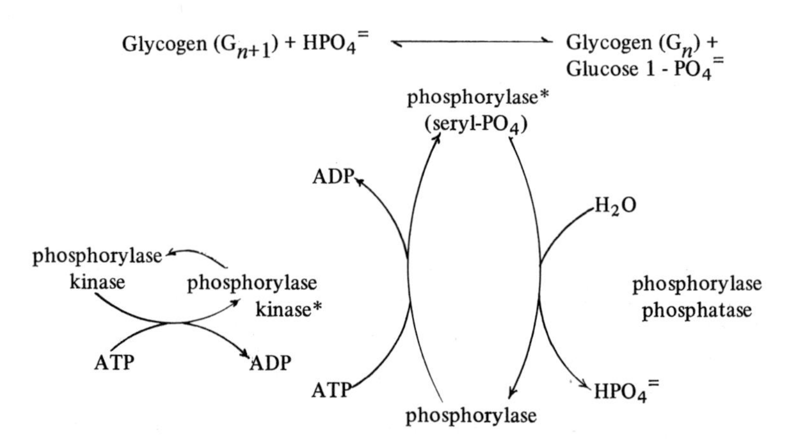

Fig. 1. Activation and inactivation of glycogen phosphorylase. An asterisk (*) indicates the active form of the enzyme.

Krebs et al. [2] showed that rabbit muscle phosphorylase kinase exists in two forms, an activated form which has activity at both pH 6.8 and 8.2 and a non-activated form which shows activity only at pH 8.2. The lack of activity at neutral pH 6.8 is ascribed to an extremely high K_m for phosphorylase for the non-activated kinase [2].

2. Phosphorylase kinase deficiency in man

Phosphorylase kinase deficiency is one of the seven well-defined glycogen storage diseases in man. These seven are:

Localized

Glucose 6-phosphatase deficiency (liver, kidney)	(Type I)
Striated muscle phosphorylase deficiency	(Type V)
Muscle phosphofructokinase deficiency	*

Generalized

Partial phosphorylase deficiency (liver, leucocytes and other tissues)	(Type VI)
Defects in the debranching enzyme system	(Type III)
Lysosomal α-glucosidase deficiency	(Type II)
Branching enzyme deficiency	(Type IV)

The first three deficiencies are localized and study of these diseases is hampered by the fact that one specific tissue must be obtained for study. In the last four diseases almost any tissue may be used and a number of investigators, including myself, have developed assays for the enzyme in blood cells (for a review see ref. 7). This allows easy repetition of enzyme assays without repeated biopsies, thus increasing reliability and also allowing genetic studies.

We have not only used leucocytes for diagnostic purposes but in a number of these diseases, the study of leucocyte enzymes has yielded important new data, not previously available from the study of other tissues.

Using leucocytes as an enzyme source we could show that the glycogen disease characterized by a partial phosphorylase deficiency had to be divided into two subtypes, i.e. type VIa where both phosphorylase and phosphorylase kinase were low [7,8], and type VIb where phosphorylase kinase was normal. The phosphorylase kinase deficiency will be further discussed in this paper. It is the mildest known glycogen storage disease and probably the

* No agreement exists on the number to be assigned to this type and we feel that it is not useful to introduce a new number for a disease when the enzymic defect is known.

most frequent. I estimate the minimum incidence at approximately 1:150,000 births. This figure was arrived at by dividing the number of patients below 10 years of age in the Netherlands by the total number of births during these 10 years in the Netherlands.

3. Phosphorylase kinase deficiency in mice

Phosphorylase kinase deficiency in mice was described by Lyon and Porter [9] after they had first observed that the muscle of the *I*-strain mice did not contain any phosphorylase *a*, while phosphorylase *b* was present in normal amounts [10].

Table 1
Comparison of phosphorylase kinase defiency.

Man	*Mice*
Sex-linked	Sex-linked
No symptoms or signs in adults, deficiency persists	No symptons or signs in adults, deficiency persists
Mild condition in children	High perinatal death rate, growth rate seems slower than normal
Phosphorylase kinase present but with high K_m for phosphorylase	Phosphorylase kinase present to small extent but easily precipitable
Affected tissues: muscle, liver, blood cells	Affected tissues: muscle, heart

It would, of course, be of great value if we could study a human disease in animals and our experiments are therefore aimed at finding the similarities and differences between the conditions in man and mice. Table 1 compares the effects and location of phosphorylase kinase deficiency in the two species.

4. Genetics

It has been clearly shown by Lyon, Porter and Robertson [11] that the genetic lesion in the *I*-strain mice is located on the X-chromosome. Matings between a kinase-deficient female mouse and a normal male mouse produced deficient male offspring and heterozygous female offspring as proven by enzyme assays and in backcrosses of the second (f_1) generation. Matings between a normal female mouse and a kinase-deficient male produced normal male offspring and all females of the second (f_1) generation were heterozygous. Our experience so far completely confirms Lyon's [11] experiments.

Huijing and Fernandes [12] were able to show that phosphorylase kinase deficiency in humans is also sex-linked in such a manner that the genetic lesion must be on the X-chromosome. This was shown by enzyme assays in leucocytes of members of five apparently unrelated pedigrees with two or more patients with this glycogen disease. The largest of these pedigrees comprises more than 200 subjects, among them 21 males in whom phosphorylase kinase assays showed a very low activity. A deficient male transmits the mutated gene to all his daughters, who are heterozygous. A heterozygous woman will transmit the mutated gene to half her sons who will show low phosphorylase kinase activity, while the other half of her sons will be normal. Half her daughters will be heterozygotes or carriers again. A much smaller pedigree with 3 deficient subjects and 2 heterozygous women studied recently by us is shown in fig. 2.

At present it would seem that all heterozygotes show asymptomatic liver enlargement during early childhood. Thus the condition can be described as a sex-linked, dominant one.

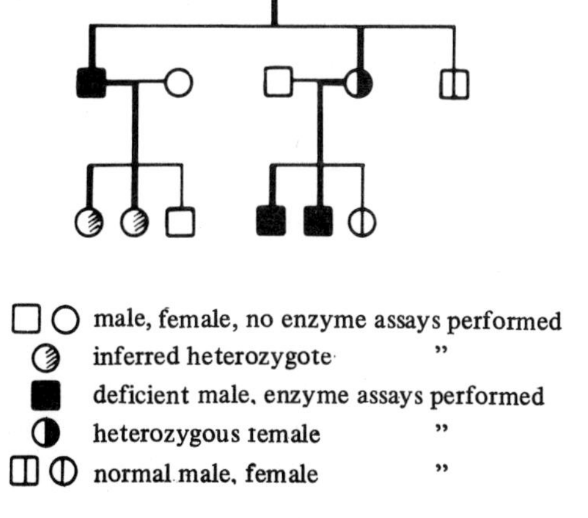

Fig. 2. Inheritance of low phosphorylase kinase activity.

5. Symptoms caused by the deficiency

No obvious symptoms or signs caused by a low phosphorylase kinase activity are seen in adult males. Six adult patients ranging from 21 to 73 years old were studied. Symptoms and signs observed in children disappear around puberty. This is also seen in one other glycogen disease, namely that caused by a deficiency in the debranching enzyme system [see e.g. ref. 13].

Similarly, there is no indication that adult mice of the *I*-strain or adult deficient hybrid mice are hampered in any way by their deficiency. In swimming tests the *I*-strain mice perform better than control mice with normal phosphorylase kinase activity [9].

However, *I*-strain mice are slow breeders. Due to a high incidence of perinatal death, litter size is small. Dr. Paul F. Fenton, who originally bred the *I*-strain mice, has the same experience (personal communication). Table 2 shows that the average litter size in seven matings was less than three whereas hybrid litters and control litters were substantially larger. It is our experience that young *I*-strain mice and offspring of a cross between a *I*-female and a C57 Black (B1) male grow slower than inbred C57 B1 control mice. This may be due to insufficient lactation of the *I*-female.

Children with a low phosphorylase kinase activity do show symptoms caused by their deficiency; these are listed in table 3.

Table 2
Litter size of inbred mouse strains or hybrids*.

Parents genotype	C57 B1 X C57 B1 XX XY	I X I xx xY	I X C57B1 xx XY	C57B1 X I XX xY
Litter genotype	XX,XY	xx,xY	xX,xY	xX,XY
Litter size	8	3		
	9	4	8	8
		1	5	9
		3	7	1
		4		
		2		
		0		

Y = Y-chromosome, X = normal X-chromosome, x = mutant X-chromosome, carrying "phosphorylase kinase deficiency".

* The figures indicate the number of young mice surviving 48 hours after birth. In several cases of litters of the I X I type it was observed that as many as 7 or 8 young were actually born but the early mortality rate was high.

Table 3
Symptoms, signs and laboratory findings in children with a low phosphorylase kinase activity.

Asymptomatic liver enlargement
Mild muscular weakness
Mild growth retardation, when untreated
Increased plasma transaminase activity, other liver function tests normal
Increased triglycerides and glycerol in blood
Very mild fasting hypoglycemia
Increased blood-lactate concentration after carbohydrate administration (persists in adults)

6. Phosphorylase kinase activity in leucocytes of patients with type VIa glycogen disease

30 male patients with type VIa glycogen disease showed an average of 14% of the normal mean activity of leucocyte phosphorylase kinase. Heterozygotes have 59% of the normal mean activity. There are two obvious explanations for the residual activity: either in the patients 14% of the normal amount of the normal enzyme is made or a normal amount of enzyme is present but each enzyme molecule shows 14% of the normal activity. Our findings, given in table 4, show that the latter explanation is the right one, for the maximal velocity, expressed as units of phosphorylase activated per mg of leucocyte protein is virtually identical in normal controls, patients and heterozygotes. K_m values for phosphorylase are at least 8 X those of comparable controls and on the average are 19 X those found in controls. In comparing these values it should be noted that the leucocytes used in these experiments were isolated several thousand miles from our laboratory and were frozen and shipped in dry ice. In frozen control leucocytes handled this way the K_m of phosphorylase kinase for phosphorylase is more variable and somewhat higher than in freshly isolated leucocytes. It is therefore best to compare values within one experiment.

The high K_m could be due to a structural change in the phosphorylase kinase itself, but it is also possible that it is due to a defect in the activation of the non-activated form of the kinase or to more rapid inactivation. The non-activated form of phosphorylase kinase can be converted into the active form by a cyclic AMP-dependent "protein kinase" [14] or by a calcium-dependent

Table 4
Kinetic properties of phosphorylase kinase in human leucocytes.

Expt.	Subject	Sex	K_m for phosphorylase (U/ml)	V_{max} (U/min/mg protein)
Normal subjects				
1	N.B.	male	62	1.6
2	F.O.	male	135	2.0
3	P.H.	male	130	1.5
3	C.R.	female	185	1.0
4	C.U.	male	62	0.9
4	A.Bn.	female	50	1.1
Patients with glycogen storage disease type VIa*				
1	L.B.	male	2000	3.0
2	T.O.	male	1667	3.0
3	H.Ke.	male	1429	2.5
3	G.K.	male	5000	2.4
4	H.H.	male	1000	2.5
4	A.Bn.	male	430	1.5
Heterozygotes				
2	M.O.	female	500	1.7
3	A.Ke.	female	830	1.7
3	I.K.	female	500	1.9
4	C.Bn.	female	182	2.0
4	M.H.	female	167	0.8
5	F.M.	female	480	1.2

* These subjects showed a low leucocyte phosphorylase activity in the absence of AMP and a low leucocyte phosphorylase kinase activity when assayed under conditions as described before [7,8].

protein [15]. An inactivating phosphatase for the active kinase is also known [16].

The "protein kinase" phosphorylates other proteins than phosphorylase kinase, e.g., casein [13] or histone [17]. Preliminary experiments using histone and [$\gamma - {}^{33}$P] ATP as substrates suggest that protein kinase in leucocytes of patients with a low phosphorylase kinase activity is normal. However these experiments will have to be repeated with non-activated phosphorylase kinase as a substrate before any conclusions can be drawn.

Also the calcium-dependent activation of phosphorylase kinase and the inactivation of activated phosphorylase kinase in tissues of patients require investigation.

7. Phosphorylase kinase activity in *I*-strain mice

Lyon and Porter [9] could not find any activity of phosphorylase kinase in muscle of the *I*-strain mice. However, we have found phosphorylase kinase activity in muscle of the *I*-strain mice. Compared with muscle of a control mouse this activity was 10-20% of normal. This kinase must be inoperative since no significant activity of phosphorylase *a* was detectable in the muscle of the *I*-strain mouse. In the control mice (C57 Black) 54-89% of the phosphorylase was in the *a* form.

There are two main differences between the experiments of Lyon and Porter [9] and ours. Lyon tried to prevent activation of phosphorylase (and the kinase) by freezing the muscles rapidly between blocks of dry ice and by other precautions. We tried to get as much activation of both enzymes as possible. The tissues were not frozen or handled with extreme speed. However this does not seem a likely explanation for our finding of phosphorylase kinase since Lyon's attempt to activate the kinase in muscle extracts was not successful. The possibility remains that in the intact muscle, kept at room temperature for a short time for weighing, etc., activation did occur.

A second difference is the fact that we assay the kinase activity in a whole 1% homogenate, whereas Lyon and Porter [9,10] use a 10,000 × *g* supernatant of a 2% homogenate. It seemed unlikely that centrifuging would be the cause of the difference since in control mice the bulk of the phosphorylase kinase stays in the 10,000 × *g* supernatant. However, when a 1% homogenate of muscle of a deficient mouse was centrifuged at 10,000 × *g* all the activity in the homogenate could be accounted for in the resuspended 10,000 × *g* precipitate, and none was in the supernatant. This interesting observation, together with the kinetic parameters of the phosphorylase kinase in the *I*-strain mice, are the subject of further investigation.

At present it is difficult to speculate to what degree the conditions in man and mouse are similar and whether the mouse will be a suitable model for study of the human disease characterised by a low phosphorylase kinase activity.

Acknowledgements

This work was supported by NIH grant AM 13359. The author wishes to express sincere thanks to Dr. Harold A. Hofmann, National Cancer Institute, Bethesda, Maryland, for generously providing the mice of the I/Hf strain. Expert technical assistance by Miss Dity Wolter is also acknowledged.

References

[1] E.H. Fischer, D.J. Graves, E.R.S. Crittenden and E.G. Krebs, J. Biol. Chem. 234 (1959) 1698.
[2] E.G.Krebs, D.S. Love, G.A. Bratvold, K.A.Trayser, W.J. Meyer and E.H. Fischer, Biochemistry 3 (1964) 1022.
[3] R.J. Delange, R.G. Kemp, W.D. Riley, R.A. Cooper and E.G. Krebs, J. Biol. Chem. 243 (1968) 2200.
[4] E. Bueding, N. Kent and J. Fischer, J. Biol. Chem. 239 (1961) 2099.
[5] T.W. Rall, W.D. Wosilait and E.W. Sutherland, Biochim. Biophys. Acta 20 (1956) 69.
[6] G. Hug, W.K. Schubert and G. Chuck, J. Clin. Invest. 48 (1969) 704.
[7] F. Huijing, in: Control of Glycogen Metabolism, ed. W.J. Whelan (Academic Press, London, 1968) p. 115.
[8] F. Huijing, Biochim. Biophys. Acta 148 (1967) 601.
[9] J.B. Lyon Jr. and J. Porter, J. Biol. Chem. 238 (1963) 1.
[10] J.B. Lyon Jr. and J. Porter, Biochim. Biophys. Acta. 58 (1963) 248.
[11] J.B. Lyon Jr., J. Porter and M. Robertson, Science 155 (1967) 1550.
[12] F. Huijing and J. Fernandes, Am. J. Human Genet. 21 (1969) 275.
[13] S. van Creveld and F. Huijing, Metabolism 13 (1964) 191.
[14] D.A. Walsh, J.P. Perkins and E.G. Krebs, J. Biol. Chem. 243 (1968) 3763.
[15] W.L. Meyers, E.H. Fischer and E.G. Krebs, Biochemistry 3 (1964) 1033.
[16] W.D. Riley, R.J. DeLange, G.E.Bratvold and E.G. Krebs, J. Biol. Chem. 243 (1968) 2209.
[17] T.A. Langan, Science 162 (1968) 579.

ACTIVE SITES OF CARBONIC ANHYDRASES

Philip L. WHITNEY
Department of Biochemistry, University of Miami School of Medicine, Miami, Florida 33152, USA

Abstract: Whitney, P.L. Active Sites of Carbonic Anhydrases. *Miami Winter Symposia*, 1, pp. 233–245. North-Holland Publishing Company, Amsterdam, 1970.

Carbonic anhydrases from several sources have many common properties. These include: a molecular weight of about 30,000, one molecule of zinc which is essential for activity, the ability to catalyze the hydrolysis of esters as well as the hydration of carbon dioxide, and inhibition by aromatic sulfonamides.

A common finding with carbonic anhydrases from erythrocytes is the presence of several isozymes. There are three major isozymes in human erythrocytes, and these have been designated A, B, and C. Although these isozymes are very similar, enzymes B and C have quantitive differences in activity and inhibition.

Human enzymes B and C are also different with respect to inhibition and reaction with some reactive inhibitors. Chloroacetyl chlorothiazide, bromoacetate, and iodoacetamide are all reversible inhibitors when activity is measured soon after adding the reagents. After longer times with the B'enzyme, these reagents proceed to react with histidines in the active site region. No such reaction occurs with C enzyme. On the other hand bromoacetate and iodoacetamide inhibit and react with enzyme A in the same way as with enzyme B.

A detailed study has been made of inhibition and reaction of bromoacetate and iodoacetamide with enzyme B. These reagents react at the 3′ nitrogen of a histidyl residue with the loss of most, but not all, of the enzymatic activity. Since the reacted enzyme retains some activity, this suggests that this histidyl residue may not be essential for enzymatic activity, even though it appears to be in the active site region.

The presence of residual activity after reaction of bromoacetate with enzyme B presents an opportunity to study the effects of a specific active site-perturbing group. Introduction of the carboxymethyl group results in significant changes in activity and inhibition. It also causes changes in spectral properties of the cobaltous enzyme, in which zinc has been replaced by a cobaltous ion. These results are of interest for correlating enzyme structure and function.

This paper reviews some functions and common properties of carbonic anhydrases, and points out some differences in isozymes of carbonic anhydrase from human erythrocytes.

Chemical modification studies employing reagents with specificity for the active site of carbonic anhydrases have shown that isozymes A and B are alike and C is quite different. Some further studies of inhibition, reaction and products of reaction of bromoacetate and iodoacetamide with enzyme B are also described. Modification by these reagents results in the introduction of a specific perturbing group in or near the active center.

1. Function and properties of carbonic anhydrases

Since the discovery of carbonic anhydrase (carbonate hydro-lyase EC 4.2.1.1) by Meldrum and Roughton in 1933 [1], many studies have been done to explore its distribution and function in nature. It has been found in many places where CO_2, HCO_3^-, CO_3^{2-}, and H^+ are involved. In many cases its activity also affects the distribution of other ions and osmotically required water. A few representative cases are given in table 1. The function has been found by studying the effects of specific inhibitors of carbonic anhydrase.

The nature of proteins with carbonic anhydrase activity has also been the subject of many studies since Keilin and Mann isolated carbonic anhydrase from bovine erythrocytes in 1940 [3]. They discovered that it was a metallo-enzyme, requiring zinc for activity, and that aromatic sulfonamides are powerful inhibitors of enzymic activity. Although the only known natural function of carbonic anhydrase is catalysis of the hydration of carbon dioxide and dehydration of bicarbonate, recent studies on the catalytic potential of the enzyme have shown that it will also catalyze the hydration of aldehydes [4] and the hydrolysis of esters [5–7], carbobenzoxy chloride [8], sulfonyl chlorides [9], fluorodinitrobenzene [10], and sultones [11].

It is of interest to see how many of these properties are found in carbonic anhydrases from different sources. Table 2 shows that each of the four given properties, are common to many carbonic anhydrases, but none is common

Table 1
Some functions of carbonic anhydrase [see ref. 2].

Location	Function
Erythrocyte	Exchange of CO_2
Kidney	Excrete H^+ instead of HCO_3^-, K^+, Na^+, H_2O
Stomach	Excrete H^+ from gastric mucosa
Eye	Excrete HCO_3^-, H_2O in aqueous humor
Nervous system	Excrete H^+, Cl^- in cerebrospinal fluid
Hen oviduct, shell fish	Excrete HCO_3^- for CO_3^{2-} source in shell formation
Gull salt gland	Excrete 0.8 M NaCl

Table 2
Some properties common to many carbonic anhydrases.

	M.W.	Essential zinc	Sulfonamide inhibition	Esterase activity
Human [12]	30,000	+	+	+
Monkey [12]	30,000	+	+	+
Bovine [3,14]	30,000	+	+	+
Horse [15]	30,000	+	+	+
Sepia officinalis [16] (Cuttlefish)	(<100,000)[a]	+	+	(–)[b]
Heliothis zea [17] (Corn ear worm)	(<100,000)[a]	(+)[c]	+	
Neiceria sicca [18]	30,000	+	+	Slow
Parsley [19]	6 × 30,000	+	(±)[d]	–
Spinach [20]	140,000	–	–	

[a]Estimates from gel filtration on Sephadex. The results are tentative since these enzymes are naturally membrane-bound and difficult to solubilize.

[b]The possibility of some slow esterase activity was not ruled out.

[c]The presence of metal was inferred from inhibition by low concentrations of metal poisons and sulfonamides.

[d]A crude assay of CO_2 hydrating activity showed weak inhibition by acetazolamide, but no inhibition was observed in assays with a stopped-flow apparatus.

to all of the enzymes. Despite the lack of universality of these properties, they are important characteristics of many proteins with carbonic anhydrase activity and may be useful for classifying enzymes from different sources.

A common finding with carbonic anhydrases from erythrocytes is the presence of several isozymes. Each isozyme has all the properties listed in table 2 for mammalian enzymes, but each has some unique properties. Three major isozymes, A,B and C, have been found in human erythrocytes. A few of their properties are given in table 3. Enzyme A appears to be identical with form B except for a difference in molecular charge which permits their separation. More careful studies have revealed nearly a dozen different isozymes, each of which is very similar to enzyme B or C [22].

Table 3
Isozymes of human erythrocyte carbonic anhydrases.

		Enzyme A	Enzyme B	Enzyme C
Relative amount [12]		1	10	2
k_{cat} (min^{-1}):	CO_2 hydrating, pH 8.8 [21]	a	12,800,000	82,500,000
	p-Nitrophenyl acetate, pH 9.0 [5]	a	130	3,300
	o-Nitrophenyl acetate, pH 9.0 [5]		270	100
K_i:	Acetazolamide [5] (μM)		0.35	0.01
	Chloride, pH 7.6 (mM)		15	500

[a]There are no good data of k_{cat} for enzyme A, but the activity of the isozyme is the same as enzyme B.
k_{cat} is maximal velocity divided by the concentration of enzymes.

2. Active site specific chemical modifications of human isozymes

Differences between human erythrocyte enzymes B and C are readily apparent from results of chemical modification studies. In these studies, reversible inhibitors were chosen or made to carry a reactive group to the active site. While bound as reversible inhibitors they could then proceed to react with amino acid side chains in or near their binding site:

$$E + I \rightleftharpoons EI \rightarrow EI^*$$

$$K_i = \frac{[E]\ [I]}{[EI]} \text{ and } t'_{1/2} = R t_{1/2}, \text{ where } R = \frac{[EI]}{[E] + [EI]}$$

$t_{1/2}$ is the half-time of reaction observed at a given concentration of inhibitor. The value of R, the degree of saturation of the enzyme with inhibitor, can be calculated from K_i. $t'_{1/2}$ is the half-time of reaction of the inhibitor with the enzyme to form EI* when the enzyme is saturated with inhibitor (R =1). Experiments of this type will first be described for enzyme B [9,23] and these results will then be compared with results for enzymes A and C.

At pH 7.6, bromoacetate inhibits enzyme B with a K_i of 3.8 mM. It then proceeds to EI* by reacting at the 3′ nitrogen of a histidyl side chain with

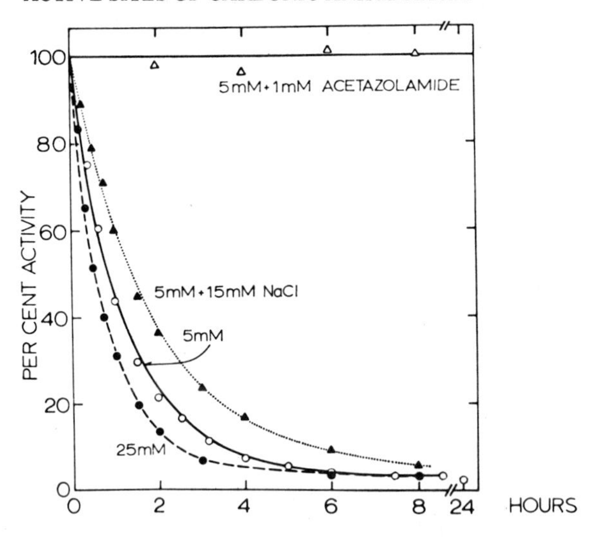

Fig. 1. Reaction of bromoacetate with enzyme B at pH 7.6 with loss of esterase activity [24, Reprinted with permission of The Journal of Biological Chemistry].

nearly complete inactivation of the enzyme as shown in fig. 1. Confirmation of this mechanism of reaction is given in table 4. The rate of inactivation is not strictly proportional to the concentration of inhibitor, but is proportional to R, as shown by the constant value of $t'_{1/2}$. Another inhibitor, chloride, competes with the binding of bromoacetate and in this way protects the enzyme

Table 4
Inactivation of carbonic anhydrase B by bromoacetate at pH 7.6 [24].

Enzyme	Bromoacetate (mM)	$t_{1/2}$ (hr)	R	$t_{1/2}$ (hr)
Native	5	0.87	0.56	0.49
	25	0.54	0.87	0.47
	5 + 15 mM NaCl	1.33	0.39	0.52
Apo-	20	27		
Co (II)	0.050	1.5	0.20	0.3

The rate of inactivation was followed by the decrease in esterase activity (The ester used in all these experiments was *p*-nitrophenyl acetate). Apoenzyme was assayed after addition of Zn (II). Enzyme concentration was 0.4 mM for native and apoenzyme and 0.035 mM for Co (II) enzyme. $t_{1/2}$ was the observed half-time of the reaction. R was calculated from K_i values of 3.8 mM for bromoacetate with the native enzyme and 0.17 mM for Co (II) enzyme. The K_i for NaCl with the native enzyme was 15 mM.

Table 5
Comparison of the effects of reactive inhibitors on human carbonic anhydrases at pH 7.6.

		Enzyme A	Enzyme B	Enzyme C
Bromoacetate:	K_i (mM)	4.4	3.8	70
	$t'_{1/2}$ (hr)	0.5	0.5	–
Iodoacetamide:	K_i (mM)	24	26	1000
	$t'_{1/2}$ (hr)	2.4	2.2	–
Chloroacetyl:	K_i (mM)		0.5	<0.5
Chlorothiazide:	$t'_{1/2}$ (hr)		6.5	–

See table 4 for experimental conditions.

from reaction with bromoacetate. Removal of zinc from the enzyme prevents the specific reaction by bromoacetate, probably because the metal is required for binding the reagent. Replacing Zn (II) with Co (II) gives an active enzyme [23] with which bromoacetate readily binds and reacts.

Other inhibitors, iodoacetate [25] iodoacetamide [24] and chloroacetyl chlorothiazide [9] also react with the B enzyme. Iodoacetamide and iodoacetate react with the same histidine as bromoacetate [25]. However, the chloroacetyl sulfonamide reacts with a second histidine which is far removed from the first histidine in terms of its position in the primary sequence of the enzyme [26], but which is probably brought quite close by the folding of the peptide chain into the native structure.

Some of these reagents have also been tried with enzymes A and C, and these results are given in table 5. The behavior of enzyme A is identical with enzyme B. However, the K_i values for enzyme C are markedly different. In addition, no specific reaction to form EI* was observed. Thus, the chemical modification studies point out marked differences between the active sites of enzymes B and C.

3. Bromoacetate and iodoacetamide with enzyme B

The rest of the paper will be devoted to a description of some of the properties of inhibition and reaction of bromoacetate and iodoacetamide with enzyme B. Fig. 2 shows the pH dependence of inhibition by these reagents. Iodoacetamide inhibits at high pH, whereas bromoacetate becomes an extremely weak inhibitor in this region, a property which is typical of anionic inhibitors of carbonic anhydrase [6,27]. The K_i for iodoacetamide increases at

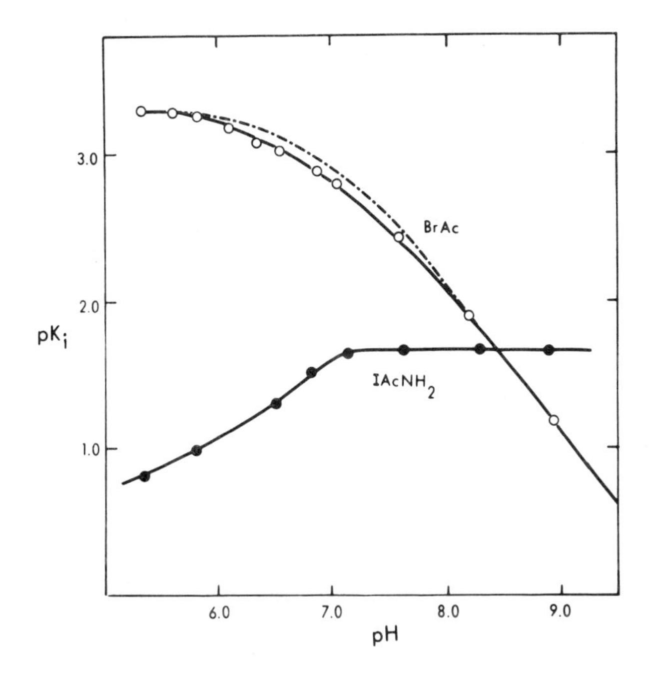

Fig. 2. Inhibition of esterase activity of enzyme B by bromoacetate and iodoacetamide in the pH region 5 to 9. The dashed line is a curve calculated for one p*K* of 6.8. The solid line through the points for bromoacetate ($BrAc^-$) was calculated with p*K* values of 7.3 and 6.4 and the K_i equal to 1.54 mM with the first group in its acidic form and 0.47 mM with both groups in their acidic forms. No theoretical curve was drawn for the results with iodoacetamide ($IAcNH_2$).

low pH (pK_i decreases) in contrast to the decrease in K_i for bromoacetate. The pH dependence of the K_i for bromoacetate does not fit the theoretical curve for one p*K* (the dashed line) but fits a theoretical curve (solid line) drawn assuming that the protonation of two groups on the enzyme promote the binding of inhibitor. The p*K* values of these two groups are 7.3 and 6.4.

The pH dependence of reaction with these two inhibitors is shown in fig. 3. The reaction with bromoacetate depends on a group with a p*K* of about 5.6. This is probably the p*K* of the reactive histidine in the presence of reversibly bound bromoacetate. This compares with a p*K* of 5.8 for the reaction with iodoacetate [25]. From this data it is not possible to estimate the p*K* of this histidine in the absence of inhibitors.

Reaction of iodoacetamide with the enzyme is essentially pH independent over this pH region. Since reaction only occurs with the unprotonated form

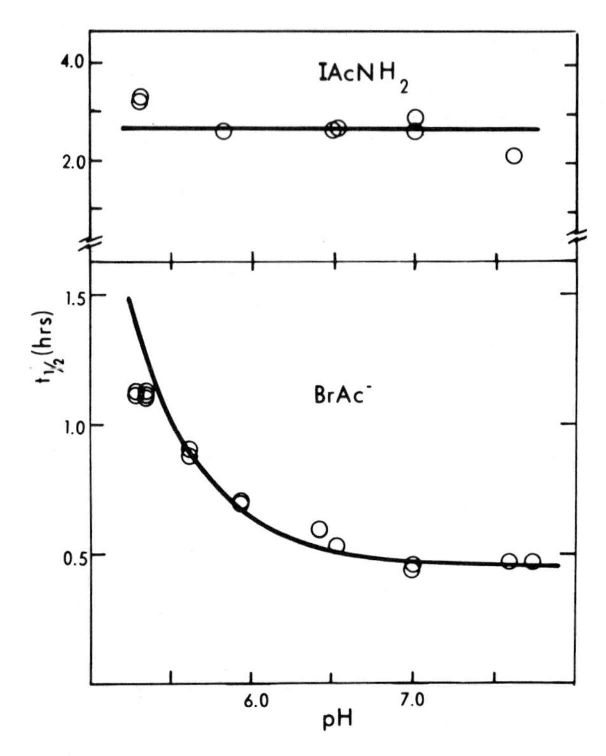

Fig. 3. Reaction of bromoacetate and iodoacetamide with enzyme B in pH region 5 to 8. The curve drawn through the points for bromoacetate (BrAc) is a theoretical curve calculated for a pK of 5.6.

of the imidazole side chain of histidine, this group must be in the unprotonated form throughout this pH region when iodoacetamide is bound. The increase K_i for this reagent at low pH would then be the result of competition with protons for binding at this imidazole group. The pK in the absence of inhibitor has not been determined due to other complicating factors.

4. Properties of carboxymethyl enzyme B

It is now of interest to look at some of the properties of EI* obtained by reaction with bromoacetate, since the added carboxymethyl group is a specific perturbant of the active site. One important property is that the modified

Table 6
Inhibition of esterase activity of native and carboxymethyl (CM) carbonic anhydrase B at pH 7.6.

	Native enzyme K_i (mM)	CM enzyme K_i (mM)
Sulfanilamide	0.0094	3.0
Acetazolamide	0.0003	0.0017
Chloride	27	140
Bromoacetate	4.4	19
Iodoacetamide	25	27

The ionic strength of the assay buffer was 75 mM, except with chloride, where the ionic strength was 400 mM.

enzyme still retains some residual activity which is not due to the presence of traces of unmodified enzyme. The most important evidence supporting this statement is that Bradbury showed that the carboxymethyl enzyme can be separated from the native enzyme by chromatography on IRC–50; the purified product still retains residual activity [25]. Secondly, several inhibitors have different K_i values for the modified enzyme, as shown in table 6. The effect of the carboxymethyl group ranges from a 300 fold increase in K_i for sulfanilamide to no effect on the K_i for iodoacetamide. It is interesting that

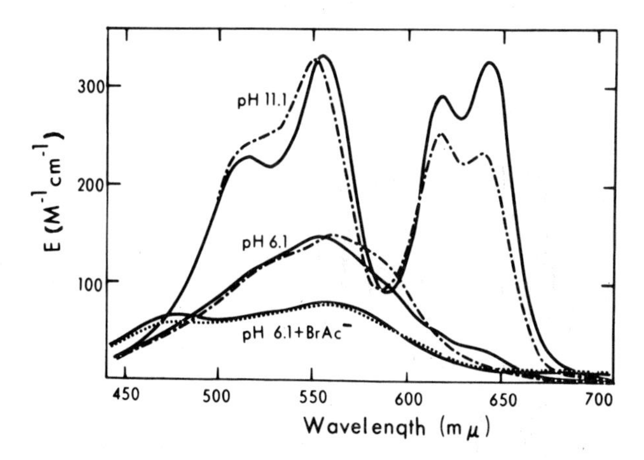

Fig. 4. Visible spectra of native (solid lines) and carboxymethyl (broken lines) Co (II) enzyme B at pH 6.1, 11.1, and at pH 6.1 with 70 mM bromoacetate (BrAc). The concentration of enzyme was 0.30–0.45 mM.

the residual activity of the modified enzyme is inhibited by bromoacetate. The activity of the carboxymethyl enzyme was titrated with acetazolamide, and the results are consistent with the conclusion that every molecule of modified enzyme is active and has the same K_i for this inhibitor [28]. The presence of residual activity suggests that the histidine which is modified by bromoacetate may not be essential for activity.

When Zn (II) is replaced by Co (II) in carbonic anhydrase, the enzyme has an absorption spectrum in the visible wavelength region which changes with pH and the addition of inhibitors [23,27,29]. The effect of adding the carboxymethyl group is shown in fig. 4, where the spectra of Co (II) enzyme and carboxymethyl Co (II) enzyme are given at pH 6.1 and 11.1, and at pH 6.1 in the presence of bromoacetate. The carboxymethyl group produces some changes in the spectrum at low pH and marked changes at high pH, especially around 530 and 640 nm. It appears that the binding of bromoacetate is very similar in both enzymes. These observations, plus the facts that the carboxymethyl enzyme has residual activity and bromoacetate inhibits this

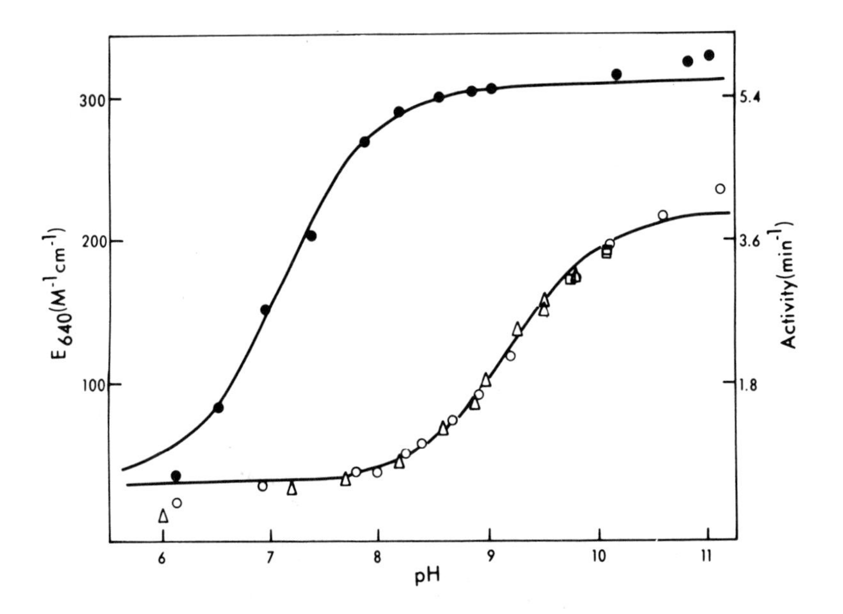

Fig. 5. pH-Dependence of absorptivity of native (solid circles) and carboxymethyl (open circles) Co (II) enzyme B at 640 nM and pH-dependence of esterase activity of carboxymethyl (Co (II) enzyme B (open triangles and squares). The curves were calculated for pK values of 7.1 for the native enzyme and 9.2 for the modified enzymes.

residual activity, suggest that bromoacetate is pulled away from its inhibitory binding site as it reacts to form EI*.

Fig. 5 shows the pH-dependence of absorption at 640 nm. The curves are calculated for p*K* values of 7.1 for the unreacted enzyme and 9.2 for the carboxymethyl enzyme. Previous results on the pH-dependence of esterase activity for the native enzyme gave a p*K* of about 7.3, which may not be significantly different from 7.1 [5,24]. The pH dependence of esterase activity for the carboxymethyl Co (II) enzyme is shown to follow the curve calculated for a p*K* of 9.2. However, even at high pH the activity for the carboxymethyl enzyme is only about one-fifth of the activity for the native enzyme. This diminished activity is reflected in the differences in the spectra of the cobalt enzymes at high pH.

It appears that the pH-dependence of esterase activity and spectra are affected alike. The increase in p*K* from 7.1 to 9.2 is probably due in large part to the perturbing effect of the negative charge of the carboxymethyl group. Although the group (or groups) responsible for this pH-dependence is not known, it may be a water molecule bound to the metal. As the pH increases, it dissociates a proton to become a metal-bound hydroxide group [29].

Comparison of the effects of the carboxymethyl group with different perturbing groups, such as the carboxamidomethyl group, is hoped to provide more information on the nature of the active site of this enzyme. The true worth of these results will become evident only upon correlation with the 3-dimensional structures calculated from X-ray diffraction data. The low resolution structure of the C enzyme is already known [30] and work on enzyme B by the same group has been progressing quite well. It seems reasonable to expect that in the near future we will be able to fit our clues together into a meaningful picture of the active site and its functions.

Acknowledgements

I wish to thank Professor John T. Edsall and Bo. G. Malmström for support and direction in these studies. I also wish to thank Dr. Guido Guidotti for many helpful suggestions and Drs. Per Olof Nyman and Sven Lindskog for numerous communications of information and ideas. I especially appreciate support for this work by Professor Frank R.N. Gurd while he sponsored me as a research associate. This work was supported by grants from the United States Public Health Service (HE-03169 to Professor Edsall; HE-05556 and GM-06308 to Professor Gurd) and the National Science Foundation (GB-1255 to Professor Edsall). I received support from a post-doctoral fellowship from the United States Public Health Service (GM-19189) and from an institutional grant for science to the University of Miami from the National Science Foundation.

References

[1] N.U. Meldrum and F.J.W. Roughton, J. Physiol. (London) 80 (1933) 113.
[2] T.H. Maren, Physiol. Rev. 47 (1967) 595.
[3] D. Keilin and T. Mann, Biochem. J. 34 (1940) 1163.
[4] Y. Pocker and J.E. Meany, Biochemistry 4 (1965) 2535;
Y. Pocker and J.E. Meany, Biochemistry 6 (1967) 239.
[5] J.A. Verpoorte, S. Mehta and J.T. Edsall, J. Biol. Chem. 242 (1967) 4221.
[6] A. Thorslund and S. Lindskog, European J. Biochem. 3 (1967) 117.
[7] Y. Pocker and D.R. Storm, Biochemistry 7 (1968) 1202.
[8] B.G. Malmström, P.O. Nyman, B. Strandberg and B. Tilander, in: Structure and Activity of Enzymes, eds. T.W. Goodwin, J.I. Harris and B.S. Hartley (Academic Press, London, 1964) p. 121.
[9] P.L. Whitney, G. Fölsch, P.O. Nyman and B.G. Malmström, J. Biol. Chem. 242 (1967) 4206.
[10] P. Henkart, G. Guidotti and J.T. Edsall, J. Biol. Chem. 243 (1968) 2447.
[11] E.T. Kaiser and K.-W. Lo, J. Am. Chem. Soc. 91 (1969) 4912.
[12] P.O. Nyman, Biochim. Biophys. Acta 52 (1961) 1.
[13] T.A. Duff and J.E. Coleman, Biochemistry 5 (1966) 2009.
[14] S. Lindskog, Biochim. Biophys. Acta 39 (1960) 218.
[15] A.J. Furth, J. Biol. Chem. 243 (1968) 4832.
[16] A.D.F. Addink, Thesis, Carbonic Anhydrase of *Sepia officinalis* (State University of Utrecht, 1968).
[17] P.L. Whitney, in: CO_2: Chemical, Biochemical and Physiological Aspects, eds. R.E. Forster, J.T. Edsall, A.B. Otis and F.J.W. Roughton (National Aeronautics and Space Administration, Washington D.C., 1969) p. 176.
[18] P.O. Nyman, Personal communication.
[19] A.J. Tobin, Ph. D. Thesis, Parsley Carbonic Anhydrase: Purification and Properties (Harvard University, 1969).
[20] C. Rossi, A. Chersi and M. Cortino, in: CO_2: Chemical, Biochemical and Physiological Aspects, eds. R.F. Forster, J.T. Edsall, A.B. Otis and F.J.W. Roughton (National Aeronautics and Space Administration, Washington D.C., 1969) p. 131.
[21] R.G. Khalifah and J.T. Edsall, Paper presented at the 3rd International Biophysics Congress held in Cambridge, Mass. Aug-Sept, 1969.
[22] S. Funakoshi and H.F. Deutsch, J. Biol. Chem. 244 (1969) 3438.
[23] S. Lindskog and P.O. Nyman, Biochim. Biophys. Acta 85 (1964) 462.
[24] P.L. Whitney, P.O. Nyman and B.G. Malmström, J. Biol. Chem. 242 (1967) 4212.
[25] S.L. Bradbury, J. Biol. Chem. 244 (1969) 2002, 2010.
[26] B. Andersson, P.O. Nyman and L. Strid, in: CO_2: Chemical, Biochemical and Physiological Aspects, eds. R.E. Forster, J.T. Edsall, A.B. Otis and F.J.W. Roughton (National Aeronautics and Space Administration, Washington D.C., 1969) p. 109.

[27] S. Lindskog, Biochemistry 5 (1966) 2641.
[28] P.L. Whitney, European J. Biochem, in press.
[29] J.E. Coleman, Biochemistry 4 (1965) 2644.
[30] K. Fridborg, K.K. Kannan, A. Liljas, J. Lundin, B. Strandberg, R. Strandberg, B. Tilander and G. Wiren, J. Mol. Biol. 25 (1967) 505.

COMPARATIVE STUDIES ON MUSCLE GLYCOGEN PHOSPHORYLASE OF SHARK AND MAN: SUBUNIT STRUCTURE, KINETIC AND IMMUNOLOGICAL PROPERTIES*

A.A. YUNIS and S.A. ASSAF

Departments of Medicine and Biochemistry, and The Howard Hughes Medical Institute, University of Miami School of Medicine, Miami, Florida 33136, USA

Abstract: Yunis, A.A. and Assaf, S.A. Comparative Studies on Muscle Glycogen Phosphorylase of Shark and Man: Subunit Structure, Kinetic and Immunological Properties. *Miami Winter Symposia* 1, pp. 246–256. North-Holland Publishing Company, Amsterdam, 1970.

Kinetic, structural and immunological studies on glycogen phosphorylase from homoiothermic and poikilothermic animals have been undertaken to better understand the role of this enzyme in glycogen metabolism. Phosphorylase has been isolated from shark muscle in crystalline, homogeneous form by ammonium sulfate fractionation and chromatography on DEAE-cellulose. Human muscle phosphorylase *b* was crystallized in the presence of AMP and Mg^{2+}. Shark phosphorylase crystallized readily in tris-EDTA buffer at pH 6.8–7.5 or in 30% ammonium sulfate. Phosphorylase *b* from both human and shark muscle underwent association in the presence of AMP and Mg^{2+} as known for rabbit muscle phosphorylase *b*. However, unlike muscle phosphorylase of rabbit, conversion of shark phosphorylase $b \rightarrow a$ was not accompanied by a doubling in molecular weight when similar protein concentrations were used in the sucrose density gradient centrifugation. Human phosphorylase *a* when freed of AMP appeared to be a slowly equilibrating mixture of dimeric and tetrameric forms as determined by sucrose density gradient centrifugation and sedimentation velocity experiments.

Kinetic studies in the direction of glycogen synthesis and glycogen degradation were performed and differences in kinetic behavior at cold temperature noted. Antisera prepared in rabbits against human and shark muscle phosphorylase exhibited a marked degree of specificity with very weak cross reactivity as determined by antigen-antibody precipitation in agar and by inhibition of enzymatic activity. Differences and similarities between muscle phosphorylase of man and of shark are discussed in the light of the extensive data available on phosphorylase from other sources.

* This work was supported by Public Health Service Grants AM 09001-06, AM 05472-05 and AM 13087-02.

Muscle glycogen phosphorylase [ph] has been isolated and crystallized from various animal species [1–6]. It exists in two forms: an unphosphorylated form called ph *b* and a phosphorylated form or ph *a*. More recently, interest has centered on the subunit structure of this enzyme (table 1).

The *a* form of most species studied thus far is a tetramer with a M.W. of 360,000. Lobster muscle ph *a*, however, remains a dimer even at cold temperature and high enzyme concentrations as both the ph *b* and *a* forms have the same sedimentation coefficient of 8.5 S [5,7]. Frog muscle ph *a* was found to sediment in a single peak with an $S_{20,w}$ of 11.2 S suggesting an equilibrium between dimeric and tetrameric species [8]. Dissociation of the frog enzyme to a dimeric form with an $S_{20,w}$ of 8.1 *S* was favored at low protein concentration and warm temperature. In an earlier study, Wang and Graves [9] observed dissociation of rabbit ph *a* but at a much lower concentration of the enzyme indicating that ph *a* from rabbit muscle favors association more than the frog enzyme. In a further study on rabbit ph *a* dissociation, it was reported that phosphorylase *a* dissociates to its more active dimeric form upon preincubation with glucose or glycogen [10,11]. Similarly frog ph *a* was also found to dissociate, but at a more rapid rate than rabbit ph *a* [8].

In the present study we have concentrated on the subunit structure and kinetic behavior of muscle ph from human (a homoiothermic species) and shark (a poikilothermic species), with special emphasis on the effect of temperature. Although human muscle ph has been previously purified and partially characterized [3,12–14], its subunit structure and kinetic behavior have not been investigated in detail.

Four times crystallized human muscle ph *b* and *a* were prepared as previously described [3]. Crystalline muscle phosphorylase from silky shark

Table 1
Summary of subunit structure of muscle phosphorylase.

	Rabbit	Human	Lobster	Rat	Frog
M.W. ph *b*	180,000	*	170,000	195,000	188,000
M.W. ph *a*	360,000	*	170,000	367,000	-
No. of subunits ph *a*	4	4	2	4	-
Dissociation of ph *a* with glucose	+	*	No	-	+
$S_{20,w}$ ph *b*	8,3	8,9	8,5	9	8,13
ph *a*	13,2	13,5	8,5	13,9	11,2

* See text

(*Carcharhinus falciformis*) was prepared as briefly outlined below:

1) Extract muscle in 1 mM EDTA, 1 mM dithiothreitol, pH 7.0.
2) Fractionate with saturated ammonium sulfate collecting the fraction between 30 and 41%. Dialyze against 2 mM tris – 1 mM EDTA, pH 7.5.
3) Repeat step 2.
4) Centrifuge at 70,000 *g* for 1 hr. Discard the sediment.
5) Subject the enzyme to column chromatography on DE-52 cellulose using a concave gradient of 40 mM tris, 5 mM EDTA, pH 7.5. Phosphorylase *a* elutes after ph *b*, each emerging in a distinct peak.

Phosphorylase *a* was also prepared by phosphorylation of ph *b* with ATP and Mg^{2+} in the presence of catalytic amounts of purified phosphorylase kinase from rabbit muscle [15]. Both ph *a* and *b* were crystallized either in ammonium sulfate or by dialyzing enzyme solutions (7–15 mg/ml) against 50 mM tris–10 mM EDTA, pH 6.8. Recrystallization was accomplished by dissolving crystals at 30° followed by recooling to 0°.

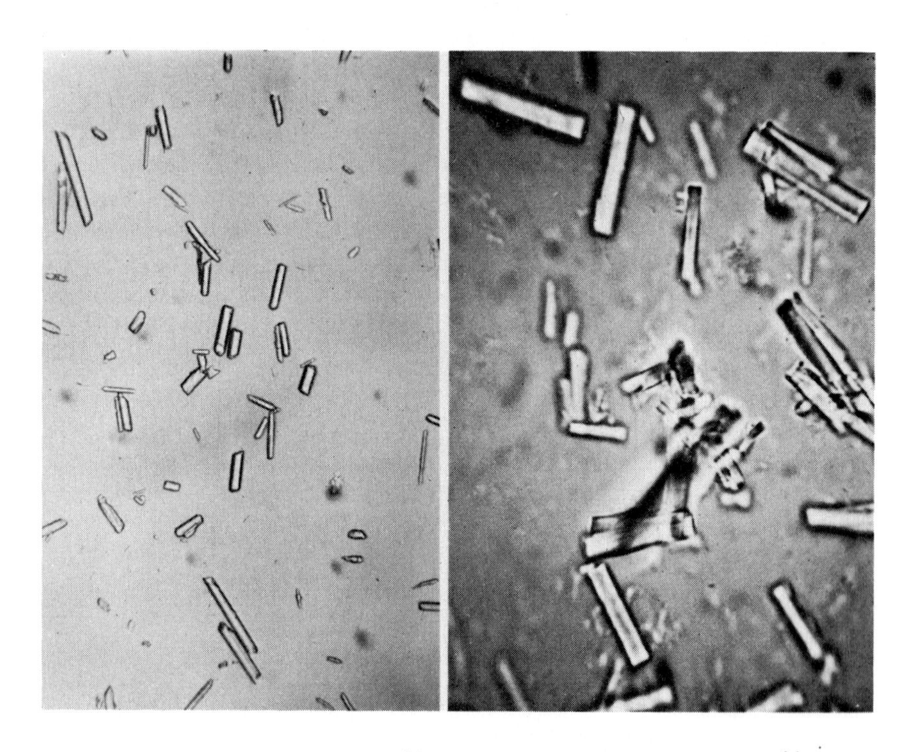

Fig. 1. Left: Shark muscle ph *a* crystals × 100. Right: shark muscle ph *b* crystals × 430.

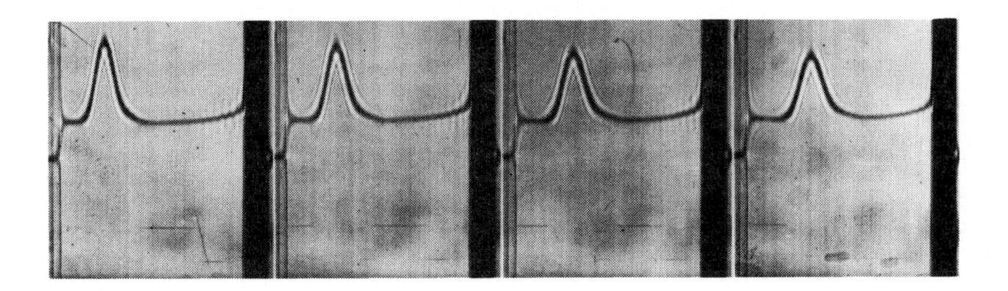

Fig. 2. Sedimentation velocity pattern of shark muscle ph *b* (6 mg/ml in 50 mM tris-HCl, 1 mM dithiothreitol, 10 mM EDTA, pH 6.9) in the model E analytical ultracentrifuge, at 40,000 rpm, and 5°, bar angle 52°. Photographs taken (from left to right) at 8 min intervals.

Shark ph *b* and *a* crystals are shown in fig. 1. The purified enzyme traveled in a single band when subjected to a polyacrylamide gel electrophoresis. It also sedimented as a homogeneous peak in the model E analytical ultracentrifuge as shown in fig. 2.

Our previous studies [3,12–14] have shown many similarities between human and rabbit muscle ph including requirement for AMP and Mg to crystallize ph *b*, sedimentation coefficients for ph *b* and *a*, amino acid sequence of the phosphorylated peptide in ph *a* and immunological reactivity with antibody. However, it was observed that while rabbit ph crystallized readily between 5 and 15°, human ph required a critical temperature of 0° for its crystallization and to prevent the formed crystals from dissolving. A possible ex-

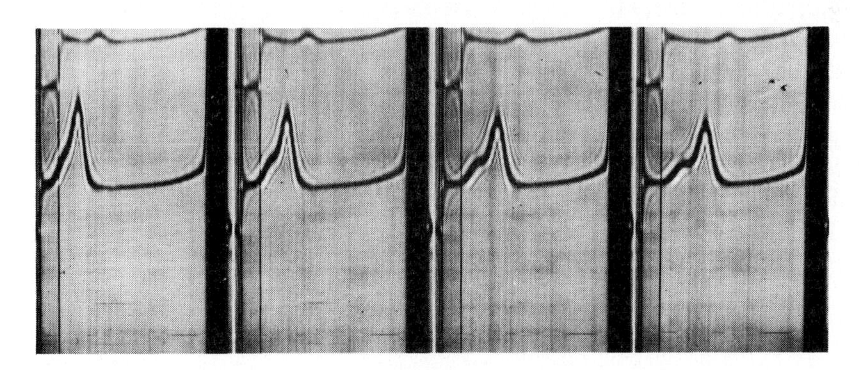

Fig. 3. Sedimentation velocity pattern of human muscle ph *a* (7 mg/ml) (lower curves) and rabbit muscle ph *a* (1.8 mg/ml) (upper curves) (conditions as in fig. 2, except bar angle 45°).

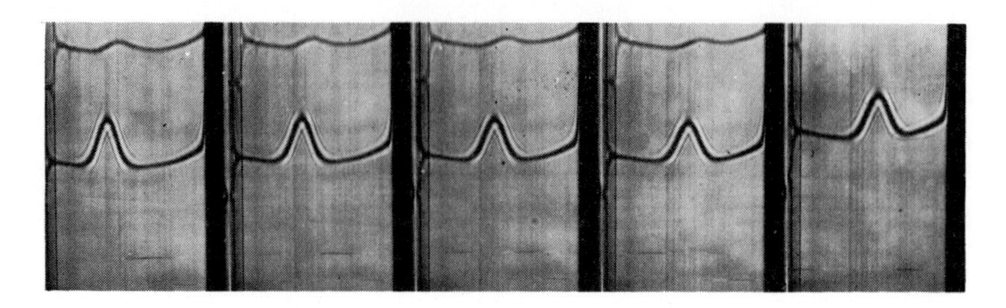

Fig. 4. Sedimentation velocity pattern of human (lower curves) and rabbit (upper curves) muscle ph *a* in the presence of 0.056 M glucose (conditions as in fig. 3).

planation for this phenomenon is suggested by our analytical ultracentrifuge experiment shown in fig. 3. This enzyme travelled in two peaks, one with an $S_{20,w}$ of 8.4 corresponding to a dimer and a heavy sedimenting peak with an $S_{20,w}$ of 13.0 corresponding to a tetramer. Rabbit ph *a* shown also in fig. 3 sedimented as one heavy component with an $S_{20,w}$ of 12.6 even though run at a low concentration (1.8 mg/ml). Glucose, which has been shown to dissociate rabbit ph *a* to dimeric species [10] caused the disappearance of the heavy peak, and only one peak for both human and rabbit ph *a* with an $S_{20,w}$ of 8.5 corresponding to the dimer was observed as shown in fig. 4.

Dissociation was also studied by ultracentrifugation in a sucrose gradient (5–20%) and the molecular weights were determined according to Martin and Ames [16]. Using crystalline bovine liver catalase as a marker (M.W. 250,000 [17]) the M.W. of human and shark ph *b* were found to be 177,000 and 180,000 respectively (fig. 5). Shark ph *a* was found to have the same molecular weight as shark ph *b*. Under these conditions human ph *a* gave a M.W. of 250,000 which is intermediate between a dimeric and tetrameric species indicating an equilibrium between these 2 states in agreement with the sedimentation velocity studies shown in fig. 3. For comparison rabbit ph *a* is also shown here. A M.W. of 335,000 was calculated for this enzyme indicating slight or no dissociation. Dimers of human ph *a* resulting from diluting the enzyme to lower concentrations were found to be more active than the tetrameric form. However, whereas human ph *a* dissociated to the dimeric form at a concentration of 2.5 mg/ml, rabbit ph *a* remains predominantly a tetramer at this concentration [18].

To better understand the role of temperature in the regulation of enzymic activity, kinetic studies were performed at 30 and 0° in the direction of glycogen synthesis.

Using glucose-1-phosphate (G-1-P) as the varied substrate at fixed saturating

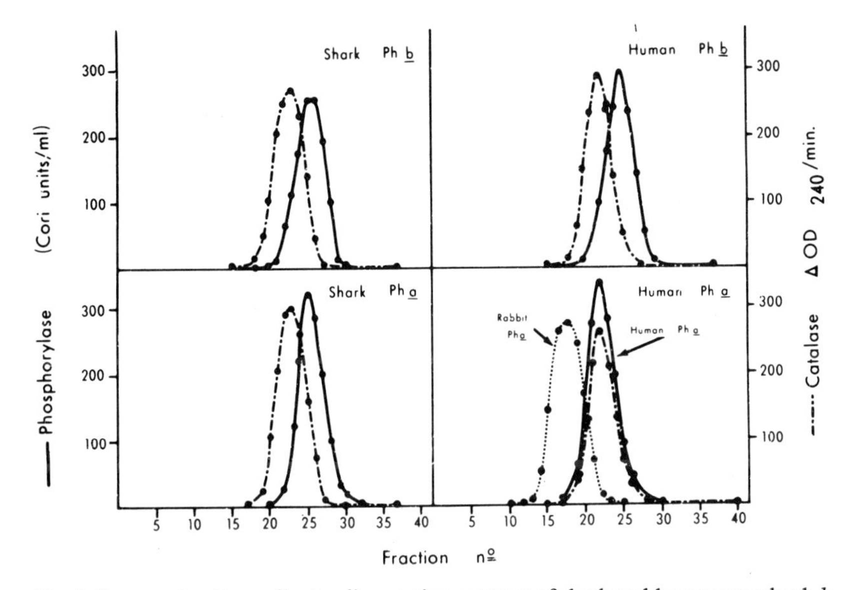

Fig. 5. Sucrose density gradient sedimentation patterns of shark and human muscle ph *b* and *a*. A 0.2 ml sample of enzyme solution containing approximately 1 mg of protein was layered on 5 ml of sucrose gradient (5–20%). Centrifugation was carried out at 35,000 rpm at 4° for 12 hr in a Spinco preparative model L ultracentrifuge, SW39 rotor. Fractions were collected and their phosphorylase activity determined. Bovine liver catalase was used as a marker (M.W. 250,000). Rabbit ph *a*, centrifuged separately, is plotted here for reference.

concentrations of glycogen (2%) and 5′-AMP (1 mM) initial velocities in μmoles/min/mg of enzyme at pH 6.8 were calculated. A double reciprocal plot of this data for both shark and human ph *b* is shown in fig. 6. As can be seen from these data, significant differences in the Michaelis-Menten constants and values of V_{max} were observed.

Whereas V_{max} for human ph *b* at 0° is 2.5 μmoles G-1-P/min/mg, it is 83 μmoles, or 33-fold higher at 30°. For shark ph *b* on the other hand, the V_{max} at 30° is similar to that of human ph *b* but at 0° it is 0.8 μmole/min/mg, or only one-third of that of human ph *b*. Although there was a great difference in the affinity of the two enzymes for G-1-P at 30° ($K_m = 1.2 \times 10^{-2}$ M for shark ph *b* vs 3×10^{-3} M for human ph *b*), at 0° the enzyme affinity for G-1-P increased for shark ph *b* ($K_m = 5 \times 10^{-3}$ M) but slightly decreased for the human enzyme.

A similar kinetic pattern was observed for glycogen (not illustrated). Also, not shown are the kinetics of human and shark ph *a* with varied G-1-P. A kinetic behavior similar to that of ph *b* was observed.

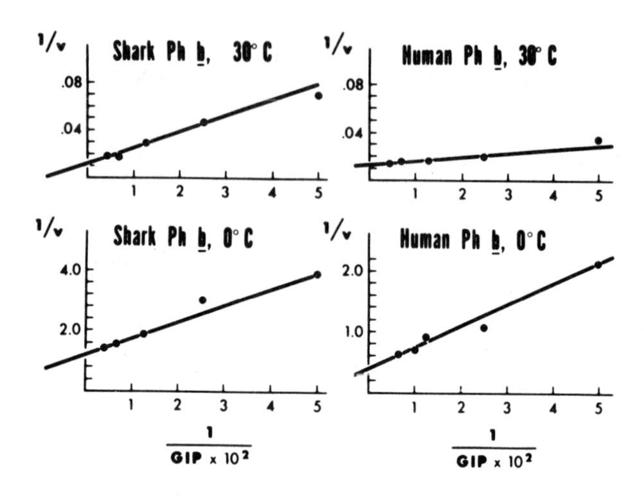

Fig. 6. Double reciprocal plot for glucose-1-phosphate at 30° and 0° for shark and human muscle phosphorylase *b* diluted in 50 mM tris, 50 mM maleate, 5 mM EDTA, 1 mM dithiothreitol pH 6.8). Final concentrations of glycogen and AMP were fixed at 2% and 1 mM, respectively.

Non-linear kinetics obtained in the direction of glycogen degradation with varying concentrations of AMP and P_i have been plotted in various forms to determine their fitness or deviation from various proposed mechanisms. It is of interest that ph *b* of both shark and human muscle were found to exhibit some activity in the absence of the allosteric effector AMP, particularly at high P_i concentration.

Immunological studies with shark and human phosphorylase revealed marked specificity with little or no cross-reaction. Antiserum prepared against shark ph *b* in roosters showed no cross reaction with human or rabbit ph *b* as shown by precipitation in agar (fig. 7).

Reaction of the antibody with the specific antigen resulted in inhibition of enzymatic activity. Thus specificity could also be evaluated by enzyme inhibition studies. The effect of rabbit anti-human and rabbit anti-shark ph antisera on the activity of human, shark, and rabbit ph is shown in fig. 8. In each case, the antibody inhibits the specific antigen with little effect on the other phosphorylases.

In fig. 9 we see the effect of rooster anti-rabbit and anti-shark ph antibody. Whereas shark ph appears to be immunologically distant from both human and rabbit ph the latter two are very closely related immunologically as shown here by inhibition of both enzymes virtually to the same extent by rooster anti-rabbit ph antibody, an observation we had made earlier [12].

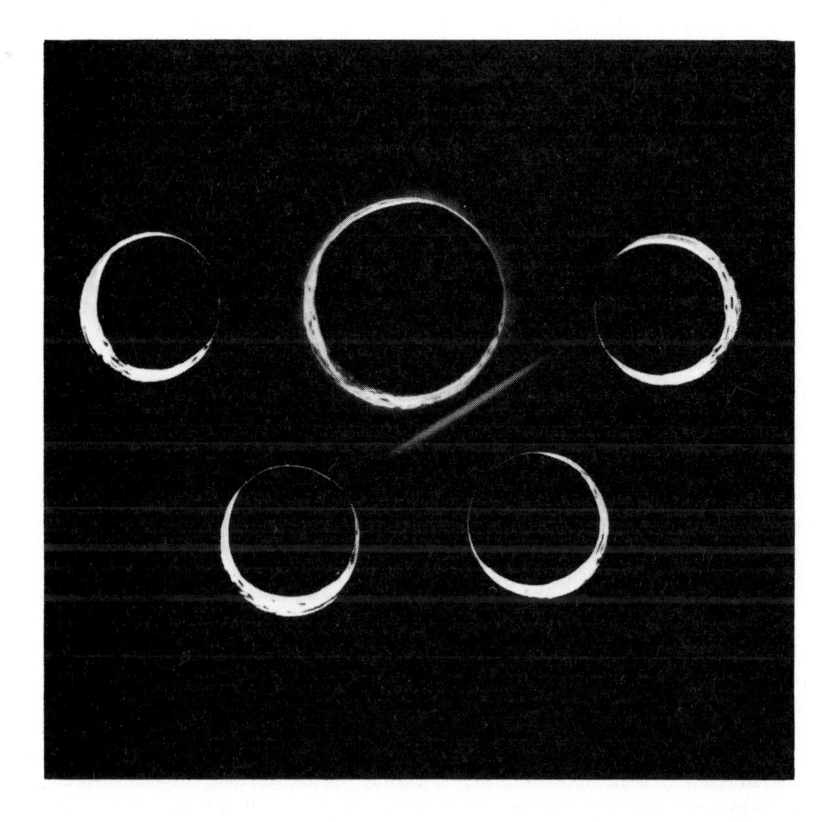

Fig. 7. Antigen-antibody precipitation in agar. Center well contained rooster anti-shark muscle ph *b* serum. From left to right: rabbit muscle ph *b*, rooster muscle ph *b*, human muscle ph *b*, and control. A precipitin line was observed only against shark ph.

From the present work we can emphasize the following points:

1. The regulation of phosphorylase activity through association-dissociation of its subunit structure is species dependent. Dissociation from tetramer to dimer appears to be very important in human, frog and rabbit muscle. This regulatory mechanism appears to be most sensitive in human muscle where dissociation seems to be most apparent.

2. The assumption that crystallization of muscle ph *b* is preceded by the formation of AMP-Mg-enzyme complex does not hold for all species. Lobster ph is crystallized in ammonium sulfate and we have seen that shark ph crys-

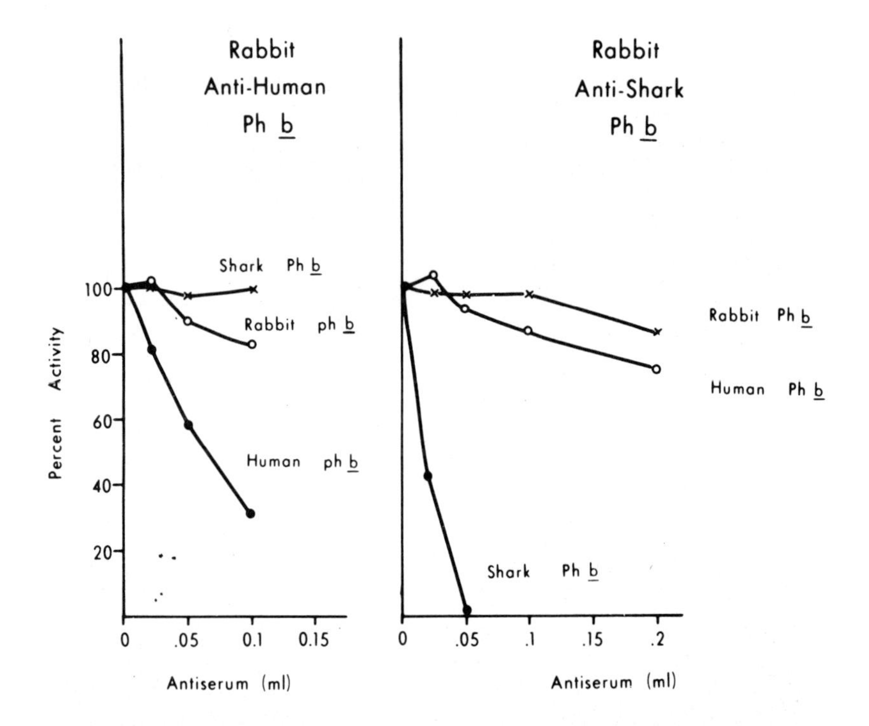

Fig. 8. Effect of rabbit anti-human muscle ph *b* serum and rabbit anti-shark muscle ph *b* serum on the activity of human, shark and rabbit muscle ph. Solutions of rabbit, human and shark enzyme were adjusted to approximately the same activity. Appropriate dilutions were made in a total volume of 2 ml of 40 mM glycerol-P–30 mM cysteine buffer pH 6.8 containing the indicated amounts of antiserum. Activity was assayed by adding 0.2 ml of enzyme solution to 0.2 ml of substrate (containing 64 mM glucose-1-phosphate and 2% glycogen) and incubating 5 min at 30°. Reaction was stopped by the addition of 2.1 ml of 5% trichloroacetic acid. Inorganic phosphate was determined on the supernatants by the method of Fiske and SubbaRow [18]. Control tubes contained equivalent amounts of normal rabbit serum.

tallizes in buffer. The possibility of the formation of a tris-enzyme complex for shark ph is currently being investigated.

3. Kinetic studies at 30° and 0° indicate that human ph is more active at cold temperatures than shark ph and that this difference is not due to increased enzyme substrate affinity.

4. Kinetics in the direction of glycogen degradation on ph *b* of human and shark muscle suggest an allosteric behavior. This data is currently being ana-

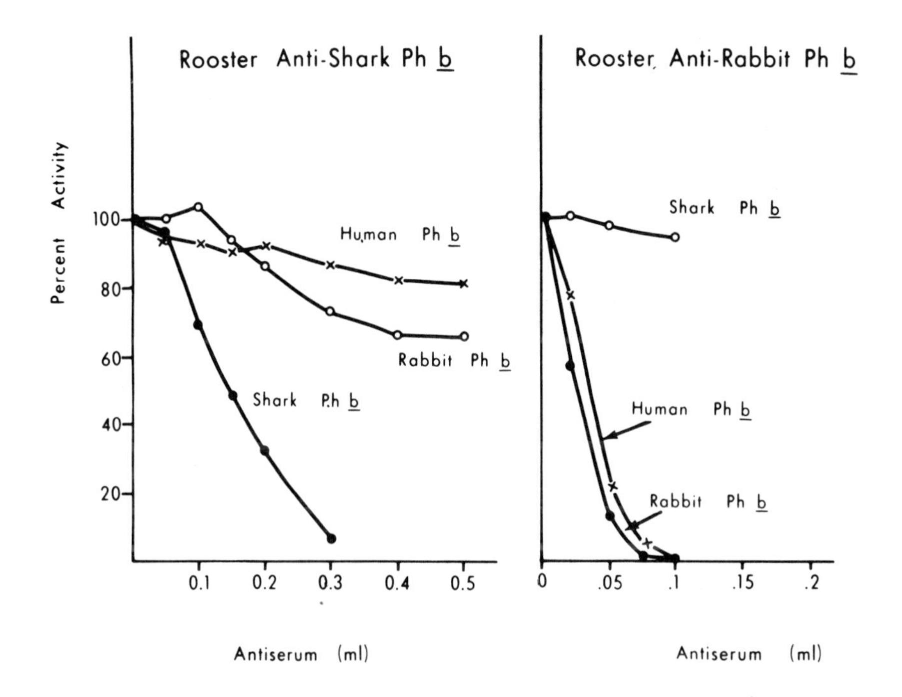

Fig. 9. Effect of rooster anti-shark muscle ph *b* serum and rooster anti-rabbit muscle ph *b* serum on the activity of human, shark and rabbit muscle ph. Reaction mixtures were prepared as described in fig. 8.

lyzed to determine its fitness or deviation from previously described mechanisms.

References

[1] C.F. Cori, G.T. Cori and A.A. Green, J. Biol. Chem. 151 (1943) 39.
[2] E.H. Fischer and E.G. Krebs, J. Biol. Chem. 231 (1958) 65.
[3] A.A. Yunis, E.H. Fischer and E.G. Krebs, J. Biol. Chem. 235 (1960) 3163.
[4] K. Hanbusa, T. Kanno, S. Adachi and H.J. Kobayashi, J. Biochem. 62 (1967) 194.
[5] S.A. Assaf and D.J. Graves, J. Biol. Chem. 244 (1969) 5544.
[6] C.L. Sevilla and E.H. Fischer, Biochemistry 8 (1969) 2161.
[7] R.W. Cowgill, J. Biol. Chem. 234 (1959) 3146.
[8] B.E. Metzger, L. Glaser and R. Helmreich, Biochemistry 7 (1968) 2021.

[9] J.H. Wang and D.J. Graves, Biochemistry 3 (1964) 1437.
[10] J.H. Wang, M.L. Shonka and D.J. Graves, Biochem. Biophys. Res. Commun. 18 (1965) 131.
[11] J.H. Wang, M.L. Shonka and D.J. Graves, Biochemistry 4 (1965) 2296.
[12] A.A. Yunis and E.G. Krebs, J. Biol. Chem. 237 (1962) 37.
[13] M.M. Appleman, A.A. Yunis, E.G. Krebs and E.H. Fischer, J. Biol. Chem. 238 (1963) 1358.
[14] R.C. Hughes, A.A. Yunis, E.H. Fischer and E.G. Krebs, J. Biol. Chem. 237 (1962) 40.
[15] E.H. Fischer and E.G. Krebs, in: Methods in Enzymology, Vol. 5., eds. S.P. Colowick and N.O. Kaplan (Academic Press, New York, 1962) p. 369.
[16] R.G. Martin and B.N. Ames, J. Biol. Chem. 236 (1960) 1372.
[17] J.B. Sumner and N.J. Gralen, Biol. Chem. 125 (1938) 33.
[18] C.Y. Huang and D.J. Graves, Biochemistry 9 (1970) 660.
[19] C.H. Fiske and Y.J. SubbaRow, Biol. Chem. 66 (1925) 375.

HOMOLOGIES OF STRUCTURE AND FUNCTION AMONG NEUROHYPOPHYSIAL PEPTIDES

W.H. SAWYER

Department of Pharmacology, Columbia University, College of Physicians and Surgeons, New York, New York, USA

Abstract: Sawyer, W.H. Homologies of Structure and Function Among Neurohypophysial Peptides. *Miami Winter Symposia* 1, pp. 257–269. North-Holland Publishing Company, Amsterdam, 1970.

Vertebrate neurohypophysial hormones are cyclic octapeptides falling into two general categories, the basic principles arginine vasotocin (AVT), arginine vasopressin (AVP) and lysine vasopressin (LVP), and the more nearly neutral principles resembling oxytocin. AVT is found in all nonmammalian tetrapods and such ubiquity suggests that it may be ancestral to the other peptides. One can fit the known principles into a rather simple scheme suggesting their possible evolution from AVT by successive point mutations, except for two oxytocin-like principles that contain serine rather than glutamine in the 4-position. These are glumitocin, in some elasmobranchs, and isotocin, in actinopterygian fishes. Mutations leading to such substitution require at least two steps, and 4-proline analogues would be likely intermediates. Several 4-proline analogues have been synthesized as a first step towards attempts at isolating and identifying such hypothetical intermediates among primitive fishes. Although neurohypophysial peptides appear to constitute a homologous series of biologically active peptides, progress toward identifying their physiological functions among nonmamalian vertebrates has been disappointing. Oxytocin has obvious value to female mammals during parturition and lactation. Its importance to nonreproducing females or to males is unknown. There is little information suggesting the nature of the endocrine functions of oxytocin-like peptides among nonmammalian vertebrates. AVT, the basic peptide in neurohypophyses of all nonmammalian vertebrates, appears to be an antidiuretic hormone in tetrapods. In fresh water actinopterygian fishes and lungfishes, however, it causes diuresis. This apparent paradox provides an interesting example of how the use to which a single peptide hormone is put may depend primarily on the osmotic relations between the organism and its environment. It is possible, however, that AVT and the vasopressins do have some homologous modes of action in fishes and tetrapods. These may concern the regulation of blood flow in specific organs. The better known effects of these antidiuretic hormones on membrane permeability may well be secondarily acquired adaptive responses that appeared during the evolution of the amphibians.

1. Introduction

The known neurohypophysial hormones form an apparently homologous group of octapeptides. Their low molecular weight, relatively uniform structure, and characteristic biological properties should make them a useful group in which to study how molecular structures and physiological functions may have changed during vertebrate evolution. The active neurohypophysial principles have been identified in a wide variety of living species. On the basis of such limited information one is tempted to try: (1) to interpret the distribution of peptides among surviving species in terms of the possible course of their molecular evolution, and (2) to correlate the evolution of these peptides with the functions they may serve in the animals that secrete them. In this paper I will discuss the limited progress that has been made in both these directions.

2. Evolution and the structures of neurohypophysial peptides

The elucidation of the structure of oxytocin and its confirmation by synthesis by du Vigneaud and his collaborators in the early 1950's stimulated a period of intense study of the chemistry of neurohypophysial peptides. In subsequent years the molecular structures of six other naturally occurring neurohypophysial principles have been determined and confirmed by synthesis [1]. These are all cyclic octapeptides closely similar in structure (table 1). The amino acids in five positions remain the same in all known active principles. Evolutionary changes have been confined to the third, fourth, and eight positions.

The neurohypophysial principles fall into two groups (table 1). One contains peptides with basic amino acids, either arginine or lysine, in the 8-position. These have potent vasopressor and antidiuretic effects in mammals. The other group are the "neutral" peptides that resemble oxytocin in having weak vasopressor and antidiuretic properties but strong oxytocic and milk-ejecting activities in mammals. The typical vertebrate neurohypophysis contains at least one basic and one neutral peptide. In mammals, at least, these can be released independently in response to appropriate stimuli.

The neurohypophysis of an adult mammal contains large amounts of oxytocin and a vasopressin. In most species this is arginine vasopressin (AVP). Among members of the pig family, however, lysine vasopressin (LVP) may occur, either alone, or with AVP. The distribution of vasopressins among individual warthogs led Ferguson [2] to conclude that they are inherited as a single

Table 1
Amino acid sequences of the known active natural neurohypophysial principles.*

A. Vasopressor-antidiuretic peptides:

1. Arginine vasopressin

Cys - Tyr - Phe - Gln - Asn - Cys - Pro - Arg - Gly - NH_2
1 2 3 4 5 6 7 8 9

2. Lysine vasopressin

Cys - Tyr - Phe - Gln - Asn - Cys - Pro - *Lys* - Gly - NH_2

3. Arginine vasotocin

Cys - Tyr - *Ile* - Gln - Asn - Cys - Pro - Arg - Gly - NH_2

B. Neutral oxytocin-like peptides:

1. Oxytocin

Cys - Tyr - *Ile* - Gln - Asn - Cys - Pro - *Leu* - Gly - NH_2

2. Mesotocin (8-isoleucine oxytocin)

Cys - Tyr - *Ile* - Gln - Asn - Cys - Pro - *Ile* - Gly - NH_2

3. Isotocin (4-serine, 8-isoleucine oxytocin)

Cys - Tyr - *Ile* - *Ser* - Asn - Cys - Pro - *Ile* - Gly - NH_2

4. Glumitocin (4-serine, 8-glutamine oxytocin)

Cys - Tyr - *Ile* - *Ser* Asn - Cys - Pro - *Gln* - Gly - NH_2

* Amino acids differing from those in the corresponding positions in the arginine vasopressin molecule shown in italics.

allelic pair and that none of the three genotypes has any selective advantage. The recent description of a mutant strain of mice that possess lysine vasopressin rather than AVP found in other strains [3] provides promising material for further studies on the genetic determination of the nature of vasopressins.

Recently reported chromatographic and pharmacological evidence indicates that fetal mammals may possess arginine vasotocin (AVT) as well as AVP [4]. AVT is the typical basic vasopressor peptide in nonmammalian neurohypophyses. Thus its appearance in the fetal neurohypophysis may represent an interesting example of "molecular recapitulation". AVT has not been detected in the adult neurohypophysis but its presence in the fetus indicates that the genetic basis for its synthesis is present among mammals although it may not be expressed in the adult.

Heller [5] first pointed out that neurohypophyses from nonmammalian vertebrates contain a principle or principles with biological activities clearly different from those of the mammalian hormones. Further progress toward identifying nonmammalian principles only became possible after the chemistry of the mammalian principles was known. Munsick et al. [6] were the first to note a striking similarity in pharmacological and chromatographic properties between the vasopressor principle from the chicken neurohypophysis and a peptide, arginine vasotocin (AVT), that had been synthesized by Katsoyannis and du Vigneaud [7] as part of their studies on the relations between peptide structures and biological activities. A pharmacological survey of other representative species suggested to us that AVT was present in all major vertebrate groups with the possible exception of the cartilaginous fishes [8]. The remarkable fact that Katsoyannis and du Vigneaud had synthesized a peptide hormone before its natural existence was recognized was soon confirmed by chemical identification of AVT in a variety of nonmammalian tetrapods and in teleost fishes. Chromatographic and pharmacological data strongly indicate that AVT also exists in cyclostomes, cartilaginous fishes, lungfishes, chondrostean and holostean fishes [1].

If the basic vasopressor peptides can be considered to belong to a homologous series it is noteworthy that AVT appears in the most primitive living

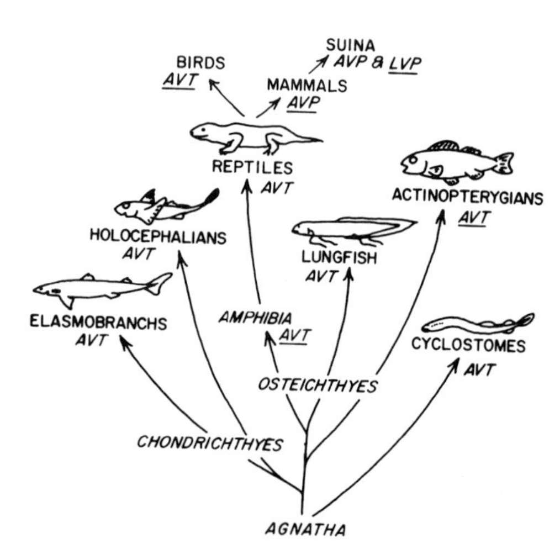

Fig. 1. The phyletic distribution of the basic vasopressor-antidiuretic peptides. AVT = arginine vasotocin, AVP = arginine vasopressin, LVP = lysine vasopressin.

vertebrates, and has persisted in all major branches of evolution (fig. 1). Only among adult mammals has it apparently been replaced as the major basic principle by AVP, and , in some species, by LVP. These changes could reflect two point mutations, one occurring early in the evolution of mammals or, more likely, among the premammalian reptiles since AVP occurs in monotremes. This resulted in the substitution of phenylalanine for isoleucine in the 3-position. The second, causing substitution of lysine for arginine presumably occurred early in the evolutionary line leading to the pigs, peccaries and the hippopotamus.

Interpretation of the distribution of neutral oxytocin-like peptides is more difficult. These principles are not as easy to isolate or identify as the basic peptides, and there appear to be more of them. Oxytocin itself has been positively identified in many mammals and in the chicken (fig. 2). It has also, surprisingly, been found in a cartilaginous fish, the holocephalian ratfish [9, 10]. The appearance of oxytocin in such distantly related species suggests that it may be a very ancient molecule, if it were present in an ancestor common to both cartilaginous fishes and mammals. Alternately, one could suggest that it arose independently in these two divergent evolutionary lines.

There is also evidence that oxytocin may occur in some lungfishes [11], amphibians [12] and reptiles [13]. Since mesotocin has been positively identified in some amphibians and reptiles [14,15], and there is strong evidence for its presence in lungfishes [11,16], it appears possible that these closely

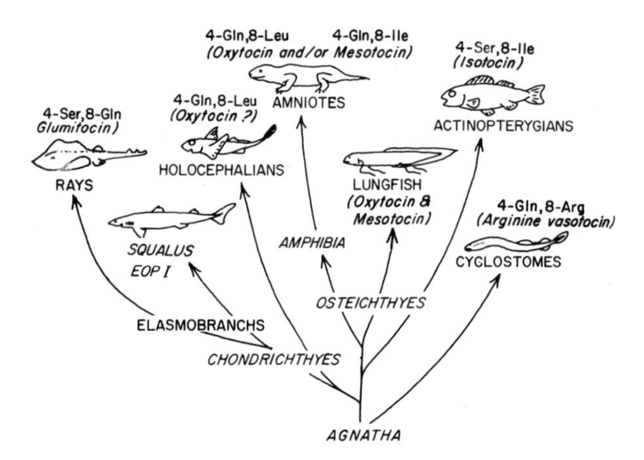

Fig. 2. The phyletic distribution of neutral principles that resemble oxytocin. No neutral principle has yet been found in cyclostomes. EOP I indicates one or more unknown oxytocin-like principles present in *Squalus* and other sharks.

related principles may coexist in some species among the cold-blooded vertebrates [10].

AVT is the only active principle yet identified in cyclostomes [17,18]. It is tempting to suggest that it is the most primitive peptide known and is thus ancestral to all others. Either oxytocin or mesotocin could arise as a result of point mutations changing a codon for arginine. However, to allow for all the substitutions that appear in the 8-position (arginine, lysine, isoleucine, leucine, glutamine) one has to postulate that two different codons for arginine have been used, and that a "silent" mutation occurred between them at some time in vertebrate evolution [19,20].

The two known natural principles containing serine pose a more difficult problem of interpretation. There is no single point mutation that could result in the substitution of serine for glutamine. Nonetheless, 4-serine peptides appear in both elasmobranchs and teleosts. One must postulate a minimum of two steps, with a 4-proline as the most probable intermediate [19,20]. It thus would be interesting to look for 4-proline peptides among the primitive fishes. Before this is possible, of course, one must know what to look for. For this reason two groups have synthesized 4-proline analogues of the known neutral neurohypophysial principles [21,22]. When assayed by the methods routinely used in finding and isolating neurohypophysial principles among fishes we find 4-proline analogues very weakly active. If more sensitive means can be found for assaying these analogues it would be of great interest to apply them to extracts of neurohypophyses from primitive fishes. If a 4-proline analogue was ancestral to both isotocin and glumitocin it must have existed prior to the separation of phylogenetic lines leading to cartilaginous and bony fishes. It is interesting to note that the surviving cyclostomes seem to have AVT but no detectable neutral principle [14,17,18]. Cyclostomes may, of course, be exceptions to the rule that the vertebrate neurohypophysis contains both neutral and basic peptides. On the other hand, if the neutral principle present contained a 4-proline its presence would be missed by the methods we used.

Glumitocin, a neutral principle containing 4-serine, has been isolated by Acher and his co-workers [23] from several skates of the genus *Raia*. Neutral principles other than glumitocin have been found among other elasmobranchs [24–27]. Despite valiant attempts none have yet been identified, largely because the principles are present in elasmobranch neurointermediate lobes in distressingly small amounts [25]. Thus we must rely on pharmacological and chromatographic characterization for the present.

The neutral principle from the spiny dogfish, *Squalus acanthias*, has been studied in some detail [18,24]. The biological properties of the neutral fraction from *Squalus* cannot be due to any of the known natural neutral principles [28]. The fraction is highly active, and its pattern of activities resemble those found in analogues with the ring of oxytocin and a neutral amino acid in the 8-position. When compared directly with available analogues of this type, some of which were synthesized by Dr. Manning and his associates [29,30] for this study, the *Squalus* fraction clearly differs. Thus its activities cannot be due to any single peptide to which it has been directly compared [28]. There is a possibility, however, that the fraction is not a single peptide and preliminary experiments indicate that it may be resolved into two subfractions, only one of which seems to be an 8-substituted oxytocin analogue. The nature of the second remains obscure. It is not glumitocin. If such separation can be repeated on a preparative scale it may make it possible to identify two heretofore unrecognized natural principles. We are hopeful that their structures can offer information that will be useful in our attempts to understand the evolution of these active peptides.

Acher [14] has attempted to interpret the evolution of neurohypophysial peptides as representing two parallel lines. One leads from AVT to the vasopressins and, as we have seen, this appears to be simple and reasonable. The other posited line moves from isotocin to mesotocin to oxytocin. One can question such a scheme on several grounds: (1) It is unlikely that lungfishes and tetrapods, (which have mesotocin) evolved from actinopterygian fishes (the only group known to have isotocin). (2) The evolutionary transition from isotocin (with a 4-serine) to mesotocin (with a 4-glutamine) requires more mutations than that from AVT to oxytocin or to mesotocin. (3) Oxytocin and other neutral 4-glutamine peptides exist in cartilaginous fishes that certainly did not evolve from actinopterygians.

Thus attempts to interpret phyletic distribution in terms of peptide phylogeny are premature. One can, however, apply the rule of parsimony and ask what would be the simplest scheme for evolution, that is, the scheme requiring the postulation of the least number of point mutations. This "game" has been played by Vliegenthart and Versteeg [19], Geschwind [20] and me, with similar, if slightly different, results (fig. 3). Such games are not highly informative but they do point out what possible intermediates we should look for that might help us define the probable course of molecular evolution among neurohypophysial peptides.

3. Evolution and responses to neurohypophysial hormones

Knowledge of the distribution of neurohypophysial peptides among the vertebrates is not very satisfying unless we can interpret this on the basis of

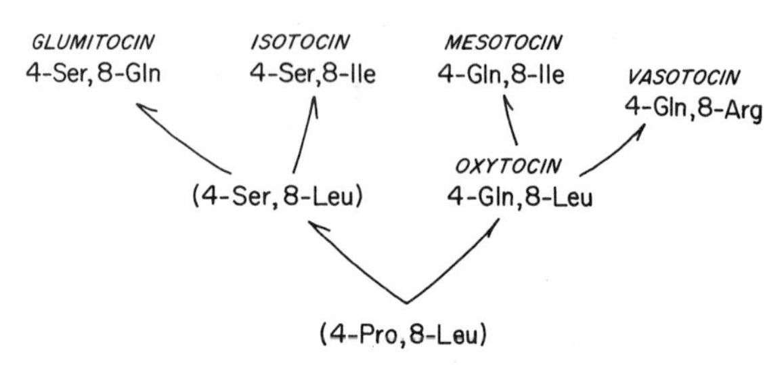

Fig. 3. A highly hypothetical scheme outlining the possible relationships among the neurohypophysial peptides from nonmammalian vertebrates. The arrows connect peptides that could be derived from one another by single point mutations. Any one of these could be primitive. The peptides in parentheses have been synthesized but are not known to exist naturally in any living vertebrates.

physiological functions. Have these peptides served similar or homologous functions throughout vertebrate phylogeny? Are they even hormones among lower vertebrates? We can only suggest answers to these questions at this time, based on fragmentary information.

A. *Neutral oxytocin-like principles*

Oxytocin is released from the neurohypophysis by female mammals during parturition and suckling. Its physiological functions as an oxytocic and milk-ejecting hormone are clear. But we have no idea of why oxytocin is secreted by the non-reproducing female or male mammal. It is not surprising, therefore, that we have no better explanation for the presence of oxytocin and its neutral analogues in nonmammalian vertebrates [26,31]. The neutral principles are relatively weak oxytocic agents except on the mammalian uterus. AVT is many times more active in causing contraction of oviducts in birds, reptiles [6] or amphibians [32] than are the neutral principles.

In some teleost fishes the neutral principle, isotocin, may increase sodium transport across the gills [33]. It is not clear that isotocin actually functions as a hormone regulating sodium balance or whether this particular response occurs in fishes other than teleosts. Perhaps the neutral oxytocin-like principles have no endocrine functions among nonmammalian vertebrates [31]. These peptides are elaborated and stored in the hypothalamo-neurohypophysial systems of many species in large quantities. It seems unlikely that such an elaborate and specialized system would have evolved to manufacture and store useless peptides. Thus our inability to assign physiological functions probably results from our failure to have looked for the right things.

B. *Basic, vasopressor peptides*

The vasopressins serve an important function in mammals as antidiuretic hormones. They act primarily to increase the permeability of the renal tubules to water. This allows osmotic reabsorption of free water and concentration of the urine. AVT appears to serve a homologous function in birds. AVT is also antidiuretic in reptiles and amphibians [34,35]. Here the peptide also causes a fall in glomerular filtration rate (GFR), the "glomerular antidiuretic" effect. Among anuran amphibians AVT increases the osmotic permeability of the skin and urinary bladder. Although such "hydro-osmotic" responses are of great interest to physiologists as models for the action of antidiuretic hormones these responses appear to be strictly limited to this single, specialized group of amphibians [31,36].

When the "antidiuretic" hormone, AVT, is administered to a goldfish or an eel in fresh water, it produces a brisk increase in urine flow and GFR [33, 37]. Thus we encounter an intriguing paradox, in which a hormone that decreased water excretion in tetrapods causes increased water excretion in teleost fishes.

In an attempt to resolve this apparent paradox we have studied a species that occupies a phyletic position somewhere between teleost fishes and amphibians. This is the African lungfish (*Protopterus aethiopicus*) [38].

When AVT is injected intravenously into a large lungfish there is a striking increase in GFR and urine flow. In these respects the response resembles that of a goldfish or eel. There is also a disproportionate increase in sodium excretion which is not seen in teleosts. Relative free water excretion does not change, indicating that AVT does not alter tubular permeability to water in any of the fish species studied. Thus the diuretic response to AVT appears to be primarily a glomerular diuresis, exactly opposite to the glomerular antidiuresis seen in cold-blooded tetrapods [39].

If one records pressures from the dorsal aorta of the lungfish one finds that AVT is an exceptionally potent vasopressor agent [38]. The increased dorsal aortic pressure is transient, however, and diuresis may persist for hours after the pressure has returned to normal levels or below them. Thus the increased GFR is not simply a result of increased renal perfusion pressure. In eels, increased GFR and urine flow may occur after doses of AVT that actually lower renal perfusion pressure [37]. Thus one must look for some cause of increased GFR other than increased dorsal aortic pressure. Possible mechanisms could be either dilatation of the afferent glomerular arterioles or constriction of efferent arterioles. The latter possibility is more intriguing for several reasons: a) Throughout the vertebrates, basic neurohypophysial peptides, if they effect blood vessels, are almost always vasoconstrictors. b) The glomerular antidiuretic effects seen in reptiles and amphibians probably reflect constriction of afferent glomerular arterioles [40]. c) Thurau (personal communication) has recently demonstrated that AVP in physiological doses causes constriction of the efferent arterioles of the juxtamedullary glomeruli of rats. This doubles GFR in these glomeruli but, since they are a small part of the total glomerular population, there is no measurable increase in total renal GFR.

Obviously much remains to be done before we can define the lungfish response to AVT in more detail. The possibility that it is due to efferent arteriolar constriction is attractive since it allows us to suggest that fish diuretic responses merely reflect the vasoconstrictor action of AVT at one particular vascular site. Thus it might be considered homologous with vasoconstrictor responses to basic peptides among tetrapods. These are the glomerular antidiuresis of reptiles and amphibians, the glomerular diuresis in specialized nephrons of the rat, and the systemic vasoconstrictor responses best seen in mammals.

If the glomerular diuretic effect is physiological we might ask its function. In fresh water fishes, exposed to continuous osmotic inflow of water from the dilute environment, a hormone that could increase water excretion would be useful. In the lungfish, however, AVT also increases sodium excretion. This might be considered disadvantageous to a free-swimming fresh water fish, to which sodium is usually a precious commodity. In our experiments we did find that at lower doses AVT often increased water excretion without a disproportionate sodium loss. Perhaps this is the range in which the fish usually uses AVT to regulate water excretion.

The lungfish, however, faces one unusual situation that does not confront most fresh water fishes. This fish can estivate by burying itself in the mud. During such periods it excretes no urine. The concentrations of urea and electrolytes in the body fluids rise to very high levels. When the fish returns to water it gains weight rapidly, as water enters the hyperosmotic body fluids. During the first few hours in water the fish undergoes an intense solute diuresis and excretes the excess salt and urea. This requires that GFR be increased rapidly from essentially nil to relatively high values. Actually, the rate of sodium loss by a lungfish in the first 24 hr after returning to water following a period of estivation [41] is approximately the same as that produced acutely by the intravenous administration of 20 ng/kg of AVT. Thus AVT may function as a diuretic and natriuretic hormone at this highly critical juncture in the life history of the lungfish. Godet [42] reported that hypophysectomized lungfish cannot excrete enough water following estivation and become edematous and die. Although one cannot rule out the possibility that this reflects the lack of some adenohypophial factor it is tempting to suggest that death results from the lack of the specific diuretic-natriuretic hormone, AVT.

4. Conclusions

Neurohypophysial peptides fall into two general classes, one basic, one neutral. Most vertebrates have at least one of each. AVT is the basic principle present in the neurohypophysis of all nonmammalian vertebrates. This appears to be an extraordinary example of evolutionary stability for a peptide molecule. The neutral principles resembling oxytocin show more variety. The information available concerning their identities and phyletic distributions is still inadequate to allow one to draw any firm conclusions concerning the molecular changes in these peptides that have occurred during vertebrate evolution.

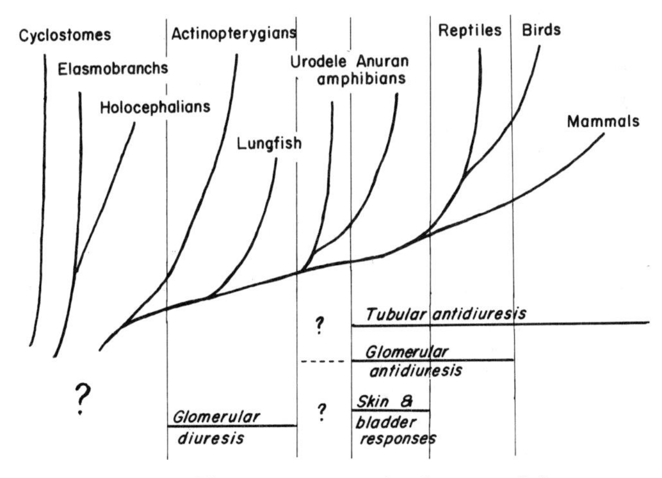

Fig. 4. Phyletic distribution of known responses related to water balance among vertebrates to vasopressor-antidiuretic peptides. The broken line indicates that antidiuretic responses have been observed in some, but not all, urodeles.

We have little idea concerning the functions of the neutral peptides except for those of oxytocin in female mammals. Thus one cannot offer any explanation for the numerous structural changes that must have occurred among these peptides. In tetrapods AVT has typical tubular antidiuretic effects but it is diuretic in some fishes (fig. 4). The glomerular diuresis produced by AVT in fishes may reflect vasoconstriction of efferent glomerular arterioles. Since vasoconstrictor responses, at one site or another, can be demonstrated in bony fishes, amphibians, reptiles and mammals, one may postulate that these are indeed homologous and primitive, whereas the more familiar effects of basic neurohypophysial peptides on water permeability first appeared as adaptations to terrestrial life among the amphibians.

Acknowledgements

Unpublished work discussed in this paper has been supported by Research Grants from the National Science Foundation (GB 4932) and the National Institute for Arthritis and Metabolic Diseases (AM 01940) and by a General Research Support Grant to Columbia University from the National Institutes of Health.

References

[1] W.H. Sawyer, in: Neurohypophysial Hormones and Similar Polypeptides, ed. B. Berde (Springer, Berlin, 1968) p. 717.
[2] D.R. Ferguson, Gen. Comp. Endocrin. 12 (1969) 609.
[3] A.D. Stewart, J. Endocrinol. 41 (1968) xix.
[4] E. Vizsolyi and A.M. Perks, Nature 223 (1969) 1169.
[5] H. Heller, J. Physiol. 100 (1941) 124.
[6] R.A. Munsick, W.H. Sawyer and H.B. van Dyke, Endocrinology 66 (1960) 860.
[7] P.G. Katsoyannis and V. du Vigneaud, J. Biol. Chem. 233 (1958) 1352.
[8] W.H. Sawyer, R.A. Munsick and H.B. van Dyke, Nature 184 (1959) 1464.
[9] W.H. Sawyer, R.J. Freer and T.-C. Tseng, Gen. Comp. Endocrin. 9 (1967) 31.
[10] B.T. Pickering and H. Heller, J. Endocrinol. 45 (1969) 597.
[11] B.T. Pickering and S. McWatters, J. Endocrinol. 36 (1966) 217.
[12] R.A. Munsick, Endocrinology 78 (1966) 591.
[13] B.T. Pickering, J. Endocrinol. 39 (1967) 285.
[14] R. Acher, Proc. Roy. Soc. (London) Ser. B. 170(1968) 7.
[15] R. Acher, J. Chauvet and M.T. Chauvet, Biochim. Biophys. Acta 154 (1968) 255.
[16] W.H. Sawyer, J. Endocrinol. 44 (1969) 421.
[17] B.K. Follett and H. Heller, J. Physiol. 172 (1964) 74.
[18] W.H. Sawyer, Gen. Comp. Endocrin. 5 (1965) 427.

[19] J.F.G. Vliegenthart and D.H.G. Versteeg, J. Endocrinol. 38 (1967) 3.
[20] I.I. Geschwind, Am. Zoologist 7 (1967) 89.
[21] W.H. Sawyer, T.C. Wuu, J.W.M. Baxter and M. Manning, Endocrinology 85 (1969) 385.
[22] J. Rudinger, O.V. Kesarev, K. Poduska, B.T. Pickering, R.E.J. Dyball, D.R. Ferguson and W.R. Ward, Experientia 25 (1969) 680.
[23] R. Acher, J. Chauvet and M.T. Chauvet, Nature 216 (1967) 1037.
[24] W.H. Sawyer, Gen. Comp. Endocrin. 9 (1967) 303.
[25] A.M. Perks, Gen. Comp. Endocrin. 6 (1966) 428.
[26] A.M. Perks, in: Fish Physiology, Vol. 2., eds. W.S. Hoar and R.J. Randall (Academic Press, New York, 1969) p. 112.
[27] B.P. Roy, Gen. Comp. Endocrin. 12 (1969) 326.
[28] W.H. Sawyer, J.W.M. Baxter, M. Manning, E. Heinicke and A.M. Perks, unpublished observations.
[29] J.W.M. Baxter, M. Manning and W.H. Sawyer, Biochemistry 8 (1969) 3592.
[30] J.W.M. Baxter, T.C. Wuu, M. Manning and W.H. Sawyer, Experientia, 25 (1969) 1127.
[31] F. Morel and S. Jard, in: Neurohypophysial Hormones and Similar Polypeptides, ed. B. Berde (Springer, Berlin, 1968) p. 655.
[32] H. Heller, E. Ferreri and D.H.G. Leathers, J. Endocrinol. 37 (1967) xxxix.
[33] J. Maetz, J. Bourguet, B. Lahlou and J. Houndry, Gen. Comp. Endocrin. 4 (1964) 508.
[34] W.H. Dantzler, Am. J. Physiol. 212 (1967) 83.
[35] W.H. Dantzler, D.P. Shaffner, P.J.S. Chiu and W.H. Swayer, Am. J. Physiol., 218 (1970) 929.
[36] P.J. Bentley, Gen. Comp. Endocrin. 13 (1969) 39.
[37] I. Chester Jones, D.K.O. Chan and J.C. Rankin, J. Endocrinol. 43 (1969) 21.
[38] W.H. Sawyer, Am. J. Physiol., in press.
[39] W.H. Sawyer, Am. J. Med. 42 (1967) 678.
[40] W.H. Sawyer, Am. J. Physiol. 164 (1951) 457.
[41] P.A. Janssens, Comp. Biochem. Physiol. 11 (1964) 105.
[42] R. Godet, Ann. Fac. Sci. Univ. Dakar 6 (1961) 183.

BIOCHEMICAL AND CONFORMATIONAL STUDIES ON GROWTH HORMONES

A.C. PALADINI, J.M. DELLACHA and J.A. SANTOME
Facultad de Farmacia y Bioquimica, Departamento de Quimica Biologica y Centro para el Estudio de las Hormonas Hipofisarias, Buenos Aires, Argentina

Abstract: Paladini, A.C., Dellacha, J.M. and Santome, J.A. Biochemical and Conformational Studies on Growth Hormones. *Miami Winter Symposia* 1, pp. 270–287. North-Holland Publishing Company, Amsterdam, 1970.

A review is made of the physicochemical and chemical properties of several growth hormones. A detailed comparison of the primary structures of human and bovine growth hormones brings out the existence of a highly homologous central region in their molecules while the *N*- and *C*-terminal chains are variable. Studies made by analytical dialysis and hydrogen exchange measurements show that the molecule of human growth hormone has greater flexibility, under physiological conditions, than all other hormones tested. It seems reasonable to conclude that the common biological activity and peculiar species-specificity shown by these hormones may be related to the homologous and less-homologous zones in their molecules, respectively.

1. Introduction

Twenty five years ago, Li, Evans and Simpson [1,2] established on a firm experimental basis the existence of a protein in bovine pituitaries with properties of a growth hormone. Since then many investigations have been performed with this and other similar proteins isolated from the anterior pituitary of a wide range of animal species [3]. It is now accepted that the presence of this hormone in the animal body is indispensable for it to achieve the adult state and endure a normal life thereafter [4,5].

Although it is clear that growth hormone exerts a regulatory role in nucleic acid and protein metabolism [6], the mechanism of the action is obscure.

It was a rather early finding that these hormones show species-specificity in their action: the rat is responsive to the hormones of several animal species and to that of the lungfish but not to the hormone from the teleostean fishes. Non-primate growth hormones are inactive in primates; bovine growth hor-

Table 1
Some physicochemical parameters of various growth hormones.

Physical or chemical parameter	Growth hormone from:									
	Man	Monkey	Ox	Sheep	Pig	Horse	Whale	Rabbit	Rat	Dog
Molecular weight $\times 10^{-3}$	21.7	23.0	21.0	20.3	22.0	$21 \rightarrow 45$	39.9	$24 \rightarrow 45$	20.5	–
Isoelectric point	4.9	5.5	6.9	6.3	6.3	6.2	6.2	6.1	5.9	6.3
S-S linkages	2	4	2	2	2	2 (per 21,000 g)	3	1.5 (per 24,000 g)	2	2
N-Terminal amino acid	Phe	Phe	Phe,Ala	Phe,Ala	Phe	Phe	Phe	Phe	Phe	Phe
C-Terminal amino acid	Phe	Phe	Phe	Phe	Phe	Phe	Phe	Phe	Phe	Phe
References	7	7	8	9	7,10	11	12	13	13,14	15

mone is effective in pigeons but not in the chicken while the guinea-pig is resistant to all growth hormones, including its own [3].

The molecular reasons for the biological inefficiency of mammalian growth hormones in man is a subject of great practical interest since it forms the basis of a rational approach to their modification for useful human use and may also be important for the understanding of their mechanism of action.

Many physical and chemical measurements have been carried out in several of these proteins (table 1) and although the results are not strictly comparable in all instances, they support, in general, the hypothesis put forward by Li [7] that the observed species-specificity in biological behaviour is related to molecular variations among the various growth hormones. Nevertheless many similarities do also exist and a symmetrical hypothesis can be stated, namely: analogous biological behaviour must have a common structural basis. Comparative structural studies together with physicochemical measurements performed under strictly comparable conditions, on as many hormones as possible, should give experimental support to one or both hypotheses.

The range of molecular weights recorded in table 1 seems to indicate great differences in size between the hormones of some species but it should be recalled that it is quite possible that the higher values correspond to aggregated preparations.

To complement the information gathered in table 1 the amino acid composition of 8 hormones is shown in fig. 1. The individual amino acid values have been expressed as percentages of the total number of residues in the molecule and are arranged in a graphical form that permits easy comparisons to be made. The sequence of amino acids in the abcissa starts with the small hydrophilic ones and ends with those of hydrophobic properties [16]. The graph makes readily apparent that all growth hormones compared have a very similar amino acid composition, the analogies in this respect being even higher than those found between two homologous molecules like the α and β chains of human hemoglobin (fig.1). In spite of this fact close examination of fig. 1 reveals at least three regions where human growth hormone differs significantly from the average composition of all the other hormones. Human growth hormone molecule has the lowest relative content of Ala, Gly, Arg and Lys and the highest of Glu, Asp and Ser. These modifications should account for the rather low isoelectric point singularly shown by human growth hormone (table 1).

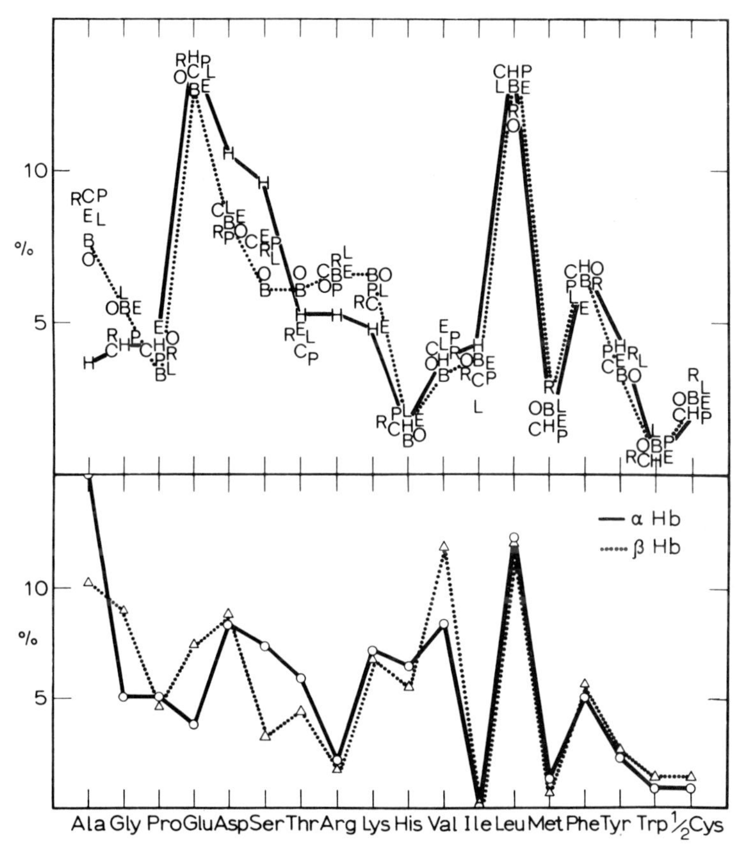

Fig. 1. Comparison of the amino acid composition of 8 mammalian growth hormones and of the α and β chains of human hemoglobin [9], presented in graphical form. The correspondence of letters and growth hormones is as follows: H, man [17]; B, ox [18]; O, sheep [9]; P, pig [10]; E, horse [11]; C, dog [15]; R, rat [13] and L, rabbit [13].

2. Do growth hormones really form a family of homologous proteins?

The similar physicochemical properties of the group of growth hormones already discussed, together with the close amino acid composition and identical *N*- and *C*-terminal amino acids are experimental evidence in favor of an affirmative answer to this question. Also the *C*-terminal sequences of 5 hormones shown in table 2 adds weight to the homology hypothesis.

A much more extensive comparison of primary structures can be now made between two of these hormones since Li et al. [17] have recently established

Table 2
C-Terminal amino acid sequence in various growth hormones.

Human: [17]	Arg - Ile - Val - Gln - Cys - Arg – Ser - Val - Glu - Gly - Ser - Cys - Gly - Phe.COOH
Bovine: [8]	Arg - Val - Met - Lys - Cys - Arg - Arg - Phe - Gly - Glu - Ala - Ser - Cys - Ala - Phe.COOH
Ovine: [9,20]	Met - Lys - Cys - Arg - Arg - Phe - Gly - Glu - Ala - Ser - Cys - Ala - Phe.COOH
Porcine: [10,19]	Arg - Val - Met - Lys - Cys - Arg - Arg - Phe - Val - Glu - Ser - Ser - Cys - Ala - Phe.COOH
Equine: [11]	Arg - Val - Met - Lys - Cys - Arg - Arg - Phe - Val - Glu - Ser - Ser - Cys - Ala - Phe.COOH

```
                      10                      20                      30
H: F P T I P L S R L F D N A M L R I S L L L I Q S W L E P V E
B: F P A M S L (A.P.V.T.E.K.A.T)- G I S - - - - E - W L F L R/A

                        40                      50
H: F A H R - L H E L A - F D T Y E E F E E A Y I - P K E Q K Y
B: - E H L E L(A.D.T.)A H F - - K E - F Z R T Y I Z P G Q R Q Y/

        60                  70                      80
H: S F L Q D P E T S L C F S E S I P T P S - N R E E T Q - K S
B:              F(T.S. -.C.-.G.E - =I.P.A.P)K E D L N E A Q E K S

           90                   100                     110
H: D L E L L R S V F A N S L V Y G A S N S D V Y D L L - K D L
B: D L E L L R S V F T N S L V F G T S D R - V Y E K N E K K L
            120                    130                     140
H: E E G I E T L M G R - L E D P S G R T G Q I F K E T Y S K F
B: E E G I - - L M - R Q L E D T G P R A G Q I L K T Q Y D K F

                150                     160                     170
H: D T N S H N D D A L L K D Y G L L Y C F R K D M D K V E T F
B: D D T M R S N D A L L K N Y G L L S C F K R N L H T K E T Y

               180
H: L R I V Q C R - S V E G S C G F *
B: L R V M K C R R F G E A S C A F *
```

Fig. 2. Alignment for maximum homology of the amino acid sequences in HGH and in BGH. The special punctuation rules and the one-letter amino acid abbreviations used by the Atlas of Protein Sequence and Structure [16] have been followed. H: HGH; B: BGH.

the complete amino acid sequence of the human protein and in our laboratory [8,24] we are presently working in the last details of this structure in BGH*.

Both sequences are shown in fig. 2 aligned side by side so as to present maximum homology. The special punctuation rules used by the Atlas of Protein Sequence and Structure [16] have been followed. Disulphide bridges occur between positions 68 and 162 and between 179 and 186 in the human hormone, and in corresponding positions in the bovine protein. Although fig. 2 shows that both molecules have several homologies spread throughout the complete sequence it is inadequate to analyze them in terms of frequency of identities, changes or deletions along the chain. This study is greatly facilitated by adoption of a graphical representation of the homology between

* *Non-standard abbreviations*: HGH: Human growth hormone; BGH: bovine growth hormone; OGH: ovine growth hormone; PGH: porcine growth hormone; EGH: equine growth hormone; ORD: optical rotatory dispersion.

both molecules that takes into account the probability of each amino acid exchange. This last information has been obtained by Dayhoff et al. [21] by the comparison of a great number of sequences within families of homologous proteins.

In fig. 3 individual amino acid changes are indicated by discrete short vertical bars placed in the plane determined by two ortogonal cartesian axes. The horizontal one shows the number of the amino acid in the sequence while in the vertical axis different levels indicate identity, mutation or deletion. In the mutation category each change is represented at one of five different levels according to its probability of occurrence.

The type and extent of the homology existing between the two proteins

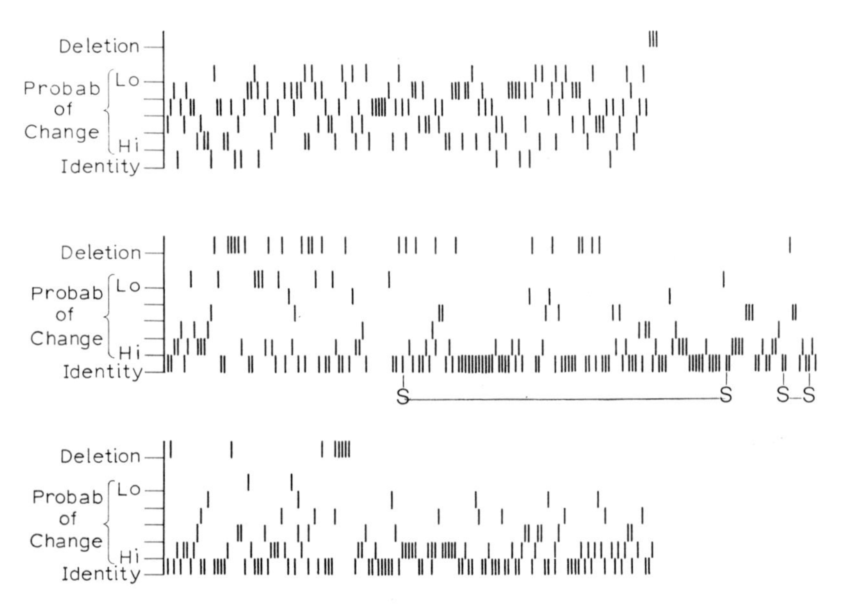

Fig. 3. Graphical representation of the homology existing between pairs of proteins. Individual amino acid changes are indicated by short vertical bars placed in the plane determined by two orthogonal cartesian axes. The horizontal one shows the number of the amino acid in the sequence while in the vertical axis different levels indicate identity, mutation or deletion. In the mutation category each change is represented at one of five different levels according to its probability of occurrence [16,21].
Top graph: β chain of human hemoglobin [22] and *S. aureus* nuclease [23].
Middle graph: Human growth hormone [17] and bovine growth hormone [24].
Bottom graph: α and β chains of human hemoglobin [22].

being compared is denoted by the horizontal and vertical spread of the bars and these characteristics as well as its regional variations are immediately apparent in fig. 3. The comparison of the sequences of human β hemoglobin with that of a totally unrelated protein gives a graph (fig. 3, top) with the bars almost uniformly distributed all over the plane as should be expected when chance alone is operating.

The pattern obtained is clearly different in the case of two homologous proteins like human α and β hemoglobins (fig. 3, bottom) where more than 70% of the bars are concentrated on the levels corresponding to identity or highly probable mutations. The graph obtained in the comparison of HGH and BGH sequences (fig. 3, middle) shows striking changes in homology as one progresses along the chain: in the first 72 amino acids (in HGH) there are 27 identities and 54 changes; in the next 90 amino acids the identities increase to 50 and there are only 31 changes; in the last 26 amino acids the proportion of identities to changes is almost one, 11 and 16, respectively. This analysis indicates that the evolutionary distance between HGH and BGH varies according to the region of the chain being compared. The middle section shows a high degree of relatedness, similar to that existing between the β and γ chains of human hemoglobin; the comparison of the *C*-terminal region indicates an intermediate homology similar to that of the α and β chains of human hemoglobin. Finally, in the *N*-terminal region both proteins have a very low degree of homology.

In the first 72 amino acids of human and bovine growth hormones, a number of radical amino acid substitutions occur and they should strongly influence the conformation of this region of the molecule.

The variability of sequence found in the *N*- and *C*-terminal sections of human and bovine growth hormones provides the only example of this phenomenon known to occur in homologous proteins outside the field of immunoglobulins where it has aroused much interest and speculation [25]. The problem is how could this variability have arisen in terms of present knowledge of genetic control of protein structure? It is assumed that this sequence variation must be the basis of antibody combining specificity and it is tempting to explain the species specificity of growth hormones on the same basis, although no direct evidence is available as yet, in either field.

We might speculate that the variable sequences are like the activation peptide for serine proteinases (chymotrypsin, trypsin, trombin, etc.). In that case the animal species insensitive to a particular growth hormone would be unable to release these peptides and uncover the active part of the hormone. Experiments with protease digested hormones pointing to the possible existence of a common active "core" [26] would seem to support the above hypothesis.

3. Conformation studies on various growth hormones

Information related to the tertiary structure and motility in solution of some of these proteins has been gathered in our laboratory by studies of the rate of dialysis through calibrated membranes and by measurement of the rate of exchange of the hydrogens in the molecule. The different growth hormones used in these experiments were prepared or obtained as follows: *human*, by the procedure of Roos et al. [27] ; *bovine and equine,* by the method of Dellacha and Sonenberg [28] ; *ovine*, by the method of Dellacha et al. [9] ; *porcine*, was preparation P-522 A kindly given to us by Professor A.E. Wilhelmi. All these proteins were biologically highly active, homogeneous in the ultracentrifuge and gave the usual patterns of pure preparations in polyacrylamide-gel electrophoresis.

4. Dialysis studies

Analytical dialysis through cellophane membranes is a procedure developed by Craig [29] which has a great sensitivity to detect molecular changes of size and shape. A monodisperse protein crosses the membrane following a first order kinetics which gives a straight line in a semi-logarithmic plot. The half-life, $t_{1/2}$, is characteristic of the rate which, in turn, depends on the size and shape of the molecule. A polydisperse protein gives rise to breaks in the curves.

Previous work from our laboratory [30] has shown that BGH has its maximum stability in solution in a glycine-hydrochloric acid buffer of pH 3.6 and I = 0.005. If the pH is lowered to 2.2 the molecule suffers a conformational

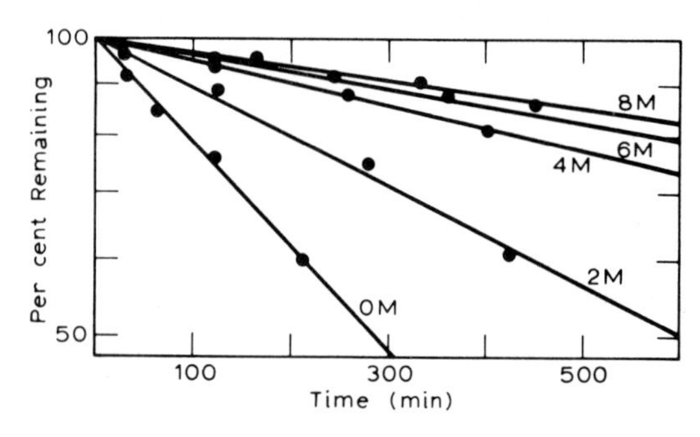

Fig. 4. Analytical dialysis. Escape patterns of BGH dissolved in 0.1 M glycine-hydrochloric acid buffer, pH 3.6, additioned of variable amount of urea, as indicated.

change that modifies its $t_{1/2}$ of dialysis from 100 min, at pH 3.6, to more than 400 min at pH 2.2. This phenomenon has been independently established by Burger et al. [31] using optical methods.

Analytical dialysis shows very clearly the gradual expansion of the molecule of BGH in solutions of urea of increasing concentrations (fig. 4). This expansion is completely reversible when the urea is eliminated as judged by the unaltered biological activity, $t_{1/2}$ of dialysis, intrinsic viscosity, elution volume in Sephadex columns and sedimentation rate.

The effect of a lyophilization step on the tertiary structure of the BGH molecule is clearly shown by dialysis (fig. 5). The change is reversible.

The complete reduction of the disulphide bridges in BGH causes an extensive modification of the tertiary structure of the molecule (fig. 6), but reoxidation under controlled conditions [32] restores the native conformation as judged by biological activity and analytical dialysis. Hydrogen exchange measurements are more sensitive and indicate that the reoxidized molecule is more permeable to the solvent.

5. Hydrogen exchange studies

Detailed information about the secondary and tertiary structure of several growth hormones was obtained by measurement of the motility of the hydrogens in their molecules [34]. In our experiments we have adopted the tracer

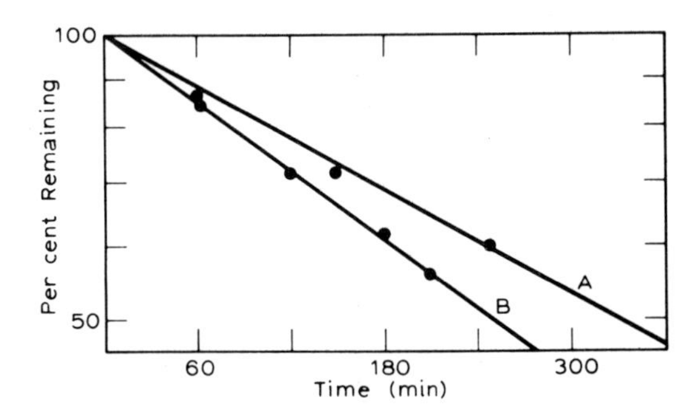

Fig. 5. Analytical dialysis; effect of lyophilization. Escape patterns obtained in 0.1 M glycine-hydrochloric acid buffer, pH 3.6, with hormones treated as follows:

Lyophilized (A), the dry hormone was dissolved in the pH 3.6 buffer immediately before the measurement.

Lyophilized and incubated (B), the solution of the dry hormone was kept 12 hr at 4° before the dialysis.

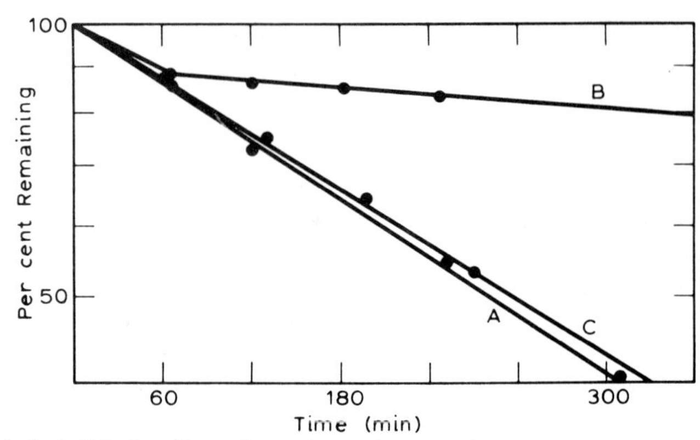

Fig. 6. Analytical dialysis, effect of complete reduction of the disulphide bridges in BGH by the method of Anfinsen and Haber [33] and controlled reoxidation. Concentration of the hormone during reoxidation: 1 mg/ml. Protein recovered: 85%; biological activity of reoxidized BGH: 100%. Native hormone (A); Reduced hormone (B); Reoxidized hormone (C).

method of Englander [35], using tritiated water. This author eliminates the excess of label by gel filtration; we have used for this purpose, the rapid counter-current dialysis apparatus of Craig [36].

Experiments in which the rate of incorporation of tritium into the protein is measured show that complete equilibration is reached in 50 to 60 hr at 4° and pH 3.6, and in about the same time at pH > 9. Accordingly 60 hr at 4° were the incubation conditions regularly used in all our experiments.

Fig. 7 shows the out-exchange data obtained with human, bovine, ovine, porcine and equine growth hormones at pH 3.6 and the exchange-out curves of HGH and BGH at pH 7.5. All these experiments were done at 22°.

The kinetic analysis [35] of the exchange-out curves permits the calculation of the size and half-life time of each class of mobile protons in the hormones studied (table 3). The same method of analysis has been applied to the exchange curves of HGH published by Squire and Ottesen [37]. These authors measured the D/H exchange by infrared spectroscopy that detects peptidic hydrogens, exclusively. Since the values obtained by them at acid pH are similar to those given by our procedure, (table 3), it can be tentatively concluded that our measurements also detect peptidic hydrogens preferentially.

The rate of exchange of these protons is measurable because they are involved in some kind of secondary or tertiary interactions within the molecule, and a likely possibility is α-helix formation.

At pH 3.6 the α-helix content in HGH has been estimated by ORD as 32%

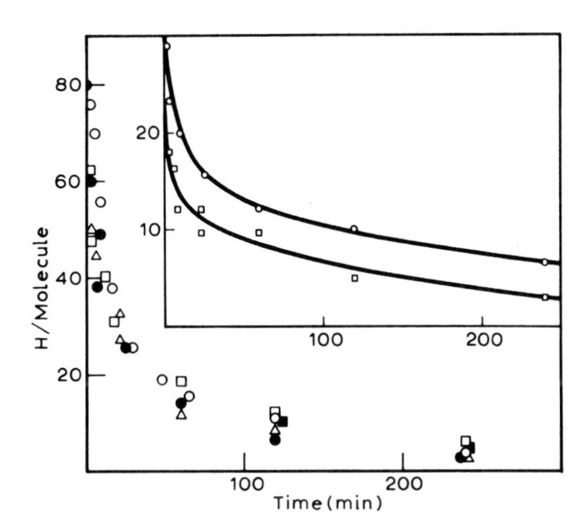

Fig. 7. Out exchange data obtained with various growth hormones.

Main graph: experiments carried out in 0.1 M glycine-hydrochloric acid buffer, pH 3.6, I : 0.005.

Inset: experiments carried out in 0.02 M sodium phosphate-0.15 M sodium chloride buffer, pH 7.5. BGH (◉); HGH (□); OGH (△); PGH (●); EGH (■).

[38] or 55% [39]. The percentage of slow peptidic hydrogens found in our experiment is 43%.

In BGH, at the same pH, the ORD measurements indicate 32% [38] or 40% [31] of α-helix, compared to our exchange measurements that give 48% of slow hydrogens. The coincidence between both set of values is very good and it is tempting to generalize its meaning and to conclude that human,

Table 3

Kinetic classes of mobile protons in human, bovine, ovine and porcine growth hormones.

Half-life time (min)	Peptidic H* pH 3.6	HGH Total H pH 3.6	HGH Total H pH 7.5	BGH Total H pH 3.6	BGH Total H pH 7.5	OGH Total H pH 3.6	PGH Total H pH 3.6
Fast	52						
2 – 8	54	46	13	56	16	30	69
18 – 70	15	17	0	15	9	19	16
170 – 180	11	14	10	12	11	8	14
Slow	46						

* Values calculated from the curves published by Squire and Ottesen [37].

bovine, ovine, porcine and equine growth hormones possibly have very similar secondary and tertiary structures, at pH 3.6 and at low ionic strength. This statement must remain tentative considering the many uncertainties that still prevail in the interpretation of exchange [34] and ORD measurements [40] in terms of protein structure.

H-Exchange measurements are usually done at acid pH because the rates are not so fast in this region, but the information obtained is only remotely connected to the physiological properties of the hormones.

At neutral pH, HGH and BGH have a similar exchange behaviour (fig. 7) but the number of measurable hydrogens is very low due to the base catalysis effect.

Squire and Ottesen [37] found that HGH at pH 9.3 exchanged completely its peptidic hydrogens in less than 30 min. We have confirmed this finding also at pH 7.5 and found the same behaviour in BGH. The greater permeability to the solvent of these molecules at neutral or alkaline pH indicates that they are more expanded in this condition and suggested a procedure to estimate the size of the area exposed: if the conformation change is reversible, incubating the hormone with tritiated water at neutral pH and then shifting the pH to 3.6, a new class of slowly exchangeable hydrogens should appear. Its size must be proportional to the area of the molecule exposed at the higher pH.

In order to attempt the measurement described we first investigated if the proteins recovered their original conformation at pH 3.6 after an incubation of 60 hr at 4° at pH 7.5. This control could be done by following the analytical dialysis behaviour of the hormones before and after the pH change. The very similar half-life time values obtained (table 4), were a sensitive indication of the reversibility of the expansion. As a further probe of the flexibility of these proteins, the extent and reversibility of the unfolding action of 10 M urea was explored in a similar way (table 4). HGH is the only hormone affected by this last treatment and the great increase in its half-life time of dialysis, after the conformation change, could be traced to an aggregation process since the preparation had a sedimentation coefficient 3.4 times greater than that of the native hormone. Nevertheless, control experiments indicated that the exchange behaviour of the aggregated protein was identical to that of the native hormone suggesting that the monomers had a native conformation.

Further evidence indicating the reversibility of the pH, or urea induced, conformation changes is shown in fig. 8: the exchange-out curves obtained after a conformation change have very similar shapes to those given by the native protein. The obvious difference is the clear indication of new classes of very slow hydrogens. The kinetic analysis of the curves discovers the same

Table 4
Analytical dialysis of growth hormone submitted to changes in conformation induced by variations of pH or 10 M urea.

Treatment	Half-life time values (min)			
	HGH	BGH	OGH	EGH
pH 7.5 → pH 3.6	121	210	174	160
Control	106	175	180	170
pH 3.6, 10 M urea	400	80	163	160
Control	106	75	180	170

The treatments are as follows:

pH 7.5 → pH 3.6: a solution of the hormone in 0.02 M sodium phosphate-0.15 M sodium chloride buffer, pH 7.5 was incubated 60 hr at 4° and then brought to pH 3.6 by addition of 20 volumes of glycine-hydrochloric acid buffer, pH 3.6, I : 0.005. The original volume was restored by ultrafiltration through a Diaflo membrane UM-10 (Amicon Corp., 21 Hartwell Ave., Lexington, Mass., USA). The analytical dialysis was performed after 12 hr at 4°.

pH 3.6, 10 M urea: a solution of the hormone in the glycine-buffer, pH 3.6 containing 10 M urea, was incubated 60 hr at 4° then dialyzed against the same buffer without urea and kept 12 hr at 4° before analytical dialysis.

Control: a solution of the hormone in the glycine buffer of pH 3.6, incubated 60 hr at 4°.

Table 5
Kinetic classes of protons in human and bovine growth hormones before, and after conformational changes induced by alkali or urea.

Half-life time (min)	H/molecule			
	HGH		BGH	
	pH 3.6	pH 9.3 → pH 3.6	pH 3.6	pH 3.6, 10 M urea
2 – 7	46	35	56	56
25 – 46	17	13	15	20
175 – 270	14	13	12	17

The meaning of the headings: pH 3.6; pH 9.3. → pH 3.6 and pH 3.6, 10 M urea is explained in the legend to fig. 8.

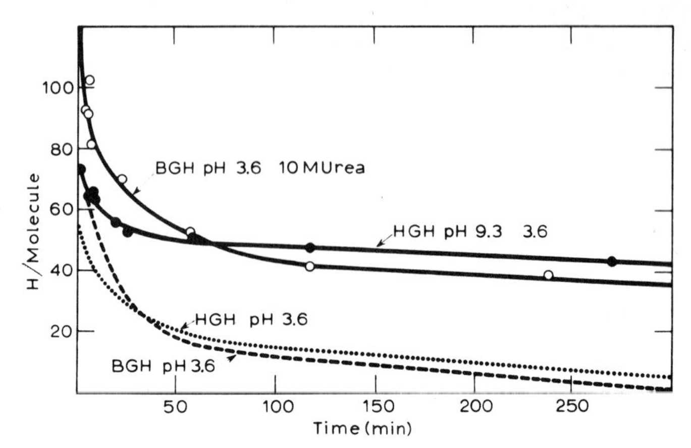

Fig. 8. Conformation changes in HGH and BGH measured by T/H exchange.

HGH, BGH, pH 3.6: exchange-out curves for these hormones dissolved in 0.1 M glycine-hydrochloric acid buffer, pH 3.6 (see fig. 7);

HGH, pH 9.3 → pH 3.6: the hormone dissolved in 0.1 M glycine-sodium hydroxide, pH 9.3 was incubated 60 hr at 4°, in the presence of 10^8 cpm/ml of tritium; at the end of this period the solution was acidified to pH 3.6 with 12 M hydrochloric acid and left 12 hr at 4° before the measurements;

BGH, pH 3.6, 10 M urea: the protein was dissolved in 0.1 M glycine-hydrochloric acid buffer pH 3.6, made 10 M in urea, plus the same amount of titrium as before. After 60 hr at 4°, the solution was dialyzed against the same buffer without urea, and left 12 hr at 4° before the measurements.

classes of measurable protons with reasonable close sizes (table 5).

With the above described evidence indicating the reversibility of the conformation changes induced in growth hormones by pH changes or 10 M urea, we could proceed to estimate the size of the new classes of slow hydrogens thus appearing. The exchange-out curves obtained before and after a conformation change become parallel to each other after about 100 min of exchange (fig. 8). The difference in ordinates between them measures the number of hydrogens trapped by the conformation change. These values have been collected in table 6. Significant differences in the degree of opening of the various molecules at physiological pH were apparent. If we admit that the newly labeled hydrogens represent peptidic protons they would correspond, in the case of HGH, to an extra exposure to the solvent, at pH 7.5, of 20% of its amino acid chain. This figure should be compared to values approximately one half to one tenth as big for the other hormones.

The greater sensitivity of HGH to unfolding was also clearly shown by the action of 10 M urea at pH 3.6: this hormone was labeled to a greater extent than all the others, indicating a considerably less tight tertiary structure. It is

Table 6
Hydrogens trapped by reversible conformation changes induced by different treatments on various growth hormones.

Exp. No.	Treatment	Unexchanged hydrogens per molecule at 120 min				
		HGH	BGH	OGH	PGH	EGH
1	pH 7.5 → pH 3.6	48	25	13	17	23
2	pH 3.6	13	11	9	7	12
	Exp.No.1 - Exp.No.2	*35*	*14*	*4*	*10*	*11*
3	pH 3.6 + 10 M urea	59	43	27	29	23
	Exp.No.3 - Exp.No.2	*46*	*32*	*18*	*22*	*11*

The treatments indicated are explained in the legend to table 4 but all the buffers had 10 cpm/ml of titrium added as titriated water.

of interest to point out that the number of hydrogens labeled by the influence of concentrated urea coincides with the total number of very slowly labeled peptidic hydrogens detected by Squire and Ottesen [37] (table 3). According to these data 10 M urea completely unfolds HGH; this is not the case for the other hormones (table 6).

The exchange characteristics of the zones of HGH and BGH exposed to the solvent at pH 7.5 could be explored in the following way: a solution of the hormone at this pH was incubated at 4° with 10^8 cpm/ml of tritium. At various times the pH of a suitable sample was lowered by addition of twenty volumes of tritiated glycine buffer of pH 3.6 and the original volume was restored by ultrafiltration, as indicated in the legend to table 4. After 12 hr at 4° free tritium was eliminated by rapid dialysis and the sample was left to out-exchange for 2 hr at room temperature before the final measurement. By this procedure it was considered that the remaining label was located in the region of the protein reversibly affected by the pH change. The number of labeled hydrogens from other regions of the molecule increases with increasing times of incubation but their maximum value cannot be higher than 13 for HGH or 11 for BGH, since these are the numbers found at pH 3.6 after 2 hr of out-exchange (fig. 7 and table 6).

The values obtained, expressed in hydrogens per molecule, were, HGH: 1 min, 10; 10 min, 17; 22 hr, 29; 55 hr, 48; BGH: 1 min, 11; 30 min, 12; 4 hr, 15; 23 hr, 17; 60 hr, 25.

These results indicate the existence of secondary and perhaps, tertiary structure, together with a small area of very fast exchange capacity, in the region of the HGH molecule uncovered at pH 7.5. This is not the case for BGH where the whole area exposed at pH 7.5 has very little organization and exchanges rapidly its hydrogens with those of the solvent.

The unique motility of HGH is the result of its peculiar amino acid composition and sequence which has a highly homologous "core" with BGH but discrepant *N*- and *C*-terminal regions. It is quite possible that a similar relationship exists between HGH and the other non-human growth hormones in which case it can be postulated that the biological activity of these proteins is a property of their common "core" while the species-specificity is controlled by the variable regions in their molecules.

Acknowledgements

The authors are obliged to the following members of the Department of Biological Chemistry who have contributed their unpublished results for this presentation: Drs. Mirtha J. Biscoglio, Silvia T. Daurat, César L. Cambiaso, Clara Peña, Edgardo Poskus, Lilia A. Retegui, Angelina V. Fontanive and Carlota E.M. Wolfenstein.

Dr. A.E. Wilhelmi generously provided pure porcine growth hormone. The able technical assistance of Dora M. Beatti, Nelly Ramos, Oscar A. Duffort, Juan P. Hecht and Héctor E. Casado, is gratefully acknowledged.

The authors are Career Investigators of the Consejo Nacional de Investigaciones Científicas y Técnicas de la República Argentina.

This work has been supported, in part, by grants from the same Institution.

References

[1] C.H. Li and H.M. Evans, Science 99 (1945) 183.
[2] C.H.Li, H.M. Evans and M.E. Simpson, J. Biol. Chem. 159 (1945) 353.
[3] I.I. Geschwind, Am. Zoologist 7 (1967) 89.
[4] Growth Hormone, ed. M. Sonenberg, Ann. N.Y. Acad. Sci. 148 (1968) Art. 2.
[5] Growth Hormone, eds. E. Muller and A. Pecile (Excerpta Medica Foundation Int. Cong. Ser., Amsterdam, 1968) Ser. 158.
[6] B.W. O'Malley, Trans. N.Y. Acad. Sci. Series II, 31 (1969) 478.
[7] C.H.Li (Excerpta Medica Foundation Int. Congr. Ser., Amsterdam, 1968) Ser. 158, p. 3.
[8] J.A.Santome, J.M.Dellacha and A.C.Paladini (Excerpta Medica Foundation Int. Congr. Ser., Amsterdam, 1968) Ser. 158, p. 29.
[9] J.M.Dellacha, M.A.Enero, A.C.Paladini and J.A.Santome, European J. Biochem. 12 (1970) 289.
[10] A.E.Wilhelmi, personal communication.

[11] L. Oliver and A.S. Hartree, Biochem. J. 109 (1968) 19.
[12] H. Papkoff and C.H. Li, J. Biol. Chem. 231 (1958) 367.
[13] S. Ellis, R.E. Grindeland, J.M. Nuenke and P.X. Callahan, Ann. N.Y. Acad. Sci. 148 (1968) 328.
[14] W.E. Groves and B.H. Sells, Biochim. Biophys. Acta 168 (1968) 113.
[15] A.E. Wilhelmi, Yale, J. Biol. Med. 41 (1968) 199.
[16] Atlas of Protein Sequence and Structure, Vol. 4, ed. M.O. Dayhoff (National Biomedical Research Foundation, Maryland 20901, 1969) p. 85.
[17] C.H. Li, J.S. Dixon and W.K. Liu, Arch. Biochem. Biophys. 133 (1969) 70.
[18] C.E.M. Wolfenstein, J.A. Santome and A.C. Paladini, Acta Physiol. Latinoam. 16 (1966) 194.
[19] J.B. Mills, Nature 213 (1967) 631.
[20] C. Pena, J.M. Dellacha, A.C. Paladini and J.A. Santome, unpublished observations.
[21] M.O. Dayhoff, R.V. Eck and C.M. Park, in: Atlas of Protein Sequence and Structure, Vol. 4, ed. M.O. Dayhoff (National Biomedical Research Foundation, Maryland, 20901, 1969) p. 75.
[22] G. Braunitzer, R. Gehring-Muller, N. Hilschmann, K. Kilse, G. Hobom, V. Rudloff and B. Wittmann-Liebold, Z. Physiol. Chem. 325 (1961) 283.
[23] H. Taniuchi, C.B. Anfinsen and A. Sodja, J. Biol. Chem. 242 (1967) 4752.
[24] J.A. Santome, J.M. Dellacha, A.C. Paladini, C.E.M. Wolfenstein, C. Pena, E. Poskus, S.T. Daurat, Z.M. Sese, A.V.F. Sanguesa and M.J. Biscoglio, in: Atlas of Protein Sequence and Structure, Vol. 4 ed. M.O. Dayhoff (National Biomedical Research Foundation, Maryland 20901, 1969) D-159.
[25] R.R. Porter, Structure of Immunoglobulins, in: Essays in Biochemistry (Academic Press, London, New York, 1969) p. 3.
[26] M. Sonenberg, M. Kikutani, C.A. Free, A.C. Nadler and J.M. Dellacha, Ann. N.Y. Acad. Sci. 148 (1968) 532.
[27] P. Roos, H.R. Fevold and C.A. Gemzell, Biochim. Biophys. Acta 74 (1963) 525.
[28] J.M. Dellacha and M. Sonenberg, J. Biol. Chem. 239 (1964) 1515.
[29] L.C. Craig, in: Advances in Analytical Chemistry and Instrumentation, Vol. 4, ed. C.N. Reilley (Interscience, New York, 1965) p. 35.
[30] J.M. Dellacha, J.A. Santome and A.C. Paladini, Ann. N.Y. Acad. Sci. 148 (1968) 313.
[31] H.G. Burger, H. Edelhoch and P.C. Condliffe, J. Biol. Chem. 241 (1966) 449.
[32] A.V. Fontanive de Sanguesa, J.M. Dellacha, J.A. Santome and A.C. Paladini, unpublished observations.
[33] C.A. Anfinsen and E. Haber, J. Biol. Chem. 236 (1961) 1361.
[34] A. Hvidt and S.O. Nielsen, Advan. Protein Chem. 21 (1966) 287.
[35] S.W. Englander, Biochemistry 2 (1962) 798.
[36] L.C. Craig and H.C. Chen, Anal. Chem. 41 (1969) 590.
[37] P.G. Squire and M. Ottesen, Biochim. Biophys. Acta 154 (1968) 226.
[38] A.H. Tashjian, L. Levine and A.E. Wilhelmi (Excerpta Medica Foundation Int. Congr. Ser., Amsterdam, 1968) Ser. 158, p. 70.
[39] T.A. Bewley and C.H. Li, Biochim. Biophys. Acta 140 (1967) 201.
[40] S. Beychok, Ann. Rev. Biochem. 37 (1968) 437.

PART 2

METABOLIC ALTERATIONS IN CANCER

ENVIRONMENTALLY INDUCED METABOLIC OSCILLATIONS AS A CHALLENGE TO TUMOR AUTONOMY*

V.R. POTTER

McArdle Laboratory, University of Wisconsin Medical School, Madison, Wisconsin, 53706, USA

Abstract: Potter, V.R. Environmentally Induced Metabolic Oscillations as a Challenge to Tumor Autonomy. *Miami Winter Symposia,* 2, pp. 291–313. North-Holland Publishing Co., Amsterdam, 1970.

Studies on *ontogeny* at the molecular level are providing new stimuli for the study of the malignant transformation at the molecular level, which in this context can be appropriately called *oncogeny*. The control of phenotypic variation in terms of enzymes associated with the maturation of parenchymal liver cells has been studied in fetal and newborn rat liver and in a series of transplantable minimal deviation hepatomas provided by Dr. Harold P. Morris. This report reviews data on nine Morris hepatoma lines in terms of the proposition that (a) whereas we no longer think of the enzyme pattern of normal liver as fixed, and (b) whereas we have a series of phenotypically diverse transplantable hepatomas that at least superficially resemble the liver phenotype at the molecular level, (c) therefore we ought to study the control factors that regulate the liver phenotype at each stage in its development and determine whether these factors produce in the hepatomas responses that will help us evaluate the hypothesis that "oncogeny is blocked ontogeny".

A set of conditions that would permit greater control of behavior and food intake in experimental rats was described and advocated in preference to *ad libitum* feeding. Diets in nearly all experiments described contained 12%, 30% and 60% protein with compensating shifts in the glucose content. Some experiments also included 0% and 90% protein (= 0% glucose).

Animals were killed at various times in each 24-hr cycle (12 hr light, 12 hr dark) in order to locate the maximum and minimum levels of enzyme activity if this was variable under standard conditions, or if it could be varied by experimental manipulation.

* Dedication: This paper was part of The Jakob A. Stekol Memorial Session in The Symposium on Metabolic Alterations in Cancer, The Second Annual Biochemistry-PCRI Winter Symposium, The University of Miami, Miami, Florida, January 19-23, 1970. The work reported in this paper was supported in part by Departmental Grant CA-07125 and Training Grant T-01-CA-5002 from the National Cancer Institute, USPHS.

A number of enzymes was studied but the present report emphasized tyrosine transaminase which reaches a peak in rat liver at 6 hr after the start of food consumption and rises to levels that are proprotional to dietary protein. Activity then falls to low levels that are fairly constant from the 12th to the 24th hr after the 8 hr period of food intake begins. Serine dehydratase and glucose-6-phosphate dehydrogenase were also reported in some experiments.

The fluctuations in tyrosine transaminase activity were correlated with fluctuations in the AIB (aminoisobutyric acid) pump, studied in terms of the concentrating capacity of a tissue relative to blood. Thus after theophylline injections, the transaminase activity rises, confirming Fuller and Snoddy, and the "pump" activity also rises. Since theophylline inhibits cyclic AMP diesterase there is an implication that cyclic AMP may be a mediator between several physiological factors and enzyme induction or blocked degradation.

Marked diversity in the activity of the mentioned enzymes was demonstrated in a series of 9 Morris hepatomas, and parallel diversities in AIB pump activity were noted.

Responses of enzyme activity and pump activity to hydrocortisone and/or glucagon demonstrated that the response of the hepatomas to these agents is not a constant feature from one hepatoma line to another, just as it is not a constant feature from one stage of liver development to another. Much more systematic work on protocols for quantitating the contributions of the factors and the nature of the phenotypic diversity in the hepatomas is needed to carry out the initial proposition and this work is in progress.

1. Introduction

There is an interesting relationship between the subject of this opening lecture in the symposium on: Metabolic Alterations in Cancer and the preceding two-day symposium on: Homologies in Enzymes and Metabolic Pathways which was opened by Professor George Wald, speaking on: Ontogeny and Phylogeny at the Molecular Level. Studies on ontogeny at the molecular level are providing new stimuli for the study of the malignant transformation at the molecular level, which in this context can be appropriately called *oncogeny* (table 1). We no longer can be satisfied with a comparison between cancer cells and the fully differentiated cells in the *adult* organism. Instead, if we are fortunate enough to have a series of cancer cells in which homologies with normal cells can be detected, such as the transplantable minimal deviation hepatomas provided by Dr. Harold Morris, we must turn our attention to the process of *ontogeny*, which is now amenable to study at the molecular level as we have seen here. A number of investigators have described isozymes and other proteins characteristic of fetal and newborn liver and differing from their corresponding proteins in adult tissue. It has turned out that many of the proteins characteristic of fetal and newborn liver have now been demon-

Table 1
Ontogeny and Oncogeny

Oncology = The study of tumors		*Ontology* = The study of beings
Oncogeny = The formation of tumors		*Ontogeny* = The formation of beings

Oncogeny as a "locked-in" stage of ontogeny	: Potter	[1]
Cancer as "dys-differentiation"	: Matsushima et al.	[2]
Cancer as "unbalanced retro-differentiation"	: Uriel	[3]

strated to occur in various lines of Morris hepatomas (references in [1]).

Because of the overlapping isozyme distribution in fetal liver and in hepatoma tissue and for other reasons, we were led to describe cancer formation or *oncogeny as a blocked ontogeny,* and to inquire into the possibility of changes in adult liver cells that would cause them to *revert* to an immature state and cause some of them to be blocked in this state and thus to be different from normal immature cells [1]. That is, if a cell or cells contained a defect or defects that would give them some of the properties of immature cells in the maturation series but in addition would prevent them from moving forward to the adult, differentiated, mature, and non-dividing state we would call them cancer cells if they continued to divide under conditions in which normal cells proceeded to the non-dividing state, regardless of the phenotypic diversity they might exhibit (fig. 1).

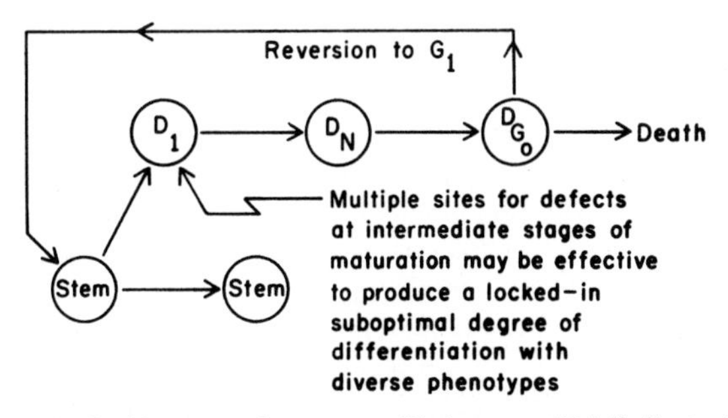

Fig. 1. *Oncogeny* as a locked-in stage of ontogeny. We interpret Uriel's "retrodifferentiation" [2] as a reversion of a differentiated G_0 cell (D_{G_0}) to a stem cell, followed by a failure in the differentiation process somewhere between the first stage (D_1) and the Nth stage (D_N).

This line of thought can be appropriately inserted between the earlier symposium and the paper on: Molecular Mechanism of Carcinogenesis by Dr. E. Farber, which follows this one, and we look forward to his report on the hyperplastic nodule in relation to the processes of oncogeny and ontogeny in liver.

So we ask, what are the characteristics of the immature partially differentiated liver cell that might be found in a minimal deviation hepatoma, and what kind of a defect or defects might freeze the cell in the transformed state? An interesting issue at this point is whether the hepatoma cells, seemingly frozen in the transformed state, are really frozen, i.e. are they irreversibly altered or could the change be reversed [4]? Although it is conceivable that at some stage the process might be reversible and some tumors might pause at the reversible stage it appears that the vast majority of neoplasms quickly progress to an irreversible state.

These introductory remarks relating to the papers in the earlier symposium lead to my major proposition in this paper, which is that (a) whereas we no longer think of the enzyme pattern of normal liver as fixed, and (b) whereas we have a series of phenotypically diverse transplantable hepatomas that at least superficially resemble the liver phenotype at the molecular level, (c) therefore, we ought to study the control factors that regulate the liver phenotype at each stage in its development and determine whether these factors produce in the hepatomas responses that will help us evaluate the hypothesis that "oncogeny is blocked ontogeny".

2. Studies on adult normal liver: ad libitum vs. controlled feeding

Any attempt to study the dynamics of rapidly changing enzyme activities in the tissues of laboratory animals must cope with the problem of variation among individual animals, since in general it is necessary to kill fairly large numbers of animals at short time intervals and to try to have enough animals to draw curves of oscillating or slowly changing enzyme activity with reasonable confidence. Prior to 1961 we instituted a system of automatic lighting with 12 hr light and 12 hr of darkness and observed very large transient increases in glucose 6-phosphate dehydrogenase activity (G6PDH)* in rat liver when rats were fed after a 2 or 3 day fast, confirming Tepperman and Tepperman [5]. We also demonstrated that the ordinary daily changes in G6PDH activity were much smaller, but the oscillation in glycogen level was very pro-

* Code numbers of enzymes: TAT, tyrosine transaminase (aminotransferase, EC 2.6.1.5), SDH, L-serine dehydratase (EC 4.2.1.13), G6PDH, glucose-6-phosphate dehydrogenase (EC 1.1.1.49).

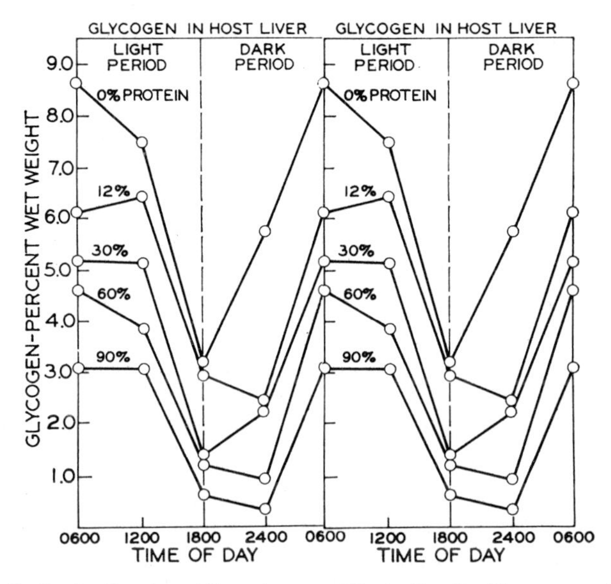

Fig. 2. Daily fluctuations in rat liver glycogen. Illustrating the 20-group protocol employing 5 protein levels and 4 killing times with data replotted to show that each average value on the curves lies between 2 other average values. The protocol is 3-dimensional but can be plotted as a family of curves in either dimension. This experiment involves only 2 rats per average shown, but there were 40 rats in the experiment. (Data plotted from Chart 6 in [6]).

nounced [5]. These experiments led to the design of an experimental protocol in which food was available only during the 12 hr dark period (18:00-06:00) and was withdrawn during the 12 hr light period (06:00-18:00) to coincide with the rats' normal nocturnal feeding habits. These experiments were begun in 1963 and first reported in 1966 [6]. Since we were interested in testing the adaptive response of minimal deviation hepatomas to environmental change we also varied the protein content of the diet to provide 5 groups of rats receiving 0, 12, 30, 60 and 90% protein, respectively, with compensating shifts in the glucose content of the diet. Tumor-bearing rats were killed at 06:00 (end of feeding period), 12:00 (middle of fasting period), 18:00 (start of feeding period) and 24:00 (middle of feeding period), thus giving a total of 20 groups based on variations in dietary protein and killing times. Since the animals were adapted to the regimen for a much longer period (44 days) than the time required for equilibrium (usually 5-10 days) every time point can be regarded as lying between 2 other time points and the data can be plotted with time as the abscissa repeating the data in order to demonstrate its oscillatory nature. The orderly nature of the experimental groups is well illustrated by the glycogen oscillations, especially when it is considered that each point is the average from only 2 animals (fig. 2 based on Chart 6 in [6]). This chart illustrates the fact that the protocol is 3-dimensional and each average value is buttressed by either 3 or 4 other average values that surround it.

Among a series of 5 enzymes studied by means of the 20-group protocol, it was noticed that tyrosine transaminase (TAT) showed a remarkable fluctuation in which the increases in the host liver between 18:00 and 24:00 were proportional to the protein content of the diet. This 1966 report was to my knowledge the first indication of a marked (i.e. 8-fold) daily fluctuation of an enzyme activity, arrived at by the application of the concept of circadian rhythms but it was stated "We have avoided the use of the term 'circadian rhythm' not because we minimize its importance but because we have not attempted to prove that the observations are due to circadian rhythms. Our studies involved the reinforcement of the normal light and dark feeding patterns by actually removing the food during the light periods, a procedure that has proved very helpful in standardizing the metabolic control systems so that hepatomas could be compared with normal liver". From that study we at once proceeded to the study of normal animals (without tumors) and to *ad libitum* feeding and again reported marked fluctuations in tyrosine transaminase activity [7,8]. Almost simultaneously with our 1967 report 3 groups independently reported similar daily fluctuations in the activity of this enzyme in normal rat liver [9–11]. Wurtman [12] has recently reviewed the literature on this enzyme under the heading of "time-dependent variations in amino acid metabolism" and has commented on the problem of defining a metabolic oscillation as circadian when it results from the complex interplay of exogenous and endogenous signals. He concludes that "A recognition of the fact that 'normal' (sic) levels of enzyme activity may vary markedly as a function of time of day is essential to the interpretation of the effect of experimental manipulation on enzyme activity". Wurtman omitted consideration of our data bearing on the relative merits of *ad libitum* feeding in comparison with controlled feeding schedules and diets, over periods long enough to permit adaptation to the regimen, and has emphasized the desirability of *ad libitum* regimens [13, unrecorded discussion].

Consideration of the arguments for and against *ad libitum* feeding will become increasingly important as attempts are made to study regulatory mechanisms in whole animals and to correlate them with findings with tissue cultures of cells showing similar responses (not discussed here, cf. [8]). When Wurtman uses the word "normal" in quotes we need to press for an answer to the question of whether a rat on an *ad libitum* diet is more "normal" than a rat that eats only at controlled times. We take the position that the properly controlled feeding schedule, which takes into consideration the rats' instinctive nocturnal feeding, is more "normal" than is *ad libitum* feeding under typical laboratory conditions. (Presumably there is no question about the fact that it is more readily standardized and manipulated.) But we emphasize that in order to have *ad libitum* conditions comparable between widely separated laboratories it would be necessary to specify the space occupied by the rats and the ambient noise and other disturbances in the same room and in adjacent rooms. Different laboratories report wide variations in eating patterns under *ad libitum* conditions. Thus in Wurtman's laboratory [14] rats on an *ad libitum* regimen ate 20% of their daily total during 12 hr of light (9:00 to 21:00) and 80% in the dark, and under reversed lighting conditions ate to approximately the same ratio (18% in the light and 82% in the dark). These figures differ considerably from those of LeMagnen and Tallon who made a detailed study of individual feeding patterns in rats [15]. Their (English) summary stated: "1) The rat, consuming a more or less constant amount of food in 24 hr, does so in meals of varying amounts, separated by irregular intervals. 2) The number of meals taken each day, and their even distribution between the 12 hr of day or night show remarkable constancy." (The mean frequency was 4 per day and 4 per night (fig. 13, ref. [15]). "3) Mean nocturnal consumption is about 50% greater than that during the day, difference being chiefly due to a nocturnal increase in the magnitude of each meal". (Items 4 and 5 in the Summary here omitted.) They reported that the average meal was 2 g in the day and 3 g in the night, and from a total of 40 rats observed for 200 days the average intake in the dark was 12.95 g and in the light it was 8.45 g or 61% in the dark and 39% in the light. This report indicated a great variation in individual rats from time to time as well as great variation in the average response of one rat compared with another rat. While the basis of the behavioral responses is of considerable interest it seems that the behavioral responses can be made more regular by standardizing one of the important links in the chain of behavioral events, namely the availability of food. We observe that marked motor activity coincides with the onset of darkness and the simultaneous availability of food, and we have shortened the period of food availability to 8 hr, during which time the animals can take several meals and maintain a continuous state of food absorption, yielding patterns of physiological activity that appear to be difficult or impossible to obtain under *ad libitum* feeding regimens [16,17].

Under conditions of regulated lighting and food availability the behavioral activity and concomitant flux of normal regulations may be brought under control because the two factors reinforce each other as *Zeitgeber*(s), triggering

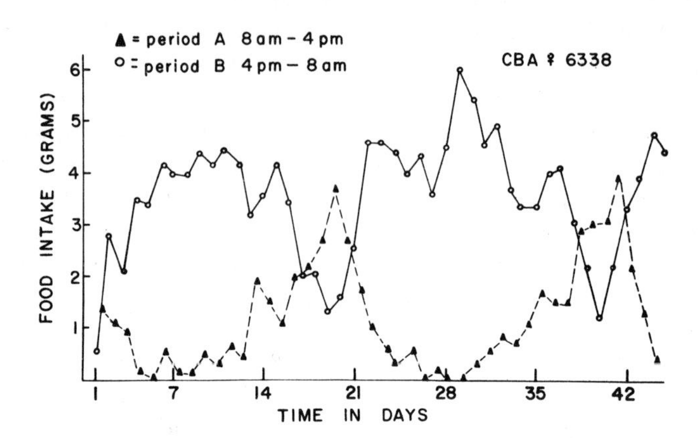

Fig. 3. Variations in the relative amounts of food consumed by mice with a hereditary retinal defect (Strain C57). Lacking perception of the daily light/dark cycle as a triggering cue, it appears that the mice operate on an endogenous biological clock that differs slightly from 24 hr. (Data courtesy of Dr. R.A. Liebelt, Houston, Texas).

a complex system of biological clocks to hold the overall system to a 24 hr cycle. An interesting deviation from the 24 cycle when mice with an hereditary retinal defect were fed *ad libitum* has been demonstrated by Liebelt (fig. 3, personal communication).

Although the present report will emphasize studies on tyrosine transaminase (aminotransferase) as a remarkably sensitive indicator of flux in the *milieu interieur* this enzyme is not unique in this sensitivity. Hamprecht, Nüssler, and Lynen [18] have observed an even more striking daily flux in the activity of hydroxymethylglutaryl coenzyme A reductase in rat liver and observed that livers from fasted rats had less than 1% as much activity as fed rats.

3. Factors affecting rat liver tyrosine aminotransferase

Reference has already been made to some of the early reports that have demonstrated the remarkable fluctuations in the activity of tyrosine transaminase (aminotransferase, TAT), and additional reports are cited in several of the references already mentioned [12–14,16]. The literature on this enzyme is now too voluminous to review in detail, and the separation of direct effects from indirect effects is still obscure (fig. 4). The problem is complicated still further by the report that TAT isoenzymes are demonstrable in

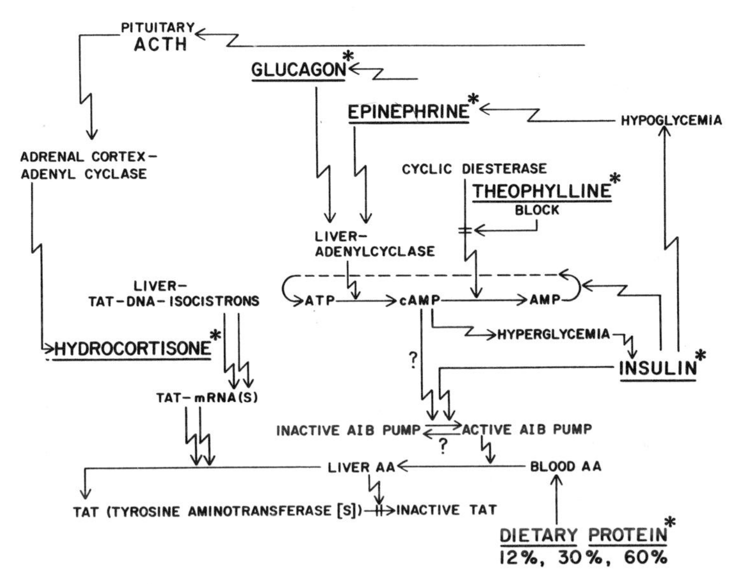

Fig. 4. Factors affecting TAT activity in liver and in hepatomas. Experimental variables are marked with an asterisk: glucagon, theophylline, hydrocortisone, dietary protein. *Isocistron* is a term not previously used but which is suggested as an appropriate term for a DNA sequence that contains the information for an *isozyme*. Alternative arrows indicate the direct observations and indirect possible mechanisms still under investigation.

terms of different responses to different inducers (Holt and Oliver, [19]). Recently, Fuller and Snoddy ([20], personal communication) have reported the induction of TAT in normal rat liver by means of theophylline, an inhibitor of the diesterase that converts cyclic AMP to 5′ AMP (see fig. 4). Since we had previously observed a correlation of TAT activity with amino acid pump activity, studied by means of radioactive AIB* (aminoisobutyric acid) we were interested to learn whether theophylline would stimulate AIB uptake as well as increased TAT activity. Experiments by Scott et al. [21] are shown in fig. 5. They show a high degree of correlation between the two parameters although we know that they can be dissociated (Scott et al. [21]; Krawitt et al. [22]). The data in fig. 5 were obtained from rats that were adapted to a feeding schedule that provided food during darkness that began at 08:30 but on the day of the experiment food was witheld. The saline controls showed no increase in TAT or in AIB uptake. The data in fig. 5 and the schematic dia-

* Non-usual abbreviations: AIB, aminoisobutyric acid; AA, amino acids.

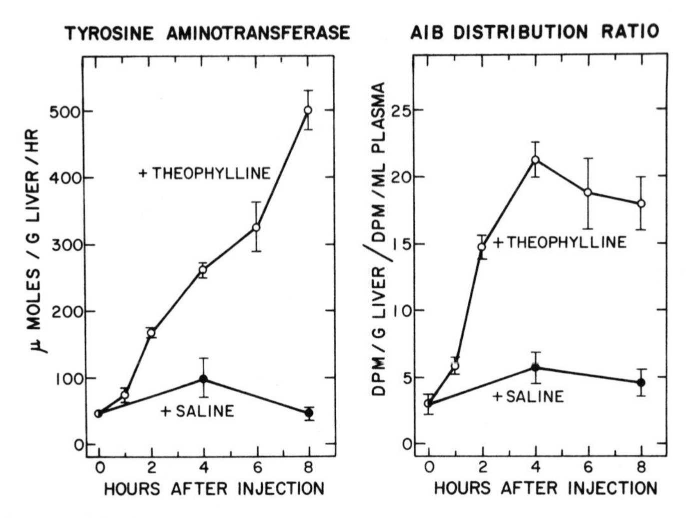

Fig. 5. Increased TAT activity and increased AIB pump activity in normal rat liver after a single injection of theophylline. The injection (theophylline 10 mg/100 g body weight in 1 ml saline or 1 ml saline only) was given at the onset of darkness 16 hr after an 8 hr feeding period, with food withheld during the period that 30% protein diet was ordinarily present. (Data from Scott et al. [21]).

gram in fig. 4 provide background for the dietary experiments reported next.

Fig. 6 shows the effect of feeding 12%, 30% or 60% protein diets according to a controlled feeding referred to as "8+16". The figure shows the fluctuations in TAT and serine dehydratase (SDH) during one half of the 24 hr cycle, namely the dark half during which the animals are fed. The fluctuation in TAT is over 8-fold in the case of the 60% protein diet, while the fluctuations in SDH are not considered significant. The TAT values remain low and relatively constant during the light period as shown in unpublished studies. The plan of the experiment shown in fig. 6 has been used to study the response of certain hepatoma strains to diet and hormonal factors in terms of enzyme activities and AIB pump activity.

4. Naturally-occurring levels of TAT, SDH and AIB pump activity in liver and in hepatoma strains

AIB can be injected into rats one or two days before the animals are sacrificed because it is not oxidized or incorporated into protein and is not rapidly excreted by the kidneys, but nevertheless is concentrated by various tissues

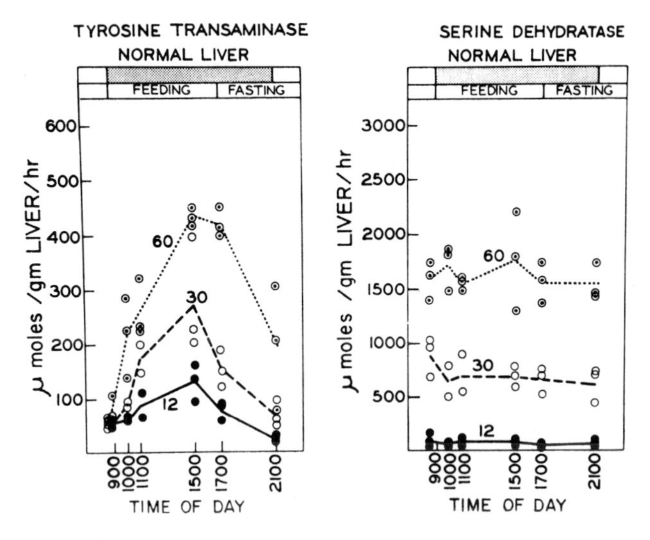

Fig. 6. Variations in levels of TAT and SDH in normal rat liver. Both time of day and percent protein in diet (indicated by the figures 60, 30 and 12) affect TAT but only the percent protein affects SDH under the conditions described. The cross-hatched bar indicates darkness *in the animal room.* Each point represents one animal and the curves are drawn through average values. Data from Charts 1A and 1B, Potter et al. [24] reprinted, with permission, from *Cancer Research.*

and hepatomas to characteristic tissue/blood ratios. Fig. 7 shows the *initial* levels in Morris hepatoma 7800, and in the liver and blood of the same animals when the AIB was injected one hr after feeding began. The hepatomas appeared to reach equilibrium sooner than the liver, and individual animals appeared to be well synchronized with others in their group. Further data are reported by Baril et al. [23].

The experiments reported in fig. 6 and in fig. 7 provide the protocols for experiments on several hundred rats representing normal non-tumor-bearing controls, and nine different lines of transplantable Morris hepatoma. Fig. 8 is a composite of published data on TAT and SDH [24] and more recent data on AIB pump activity [23]. Two of the hepatoma lines were karyotypically normal (9618A and 9633), four lines had a normal number of 42 chromosomes but with slight abnormalities (9121, 7794A, 9098, 7800), hepatoma 9108 had 42 chromosomes with some abnormalities, hepatoma 7794B had 44 chromosomes, and hepatoma 5123C had 90-96 chromosomes when last studied [25]. The AIB was injected one day before the animals were killed. Not shown are values for the normal and host livers which were in general quite similar to

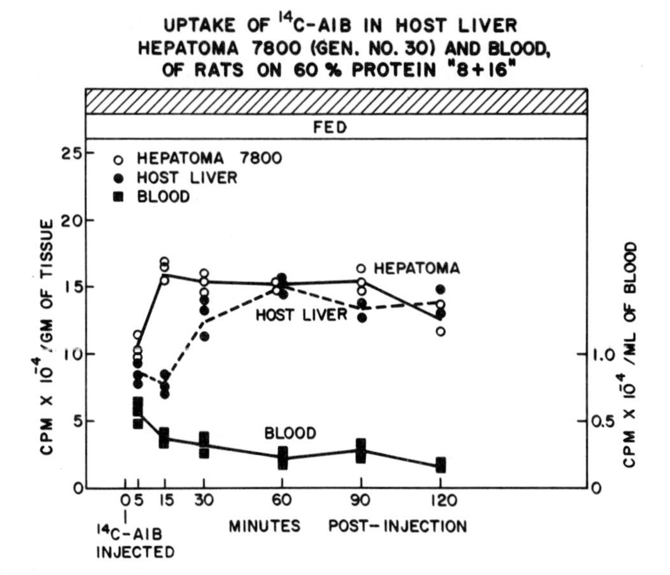

Fig. 7. Variations in levels of AIB radioactivity in blood, liver and in hepatoma 7800. Initial values following a single subcutaneous injection of AIB which was demonstrated to be concentrated in the tissues (pumped) but not metabolized. Each animal yielded 3 points on the chart: hepatoma, liver and blood values. In subsequent experiments the injection was 20-24 or more hr before sampling. Data from Chart 1, Baril et al. [23] reprinted, with permission, from *Cancer Research*.

the values given in fig. 6 for TAT and SDH which if plotted in fig. 8 would run diagonally from near 50 units to maximal values of 500 and 1700, respectively, at 60% protein, (see also the control values in fig. 9, next section) in contrast to the hepatoma values which in all cases deviate from normal and in most cases show no correlation to dietary protein. The AIB pump values for normal liver and host liver were very similar and ranged from 24 to 35 on the scale shown in fig. 8, in which hepatoma 5123C showed a tissue/blood ratio of about 140. Fig. 8 shows a rough parallelism between AIB pump activity and enzyme activity in the hepatomas and it may be noted that there is also a rough parallelism with growth rate, with rapid growth in 5123C and very slow growth in 9618A. I have attempted to plot SDH and TAT activity against AIB pump activity and found that the hepatomas show little enzyme response to AIB pump activity except in the case of 5123C, while normal liver showed wide variations in enzyme activity within a fairly narrow range of pump activities. Now that we are able to vastly increase the pump activity in normal

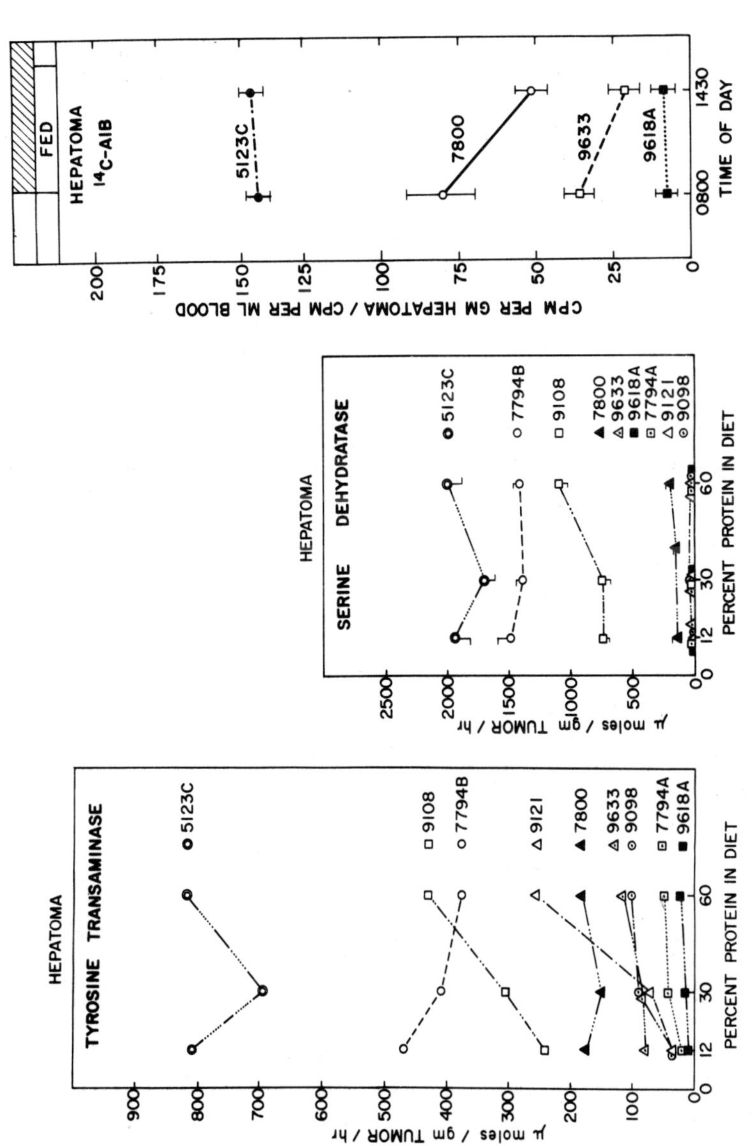

Fig. 8. Phenotypic diversity in Morris hepatomas. Data on TAT and SDH from Charts 6B and 7B, Watanabe et al. [26], and data on AIB radioactivity from Chart 3, Baril et al. [23], reprinted, with permission, from *Cancer Research*, where details are recorded. TAT data are averages from 3 rats.

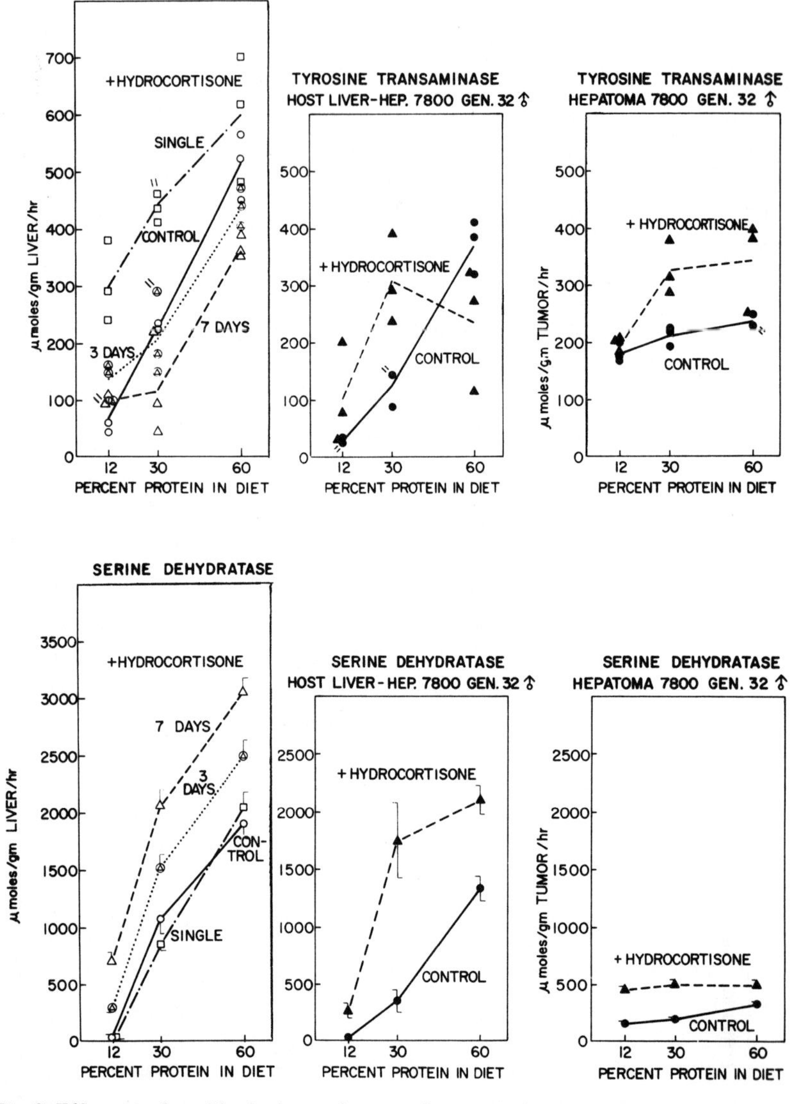

Fig. 9. Effects produced by hydrocortisone on 3 enzymes in normal liver, Morris hepatoma 7800, and 7800 host liver as a function of dietary protein. The rats bearing the hepatoma were treated with hydrocortisone for 0 and 7 days and the normal rats were treated for 0, 1, 3, and 7 days. Data from Charts 4, 5, 6, 9A, 9B, 11A, 11B, 13A, and 13B, Watanabe et al. [26], reprinted, with permission, from *Cancer Research.*

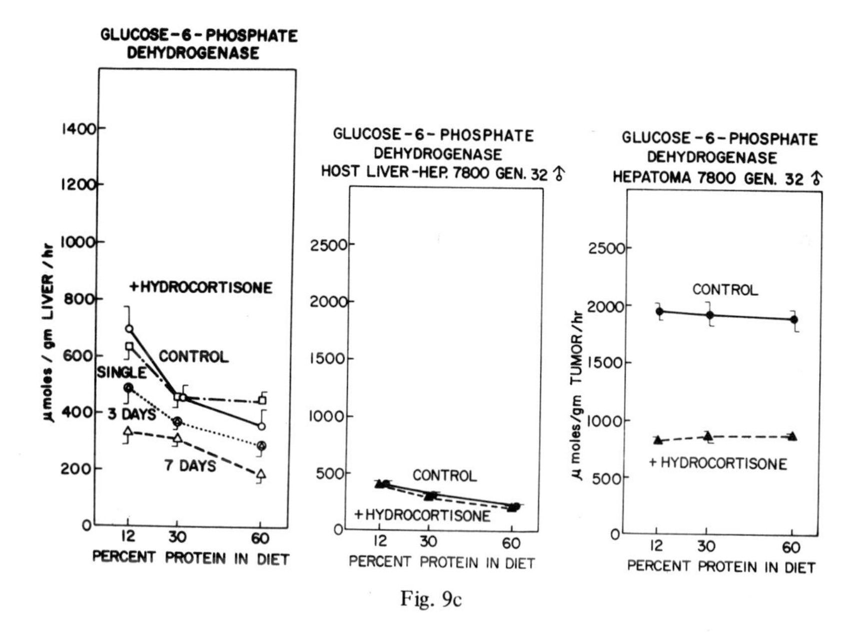

Fig. 9c

liver with theophylline (fig. 5) it will be desirable to test the response of the hepatomas to similar manipulation. This has not yet been done. However, several diverse responses by various hepatomas have been obtained with hydrocortisone and glucagon (next section).

5. Comparative studies on the responses of liver and of hepatomas to hydrocortisone and glucagon

It has proved to be helpful to study several enzymes on the same samples of tumor and liver and to study the response to hormones with all 3 levels of dietary protein used in the previous experiments. Fig. 9 shows the values for TAT, SDH, and G6PDH in normal liver, host liver and in hepatoma 7800 in animals on 3 levels of dietary protein at various times after hydrocortisone (3 mg per 100 g body weight) [26].

Values for individual animals are given in the case of tyrosine transaminase, because in each case data from 6 of the 9 animals were discarded for the reason that only the animals killed 6 hr after feeding reflect the protein intake (see fig. 6, TAT) while for the other two enzymes all 9 rats gave data that could be averaged and reported with the standard error (see fig. 6, SDH).

The normal rats were tested after 0, 1, 3 and 7 days of 2 × daily injections of hydrocortisone and each enzyme gave a characteristic response in relation to dietary protein, relative to day zero. For tyrosine transaminase the changes were: Up at 1 day; back to day zero levels, except on 12% protein, at 3 days; and further decrease below day zero levels, except on 12% protein, at 7 days. For SDH the changes were: no change at 1 day; up at 3 days; further up at 7 days. For G6PDH the changes were: no change at 1 day; down at 3 days; further down at 7 days. The reciprocal changes between serine dehydratase and G6PDH have been thoroughly documented previously [24] and are here seen recapitulated in relation to hydrocortisone [26]. Even more striking are the reciprocal changes in these two enzymes in hepatoma 7800 in rats treated with hydrocortisone (see below).

Included in fig. 9 are experiments with hepatoma 7800 with the hydrocortisone protocol based on the data from normal rats shown in the same figure. The consistency of the responses at the different levels of dietary protein are quite remarkable in the case of SDH and G6PDH and the increase and decrease respectively confirms the rule of reciprocity previously noted [24]. Yet the difference between the hepatoma and the host liver or normal liver is striking. The hepatoma is not completely autonomous or oblivious of the effects of protein and/or hydrocortisone but it remains on the whole non-responsive, i.e. non-adaptive, to changes in dietary protein, whereas the normal liver responds in a way that relates to its normal gluconeogenic function.

6. Effect of glucagon on cortisonized normal rats

From experiments with hydrocortisone alone, we moved to experiments in which cortisonized animals were treated with one injection of glucagon and killed 5 hr later. In addition to enzyme assays, the AIB pump activity was assayed in the same animals, which were injected with tracer doses of AIB about 24 hr previously. Data from normal animals adapted to 30% protein are shown in fig. 10. It may be noted that the effect of hydrocortisone alone is to produce increased TAT activity at 1 day, returning to near normal at 3 days, and to depressed levels at 7 days, confirming the experiment in fig. 9, but when a single injection of glucagon was given a few hours before the killing time there was a significant stimulation of the control or depressed values at 0, 3 or 7 days but no significant elevation of the already elevated value at 1 day. It may also be noted that the AIB pump activity was also stimulated at 0, 3 or 7 days but not at 1 day. We concluded that the glucagon effects are significant despite the fact that only 3 animals received glucagon in each experiment be-

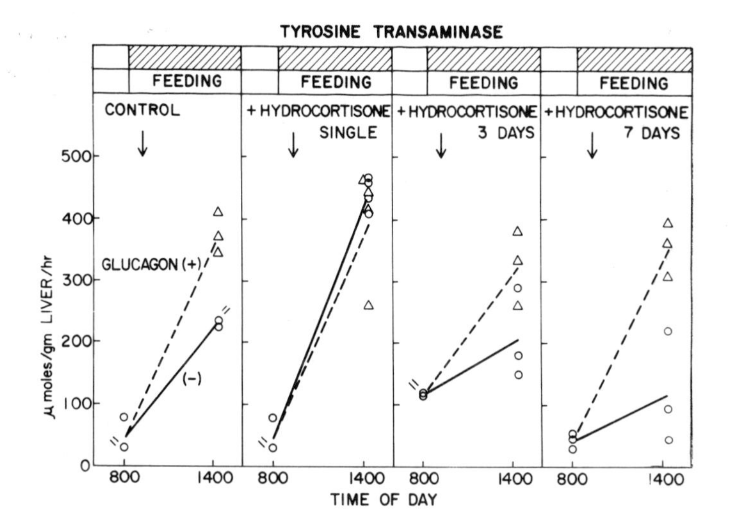

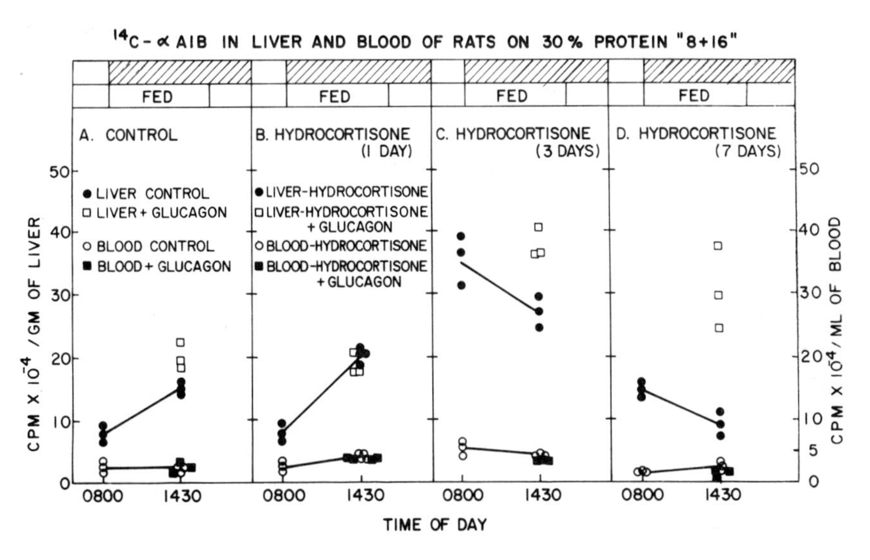

Fig. 10. Correlations between TAT activity and AIB pump activity in normal rat liver. Glucagon was administered to rats after 0, 1, 3 and 7 days of hydrocortisone treatment. Enzyme data from Chart 2, Watanabe et al. [26]. AIB data on the same animals from Charts 5A,B,C and D, Baril et al. [23]. The data shown are for animals adapted to 30% protein diets. Data for other enzymes and for animals fed 12% or 60% protein are in the references cited. The glucagon was injected at one hour after the onset of feeding as shown by the arrows in the upper chart. Data reprinted, with permission, from *Cancer Research.*

cause there were actually 36 animals that received glucagon and the overall results from the 12 groups of 3 animals each (0, 1, 3 and 7 days of cortisone and 12%, 30% and 60% protein in diets) were remarkably consistent with the above summary [23,26].

7. Effect of glucagon and cortisone on hepatomas 9618A and 9633

The normal rat liver diploid number of 42 chromosomes prevails from birth to a transition point at about 50 g body weight when diploid and tetraploid cells are about equal in number. By the time the rats reach 100 g body weight the diploid cell fraction has stabilized at about 10% (Naora, [27]). Among the Morris hepatomas with 42 chromosomes we are quite interested in comparing 7800, 9618A and 9633 because from a qualitative standpoint the presence or absence of tyrosine transaminase and serine dehydratase in fetal liver, adult liver, and in the 3 hepatomas can be represented as a spectrum as shown in table 2. This table represents a simplistic overview of data from many experiments at three levels of dietary protein, and the (-) values apply to the highest protein levels as well as the lowest levels. Although the (-) levels of enzyme are not zero, they cannot be induced to higher levels by high protein diets. We do not know whether the (-) values in fetal liver can be increased to high values by a high protein diet because the experiment cannot be done (i.e. giving a high protein diet to the pregnant rat is not effective but it is not a valid experiment). However, we do know that tyrosine transaminase can be induced in fetal liver by glucagon administered to the fetal rat *in utero* as shown by Greengard and Dewey [28]. We therefore proceeded to test whether glucagon could induce serine dehydratase or tyrosine transaminase and/or AIB pump activity in cortisonized animals and in untreated animals bearing hepatoma 9618A, which for the two enzymes is described as (-,-) in table 2. Fig. 11 shows

Table 2
Qualitative description of two enzyme activities in 3 hepatomas, and in fetal and adult liver.

Enzyme	Tissue: liver or hepatoma				
	Fetal liver	9618A	9633	7800	Adult liver
Tyrosine transaminase	-	-	+	+	+
Serine dehydratase	-	-	-	+	+

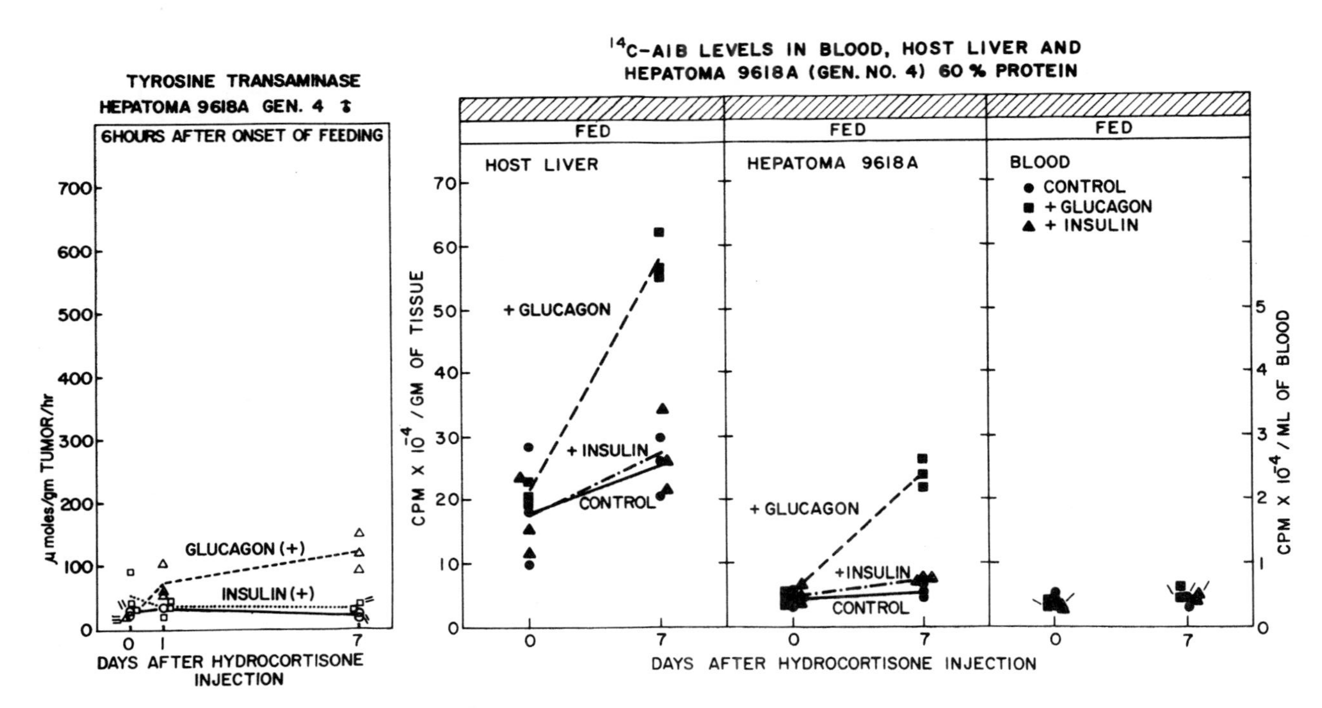

Fig. 11. Correlation between TAT activity and AIB pump activity in Morris hepatoma 9618A. Animals were killed six hours after feeding began and at 3 hr after glucagon was administered to rats after 0 or 7 days of hydrocortisone treatment. Enzyme data from Chart 27B, Watanabe et al. [26], and AIB data on the same animals from Chart 11, Baril et al. [27]. The animals were adapted to a 60% protein diet. Data reprinted, with permission, from *Cancer Research*.

that glucagon had no effect in untreated animals but had a stimulating effect on tyrosine transaminase and AIB pump activity in cortisonized animals [23, 26]. Not shown are data on serine dehydratase showing no induction. Subsequent experiments reported by Potter et al. [29] showed that 2 glucagon injections at zero and 2 hr into the feeding period in cortisonized animals gave marked increase in tyrosine transaminase activity at 3 hr into the feeding period with activity still maintained at the 6th hr, but with no elevation in SDH at any time. This observation was referred to as "Induction of a previously non-inducible enzyme" [29], and it has an obvious bearing on the interpretation of data presented as (-) as in table 2. Evidently we must regard as open the possibility that any given enzyme that is phenotypically very negligible in activity is not necessarily deleted from the genotype [29]. It is noteworthy that there was a good correlation between AIB pump stimulation and increased activity for tyrosine transaminase.

From these experiments with hepatoma 9618A we proceeded to hepatoma 9633 which for the 2 enzymes was reported as (+,-) in table 2. The experiment is reported in fig. 12. Glucagon had no effect on AIB pump, tyrosine transaminase, or serine dehydratase; as in the case of hepatoma 9618A, the effect of glucagon was internally consistent in the relation between the AIB pump and TAT.

Further attempts to induce serine dehydratase in hepatoma 9618A, hepatoma 9633, and other hepatomas that are (-) for SDH are in progress. Since SDH has also been found to occur in distinct isozymic forms in proportions that vary with conditions of induction (Inoue and Pitot, [30]) it is clear that further studies on the role of inducing factors must consider the existence of the isozymic forms.

8. Discussion

Obviously these experiments have been quite exploratory as far as the introductory proposition is concerned. However, from these experiments have come a working hypothesis as to the interacting control factors (fig. 4). An experimental protocol for testing the proposition that there is a complementary relationship between the amino acid pump phenomenon and the corticosteroid status of the rat with respect to the TAT activity in rat liver or in hepatomas, and the possible role of adenylcyclase and cyclic AMP in this relationship is under study in our laboratory (Dr. Fred Butcher, unpublished). We believe that the use of controlled feeding schedules is an essential part of this protocol [17]. From fig. 8 and from an earlier reference [6] it is evident

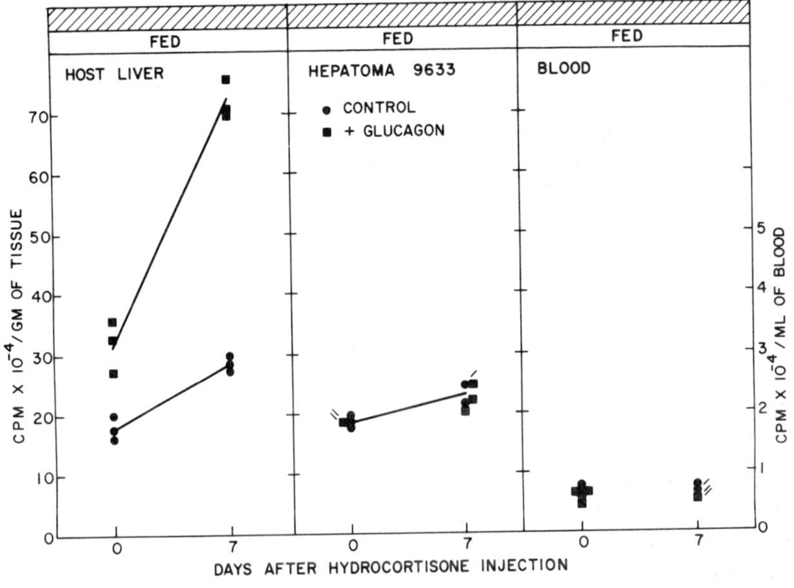

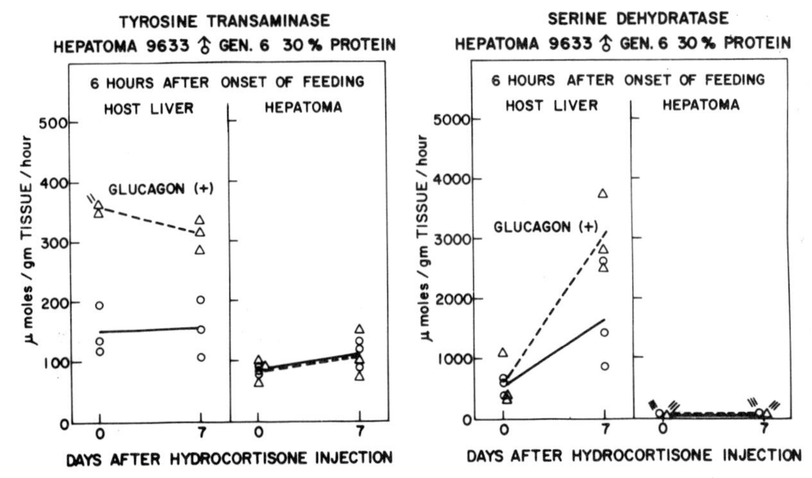

Fig. 12. Correlations between TAT activity and AIB pump acitivity in host liver and in Morris hepatoma 9633. Animals were killed six hours after feeding began and 3 hr after glucagon was administered to rats after 0 or 7 days of hydrocortisone treatment. The animals were adapted to a 30% protein diet. SDH response in the hepatoma was zero under all conditions. Enzyme data from Charts 23 and 24, Watanabe et al. [26], and AIB data from Chart 1C, Baril et al. [24]. Data reprinted, with permission, from *Cancer Research.*

that there are several hepatoma lines with SDH and TAT activities between hepatoma 7800 and hepatoma 5123C. These include hepatoma 9108 and hepatoma 7794B (fig. 8) and hepatoma 7793 [6]. None of these 5 hepatoma lines have been studied in terms of simultaneous measurements of AIB pump and TAT activity and it would seem desirable to study the action of glucagon and/or theophylline in cortisonized or adrenalectomized control and tumor-bearing rats in these hepatoma lines, in which the enzyme activity is present in sufficient amounts to be modulated in either direction. In the present report we have shown the marked effects of hydrocortisone on hepatoma 7800 (fig. 9) and have omitted the equally clear-cut effects in the opposite direction produced by adrenalectomizing rats bearing hepatoma 5123C [26]. We have shown that hydrocortisone was ineffective in hepatoma 9618A and in hepatoma 9633 while glucagon given to cortisonized animals could produce a stimulation of AIB pump activity and TAT activity in hepatoma 9618A but not in hepatoma 9633 [26]. Thus the response to glucagon and hydrocortisone is not a constant feature from one hepatoma to another, just as it is not a constant feature from one stage of liver development to another [28].

Experiments outlining a systematic method for quantitating these complex interrelationships are in progress and will be reported elsewhere. After this manuscript was completed we noted the highly relevant paper by Wicks, Kenney, and Lee [31] in which the induction of tyrosine transaminase with dibutyryl cyclic AMP in intact and adrenalectomized rats was reported, with comparisons that included glucagon, hydrocortisone, insulin, and theophylline in various combinations. However, these workers did not study AIB pump activity and hence did not observe data as presented in fig. 5 above.

Acknowledgement

This work was made possible by the cooperation of Dr. H.P. Morris, Dr. H.C. Pitot, Dr. M. Watanabe, Dr. E.F. Baril, Dr. D.F. Scott, Dr. Fred R. Butcher and Mr. R.E. Reynolds who are co-authors of various cited reports published and unpublished.

References

[1] V.R. Potter, Can. Cancer Conference 8 (1969) 9.
[2] T. Matsushima, S. Kawabe, M. Shibuya and T. Sugimura, Biochem. Biophys. Res. Commun. 30 (1968) 565.

[3] J. Uriel, Pathologie-Biologie 17 (1969) 877.
[4] Armin C. Braun. The Cancer Problem: A Critical Analysis and Modern Synthesis (Columbia University Press, New York, 1969).
[5] V.R. Potter and T. Ono, Cold Spring Harbor Symp. Quant. Biol. 26 (1961) 355.
[6] V.R. Potter, R.A. Gebert, H.C. Pitot, C. Peraino, C. Lamar Jr., S. Lesher and H.P. Morris, Cancer Res. 26 (1966) 1547.
[7] V.R. Potter, R.A. Gebert and H.C. Pitot, Advan. Enzyme Regulation 4 (1966) 247.
[8] V.R. Potter, M. Watanabe, J.E. Becker and H.C. Pitot, Advan. Enzyme Regulation 5 (1967) 303.
[9] R.J. Wurtman and J. Axelrod, Proc. Natl. Acad. Sci. U.S. 57 (1967) 1594.
[10] M. Civen, B. Lurich, B.M. Trimmer and C.B. Brown, Science 157 (1967) 1563.
[11] G.E. Shambaugh, D.A. Warner and W.R. Beisel, Endocrinology 81 (1967) 811.
[12] R.J. Wurtman, Advan. Enzyme Regulation 7 (1969) 57.
[13] V.R. Potter, E.F. Baril, M. Watanabe and E.D. Whittle, Federation Proc. 27 (1968) 1238.
[14] M.J. Zigmond, W.J. Shoemaker, F. Lavin and R.J. Wurtman, J. Nutr. 98 (1969) 71.
[15] J. LeMagnen and S. Tallon, J. Physiol. (Paris) 58 (1966) 323.
[16] M. Watanabe, V.R. Potter and H.C. Pitot, J. Nutr. 95 (1968) 207.
[17] E.F. Baril and V.R. Potter, J. Nutr. 95 (1968) 228.
[18] B. Hamprecht, C. Nüssler and F. Lynen, FEBS Letters 4 (1969) 117.
[19] P.G. Holt and I.T. Oliver, FEBS Letters 5 (1969) 89.
[20] R.W. Fuller and H.D. Snoddy, Biochem. Pharmacol. 19 (1970) 1518.
[21] D.F.Scott, F.R.Butcher, R.D.Reynolds and V.R.Potter, in: Biochemical Responses to Environmental Stress, ed. I.A.Bernstein (Plenum Press, New York) in press.
[22] E. Krawitt, E.F. Baril, J.E. Becker and V.R. Potter, Science 169 (1970) 294.
[23] E.F. Baril, V.R. Potter and H.P. Morris, Cancer Res. 29 (1969) 2101.
[24] V.R. Potter, M. Watanabe, H.C. Pitot and H.P. Morris, Cancer Res. 29 (1969) 55.
[25] P.C.Nowell, H.P. Morris and V.R. Potter, Cancer Res. 27 (1967) 1565.
[26] M. Watanabe, V.R. Potter, H.C. Pitot and H.P. Morris, Cancer Res. 29 (1969) 2085.
[27] H. Naora, J. Biophys. Biochem. Cytol. 3 (1959) 949.
[28] O. Greengard and H.K. Dewey, J. Biol. Chem. 242 (1967) 2986.
[29] V.R. Potter, R.D. Reynolds, M. Watanabe, H.C. Pitot and H.P. Morris, Advan. Enzyme Regulation 8 (1970) 299.
[30] H. Inoue and H.C. Pitot, Advan. Enzyme Regulation 8 (1970) 289.
[31] W.D. Wicks, F.T. Kenney, K.-L. Lee, J. Biol. Chem. 244 (1969) 6008.

STUDIES ON THE MOLECULAR MECHANISMS OF CARCINOGENESIS*

E. FARBER

Department of Pathology, University of Pittsburgh School of Medicine, Pittsburgh, Pennsylvania 15213, USA

Abstract: Farber, E. Studies on the Molecular Mechanisms of Carcinogenesis. *Miami Winter Symposia* 2, pp. 314–334. North-Holland Publishing Co., Amsterdam, 1970.

Newer knowledge developing in the past ten years is placing the process of carcinogenesis in a new perspective and offers to play an important role not only in the elucidation of the biochemistry of cancer but also in laying a scientific foundation for the prevention of cancer. Although advances have been significant with all three groups of carcinogens, chemicals, viruses and radiation, this presentation will be limited to the first.

Many chemical carcinogens are only carcinogenic after they are converted to highly reactive derivatives by the host tissues. The liver, although very potent in the metabolism of many potential carcinogens, is by no means the only active tissue. Many different organs or tissues can generate active carcinogens which appear to be in each case either an alkylating or arylating agent. The active carcinogens in turn interact with many of the hosts' cellular macromolecules to alter their information content or metabolic properties.

In the case of the liver, following this initial mutual interaction, innumerable islands of a new population of liver cells appear with biochemical characteristics different from both those of the initial tissue and of the subsequent cancer. This population has acquired the ability to grow under conditions in which the normal liver does not. In the liver, such growth has now been found to occur also *in vitro* with hyperplastic nodules induced by aflatoxin, ethionine or acetyl-aminofluorene (AAF). This growing population which appears to be a precursor of cancer, is however, not cancer. A most impressive property of this new population is the great similarity in the structure and function of these cells even when induced by many different agents with widely different chemical structures. This contrasts sharply with the primary cancer in which there is an extreme degree of heterogeneity. Thus, any biochemical hypothesis of carcinogenesis must explain this sequence of cell populations-relativelyhomogeneous initial population followed by a very

* The author's research included in this report was supported in part by grants from the American Cancer Society and the Beaver County Cancer Societey and by U.S. Public Health Service Research Grant CA-06074 from the National Cancer Institute, AM-05644 from the Institute of Arthritis and Metabolic Diseases, and GM-135 (a training grant) and GM-10269 from the National Institute of General Medical Sciences, and by a contract (PH-4364505) from the National Cancer Institute.

homogeneous growing cell population (minimum deviation?) followed by a very heterogeneous cancer cell population.

With one carcinogen so far examined, AAF, the intermediate growing cell hyperplastic population has obvious alterations in DNA and carbohydrate metabolism and they will be described.

This new information concerning the properties of different cell populations during carcinogensis offers new possibilities for the formulation of hypotheses of biochemical mechanisms of carcinogenesis which can begin to be subjected to experimental study and may hopefully lead to new avenues of exploration into the biochemistry of carcinogenesis.

1. Introduction

It is now clear that the process leading to the appearance of malignant neoplasia or cancer in any tissue or organ consists of a progressive and sequential series of *discontinuous steps* each associated with a new cell population. Although it was thought at one time, not so long ago [1], that the conversion from normal to neoplastic was a rapid single event, it is now well established that such is not the case. However, I should emphasize that any attempt to designate the *number* of steps, e.g. two, in carcinogenesis by any agent, is obviously premature and without a factual foundation. What is meant is more than one. The acceptance of a stepwise sequence, rather than a single step, between normal and cancer implies that the various aberrations which, in composite, we designate as cancer are not a single unit of change but are a conglomerate of many changes.

It is the conviction and belief of biochemists in cancer research that an analysis of the essential biochemical or metabolic derangements associated with cancer is a feasible as well as worthwhile undertaking. It would appear that there are in general two types of approaches: (a) a pragmatic and empirical one designed to discover some reproducible biochemical difference between normal and cancer which can be exploited for cancer therapy today, and (b) a more fundamental one, the purpose of which is to understand the essence of neoplasia with the hope that such knowledge will lead to more rational ways to prevent or interrupt carcinogenesis or to reverse or cure an existing cancer. The first I consider more development than research. Although I do not criticize such a pragmatic approach, it needs no further discussion at this time.

The second is an intellectual pursuit which many believe will contribute both to a more rational handling of human cancer and to the body of knowledge in basic cellular biology. However, I am not convinced that the models now used are amenable to the analysis and understanding desired. Since any

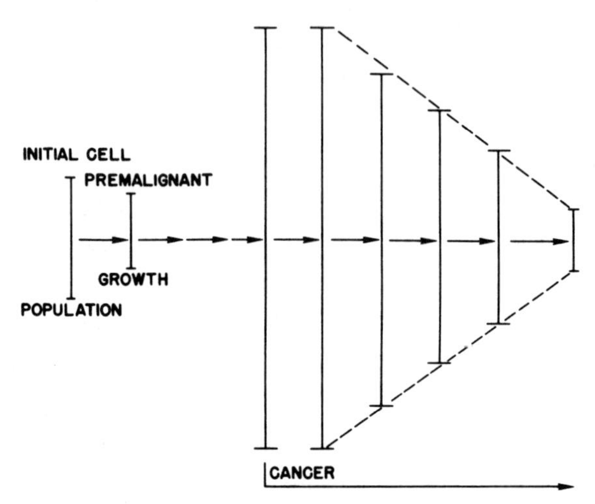

Fig. 1. Diagrammatic representation of cell populations and their spectrum. The number of populations is arbitrary and is not meant to represent any experimentally-established value. The height of each line is a rough guide to the spectrum of cell variation seen. As the cancer "ages", it tends toward a less diverse spectrum.

cancer, regardless of its appearance, has multiple deviations from the normal if it is transplantable or if it invades, metastasizes or has obvious cytologic or chromosomal abnormalities, it is evident that it is not susceptible to scientific analysis unless the number of known and obvious variables can be decreased towards one. Since it appears that each of the deviations mentioned are acquired seriatim during carcinogenesis and that each has a quantitative spectrum, it seems to me that our major hope of understanding a cell already classified as cancer is to dissect each biological aberration separately from the others and perhaps put them all back together to form a composite and integrated picture. This is today not possible by studying any cancer, especially since we have no way of deciding where any one neoplasm falls in the sequence between the first and "last" cancer (fig.1).

It is obvious, therefore, that any attempt to understand the biochemical or metabolic basis for each of the known biological properties of cancer can only hope to succeed if models are available or developed in which each of the major variables can be controlled. In my view, this is only possible by focussing our attention on that portion of the carcinogenic process (figs. 1 and 2) between the initial cell population and the earliest cancer cell population.

Hopefully, the biochemical basis for persistent growth as well as for other attributes of the ultimate cancer cell may be analyzable by isolating and study-

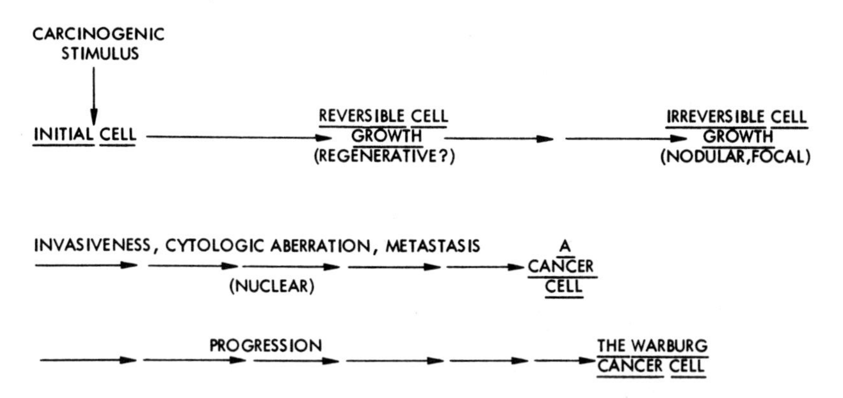

Fig. 2. Diagrammatic representation of the sequence of changes from an initial target cell to an advanced cancer cell ("The Warburg cancer cell"). The number of arrows shown is arbitrary and is not meant to represent any known value. The "Warburg Cancer Cell" is theoretically the cell approached by many malignant neoplasms that are allowed to progress for a relatively long time.

ing each of the various cell populations that appear in sequence or in parallel between the initial cell and the first primary cancer. In other words, we must spend more time developing and refining models designed to analyze metabolically each of the major discrete steps in the process.

I hasten to add that the cancer cell *per se* is becoming increasingly useful as a biological model for the study of the synthesis and control of discrete macromolecules, such as enzymes, hormones or immunoglobulins. However, at present it is difficult to see how this can be used for the study of the essentials of the chemistry of neoplasia, until means are found to selectively turn on and off the different biological behavioral parameters of the cell.

After this preamble, I would now like to present a few highlights of our current knowledge and concepts relating to two more or less discrete phases of carcinogenesis, (1) the initial mutual interaction between carcinogen and original cell and (2) a new cell population that grows out fairly early in the carcinogenic process and to extrapolate from these to a working hypothesis of the biochemistry of carcinogenesis.

2. The initial mutual interaction between carcinogen and original cell population

Carcinogenic stimuli can be conveniently divided into two major groups:
A. Agents containing information translatable as such by the cell, e.g. viruses.

B. Agents not containing translatable information. e.g. chemicals, radiation.

I shall concentrate almost exclusively on chemicals in this presentation, although some unavoidable but passing reference to viruses will be made.

2.1. *Action of cells on carcinogenesis*

The chemical carcinogens are divisible into two large groups (fig. 3) – those with no apparent requirement for metabolic activation or conversion and those inactive until converted to chemically or metabolically active derivatives, called proximate or ultimate carcinogens [cf. 2,3].

The first group, the chemically active carcinogens, is an expanding group of alkylating agents that are able to effect many chemical reactions under the conditions existing in the living tissues as well as *in vitro* in model chemical systems. Some of the most interesting are the nitrosamides and alkane sulfonates which are able to induce a wide variety of cancers in many different organs, often following a single injection or application [cf. 4, 5]. Since the biological lifetime of many of these is very short and yet they are effective in inducing cancer, their intensive study offers the real hope of separating the irrelevant from the possibly relevant reactions as regards carcinogenesis (see below).

The most significant progress relates to work during the past 10 years or so on compounds in group B (fig. 3). The outstanding work of the Millers and their collaborators in Madison on aromatic amines [2, 3], the pioneer work of Magee and associates in England on nitrosamines [4], the work of Stekol and coworkers [6] and of our group [7, 8] on ethionine and the recent studies of Gelboin and associates [9, 10] and of Grover and Sims [11] on

A. COMPOUNDS WITH NO APPARENT REQUIREMENT FOR METABOLIC ACTIVATION OR CONVERSION

e.g. β-PROPIOLACTONE, EPOXIDES, NITROSAMIDES, ALKANE SULFONATES, ETC.

B. COMPOUNDS WHICH REQUIRE METABOLIC CONVERSION TO ACTIVE DERIVATIVES

1. PARTIAL DEPENDENCE UPON ENZYMES (UNTIL A CHEMICALLY ACTIVE FORM IS GENERATED)

   e.g. AROMATIC AMINES, NITROSAMINES, POLYCYCLIC AROMATIC HYDROCARBONS

2. COMPLETE DEPENDENCE UPON ENZYMES ?

   e.g. ETHIONINE

Fig. 3. Classification of chemical carcinogens.

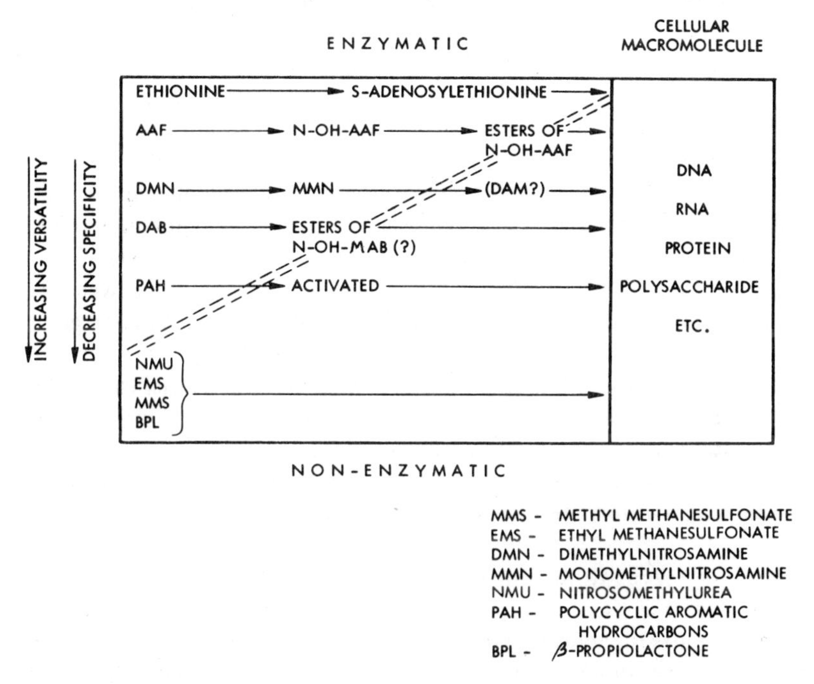

Fig. 4. Diagrammatic representation of some of the steps in the interactions between the selected carcinogens and various cell macromolecules. The reactions above the double broken lines are enzymatic, whereas those below the lines are considered to be non-enzymatic. AAF, acetylaminofluorene; DAM, diazomethane. The term versatility refers to the size of the spectrum of tumors induced while specificity refers to target cell.

polycyclic aromatic hydrocarbons are laying a solid foundation in the understanding of how cells act on chemical carcinogens and how the active derivatives in turn interact with cellular macromolecules.

A few of these are highlighted in fig. 4 where the spectrum of interaction between the initial cell population and various carcinogens is shown. Certain features stand out clearly.

(a) *N-Hydroxylation of aromatic amines.* The discovery by the Millers of this new biochemical reaction has opened up a new approach to the study of the metabolism of carcinogens. With many different aromatic amines such as derivatives of aminofluorene, azo dyes, naphthylamine and aminobiphenyl as well as others [2, 3] there is increasing evidence to suggest that metabolic conversion to a more active proximate or ultimate carcinogen is a prerequisite for carcinogenicity. This is best illustrated by acetylaminofluorene (AAF), a compound which induces cancer in the liver, mammary gland, ear duct and under special circumstances, the urinary bladder [12]. AAF also has a sharply defined species specificity. More recently, there is increasing evidence that the ultimate carcinogens are esters of the N-OH-AAF (sulfate, phosphate, glucuronidates, acetate, etc.) [2, 3, 13–17]. These are highly reactive *in vitro* with many nucleophilic groups including amino acids, proteins and nucleic acids. They are also much more carcinogenic than is the parent compound and often fail to show the tissue and species specificity of the latter.

The *N*-hydroxylation is effected by microsomal enzymes in the liver. The liver microsomal system is also active in ring hydroxylation, a reaction generally leading to inactivation of the carcinogenicity of AAF. Thus, in the case of the liver, the metabolism of the parent AAF occurs by at least two mechanisms, ring- and *N*-hydroxylation, one generating less active or inactive derivatives and the other highly active. It would seem that some inhibitors of AAF-induced liver cancer, such as 20-methylcholanthrene, preferentially stimulate the ring hydroxylation so as to greatly enhance the inactivation of AAF.

Since the enzymatic makeup of cells differ so much, it is probable that the high degree of tissue and species specificity for cancer induction by the aromatic amines is a reflection of the tissue distribution of the hydroxylating and esterifying enzymes. Hopefully, as these become better understood, one can anticipate an increasing ability to modify the metabolism of these carcinogens and thereby interrupt the carcinogenic process at will.

(b) *Oxidative dealkylation.* Although the nitrosamines are subject to a variety of metabolic reactions which produce at a minimum aldehydes, nitrous acid, hydroxylamines and hydrazines [4], it appears that the most important reaction is probably oxidative dealkylation, generating unstable monosubstituted nitrosamines [4]. These are considered to be potentially potent alkylating or arylating agents capable of reacting with many of the cell's nucleophilic groups. The probable inactivity of the parent compounds, the metabolic transformation to potent electrophilic reagents, the localization of the enzymes to the microsomes, at least in the liver, the high degree of tissue specificity for cancer induction by the various nitrosamines and the augmented carcinogenic potency and versatility with suggested ultimate carcinogens [18–21] suggest an unusual parallelism between some of the aromatic amines and some nitrosamines.

(c) *Other metabolic reactions. Ethionine,* the methionine analogue, is much more restricted in its carcinogenic spectrum in that the liver is the only or major target. Although this amino acid is susceptible to several metabolic attacks each of which mimics that of methionine, it would appear that the most likely reaction related to cancer is the enzyme-catalyzed interaction with ATP to produce *S*-adenosylethionine (SAE) [7]. Like *S*-adenosylmethionine (SAM), SAE is utilized for many different enzymatic transalkylations, even though the spectrum is considerably narrower than with SAM. However, there is one reaction which appears to be unique for SAE, production of 7-ethyl guanine [22] (see below). Whether this alkylation is enzymatic or non-enzymatic remains to be established.

Polycyclic aromatic hydrocarbons, although the earliest chemical carcinogens to be identified, have resisted clarification of their metabolism for a long time. Recent work by Grover and Sims [11] and by Gelboin and associates [9, 10] have offered the first clear-cut evidence that at least some polycyclic hydrocarbons must be metabolized before they interact with protein or DNA. Although various hydroxylated derivatives have been known for many years, virtually all have proven to be inactive or less active or no more active than the parent compound. Diamond and Gelboin [10] have recently reported than alpha-naphthoflavone inhibits the metabolism of 3,4-benzopyrene and of 7,12-dimethylbenz(a)anthracene in hamster embryo cell cultures and protects the cells against the depression in growth rate induced by these hydrocarbons. Alpha-naphthoflavone inhibits the aryl hydrocarbon hydroxylase activity in homogenates of induced hamster embryo cells and in liver microsomes from rats injected with polycyclic aromatic hydrocarbons. Thus, new possibilities for the clarification of the metabolism of this important and ubiquitous group of carcinogens now become available.

It is evident that this first phase of carcinogenesis, the action of cells on carcinogens, often represents an example of what might be called *toxic synthesis,* akin to *lethal synthesis* and *lethal incorporation* of Peters [23, 24]. Thus, the host plays an active role not only in responding to a carcinogenic environmental hazard but also in *actually generating the hazard.*

2.2. *Interaction of carcinogens with cell constituents*

As summarized in fig. 4, the majority of chemical carcinogens, either as such ("direct") or after suitable metabolic conversion ("indirect") interact with many cellular macromolecules. Since the original discovery of firm binding of azo dyes to liver protein by Miller and Miller in 1947 [25], many different oncogenic chemicals have been found to interact with cell proteins of different organs or tissues. Although quantitative variations, sometimes quite large, do occur with different cell proteins [26–28], it is impossible at this time to make a judgment concerning relative importance of any one of these to carcinogenesis. An added component in this complexity is the number of amino acids that are being found to bind to various types of carcinogens *in vivo* or *in vitro* [cf. 3]. Among these are methionine, cysteine, tryptophan and tyrosine.

In addition to proteins, it is also clearly established that many different carcinogens interact firmly with various RNAs and DNA. In some cases, the major chemical form of the product has been established. For example, with both direct (nitromethylurea and methyl methanesulfonate) and indirect (dimethylnitrosamine) methylating agents, the *major* product is 7-methylguanine in both RNA and DNA [4, 29] (fig. 5). However, other products, 1-methyladenine, 3-methyladenine and 1-methylcytosine have also been found in both types of liver nucleic acids in small but measurable amounts [30]. In the case of RNA, 7-methylguanine is a normal component of rat liver RNA including microsomal and soluble RNA [31]. However, 7-methylguanine has *not* been found in DNA from liver or other tissues, despite intensive searches [29, 31, 32]. These have included experiments using relatively small amounts (0.5 to 1 mg) [31] and large amounts (120 mg) [32] of methionine labeled in the methyl group. Thus, the 7-methylguanine appears to be an unusual or abnormal component of DNA.

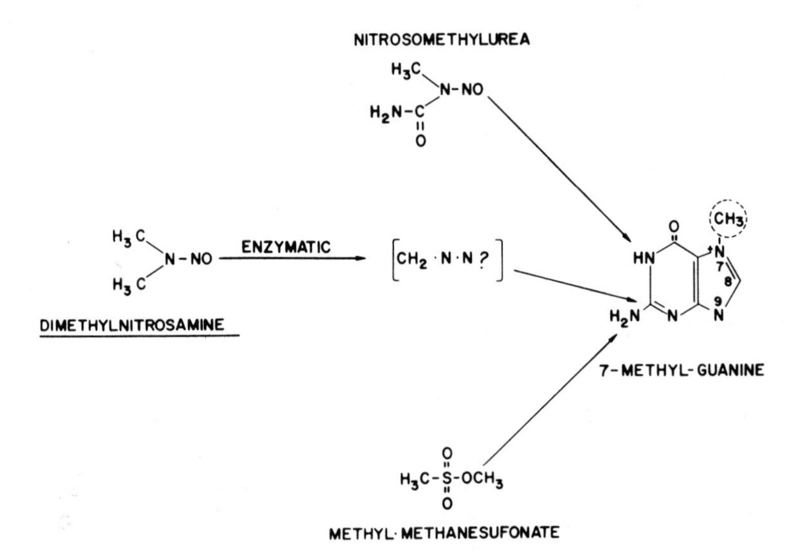

Fig. 5. 7-Methylguanine generated by "direct" (nitrosomethylurea and methyl methanesulfonate) and "indirect" (dimethylnitrosamine) methylating carcinogens.

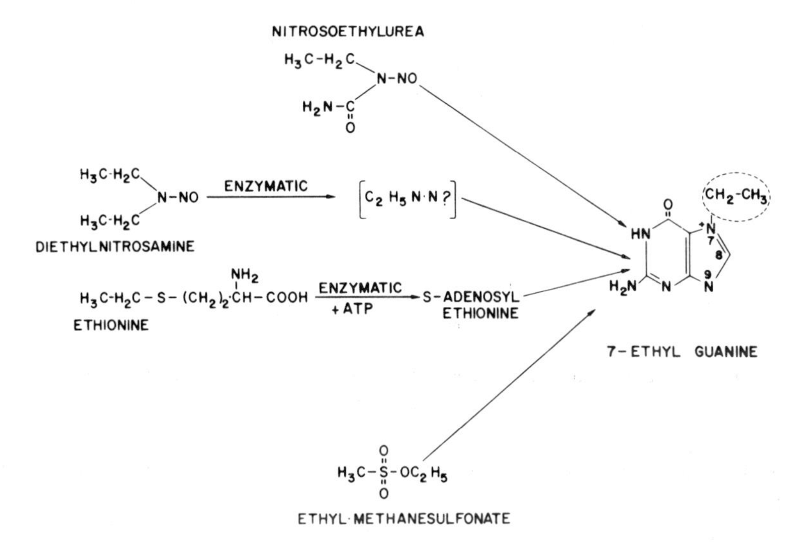

Fig. 6. 7-Ethylguanine generated by "direct" (nitrosoethylurea and ethyl methanesulfonate) and "indirect" (diethylnitrosamine and ethionine) ethylating carcinogens.

A similar pattern appears to pertain with carcinogenic ethylating agents (fig. 6). Both direct (nitrosoethylurea and ethyl methanesulfonate) and indirect (diethylnitrosamine and ethionine) ethylating compounds generate 7-ethylguanine in DNA as well as probably ethylate RNA and protein [4, 22, 32, 33]. Again, since no 7-methylguanine has been found in DNA, the 7-ethyl is a totally new compound with no normal analogue. With ethionine, this principle does not seem to hold with either transfer RNA [34] or nuclear protein [28] in rat liver and probably also not with alkylated cytosine in DNA in mouse liver [35]. In each of these situations, the ethylated compounds so far found are analogues of the normally-occurring methylated one, although as expected, the relative amounts of each of the various ethylated derivatives are different than with the methylated. Thus, there is no chemical evidence to suspect mechanisms of ethylation by ethionine of RNA, protein and DNA cytosine by other than the enzymatic routes used for methylation by methionine. However, this judgment must be reserved until the detailed mechanisms have been worked out. Another active carcinogen that alkylates the *N*-7 of guanine of DNA and of RNA as well as reacts with protein is β-propiolactone (fig. 7) [36]. Thus, three groups of active carcinogens generate substitution in the same position in the guanine of DNA and RNA. This type of alkylation was first discovered by Brookes and Lawley in 1961 [37] with bifunctional

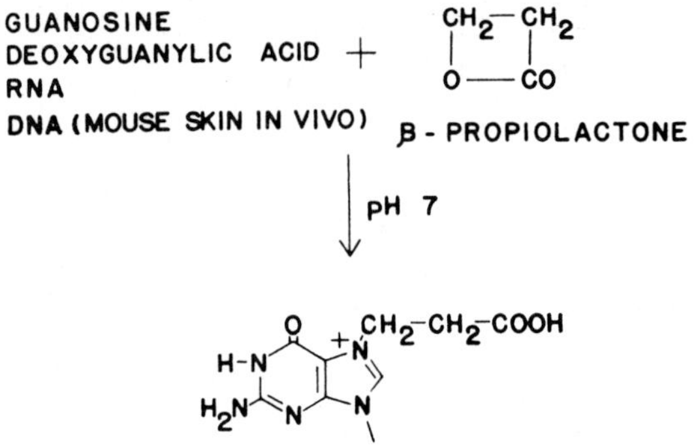

GUANOSINE
DEOXYGUANYLIC ACID
RNA
DNA (MOUSE SKIN IN VIVO) + β-PROPIOLACTONE
pH 7

Fig. 7. Interaction of β-propiolactone with guanine derivatives to generate 7-(2-carboxyethyl)-guanine.

alkylating agents which on the whole are very weak and non-versatile carcinogens.

An additional type of interaction with nucleic acids has been found with AAF. This carcinogen, after suitable activation, is able to react rapidly with DNA, RNA and their constituent nucleosides or bases to produce an aryl substitution at C-8 of guanine (fig. 8) [4, 38]. Although this is apparently not the only site of interaction, it is the major one. It appears that the interaction with DNA and RNA are different, since the former contains predominantly 8-(*N*-2-fluorenylamino)-guanine while the latter has as its major product the acetamide form (8-(*N*-2-fluorenylacetamido)-guanine) [39, 40]. This might be due to the nature of the activated N-OH-AAF – the glucuronide favoring reaction with DNA while other esters have more affinity for RNA [17].

Not only is there alkylation or arylation of purines by carcinogens but there is evidence also for reactions with pyrimidines. With ethionine, there is evidence for 5-ethylcytosine in place of 5-methylcytosine in mouse liver DNA [35]. Some labeling of a probable pyrimidine nucleotide has also been

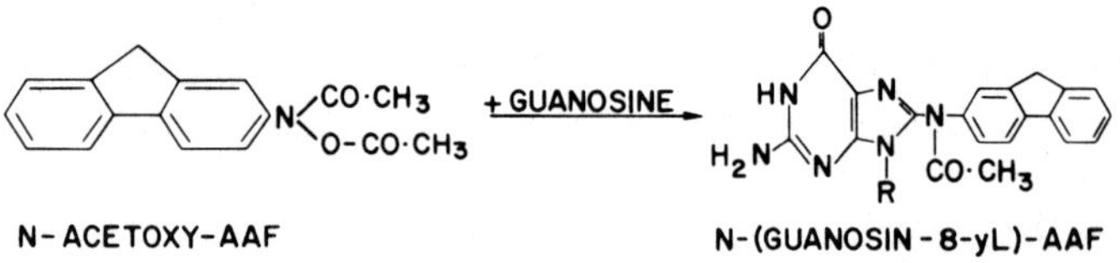

N-ACETOXY-AAF + GUANOSINE → N-(GUANOSIN-8-yL)-AAF

Fig. 8. Interaction of *N*-acetoxy-AAF with guanosine to generate the arylated derivative at C-8.

found in rat liver [41]. In the case of urethane, reaction with RNA appears to generate the ethyl ester of cytosine-5-carboxylic acid [42].

Thus, there is increasing evidence for a whole array of specific reactions between carcinogens and various cell macromolecules. Judging by the nature of the activated molecules being generated from some carcinogens, it is highly probable that virtually all nucleophilic groups, regardless of how strong or how weak, will be found to react with at least some carcinogens. This increases the complexity of the initial mutual interaction between cell components and chemical carcinogens and makes the choice of relevance to carcinogenesis much more difficult.

2.3. *What chemical interactions are relevant to carcinogenesis?*

It goes without saying that this is a most difficult question to discuss at this stage in our knowledge about cells. The first prerequisite is an understanding of the determinants of the normal variations in cell growth during the life history of the whole organism. How is an organism able to express and repress in an orderly fashion the information in its genome at all the different stages in its development and in its adult life? A rational conceptual framework for this, based on knowledge and not on speculation, would appear to be a necessity. And yet, it is not unlikely that the intensive study of carcinogenesis may itself play an important part in generating this knowledge. How should one attempt to approach this problem? It seems to me there are at least two basic approaches – a *prospective* and a *retrospective*. By prospective, I mean the selective elimination or suppression of all but one kind of interaction between a carcinogen and a molecular species in target cells. Alternatively, the synthesis or discovery of carcinogens capable of only one chemically specific interaction with a particular chemical component of the host falls naturally into this category. Be retrospective, I refer to the detailed analysis of the first cell population derived from the original target cell in the hope that only the relevant chemical alterations will continue to persist and begin to express themselves. Each approach has been used with some success and I would now like to discuss each one in turn.

However, before doing so, one major problem must be mentioned. Carcinogenesis could theoretically be triggered or initiated either by the specific *selection* of already different cells normally present in the host cell population or by the *induction*, more or less in random fashion, of changes on a probability basis without any built-in bias. This important problem [8] has been discussed previously by Prehn [43] and awaits a definitive solution. Subject to the possibility of selection as a major factor, the remainder of the discussion will be focussed on induction as the main phenomenon in carcinogenesis.

2.4. *Prospective approach*

It is already evident from work in various laboratories that alkylation of DNA or RNA, if important at all, is insufficient by itself to initiate carcinogenesis under some circumstances [e.g. 8, 29, 44]. This comes as no surprise, since it is very well documented that in some organs such as the liver, one must administer a carcinogen for usually 8 to 10 weeks to induce a relatively high incidence of cancer and single doses rarely are carcinogenic. Probably, in this organ, additional types of effects, such as induction of cell proliferation, may be needed in this non-proliferating cell population in order to set in motion the carcinogenic process [45, 46]. There are many studies to suggest interaction with DNA correlates best with carcinogenicity with several carcinogens [4, 47]. In addition, some studies have stressed RNA rather than DNA as the likely target for carcinogens [40, 48]. It may be worthwhile to point out, however, that attempts to draw anything but inexact conclusions from such correlations suffer from at least one theoretical handicap – there are thousands of known molecular species of proteins in any type of cell, at least dozens of known types of RNA and very few discrete DNA molecules (perhaps one per chromosome). Since interference with one single key protein could have far-reaching consequences on the activity and control of a cell, and since there is no known way to evaluate the significance of low levels of interaction of a carcinogen with such a protein, great caution must be used in arbitrarily ruling out "protein binding" or "RNA binding" on the basis of present day "negative" correlations.

An approach of great interest is the use of agents which inhibit or enhance the carcinogenic action of a chemical. The most convincing and clearcut example of this is the complete suppression of liver cancer induction by azo dyes, AAF and *m*-toluenediamine by 20-methyl-cholanthrene [see 49]. Unfortunately, this model does not help us to answer the question of relevance, since it appears to effectively suppress the generation of proximate or ultimate carcinogens by enhancing the inactivation or degradation of the parent carcinogen [2,3]. Yet, it does add great weight to the presumptive conclusion for a role of metabolism in the generation of active derivatives. The modification of the carcinogenic action of 2-AAF by acetanilide [50, 51] may operate through a similar overall mechanism, even though competition for the same enzyme or competition for available tissue sulfate by a major metabolite, *p*-hydroxyacetanilide [52] rather than enzyme induction may be the pathway operating. Another interesting example is that of alpha-naphthoflavone which inhibits the toxicity of some polycyclic aromatic hydrocarbons, probably again through an effect on microsomal hydroxylases [10]. Its possible influence on carcinogenicity will be awaited with interest.

These are but some examples of many in which modification of the metabolism of a carcinogen has a profound influence on its biological effects. Unfortunately, no agent has yet been found to selectively suppress or enhance one or more types of interaction with cellular macromolecules and thereby allow only one interaction to proceed. Very recently, administration of sulfate compounds to animals receiving N-OH-AAF appears to increase protein and RNA binding much more so than DNA binding [52]. One is therefore encouraged to think that the search for selective suppressors may be rewarding. The possibility of preparing various nucleophilic compounds that compete selectively with DNA, RNA, protein, etc. for electrophilic carcinogens may well open up new horizons in the study of carcinogenesis.

2.5. *Retrospective approach*

As in the case of other retrospective activities of man, such as in epidemiology, looking back has disadvantages as well as advantages. No doubt, as in epidemiology, the opportunity to use both directions of focussing on a target offers the most hope.

I would now like to review briefly some newer developments in this phase of liver carcinogenesis which have already proven very useful to us and which may be exploited profitably in many directions. In this work, I must single out particularly a colleague of mine, Dr. Epstein, whose contributions have been unusual.

As a result of earlier work on the factors that may modify liver carcinogenesis induced by ethionine [7, 53], it became apparent that one could modify the cellular populations appearing during the induction of liver cancer in such a way as to offer an interesting model for the analysis of some steps in the process. What we were after was a model that satisfied the following minimal citeria [7, 8]: (a) a cell of origin of cancer could be identified; (b) this precursor cell population occurs in a localized lesion sufficiently large to enable gross identification and isolation for a wide variety of metabolic as well as morphologic and cultural studies; and (c) the cell population is uniform in genotype and phenotype. Several such potential models are known – the hyperplastic nodules in the liver induced with virtually every hepatic carcinogen [7, 54], the plaque or hyperplastic nodule in the mammary gland of mice or rats induced by viruses or polycyclic aromatic hydrocarbons [cf. 55], papillomas of the skin induced by many different chemicals, and so-called conditional neoplasms of endocrine glands which can be induced by a variety of procedures [56].

We have concentrated our efforts so far on the liver and its hyperplastic nodules induced by AAF, ethionine or more recently aflatoxin B_1. By suitable dietary manipulation, we are now able to induce nodules up to 2 or 3 cm in diameter in well over 50% of the rats.

These nodules are composed of a new liver cell population that begins to grow out fairly early in liver carcinogenesis with virtually every hepatic carcinogen. When sufficiently large, one can often see unequivocal liver cell cancer arising within the confines of the nodule with no evidence of cancer elsewhere in the liver. Thus, the hyperplastic nodule is definitely one precursor for liver cancer. The population in the nodule is almost all liver cells with a sprinkling of bile ducts and vascular lining cells. Thus the nodules are at least as homogeneous as is normal liver from the point of view of cell composition. The liver cells are also remarkably uniform as judged by light and electron microscopy [57]. They also show unusually reproducible biochemical and cultural changes, regardless of which one of the three carcinogens is used.

The properties of this first identifiable growing population of hepatocytes, in so far as they have been studied, are as follows. *It should be emphasized that the hyperplastic nodules show the following changes many weeks after the carcinogen is removed from the diet.*

A. *Carbohydrate metabolism* [54]

(i) Decreased breakdown of glycogen on fasting or after glucagon with normal glycogen synthesis.

(ii) Progressive decrease in glucose-6-phosphatase, and glycogen phosphorylase with no change in glycogen synthetase (UDPglucose: glycogen α-4-glucosyltransferase).

(iii) Preliminary work with Dr. E. Bueding suggests that the glycogen size spectrum may be considerably larger than in the control or surrounding liver.

Probably a similar biochemical change is seen with virtually every hepatic carcinogen and perhaps also in the kidney with a carcinogenic nitrosamine [58].

B. *Proliferation in situ* [57, 59]

(i) The nodule cell population frequently contain annulate lamellae [57], cell organelles associated with cell proliferation.

(ii) The labeling index (percent labeled hepatocytes) is 10 to 20 times higher in the nodule than in the surrounding liver [59].

C. *Proliferation of smooth endoplasmic reticulum* [57].

(i) This change is striking and often persists in the cancer.

D. *Altered glycogen structure* [60]

(i) Glycogen in the hyperplastic nodules contains bound AAF derivatives as determined by (a) gas chromatography, (b) spectrophotometry, and (c) mass spectrographic analysis. The bound AAF derivative persists also in the glycogen in the cancer many months after AAF is removed from the diet but is absent from the surrounding liver.

(ii) The ultraviolet adsorption spectrum of nodule glycogen in the presence of I_2 – KI is unstable, probably due to the bound AAF.

(iii) Glycogen in nodule does not stain normally with lead hydroxide as observed in the electron microscope [57].

E. *Altered DNA in hyperplastic nodule* [45, 60]

(i) The nodule DNA shows absorption in the range 300 to 340 nm, very similar to that observed with double-stranded liver or thymus after reaction *in vitro* with *N*-acetoxy-AAF. This is absent from the DNA from the surrounding liver.

(ii) The nodule DNA, but not surrounding liver DNA, has a small population (2-3%) with a greater bouyant density (in CsCl) than the bulk of the liver DNA.

(iii) By electron microscopy, the nodule DNA shows branching and puddles not seen in DNA from normal liver, from surrounding liver or from regenerating liver at two different times after partial hepatectomy (17 and 23 hr).

(iv) Preliminary results with Dr. Walter Troll suggest that the T_m (melting curve) of nodule DNA is as much as 10° less than that of DNA from the control or the surrounding liver.

F. *Growth in vitro* [61]

(i) In every experiment to date (over 30), cells from hyperplastic nodules without cancer grow on plastic in tissue culture *in vitro* under conditions in which the control or surrounding liver show no evidence of such growth. The pattern of growth is quite different than that seen with liver cell cancer. Growth has been found regularly with nodules induced by AAF, ethionine or aflatoxin B_1.

(ii) The growing cells differentiate into ducts when grown on a collagen-coated sponge.

(iii) The cells in culture resemble epithelial cells and some contain rat serum albumen.

Thus, the evidence points to the proliferating cells being liver hepatocytes and not connective tissue.

1.6. *Implications*

The most interesting new information is (a) the *persistence* of at least one carcinogen (others have not yet been examined in detail) throughout the whole process of carcinogenesis bound to a macromolecule in a select population of non-cancerous liver cells and appearing in the same macromolecule

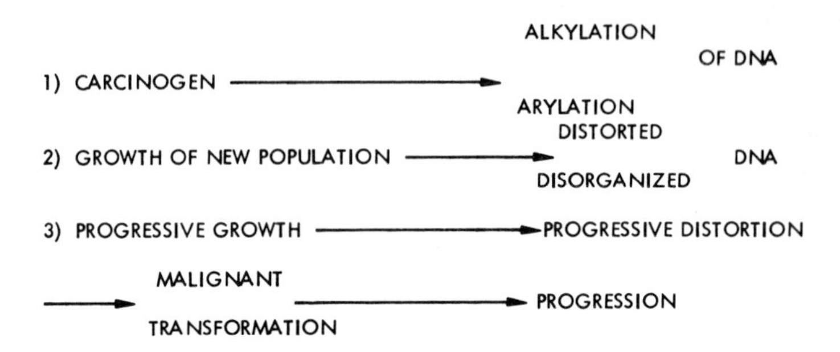

Fig. 9. Diagrammatic representation of proposed steps in a working hypothesis of chemical carcinogenesis.

in the cancer; (b) the *persistence* of bound carcinogen in DNA in the non-cancerous hyperplastic nodule with obvious distortion and disorganization of the DNA; (c) the selection of an early population of liver cells for *growth* without evidence of other characteristics of cancer; (d) the obvious distortion of DNA from the growing non-cancerous liver cell population as contrasted to the DNA from the surrounding non-hyperplastic liver; (e) the unusually *uniform appearance* of the hepatocytes in growing hyperplastic nodule as contrasted with the very heterogeneous appearance of the cells in the primary cancer [57, 62]; and (f) the remarkable *similarity* between the *hyperplastic nodules* induced by many different hepatic carcinogens.

These considerations suggest the following working hypothesis (figs. 2 and 9).

The AAF, after appropriate activation and conversion to a reactive ester (sulfate?, etc.) of N-OH-AAF, reacts with several macromolecules including DNA, RNA, protein and glycogen. Within a relatively short time (one to a few weeks), some liver cells respond by cell multiplication. Initially, the multiplication is dependent upon the continued presence of the dietary AAF and thus may be basically a regenerative response to cell injury induced by the AAF. However, after a certain period of AAF administration (8-10 weeks), multiple foci of a new population of liver cells persist and grow in the absence of any further intake of AAF. This new population has now acquired the ability to grow *in vitro* under conditions where liver normally does not. Thus growth, without other stigmata of cancer, has been "turned on". This population is referred to as nodular hyperplasia, each nodule being presumably a clone. This growing population is different from the self-limiting proliferation seen in response to an obvious stimulus such as partial hepatectomy or death of liver cells. The biochemical and molecular properties that are associated

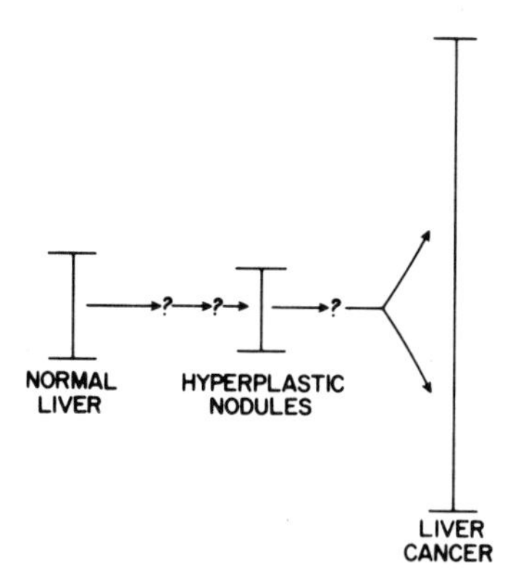

Fig. 10. Diagrammatic representation of 3 cell populations, original target cells (liver), hyperplastic nodules and liver cancer. The height of the lines is meant to represent the variations in the cell structure and function seen with each population.

with this new growing cell population must be very restrictive, since many different types of chemicals, all carcinogenic, induce hyperplastic nodules which are remarkably similar to each other (fig. 10).

The altered DNA in the surrounding non-proliferating liver cells, containing scattered sites of arylation with no disorganization of the secondary structure of the double-stranded DNA, is presumably enzymatically repaired in such a manner as to yield essentially normal liver nuclear DNA. In contrast, in the proliferating cells located in the hyperplastic nodules, the DNA has been replicated before sufficient time or other conditions have allowed full excision and repair of the altered DNA. The presence of the aryl group at certain sites could conceivably induce an altered or distorted replication of the DNA. Some precedent for this is offered by the work of Kornberg and associates [63–65] who found that *E. coli* DNA polymerase synthesized branched molecules under some experimental circumstances.

Once initiated in a proliferating cell population, the branching and distortion of the DNA would become increasingly greater as the cells continued to proliferate. Such a phenomenon could play an important role in the progressive nature of the changes in the biological behavior of the premalignant cell leading to the stepwise appearance of the discrete properties characteristic of cancer.

This formulation may not be unique to AAF. Dimethylaminoazobenzene or its active derivatives bind in a covalent-like manner to liver nucleic acids [66–68], the binding to DNA persists for months after termination of exposure to the carcinogen [69] and a non-carcinogenic derivative becomes carcinogenic when combined with partial hepatectomy [46]. These findings, coupled with the observations concerning cell populations during azo-dye carcinogenesis [70] suggest that the hypothesis outlined may be applicable, in principle, to carcinogenesis with chemicals other than AAF.

It would also seem that some examples of viral carcinogenesis may fit into this overall picture. As pointed out recently by Dulbecco [71], the rapidity of the transformation induced by viruses may be, in part at least, due to the use of *in vitro* cell populations which have been selected for proliferation. Cell proliferation is needed for initiation of transformation by SV 40 [72]. Therefore, it may be that in some systems, the carcinogen must have the capability of somehow inducing or stimulating cell proliferation, either directly or indirectly, before any molecular lesion, such as altered DNA, can be "fixed" or "imprinted" on the cell.

The essence of the above formulation would be the following: (a) chemically altered DNA; (b) inadequate or faulty repair; (c) cell proliferation in the face of unrepaired DNA and in the absence of exogenous carcinogen. One cannot help thinking that the altered DNA need not be the direct gain or loss of some specific discrete piece of information but rather the *disorganization* of the genomic information. Conceivably, the viral genome, by virtue of its unique structure, can disorganize the host genome in a predictable manner such that malignant transformation occurs rapidly along very narrow lines. The failure so far to find any specific piece of information in some viruses which is responsible for cancer induction is consistent with this. Thus chemicals and viruses could have a more or less common pathway if allowance is made for the relative randomness of a chemical interaction compared to the non-random interaction with a DNA virus.

Finally, one should stress that the above formulation is focussed on host DNA because that is the only type of macromolecule today which appears to have the necessary prerequisite for the induction of an heritable change in cell behavior. However, this is no doubt merely a reflection of our ignorance. The future could well develop experimentally sound bases for an important role for altered RNA or protein molecules in an heritable change in a cell. At that time, the knowledge that is developing concerning the mutual interaction between chemical carcinogens and target cells will no doubt be ready to be exploited for new and more imaginative concepts of carcinogenesis.

Acknowledgement

I would like to express my very sincere thanks to Dr. Sheldon Epstein for many helpful discussions.

References

[1] S. Bayne-Jones, R.G. Harrison, C.C. Little, J. Northrop and J.B. Murphy, Public Health Rept. 53 (1938) 2121.
[2] E.C. Miller and J.A. Miller, Pharmacol. Rev. 18 (1966) 805.
[3] J.A. Miller and E.C. Miller, The Jerusalem Symposium on Quantum Chemistry and Biochemistry, I. Israel Academy of Sciences and Humanities, 1969.
[4] P.N. Magee and J.M. Barnes, Advan. Cancer Res. 10 (1967) 163.
[5] H. Druckrey, R. Preussmann, S. Ivankovic, D. Schmahl, J. Afkham, G. Blum, H.D. Mennel, M. Muller, P. Petropoulos and H. Schneider, Z. Krebsforsch. 69 (1967) 103.
[6] J.A. Stekol, Advan. Enzymol. 25 (1963) 369.
[7] E. Farber, Advan. Cancer Res. 7 (1963) 383.
[8] E. Farber, Cancer Res. 28 (1968) 1859.
[9] H.V. Gelboin, Cancer Res. 29 (1969) 1272.
[10] L. Diamond and H.V. Gelboin, Science 166 (1969) 1023.
[11] P.L. Grover and P. Sims, Biochem. J. 110 (1968) 159.
[12] E.K. Weisburger and J.H. Weisburger, Advan. Cancer Res. 5 (1958) 331.
[13] J.A. Miller and E.C. Miller, Progr. Exptl. Tumor Res. 11 (1969) 273.
[14] C.M. King and B. Philips, Science 159 (1968) 1351.
[15] J.R.DeBraun, J.Y. Rowley, E.C. Miller and J.A. Miller, Proc. Soc. Exptl. Biol. Med. 129 (1968) 268.
[16] C.C. Irving, R.A. Veazey and J.T. Hill, Biochim. Biophys. Acta 179 (1969) 189.
[17] C.C. Irving, R.A. Veazey and L.T. Russell, Chem. Biol. Interactions 1 (1969) 19.
[18] R. Schoental and P.N. Magee, Brit. J. Cancer 16 (1962) 92.
[19] R. Schoental, Nature 199 (1963) 190.
[20] R. Schoental, Nature 208 (1965) 300.
[21] R. Schoental, Nature 209 (1966) 726.
[22] J.A. Stekol, in: Transmethylation and Biosynthesis of Methionine, eds. S.K. Shapiro and F. Schlenk (University of Chicago Press, Chicago, 1956) p. 235.
[23] R.A. Peters, Biochemical Lesions and Lethal Synthesis (Pergamon Press, Oxford, 1963).
[24] R.A. Peters, Brit. Med. Bull. 25 (1969) 223.
[25] E.C. Miller and J.A. Miller, Cancer Res. 7 (1947) 468.
[26] S. Sorof, E.M. Young and M.G. Ott, Cancer Res. 18 (1958) 33.
[27] B. Ketterer, P. Ross-Mansell and J.K. Whitehead, Biochem. J. 103 (1967) 316.
[28] M. Friedman, K.H. Shull and E. Farber, Biochem. Biophys. Res. Commun. 34 (1969) 857.
[29] P.F. Swann and P.N. Magee, Biochem. J. 110 (1968) 39.
[30] P.D. Lawley, P. Brookes, P.N. Magee, V.M. Craddock and P.F. Swann, Biochim. Biophys. Acta 157 (1968) 646.

[31] V.M. Craddock, S. Villa-Trevino and P.N. Magee, Biochem. J. 107 (1968) 179.
[32] P.F. Swann, A. Hawkes, E. Farber and P.N. Magee, unpublished observations.
[33] P.F. Swann and P.N. Magee, personal communication.
[34] L. Rosen, Biochem. Biophys. Res. Commun. 33 (1968) 546.
[35] A.L. Kaye, personal communication.
[36] W.N. Colburn and R.K. Boutwell, Cancer Res. 26 (1966) 1701.
[37] P. Brookes and P.D. Lawley, Biochem. J. 80 (496) 1961.
[38] E. Kriek, Biochem. Biophys. Res. Commun. 20 (1965) 793.
[39] E. Kriek, Biochim. Biophys. Acta 161 (1968) 273.
[40] E. Kriek, Chem. Biol. Interactions 1 (1969) 3.
[41] E. Farber, J. McConomy, B. Franzen, F. Marroquin, G.A. Stewart and P.N. Magee, Cancer Res. 27 (1967) 1761.
[42] E. Boyland and K. Williams, Biochem. J. 111 (1969) 121.
[43] R.T. Prehn, J. Natl. Cancer Inst. 32 (1964) 1.
[44] R. Schoental, Biochem. J. 114 (1969) 55P.
[45] S.M. Epstein, E.L. Benedetti, H. Shinozuka, B. Bartus and E. Farber, Chem.-Biol. Interactions 1 (1969) 113.
[46] G.P. Warwick, Intern. J. Cancer 3 (1967) 227.
[47] P. Brookes, Cancer Res. 26 (1966) 1994.
[48] I.B. Weinstein, Cancer Res. 28 (1968) 1797.
[49] N. Ito, Y. Hiasa, Y. Konishi and M. Marugami, Cancer Res. 29 (1969) 1137.
[50] R.S. Yamamoto, R.M. Glass, H.H. Frankel, E.K. Weisburger and J.H. Weisburger, Toxicol. Appl. Pharmacol. 13 (1968) 108.
[51] P.H. Grantham, L. Mohan. R.S. Yamamoto, E.K. Weisburger and J.H. Weisburger, Toxicol. Appl. Pharmacol. 13 (1968) 118.
[52] J.R. DeBraun, J.Y.R. Smith, E.C. Miller and J.A. Miller, Science 167 (1970) 184.
[53] E. Farber and H. Ichinose, Acta Unio Intern. Contra Cancerum 15 (1959) 152.
[54] S.M. Epstein, N. Ito, L. Merkow and E. Farber, Cancer Res. 27 (1967) 1702.
[55] K.B. DeOme, L.J. Faulkin, H.A. Bern and P.B. Blair, Cancer Res. 19 (1959) 515.
[56] J. Furth, Cancer Res. 23 (1963) 21.
[57] L.P. Merkow, S.M. Epstein, B. J. Caito and B. Bartus, Cancer Res. 27 (1967) 1712.
[58] P. Bannasch and U. Schacht, Virchows Arch. Abt. B. Zellpathol. 1 (1968) 95.
[59] M. Manugami, S.M. Epstein and B. Bartus, unpublished work.
[60] Epstein, S.M., J. McNary, B. Bartus and E. Farber, Science 162 (1968) 907.
[61] M. Slifkin, L.P. Merkow, M. Pardo, S.M. Epstein, J. Leighton and E. Farber, Science 167 (1970) 285.
[62] L.P. Merkow, S.M. Epstein, E. Farber, M. Pardo and B. Bartus, J. Natl. Cancer Inst. 43 (1969) 33.
[63] C.L. Schildkraut, C.C. Richardson and A. Kornberg, J. Mol. Biol. 9 (1964) 24.
[64] C.C. Richardson, R.B. Inman and A. Kornberg, J. Mol. Biol. 9 (1964) 46.
[65] M. Goulian and A. Kornberg, Proc. Natl. Acad. Sci. U.S. 58 (1967) 1723.
[66] F. Marroquin and E. Farber, Proc. Am. Assoc. Cancer Res. 4 (1963) 41.
[67] J.J. Roberts and G.P. Warwick, Intern. J. Cancer 1 (1966) 179.
[68] C.W. Dingman and M.B. Sporn, Cancer Res. 27 (1967) 938.
[69] G.P. Warwick and J.J. Roberts, Nature 213 (1967) 1206.
[70] R. Daoust, Can. Cancer Conf. 5 (1963) 225.
[71] R. Dulbecco, Science 166 (1969) 962.
[72] G.J. Todaro and H. Green, Proc. Natl. Acad. Sci. U.S. 55 (1966) 302.

A STEREOCHEMICAL APPROACH TO THE ACTIVE SITE OF GLUTAMINE SYNTHETASE

Alton MEISTER*

Department of Biochemistry, Cornell University Medical College,
New York City, New York 10021, USA

Abstract: Meister, A. A Stereochemical Approach to the Active Site of Glutamine Synthetase. *Miami Winter Symposia* 2, pp. 335–351. North-Holland Publishing Company, Amsterdam, 1970.

Brain glutamine synthetase interacts with L- and D-glutamate, α-methyl-L-glutamate, *threo*-β-methyl-D-glutamate, *threo*-γ-methyl-L-glutamate, β-glutamate, *cis*-1-amino-1, 3-dicarboxycyclohexane, L-methionine-*S*-sulfoximine, L- and D-methionine sulfone, but not with a number of closely related compounds. These and other findings have led to an hypothesis about the conformation and orientation of L-glutamate, ammonia, and the terminal phosphate moiety of adenosine triphosphate at the active site. The conclusions have been derived in large part from study of molecular models and are supported and extended by a new computer approach, which has been used in mapping the active site of brain glutamine synthetase.

In this talk I will attempt to review some of the work which has been done in our laboratory during the last few years on the structure and function of brain glutamine synthetase. I will deal especially with those findings which have made it possible for us to begin to think about the three-dimensional orientations of the substrates on the active site of the enzyme. It should be borne in mind that at this time we do not have any information about the chemical structure of the active site of the enzyme so that virtually everything that we have attempted to deduce about the relative positions of the substrates on the active site has of necessity come from other data, especially observations on the ability of the enzyme to interact with a variety of substrates and inhibitors. Many of these findings have been reviewed recently [1,2] and have been extended by means of a computer approach [3].

* The author wishes to acknowledge the support of the National Institutes of Health, Public Health Service, the National Science Foundation, and the John A. Hartford Foundation, Inc.

The synthesis of glutamine takes place according to the following reaction:

$$\text{L-glutamate} + NH_3 + \text{ATP} \xrightleftharpoons{Mg^{2+}} \text{L-glutamine} + \text{ADP} + P_i$$

When hydroxylamine is substituted for ammonia, the product is L-γ-glutamyl hydroxamate. The enzyme also catalyzes several other reactions [see 1,2], one of which will be discussed below.

We have used the glutamine synthetase from sheep brain in our studies; the enzyme has been isolated in homogeneous form from this source [4,5] and also from human brain [6]. The sheep and human enzymes are embarassingly similar; indeed the glutamine synthetase of peas closely resembles that of brain!** Brain glutamine synthetase has been shown to have a molecular weight of about 500,000 and to be composed of eight apparently identical subunits arranged in a manner analogous to the corners of a cube [7, 8]. The native octameric enzyme may be readily dissociated reversibly to a tetramer, and further irreversible dissociation to monomer has also been observed [9]. There is good evidence, as will be discussed below, that each subunit possesses an active site capable of combining with substrate.

There is now substantial evidence that enzyme-bound γ-glutamyl phosphate is formed as an intermediate in glutamine synthesis; the details of experiments that support this conclusion have been reviewed previously [1,2], and additional evidence in accord with this idea will be presented here. A number of years ago we carried out experiments designed to detect the formation of an activated glutamate intermediate [10]. Thus, we incubated glutamate, ATP, magnesium ions, and enzyme in the absence of ammonia in the hope that an activated glutamate intermediate might be formed on the enzyme. Since it is well known that γ-glutamyl compounds exhibit a tendency to cyclize to form pyrrolidone carboxylic acid, we reasoned that a highly activated γ-glutamyl derivative might cyclize quite rapidly. In accord with this expectation we found that glutamine synthetase does indeed catalyze the formation of pyrrolidone carboxylic acid in the absence of ammonia according to the following

** In contrast, the glutamine synthetase of *Escherichia coli* differs from that of brain (and peas) in the number [12] and arrangement of subunits, amino acid composition, specificity, ability to be inhibited by various allosteric effectors and to undergo adenylation, and in other respects (see Stadtman [24]).

equation:

$$\mathrm{ATP} + \begin{array}{c} \mathrm{COOH} \\ | \\ \mathrm{CH_2} \\ | \\ \mathrm{CH_2} \\ | \\ \mathrm{CHNH_2} \\ | \\ \mathrm{COOH} \end{array} \xrightarrow{Mg^{2+}} \begin{array}{c} \mathrm{H_2C}\text{———}\mathrm{CH_2} \\ | \qquad\quad | \\ \mathrm{O{=}C} \diagdown \; \diagup \mathrm{CHCOOH} \\ \mathrm{N} \\ \mathrm{H} \end{array} + \mathrm{ADP} + \mathrm{P_i}$$

The formation of pyrrolidone carboxylic acid is consistent with the prior formation of enzyme-bound γ-glutamyl phosphate. γ-Glutamyl phosphate exhibits a marked tendency to undergo spontaneous cyclization; this made it impossible for us to isolate the acyl phosphate itself. At this point it should be mentioned that the enzymatic cyclization of glutamate takes place at approximately equal rates with L- and D-glutamate. Furthermore, it was known from earlier studies [11] that glutamine synthetase can catalyze the synthesis of both L- and D-glutamine (and the corresponding hydroxamic acids). The ability of glutamine synthetase to act on both D-glutamate and L-glutamate indicates that the amino group of the substrate does not need to be in a specific position, and we therefore prepared the amino acid β-glutamic acid (3-aminoglutaric acid), in which the amino group is attached to the third possible position of the glutarate carbon chain [12]. The observation that β-glutamate is a good substrate for glutamine synthetase was of particular importance because the β-aminoglutaryl phosphate intermediate that might be postulated in the enzymatic synthesis of β-glutamine would be expected to be rather stable as compared to γ-glutamyl phosphate. Subsequent studies with chemically synthesized β-aminoglutaryl phosphate indeed showed that glutamine synthetase can use this compound in the presence of ADP for the synthesis of ATP. This observation gave strong support to the belief that γ-glutamyl phosphate is an intermediate in the synthesis of glutamine.

An incidential finding in the experiments on β-glutamate was that this compound, which does not possess an asymmetric carbon atom, is converted only to D-β-glutamine [13]. The strict stereospecificity of the enzyme on β-glutamate thus contrasts markedly with its rather optically unspecific activity towards glutamate. Furthermore, it was found at about the same time that glutamine synthetase acts with strict L-specificity on α-methylglutamate [14].

These observations presented us with something of a puzzle. It was curious that glutamine synthetase could act on both isomers of glutamate, with strict D-specificity on β-glutamate, and with equally strict L-specificity on α-methyl-glutamate. In an effort to explain these observations we began to build models of the substrates. We made several assumptions. Thus, since glutamine synthetase does not interact appreciably with amino acids that have only one carboxyl group, we assumed that the enzyme has sites for both carboxyl groups of glutamate. Second, since compounds that lack an amino group are not substrates, we postulated that the enzyme has a binding site for the glutamate amino group. Third, since aspartate does not interact with the enzyme to an appreciable extent, we assumed that the carboxyl groups of glutamate are oriented on the enzyme at a distance greater than the maximum intercarboxyl distance possible for aspartate. In addition, we postulated that the same relatively fixed sites on the enzyme combine with the respective functional groups of each of the substrates.

Accordingly we may consider the fully extended conformations of L- and D-glutamate in which the carboxyl groups of these molecules are as far apart as possible (fig. 1). Although the carboxyl carbon atoms of each model can be brought to the same relative positions in space, it is obvious that the amino groups of these isomers are in quite different positions and thus it is difficult to explain the enzymatic susceptibility of D-glutamate if this molecule is assumed to be oriented on the enzyme in the same manner as L-glutamate. However, if the model of D-glutamate is rotated through an angle of 69° around an axis passing through carbon atoms 1, 3 and 5, the nitrogen atom of D-glutamate will be brought to a position in space which is identical to that of the nitrogen atom of L-glutamate (fig. 2). If the L- and D-glutamate molecules re-

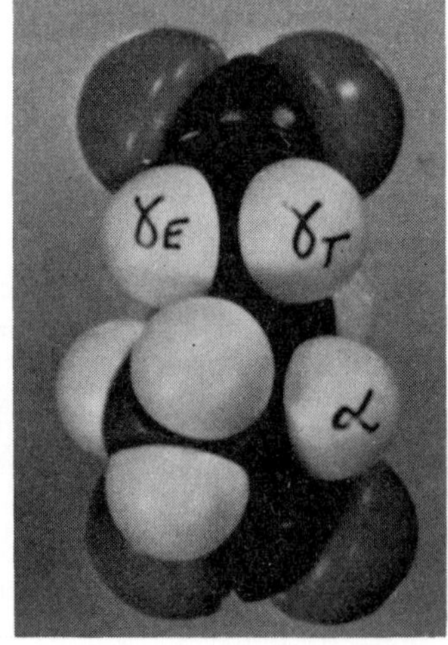

Fig. 1. Models of L- (left) and D- (right) glutamic acid (see the text).

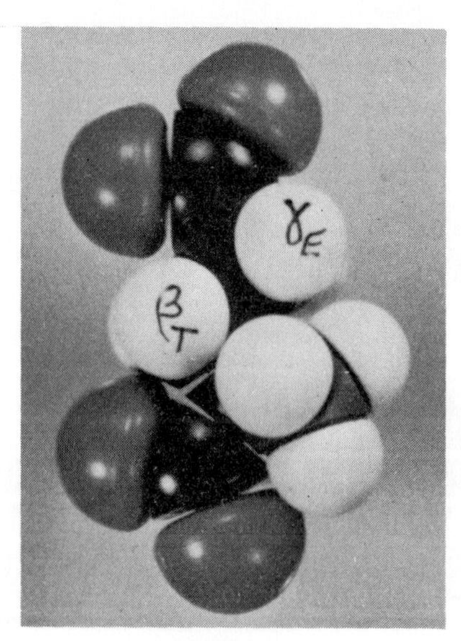

Fig. 2. The model of D-glutamic acid has been rotated to the right from the position shown in fig. 1.

presented in the models shown in fig. 1 (left) and fig. 2 are assumed to be resting in the orientations shown on the active site of the enzyme, it becomes evident that the amino and carboxyl groups of the substrates could attach to the same binding sites on the enzyme. It will be noted that the α-hydrogen atom of L-glutamate is on the side of the molecule directed away from the enzyme and thus can be seen in fig. 1. On the other hand, the α-hydrogen atom of D-glutamate is on the undersurface of the model and thus close to the enzyme (fig. 2). When the α-hydrogen atom of D-glutamate is replaced by a methyl group this might therefore be expected to interfere with attachment of α-methyl-D-glutamate to the enzyme; however, similar substitution of the α-hydrogen atom of L-glutamate would not since the methyl group would project away from the enzyme.

We subsequently carried out studies with the several β-methyl and γ-methyl derivatives of glutamic acid [15,16] and found that of the ten possible methyl-substituted glutamates only three are substrates for glutamine synthetase. These are *threo*-β-methyl-D-glutamate, *threo*-γ-methyl-L-glutamate, and as stated above, α-methyl-L-glutamate. It will be noted that the three susceptible methyl glutamates are all substituted on the left-hand side of the respective molecules (as shown in fig. 1 (left) and fig. 2). Furthermore, methyl substitutions on the undersurfaces (assumed to be in contact with the enzyme) of these molecules (fig. 3) resulted in loss of enzymatic susceptibility.

Fig. 3. Models of L- (left) and D- (right) glutamic acid; undersurfaces of models shown in fig. 2.

A summary of the specificity of glutamine synthetase is given in table 1. Consideration of the structures of the three enzymatically susceptible methyl-substituted glutamates suggested the attractive possibility of constructing a cyclohexane ring consisting of carbon atoms 2, 3, and 4 of L-glutamate and a chain of 3-carbon atoms attached to carbon atoms 2 and 4 of glutamate [17]. If one examines a model of this cyclohexane amino acid (*cis*-L-1-amino-1, 3-dicarboxycyclohexane) (fig. 4) it may be seen that this compound can exist in a form possessing a relatively rigid 5-carbon chain oriented in a manner which is identical to that of the extended conformation of L-glutamate. Furthermore, the portion of the cyclohexane ring which is not identical to the 5-carbon chain of glutamate lies wholly in the region in which methyl group

Table 1
Summary of the optical specificity of glutamine synthetase.

Substrate	Optical specificity
Glutamate	L and D
α-Methyl-glutamate	L
β-methyl-glutamate (*threo*)	D
γ-Methyl-glutamate (*threo*)	L
β-Glutamate	D

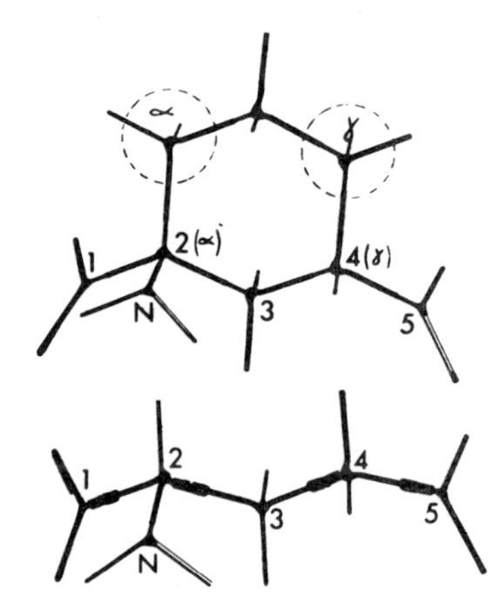

Fig. 4. Dreiding models of L-glutamic acid (below) and *cis*-L-1-amino-1,3-dicarboxycyclohexane (above).

Table 2
Relative activities of various substrates of ovine brain glutamine synthetase with ammonia and hydroxylamine.

Amino acid substrate	Relative V_{max}	
	NH_2OH	NH_3
L-Glutamate	100	100
D-Glutamate	54	27
α-Methyl-L-glutamate	67	75
threo-β-Methyl-D-glutamate	46	2.2
threo-γ-Methyl-L-glutamate	63	27
threo-β-Hydroxy-D-glutamate[a]	48	22
threo-γ-Hydroxy-L-glutamate	89	100
threo-γ-Hydroxy-D-glutamate	1.6	<0.08
erythro-γ-Hydroxy-L-glutamate	64	81
erythro-γ-Hydroxy-D-glutamate	29	38
β–Glutamate	46	18
L-α-Aminoadipate	22	0.63
D-α-Aminoadipate	11	0.19
cis-L-1-Amino-1,3-dicarboxycyclohexane	102	20.0

[a]The values for the *erythro*-L, *erythro*-D, and *threo*-L isomers of β-hydroxyglutamate (with NH_2OH) are, respectively, 3, 1.5, and 6. From [1].

substitutions can be made without loss of enzymatic susceptibility. The synthesis of this compound was carried out in our laboratory by Jerald Gass, and he found it to be an excellent substrate for the enzyme. The relative activities of various substrates with sheep brain glutamine synthetase are summarized in table 2. The high reactivity of the cyclohexane amino acid, which was tailor-made for glutamine synthetase, provides considerable support for the hypothesis developed above that L-glutamate combines with the enzyme in the extended conformation.

I would now like to turn to another aspect of the enzyme, namely its inhibition by certain methionine derivatives. It had been known for some time that glutamine synthetase is inhibited by methionine sulfone and methionine sulfoximine (for structures see fig. 5), but these inhibitions had not been studied in very great detail. We found that the inhibition by methionine sulfone was reversible but that methionine sulfoximine inhibited the enzyme irreversibly [18,20]. Inhibition by methionine sulfoximine was found to require ATP and magnesium ions and to be associated with the cleavage of ATP to ADP. Furthermore, it was shown that the rate at which irreversible inhibition by methionine sulfoximine is established is decreased in the presence of glutamate and decreased much more when both glutamate and ammonia are present. Finally, it was found that the irreversible inhibition of glutamine synthetase by methionine sulfoximine is associated with the binding of 8 molecules each of methionine sulfoximine and ADP to each molecule of enzyme. The bound methionine sulfoximine was shown to be methionine sulfoximine phosphate. This new compound was subsequently synthesized chemically and it was demonstrated that the phosphate moiety is attached to the sulfoximine

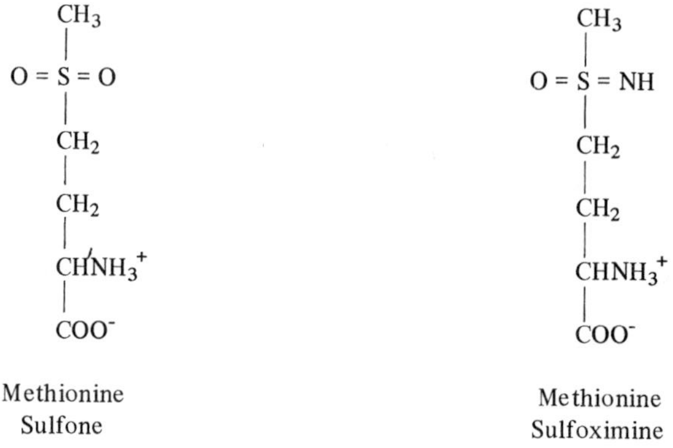

Fig. 5.

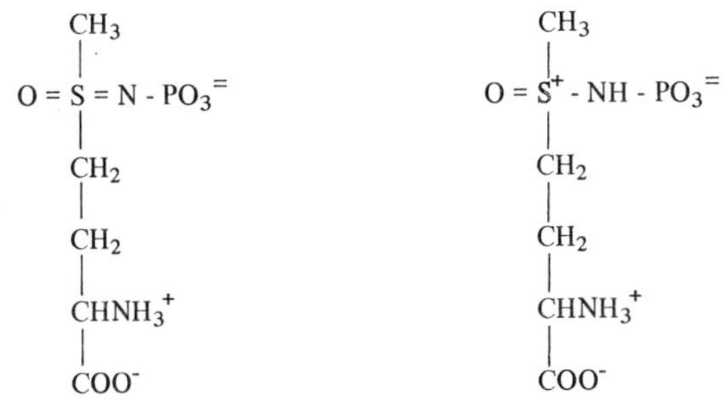

$$CH_3-S(=O)(=N-PO_3^{=})-CH_2-CH_2-CHNH_3^+-COO^- \qquad CH_3-S^+(=O)(-NH-PO_3^{=})-CH_2-CH_2-CHNH_3^+-COO^-$$

Fig. 6. Possible structures of methionine sulfoximine phosphate.

nitrogen atom. Methionine sulfoximine phosphate is relatively stable to strong mineral acid as compared, for example, to creatine phosphate and arginine phosphate. It is of interest that methionine sulfoximine phosphate is cleaved by a variety of phosphatases to methionine sulfoximine and inorganic phosphate. Possible structures of methionine sulfoximine phosphate are given in fig. 6. The formation of methionine sulfoximine phosphate under these conditions is in accord with the acyl phosphate hypothesis; indeed, this phosphorylation reaction appears to reflect an aspect of the normal catalytic reaction. In the course of this work it was found that D-methionine sulfoximine does not inhibit the enzyme and, further, that only one of the L-methionine sulfoximine isomers, i.e., L-methionine-*S*-sulfoximine, inhibits glutamine synthetase [21]. Of considerable related interest is the finding that only this isomer of methionine sulfoximine produces convulsions when administered to mice [22].

Studies on the inhibition of glutamine synthetase by methionine sulfone revealed that this compound undergoes cyclization when incubated with the

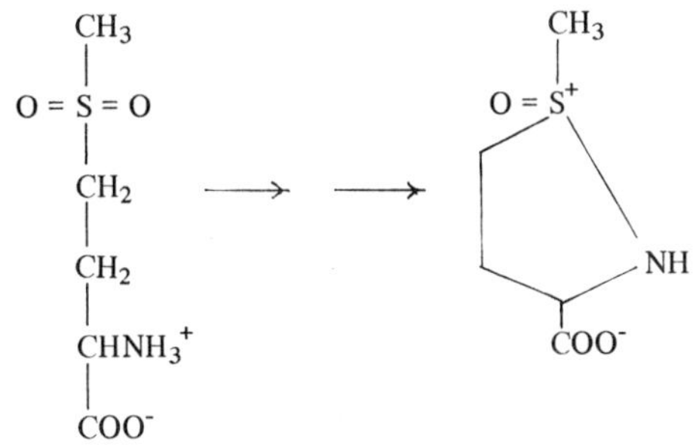

$$CH_3-S(=O)(=O)-CH_2-CH_2-CHNH_3^+-COO^- \longrightarrow \longrightarrow \text{cyclic } CH_3-S^+(=O)\text{-}NH \text{ ring with } COO^-$$

Fig. 7. Postulated reaction of methionine sulfone (see the text).

enzyme (fig. 7); presumably,this reaction, which is associated with cleavage of ATP, is analogous to the cyclization of glutamate to pyrrolidone carboxylate catalyzed by glutamine synthetase.

In an effort to integrate and further extend the observations on the specificity and inhibition of glutamine synthetase described above, a computer approach to the mapping of the active site of glutamine synthetase has been used [3]. In this work we were guided by the assumptions given above which led to the conclusion that L-glutamate attaches to the active site of the enzyme in an extended conformation and in an orientation in which the α-hydrogen atom is directed away from the enzyme. The computer study was made by Jerald Gass and this work represents a portion of his doctoral dissertation. The three-dimensional coordinates of the amino acid substrates and inhibitors were calculated with an IBM 360/40 computer and plotted as stereographs on a Calcomp 1627 on-line plotter using programs written in Fortran IV. Standard tetrahedral angles and standard bond lengths were used. The program was designed to manipulate the coordinates of the molecules; for example, rotation, translation, rotation of part of a molecule around a given bond, and finding the position of an atom to a given point is at a minimum or maximum. In the course of this study many stereographs were obtained; only several of these will be reproduced and discussed here. The complete program and the other stereographs are available elsewhere [3, 23].

The stereograph shown in fig. 8, indicates that when L-glutamate is reflected across a plane containing the α- and γ-carboxyl carbon atoms and the α-amino

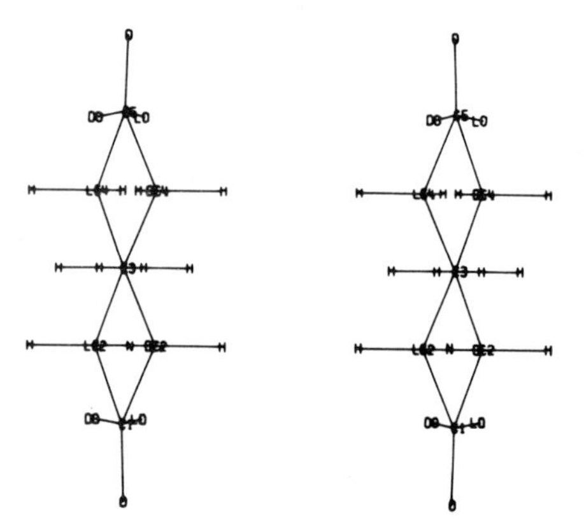

Fig. 8. Stereograph of D- and L-glutamic acid in an extended conformation. The reflecting plane passes through the C-1, C-3, and C-5 carbon atoms and the C-2 nitrogen atom.

nitrogen atom, D-glutamate is obtained. Thus, carbon atoms 1 and 5, and the nitrogen atom are in the reflecting plane, are common to both molecules, and can bind to the same sites on the enzyme. It will be seen that the three hydrogen atoms which extend far to the left are the α-L- , *threo*-γ-L- , and *threo*-β-D-hydrogen atoms. Only these hydrogen atoms can be substituted by methyl groups without loss of enzymatic susceptibility. It will also be noted that it is possible to place one, but not both, of the oxygen atoms of each γ-carboxyl group in the reflecting plane. It seems likely that the phosphate group of γ-glutamyl phosphate is bound to a specific site on the enzyme and that the phosphate groups of L-γ-glutamyl phosphate and D-γ-glutamyl phosphate occupy the same site on the enzyme. We therefore designate the oxygen atom in the reflecting plane as the oxygen to be phosphorylated; it turns out that this choice is not arbitrary. We will define the oxygen atom lying in the reflecting plane as the "phosphorylation site" (OP-site) and the position of the remaining oxygen of the γ-carboxyl group of L-glutamate as the "oxygen binding site" (OB-site). These sites are designated as "ENZ" in the other stereographs.

In considering the structure of the enzyme-bound tetrahedral addition compound presumably formed in the reaction of ammonia with γ-glutamyl phosphate, it appears reasonable to us to assume that the oxygen atoms and phosphate group of the tetrahedral intermediate are very close to their respective positions in the planar acyl phosphate. Thus, the conformation of the tetrahedral intermediate shown in fig. 9 is one in which the distances between its

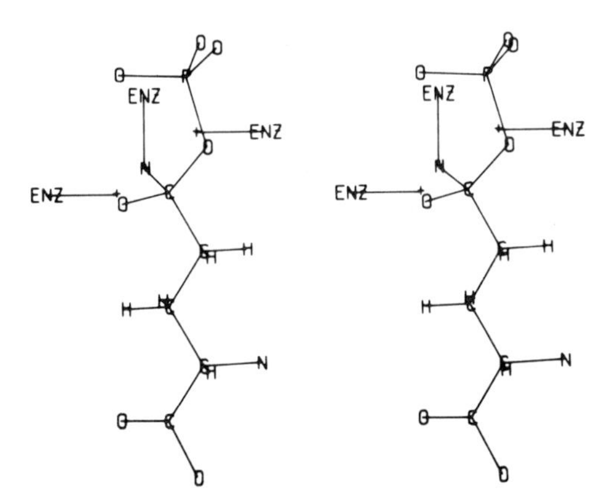

Fig. 9. The tetrahedral addition compound formed by the reaction of L-γ-glutamyl phosphate with ammonia.

two oxygen atoms and the two oxygen reference points (OP and OB) on the enzyme are at a minimum. The oxygen atoms of the tetrahedral intermediate are 0.42 Å and 0.25 Å from the OP site and OB site, respectively. We may also define a third point on the enzyme (the central "ENZ" in fig. 9). It seems likely that ammonia binds to the enzyme at a position close to the nitrogen atom in the tetrahedral intermediate and we therefore designate the nitrogen atom in fig. 9 as the ammonia binding site of the enzyme. Stereographs of L-glutamate and D-glutamate were drawn by the computer in which perpendiculars were erected to the plane containing the carbon atom and the two oxygen atoms of the carboxyl groups. With L-glutamate the position of the nitrogen atom in the tetrahedral intermediate is only 4.3° from the perpendicular to the C-5 carbon atom. On the other hand, the line perpendicular to the C-5 carboxyl group of D-glutamate makes an angle of 39° with the perpendicular to the C-5 carboxyl group of L-glutamate. It would be expected that the rate of reaction would be influenced by this angle and thus that D-glutamate would react with ammonia more slowly than L-glutamate. This is in general agreement with experimental observation; thus the relative rates of L- and D-glutamine synthesis are: L, 100; D, 27. Stereographs of *threo*-β-methyl-D-glutamate and *threo*-γ-methyl-L-glutamate were also drawn and studied. The relative rates of amide synthesis observed with these compounds as compared to those found with L- and D-glutamate are closely correlated with the steric hindrance demonstrated in quantitative terms in the stereographs. Other structure-rate relationships are also made possible by the computer approach.

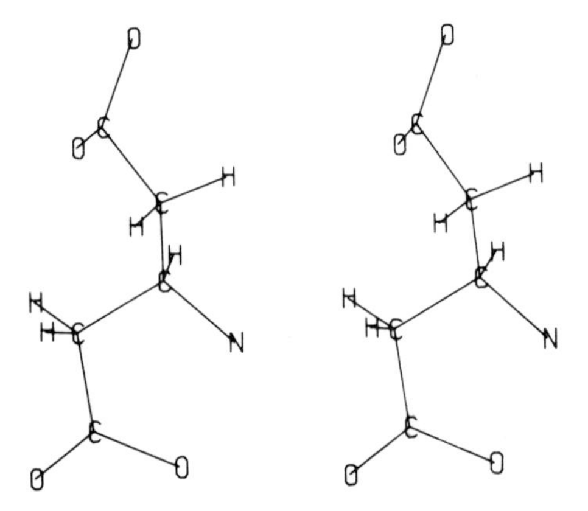

Fig. 10. β-Glutamic acid in a conformation which would lead to the formation of D-β-glutamine.

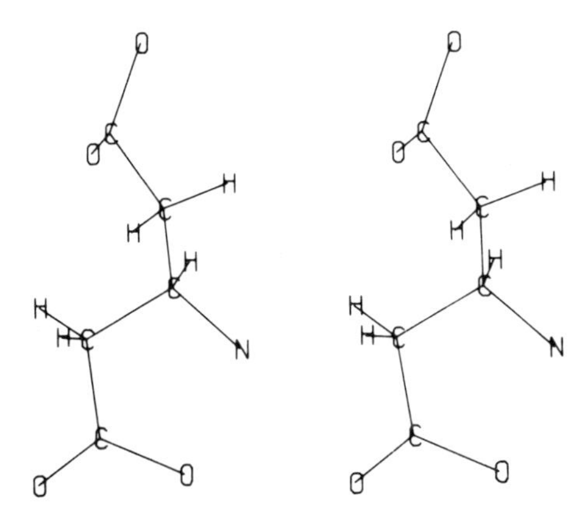

Fig. 11. β-Glutamic acid in a conformation which would lead to the formation of L-β-glutamine.

Stereographs of β-glutamate were also constructed; β-glutamate can attain a conformation in which the distance between C-1 and the amino nitrogen atom is only 0.09 Å greater than the distance between the C-1 and the amino nitrogen atom of L-glutamate. It is therefore possible for β-glutamate to bind to the enzyme with its C-1 carboxyl carbon atom coincident with that of L-glutamate and its 3-amino nitrogen atom only 0.09 Å from the 2-amino binding site. Fig. 10 shows the the conformation of β-glutamate which would lead to the formation of D-β-glutamine. Note that C-4 carbon atom extends upward. In fig. 11, the conformation of β-glutamate which would result in the formation of L-β-glutamine is shown; in this stereograph the C-4 carbon atom extends downward. The studies on the various methyl substituted glutamates have indicated that there is steric hindrance by the enzyme in this direction; therefore β-glutamate would not be expected to bind as shown in fig. 11, and hence the only conformation in which β-glutamate can bind to the enzyme is that shown in fig. 10 which leads to the formation of D-β-glutamine.

The sulfone oxygen atoms of L-methionine sulfone can be placed at minimum distances from the OP and OB enzyme sites in a manner similar to that carried out with the tetrahedral intermediate; see the stereograph shown in fig. 12. The distances between the sulfone oxygen atoms and the OP and OB sites are, respectively, 0.22 and 0.06 Å. It is of note that the methyl group of L-methionine sulfone is very close to the position on the nitrogen atom of the L-tetrahedral intermediate and it would therefore appear that the methyl group of L-methionine sulfone must lie very close to the ammonia binding

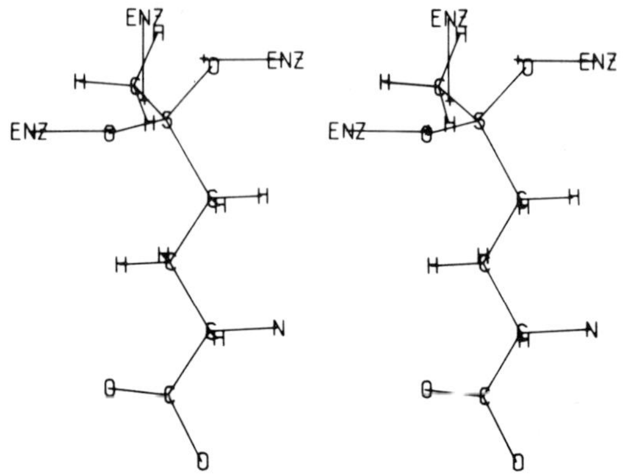

Fig. 12. L-Methionine sulfone.

site. This suggests that the site that binds ammonia is in a hydrophobic region of the enzyme; this supports the view that unionized ammonia rather than ammonium ion is bound to the enzyme.

In considering the orientation of L-methionine-*S*-sulfoximine on the enzyme, a number of possibilities were considered. The results of these studies make it seem extremely unlikely that the sulfoximine nitrogen atom binds to the ammonia binding site of the enzyme. The most satisfactory arrangement is one in which the sulfoximine oxygen atom of L-methionine-*S*-sulfoximine

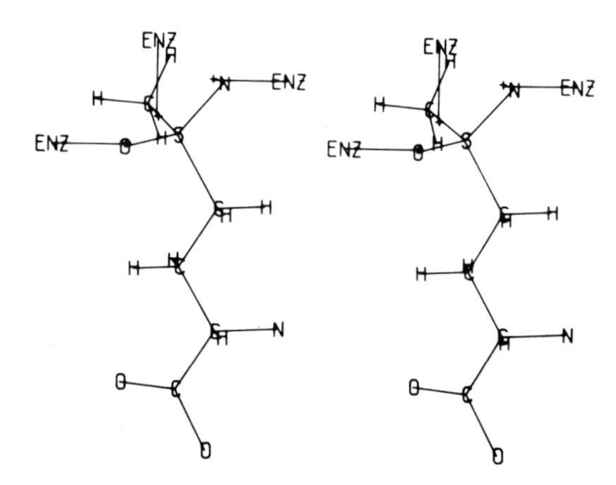

Fig. 13. L-Methionine-*S*-sulfoximine.

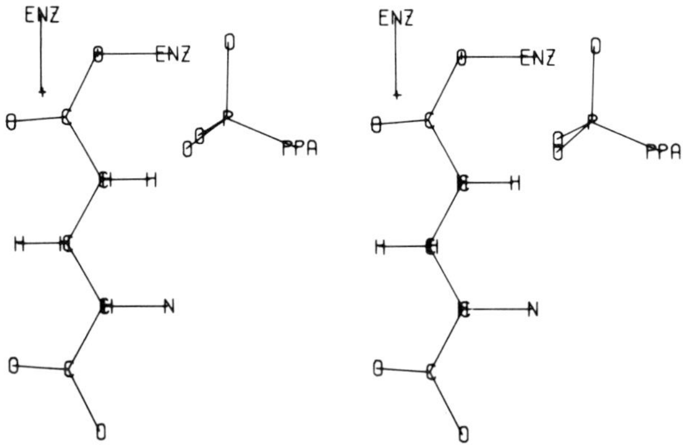

Fig. 14. L-Glutamic acid and ATP; see text.

binds to an oxygen site on the enzyme and the methyl group binds to the ammonia binding site in a manner analogous to that shown in fig. 12 for L-methionine sulfone. In this arrangement (fig. 13), L-methionine-*S*-sulfoximine is bound so that the sulfoximine nitrogen atom is in a good position to be phosphorylated; the sulfoximine nitrogen atom is 0.24 Å from the OP site, the sulfoximine oxygen atom is 0.06 Å from the OB site, and the methyl carbon atom is 0.02 Å from the position of the methyl carbon atom of L-methionine sulfone shown in fig. 12. These considerations appear to offer an explanation for the observed phosphorylation of the sulfoximine nitrogen atom of L-methionine-*S*-sulfoximine. It is conceivable that L-methionine-R-sulfoximine might bind in a manner analogous to that postulated for that of L-methionine-*S*-sulfoximine; here the positions of the sulfoximine oxygen and nitrogen atoms would be reversed and the sulfoximine oxygen atom would be close to the OP site. Since L-methionine-R-sulfoximine does not bind to the enzyme it seems most likely that the sulfoximine nitrogen atom of the R-isomer is repelled by the OB site; however, other explanations are possible. In any event it is difficult to see how the nitrogen atom of L-methionine-R-sulfoximine could be phosphorylated, and indeed the experimental evidence indicates that this reaction does not occur.

The available evidence is therefore in accord with an active site to which one part of the substrate or substrate analog can attach. The remainder of the molecule, which may have relatively bulky substituents, extends away from the enzyme. The computer calculations allow the selection of the phosphorylation site by virtue of the coincidence in the positions of one of the γ-carboxyl oxygen atoms of L- and D-glutamate. The calculations also permit designation of an ammonia binding site on the enzyme derived from a calculated position of the nitrogen atom of the tetrahedral intermediate. The computer approach also provides detailed steric explanations for the more rapid synthesis of L-glutamine as compared to D-glutamine and for the relatively low rates of amide synthesis from *threo-β*-methyl-D-glutamate and *threo-γ*-methyl-L-glutamate. Explanations for the specificity of inhibition by the methionine derivatives are also possible. We may conclude that methionine sulfone and methionine sulfoximine are analogs of the tetrahedral intermediate rather than of glutamate. Indeed the arrangement of the atoms about the sulfur atoms of the methionine derivatives is essentially tetrahedral and closely resembles the geometry of the tetrahedral intermediate. The computer approach also makes possible other speculations about the mechanism of action of glutamine synthetase; these are discussed in detail elsewhere [3,23]. Thus, the computer method applied here appears to provide new insight into both the steric relationships between the reactants at the active site of the enzyme and aspects of the reaction mechanism itself. Of course, the conclusions depend upon the assumptions made. However, a number of the findings appear to be susceptible to experimental test. The available data can be used to design an active site inhibitor that might react so as to form a covalent linkage with the enzyme. This would pave the way to a detailed understanding of the structure of the enzyme at the active site. It will be of particular interest to integrate the present data with information about the chemical structure of the active site when such information becomes available. To our knowledge this work on glutamine synthetase appears to represent the first attempt to "design" an active site by a mathematical approach of this type. Such an approach might be particularly useful in studying enzymes whose detailed structures are already known.

References

[1] A. Meister, in: On the Synthesis and Utilization of Glutamine. Harvey Lectures, Series 63 (1969) 139 (Delivered Feb. 15, 1968).

[2] A. Meister, Advan. i. Enzymol. 31 (1968) 183.

[3] J.D. Gass and A. Meister, Biochemistry 9 (1970) 1380.

[4] V. Pamiljans, P.R. Krishnaswamy, G. Dumville and A. Meister, Biochemistry 1 (1962) 153

[5] R.A. Ronzio, W.B. Rowe, S. Wilk and A. Meister, Biochemistry 8 (1969) 2670.

[6] S. Wilk (1969), unpublished observations.

[7] R.H. Haschemeyer, Abstracts 148th ACS Meeting, Sept. 1965 (Atlantic City, New Jersey).

[8] R.H. Haschemeyer, Abstracts 152nd ACS Meeting, Sept. 1966 (New York, New York).
[9] S. Wilk, A. Meister and R.H. Haschemeyer, Biochemistry 8 (1969) 3168.
[10] P.R. Krishnaswamy, V. Pamiljans and A. Meister, J. Biol. Chem. 237 (1962) 2932.
[11] L. Levintow and A. Meister, J. Am. Chem. Soc. 75 (1953) 3039.
[12] E. Khedouri, V.P. Wellner and A. Meister, Biochemistry 3 (1964) 824.
[13] E. Khedouri and A. Meister, J. Biol. Chem. 240 (1965) 3357.
[14] H.M. Kagan, L.R. Manning and A. Meister, Biochemistry 4 (1965) 1063.
[15] H.M. Kagan and A. Meister, Biochemistry 5 (1966) 725.
[16] H.M. Kagan and A. Meister, Biochemistry 5 (1966) 2423.
[17] J.D. Gass and A. Meister, Biochemistry 9 (1970) 842.
[18] R. Ronzio and A. Meister, Proc. Natl. Acad. Sci. U.S. 59 (1968) 164.
[19] R.A. Ronzio, W.B. Rowe and A. Meister, Biochemistry 8 (1969) 1066.
[20] W.B. Rowe, R.A. Ronzio and A. Meister, Biochemsitry 8 (1969) 2674.
[21] J.M. Manning, S. Moore, W.B. Rowe and A. Meister, Biochemistry 8 (1969) 2681.
[22] W.B. Rowe (1969), unpublished observations.
[23] J.D. Gass, Doctoral Dissertation, Cornell University Medical College (1970).
[24] E.R. Stadtman, Advan. Enzymol. 28 (1966) 41.

THE STRUCTURE OF THE HEAVY CHAINS OF IMMUNOGLOBULINS AND ITS RELEVANCE TO THE ANTIBODY SITE

R.R. PORTER

Department of Biochemistry, University of Oxford,
South Parks Road, Oxford, England

Abstract: Porter, R.R. The Structure of the Heavy Chains of Immunoglobulins and its Relevance to the Antibody Site. *Miami Winter Symposia* 2, pp. 352–360. North-Holland Publishing Company, Amsterdam, 1970.

The advances in knowledge of the structure of immunoglobulins during the last year have been largely concerned with sequence data on the heavy chains. Improvements in techniques of affinity labelling of the combining sites of antibodies using an aryl nitrene have also been reported. These results will be discussed and the argument put forward that the sequence in the variable areas of the *N*-terminal section of both heavy and light chains may be essentially random. Hence the sequences of specific antibodies in these sections may be little more homogeneous than those found in preparation of inert immunoglobulin. The only exceptions to be expected would be in the sequence of antibodies of exceptionally high affinity for the antigen.

1. Introduction

The four polypeptide chain structure of immunoglobulins is well known as is the presence at the *N*-terminal end of each chain of a section which varies in sequence for each myeloma protein. A similar phenomenon of variable and constant sections occurs in immunoglobulins from sera of normal individuals. That the wide range of affinity for antigen shown by antibodies is due to these variable sequences controlling the structure of the combining site seems obvious though it is not yet proven.

I propose to discuss the recent advances in knowledge of the structure of the immunoglobulins and during the last year there has been a rapid appearance of information on the primary structure of the heavy chains of myeloma proteins delineating particularly the nature and extent of the variability of the *N*-terminal end. This however gives only indirect evidence on the structure and position of the combining site and the progress which has been made on direct localization of the combining site will also be described. Finally, the evolving ideas about the nature of the site will be discussed.

In the last twelve months, sequences have been published of complete or substantial sections of the heavy chains of 3 γ chains [1–3], and 1 μ chain [4], all from human myeloma proteins. Shorter sequences have also been published from two other γ chains [1,5], and the *N*-terminal sequences of several μ chains [6].

The conclusions which can be drawn from this data are as follows:

1) The variable section of the heavy chain is of the same order of size though slightly larger than that of the variable section of the light chain. Fig. 1 summarizes the known sequences around the switch from the variable to constant sections. The numbering is taken from that of the Daw $\gamma 1$ chain and it varies from chain to chain because of the presence of insertions and deletions to the variable section. One of each is apparent in this region and others occur nearer the *N*-terminal. The heavy chain of Cor also shows what is at present a unique feature–carbohydrate in the variable section attached to the aspartic acid residue 62. Carbohydrate has been found also attached to the variable section of several Bence-Jones proteins but its significance in either chain is unknown.

Though the variable section of these γ chains is rather larger (116 residues in Daw) than in the κ and λ chains (105–109) the distance relative to the *C*-terminal half cystine of the cystine bridge in the variable section is constant as shown in fig. 2.

2) The presence of families or subgroups in the variable parts of both the κ and λ chains has been demonstrated clearly by the substantial data which has accumulated [7,8]. The same phenomenon appears to occur in the heavy chains though based on much more limited evidence. The only striking difference between the variable sections of the heavy and light chains is that the subgroups of κ and λ chains are distinct while those of γ chains appear to be shared by the μ chains. Again the evidence comes only from three full and one partial sequence of γ chains and the sequence of the first 105 residues of one μ chain, but the latter shows 70% homology with two of the γ chains and only 30% with the other two suggesting rather strongly the sharing of subgroups. The significance of this lies in that if confirmed by further data, the most obvious explanation would be that there are genes coding for the variable section and others coding for the constant portion and that at some point translocation occurs to fuse the two genes. The failure to share subgroups between the κ and λ chains may arise because the genes for the constant sections are not closely placed [9] on the chromosome while those of the heavy chains have been shown to be closely linked [10].

```
                100                                  105                 110                115
Daw.  Tyr-Tyr-Cys-Ala-Arg-Ser-Cys-Gly-Ser-Gln————————Tyr-Phe-Asp-Tyr-Trp-Gly-Gln-Gly-Ile-Leu-Val-Thr-Val

Cor.  Tyr-Tyr-Cys-Ala-Arg-Ile-Thr-Val-Ile-Pro-Ala-Pro-Ala-Gly-Tyr-Met-Asp-Val-Trp-Gly-Arg-Gly-Thr-Pro-Val-Thr-Val
Eu.   Tyr-Phe-Cys-Ala-Gly-Gly-Tyr-Gly-Ile————————————Tyr-Ser-Pro-Glu-Glu-Tyr-Asn-Gly-Gly-Leu-Val-Thr-Val
He.                                                  Ala-Phe-Asr-Val-Trp-Gly-Glx-Gly-Thr-Lys-Val-Ala-Val
Ou.   Tyr-Tyr-Cys-Ala-Arg-Val-Val-Asn-Ser
```

Fig. 1. Extent of the variable section of human myeloma heavy chains. Daw, Cor, Eu and He are $\gamma 1$ chains and Ou is a μ chain. Note the insertions and deletions shown here relative to the sequence of the Daw γ chain. There is identity of sequences from residue 117.

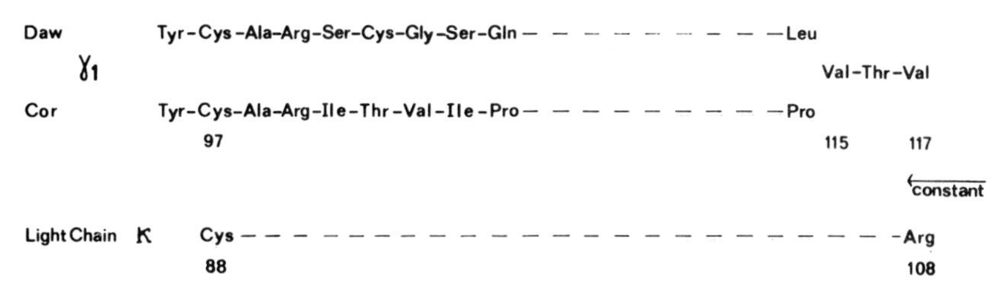

Fig. 2. Extent of the variable region of the γ and κ chain of human myeloma proteins relative to the variable region intrachain disulphide terminating at residue 97 in the γ chain and 88 in the κ chain.

3) Milstein [7] has drawn attention to the region of exceptionally high variability in the human κ chains noting particularly the sections immediately after the two half cystines which form the intra disulphide bond of the variable region, i.e., positions about 30 and 90. A similar phenomenon is apparent in the heavy chains and again suggests that these two areas which must be adjacent to each other in the molecule may be of particular importance in forming the antibody combining site.

The nature and degree of variability in the heavy and light chains appears to be identical and obviously leads to the conclusions that each chain plays an equivalent role in forming the combining site if the assumption is made, as it surely must, that the variability in the *N*-terminal section provides the structural basis for the wide range of specific affinities of antibodies.

Affinity labelling

It is clear that while all the sequence studies may give a lead as to the most likely structures to be concerned in the combining site and that they are indeed an essential prerequisite to other studies, they cannot by themselves define the size, shape and basis of affinity of the site.

The most direct chemical approach is by the affinity labelling, a technique introduced by Singer, Wofsy and colleagues [11]. In this method, antibody against a haptene such as DNP-lysine is allowed to combine with such a molecule into which a reactive group has been introduced. The specific affinity of the antibody for the haptene concentrates the haptene in the combining site and the subsequent interaction with the reactive group leads to covalent binding of the haptene to an amino acid in or near the combining site (fig. 3).

AFFINITY LABELLING (Singer)

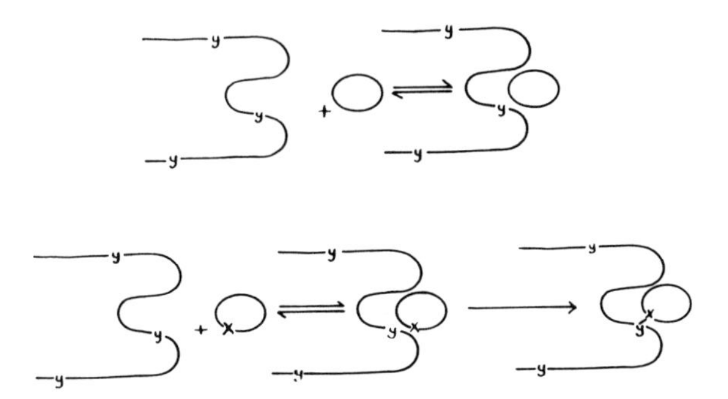

Fig. 3. Principle of affinity labelling of antibodies with a haptene containing a reactive group. Upper: interaction of antibody and haptene. Lower: antibody interacting with haptene containing a reactive group x and forming a covalent x–y bond.

Singer et al. [12] used antisera prepared in rabbits against, for example, benzene arsonate and affinity labelled with *p*-(arsonic acid)-benzene diazonium fluoroborate. In a recent paper, Thorpe and Singer [13] have reported the isolation of several labelled dipeptides from both heavy and light chains of rabbit and mice antibody to the 2,4-dinitrophenyl group using ^{3}H *m*-nitrobenzedizonium fluoroborate as the affinity labelling reagent. In the rabbit γ chain, the major peptide isolated was Thr-nitrobenzene-Tyr from the heavy chain and Val-nitrobenzene-Tyr from the light chain. In the mouse γ chain serine, threonine and glycine were found with the labelled tyrosine and in the mouse λ chain, aspartic acid together with serine and threonine predominated. The authors suggested that in each case the labelled tyrosine residue was likely to be that immediately after the second half cystine of the intra disulphide bond in the variable section of the chains but clearly much more extensive sequence data is necessary to establish this point.

An alternative reagent, diazoketone, has been proposed by Converse and Richards [14]. This has the advantage that it will react with the antibody only after ultraviolet irradiation. Hence the nonspecific reaction found with the diazo group in Singer's work is reduced considerably and the reactive ketene or carbene derivatives produced by irradiation should react with a wider range of amino acid than the tyrosine, histidine and lysine to which the diazo group is confined. With these reagents nearly half the sites of rabbit anti-DNP antibodies were blocked covalently as judged by a difference titration and the label was confined to the FAB fragment. The ratio of label in the heavy/light

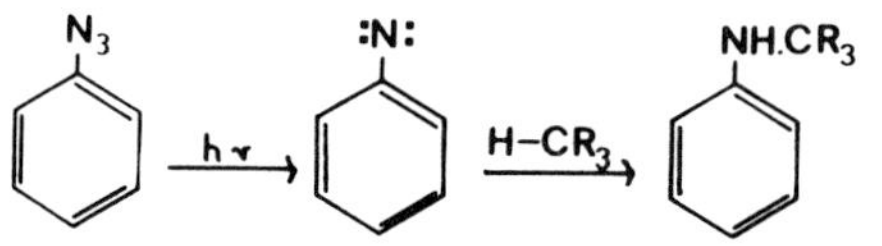

Fig. 4. Activation of aryl azide by light to give a nitrene able to insert into a carbon chain.

chains was 4:1. A third reagent has been proposed by Fleet, Porter and Knowles [15], an aryl nitrene generated by irradiation of an aryl azide (fig. 4). This has the advantage that the aryl azide can be used as a determinant group in the haptene used for immunization and hence the reactive group will be in, rather than adjacent to, the combining site. It also has the advantage that the wavelength of the light used for irradiation is > 400 nm and hence the risk of destruction of aromatic amino acids is less than with the diazo-ketone reagent. It is also probable that irradiation of the azide will give predominantly the nitrene which will insert into any-C–H bond and hence any amino acid. With this reagent all the combining sites appeared to be blocked but only 50–60% were labelled covalently as judged by the incorporation of the nitroazide phenyl-^{3}H-lysine used as haptene (fig. 5). There was a 2.5:1 ratio of label in heavy/light chain. Only the heavy chain has been studied further and preparation of the *N*-terminal half by cyanogen bromide splitting showed 90% of the label to be in this section. The identification of the labelled peptides is continuing but no characterization has been achieved so far.

It is essential that the sequence of the peptide chain likely to react with the affinity labelling reagent should be known in order that the position of the labelled peptide, when isolated, can be placed unequivocally. Working with IgG of normal rabbit serum or that of any other serum offers considerable difficulty in that there will be mixed sequences arising from the presence of different subclasses of the constant region and different subgroups of the vari-

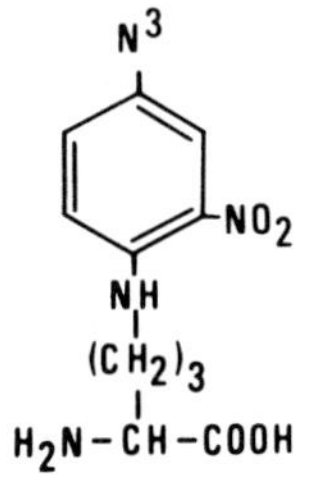

Fig. 5. The affinity label nitro azide phenyl lysine (NAP-lysine).

able region in addition to the inherent variability within each subgroup.

Surprisingly, a majority sequence of most of the *N*-terminal 250 residues of the heavy chain of rabbit IgG has been found, probably because one subclass and one subgroup predominates and constitutes some 60–80% of any preparation from the serum of rabbits of a single allotype. Fruchter et al. [16] and O'Donnell et al. [17] have reported recently sequences of the rabbit γ chain from 1-34, 84-99 and 109-253 together with the arrangement of the inter and intra disulphide bonds. The sequence 34-50 has since been completed and peptides have been isolated which may account in a large part for the 50-84 region, but no satisfactory sequence for 100-110 has been found. On the contrary, such evidence as is available suggests a high degree of variability in this latter section and an attempt is being made at present to evaluate quantitatively the degree of complexity from 100-110 in the γ chain in both purified antibodies and inert IgG. This position coincides with the region of high variability apparent in the sequence studies of human myeloma proteins referred to earlier, though the second highly variable area about 30-33 suggested by work with myeloma proteins, has in the γ chain from normal rabbit IgG given a comprehensible sequence.

What conclusion concerning the structure of antibody combining sites can be drawn from these results?

The sequence studies of myeloma proteins suggest that there is one short section of exceptional variability from residue 100-105 and possibly a second from 30-33 in the γ chain and comparable sections in the κ and λ chains. Other differences in the 80-90 residues of the variable sections, within one subgroup, are much more restricted and might be only compensatory changes necessary to maintain the overall steric structure, by accommodating the complete range of residues which appear to occur in this very unstable section of the variable region. The evidence from affinity labelling is incomplete but the preliminary evidence of Thorpe and Singer [13] is compatible with the view that this unstable section (residues 100-110) may contain the combining site. Our preliminary evidence from sequence studies of anti-DNP antibodies and inert IgG suggests that the results with the myeloma proteins are reflected in the γ chains of normal immunoglobulins and that exceptional instability of sequence from 100-110 occurs here also. It goes further and suggests that there may be considerable variability of sequence in this section even in purified antibodies to the DNP-haptene. If this is correct we may have IgG molecules with 5^{20} to 10^{20} different sequences in this very unstable region able to combine with a sufficiently high affinity (say $< 10^4$) as to be recognizable as antibodies for an equivalent number of different antigen structures. The X-ray crystallographic studies of lysozyme, ribonuclease and chymotrypsin have shown the ease with which a cleft in a protein molecule can bind specifically to a polysaccharide, a nucleic acid, and a polypeptide by varying the amino acid residue lining the cleft. It is probable that the antibody combining site will bind by a similar mechanism and it seems likely that the relevant residues controlling the specificity may be provided by the very unstable section of either or both of the peptide chains.

What distinguishes the antibody and enzyme sites is that the very great range of specificity common to both is achieved in the case of antibodies within a strictly limited overall structure. As has been demonstrated frequently, immunization leads to a range of antibody molecules each with a significant but different affinity for a given haptene such as DNP-lysine. This appears to be reflected in the sequence studies where, with techniques able to detect only sequences occurring in at least say 10-20% in a given preparation, the variety in the 100-110 region appears to be as complex in the anti-DNP preparation as in inert IgG. If antibody with only very high affinity (say 10^9) for a given haptene is studied, much more homogeneity may be found as the high affinity is likely to depend on exactness of fit which would be expected to necessitate a much more restricted sequence at the combining site and subsequently at the compensatory positions. In a preparation of anti-DNP antibody of specific affinity of 10^4 to 10^6, however, there may well be several thousand types of molecule each with distinct sequences,but able to show the requisite affinty for the haptene.

This type of mechanism would imply that the diversification leading to cells synthesizing the different sequences in the variable section of the immunoglobulins is indeed random. Though the individuals of one species may be capable of synthesizing say 10^4 immunoglobulins of different sequence, but all able to show demonstrable affinity for a given haptene, such as DNP-lysine, in any single individual at one time, there may be present cells able to synthesize only 10^3 such molecules and the method and route of immunization may be such that only half this number may be stimulated to division and synthesis. If this picture is correct the diversity of anti-DNP antibody will vary with each preparation used. Technically it may well be difficult to distinguish, chemically, preparations of IgG which combine with DNP-lysine and those which do not as it will depend on the evaluation of the relative complexity of two mixtures. Similarly myeloma proteins showing affinity for DNP-lysine may differ so much in structure that only statistical assessment of data from considerable numbers of such proteins will be meaningful.

The easiest conclusion to test experimentally, if there is essentially random production of very large numbers of possible sequences in the parts of the variable sections of the peptide chains which control the combining specificity, should be a comparison of the high affinity and low affinity antibodies. If the idea is correct, it should be apparent that the high affinity antibody is more homogeneous than low affinity antibody whatever their source. Such a comparison would be easier than an attempt to assess the relative structure of similarly complex mixtures.

References

[1] G.M. Edelman, B.A. Cunningham, W.E. Gall, P.D. Gottlieb, W. Ruithauser and M.J. Waxdal, Proc. Natl. Acad. Sci. U.S. 63 (1969) 78.
[2] E.M. Press and N.M. Hogg, Nature 223 (1969) 807.
[3] E.M. Press and N.M. Hogg, Biochem. J., (1970), 117 (1970) 641.
[4] M. Wikler, H. Köhler, T. Shinoda and F.W. Putnam, Science 163 (1969) 75.
[5] C.E. Fisher, W.H. Palm and E.M. Press, FEBS Letters 5 (1969) 20.
[6] G. Bennett, Biochemistry 7 (1968) 3341.
[7] C. Milstein, FEBS Letters 2 (1969) 307.
[8] B. Langer, M. Steinmetz-Kayne and N. Hilschman, Z. Physiol. Chem. 345 (1968) 945.
[9] R.G. Mage, G.O. Young, J. Reynek, R.A. Reisfeld and E. Apella, 17 Colloquium on Protides of the Biological Fluids (Bruges, 1969) in press.
[10] H.G. Kunkel, W.K. Smith, F.G. Joslin, J.B. Natvig and S.D. Litwin, Nature 223 (1969) 1247.
[11] L. Wofsy, H. Metzger and S.J. Singer, Biochemistry 1 (1962) 1031.
[12] S.J. Singer, L.I. Slobin, N.O. Thorpe and J.W. Fenton, Quant. Biol. 23 (1967) 99.
[13] N.O. Thorpe and S.J. Singer, Biochemistry 8 (1969) 4523.
[14] C.A. Converse and F.F. Richards, Biochemistry 8 (1969) 4431.
[15] G.W.J. Fleet, R.R. Porter and J.R. Knowles, Nature 224 (1969) 511.
[16] R.G. Fruchter, S.A. Jackson, L.E. Mole and R.R. Porter, Biochem. J. 116 (1970) 249.
[17] I.J. O'Donnell, B. Frangione and R.R. Porter, Biochem. J. 116 (1970) 261.

IMMUNOGLOBULIN STRUCTURE: PRINCIPLES AND PERSPECTIVES

F.W. PUTNAM

Department of Zoology, Indiana University, Bloomington, Indiana 47401, USA

Abstract: Putnam, F.W. Immunoglobulin Structure: Principles and Perspectives, *Miami Winter Symposia* 2, pp. 361–381. North-Holland Publishing Company, Amsterdam, 1970.

Immunoglobulins have a symmetrical tetrachain structure composed of a pair of heavy chains disulfide-bonded to each other and to a pair of light chains. Each chain is divided into subunits (two in the light and four in the heavy), and each subunit contains 110–115 residues including a difulside loop joining about 60 residues. In the NH_2-terminal subunit (the v unit) both heavy and light chains exhibit great variability in amino acid sequence which produces immunoglobulin heterogeneity and is related to antibody specificity. The COOH-terminal sequence is constant and specific for each kind of chain and is given by one c unit in light and by three c units in most heavy chains. Light and heavy chains show striking homology, and certain residues are strongly conserved such as the disulfide bridges.

These basic concepts of immunoglobulin structure have been extended by amino acid sequence analysis of macroglobulins (γM). The v unit of one γM (Ou) has 73% homology with that of one γG (Daw) though both have only about 30% homology with the v unit of another γG (Eu). These results and additional data on other μ and γ chains indicate that the NH_2-terminus of heavy chains is highly variable like that of light chains but may not be specific for the class of heavy chain. Furthermore, the κv or λv subunits of other light chains are as related in amino acid sequence to the v unit of the Ou μ chain (μv) as is the κv unit from the Ou light chain. Although γ1 and μ heavy chains have similar structural features, the c regions show surprisingly low identity in amino acid sequence. These results suggest a common evolutionary origin of heavy and light chains from a primitive subunit. The diversity of present-day light and heavy chains is attributed to evolutionary mutations that gave rise to many different immunoglobulin genes.

1. Introduction

Antibodies are remarkable both for their exquisite specificity and their infinite variety. They are members of the large family of proteins once called

γ-globulins and now called immunoglobulins. Indeed, immunoglobulins are defined as proteins that are either endowed with known antibody activity or are related to antibodies in chemical structure and hence in antigenic specificity [1]. The large size and the usual heterogeneity of immunoglobulins hinders determination of their structure by present methods of protein chemistry. However, in multiple myeloma, a tumor of the plasma cells which are normally a site of antibody biosynthesis, and in the related disease macroglobulinemia large amounts of homogeneous immunoglobulins are secreted into the serum, and incomplete immunoglobulins (Bence-Jones proteins) may be excreted in the urine. These abnormal proteins lack demonstrated antibody activity but they are classified as immunoglobulins because of their similarity in site of synthesis, in polypeptide chain structure, and in antigenic specificity [2]. They may be thought of as antibodies in search of an antigen. Indeed some of them show specific combining ability with synthetic haptens such as DNP-derivatives. The homogeneity of these abnormal proteins permits their amino acid sequence analysis, unlike normal γ-globulins and unlike the natural mixture of antibodies. However, the abnormal proteins, though serving as a model for structural study of antibody γ-globulins, appear to differ in amino acid sequence for each patient.

Because such almost infinite variability in primary structure is unique to the immunoglobulins, it is thought to be related to antibody specificity. Though indirect, this is still the best evidence for the hypothesis that specific antibodies differ from each other in amino acid sequence. This is the crux of the antibody problem and leads to the paradox: either protein structure is not invariant and is not uniquely determined by pre-existing genes, or there must be already present in the genetic makeup of vertebrate animals sufficient immunoglobulin genes to provide for an almost unlimited number of antibodies. To solve this problem knowledge is needed of the exact structure of immunoglobulins; at present this is best obtainable by study of the abnormal globulins.

2. Polypeptide chain structure of the immunoglobulins

Immunoglobulins are a family of proteins that are related through their polypeptide chain structure and the presence of common antigenic determinants located in their two kinds of polypeptide chains (heavy and light) [2,3]. These determinants are present in both normal and abnormal globulins and may be detected by use of antisera against the individually specific serum globulins produced by patients with multiple myeloma or macroglobulinemia and

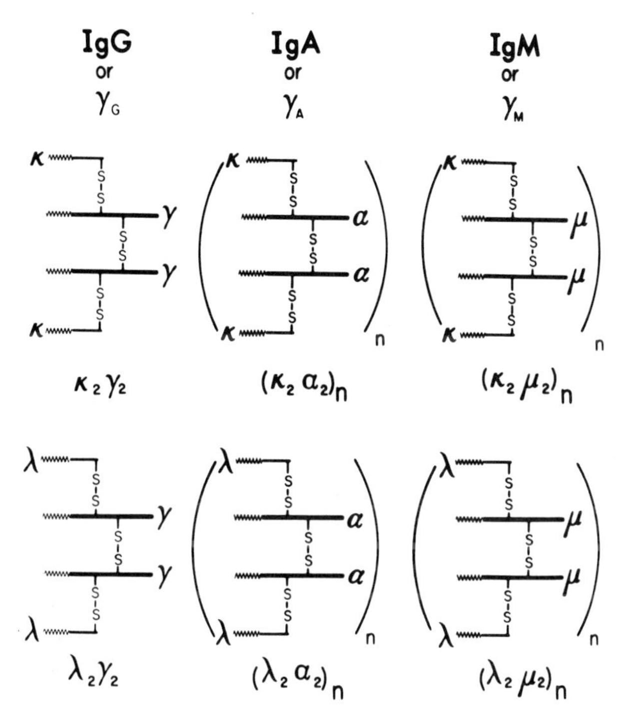

Fig. 1. The tetrachain polypeptide structure of the three major classes of immunoglobulins (IgG, IgA and IgM or γG, γA and γM), all of which are present in normal sera. The light chains are denoted κ and λ and the heavy chains, γ,α, or μ. The chain formula is given under each of the six subgroups. Most IgA and IgM globulins are polymeric; the tetrachain subunits are joined by an intermolecular disulfide bond between cysteine residues at an unknown position in the heavy chains. The zigzag lines indicate the locus of the variable amino acid sequence of the heavy and light chains. From Putnam and Köhler [3].

by antisera to the Bence-Jones proteins excreted in the urine by these patients. This led to an antigenic classification of immunoglobulins into the three major classes (designated γG, γA, and γM, or alternatively, IgG, IgA and IgM) [1], which are illustrated in fig. 1, and into two minor classes (γD and γE or IgD and IgE). This classification correlates with physicochemical properties such as electrophoretic mobility, chromatographic behavior, sedimentation coefficient, and molecular weight; it is now known to be based on characteristic differences in amino acid sequence of the "constant" regions of the light and heavy polypeptide chains.

All complete immunoglobulins have a tetrachain polypeptide structure, the main principles of which are illustrated in fig. 1 for the three major classes, γG, γA, and γM. The two minor classes, γD and γE, are believed to have a similar structure, but thus far only λ chains have been associated with γD. Each immunoglobulin molecule has a pair of identical light chains of molecular weight about 23,000 and a pair of identical heavy chains about twice as large.

Despite the similarities emphasized in fig. 1, the three classes shown, and even their various subclasses, may differ in such important features as the number and location of the interchain disulfide bonds and the degree of polymerization, as well as in the primary structure of their class-specific chains. For each of the three major classes there are two antigenic types κ and λ leading to two subgroups for each class. The subgroups are based on the nature of the light polypeptide chain and are detectable by their specific reaction with antisera to the κ and λ types of Bence-Jones proteins, respectively. The Bence-Jones protein excreted by a patient with multiple mueloma or macroglobulinemia is identical in antigenic type and in amino acid sequence to the light chain of his serum immunoglobulin and is related in structure to all normal light chains of the same antigenic type. Hence, the terms Bence-Jones protein and light chain are often used almost interchangeably.

The heavy chains carry the antigenic determinants characteristic of the class and usually contain all the carbohydrate. They differ in primary structure and are designated the γ, α, and μ chains, for γG, γA, and γM, respectively. Thus, six principal polypeptide chain formulas are possible; these are given in fig. 1. Hybrid chain combinations are not thought to occur. Within these subgroups, however, many subclasses of immunoglobulins may occur both for heavy and light chains. These subclasses may be isotypic, allotypic, or idiotypic.

Isotypes are present in the serum of all normal individuals of the same species; in man they include γG1, γG2, γG3, and γG4 which are based on the four subclasses of the γ heavy chain, $\gamma 1$, $\gamma 2$, $\gamma 3$, and $\gamma 4$ [4]. Allotypes differ among individuals of the same species and are inherited in a Mendelian fashion. About 20 allotypic Gm factors that are serologically detectable are associated with the γ chains of man and higher primates [5]. Similar allotypic markers are being found in α and μ chains. Allotypic markers are expressed at a molecular level through amino acid sequence differences in both heavy and light chains. For example, in human κ light chains the Inv allotypic marker is associated with an interchange of leucine and valine at position 191. Idiotypic variants are more difficult to define. Examples are the Bence-Jones proteins from different patients; no two of these proteins have yet been shown to be identical in antigenic determinants or amino acid sequence. Similar individual antigenic specificity is found in some antibodies.

3. Heterogeneity of the immunoglobulins

The continuously expanding subclassification described above is simply a reflection of the remarkable heterogeneity of the immunoglobulin system in normal individuals capable of producing the full spectrum of antibodies. Normal γ-globulin and even purified antibodies are heterogeneous by whatever criterion examined. This heterogeneity is manifested in many ways; for example by the multiple banding in gel electrophoresis of normal light chains and by the ambiguity in amino acid sequence of the NH_2-terminal regions of normal light and heavy chains. For this reason amino acid sequence determination of normal immunoglobulins and even of purified antibodies up to now has not been feasible. In contrast, the immunoglobulins produced by individual patients with multiple myeloma and macroglobulinemia are homogeneous and thus are suitable for sequence determination. Indeed, in the 5 years since we first showed the Bence-Jones proteins differ individually in sequence [6], many laboratories have entered the field. Now, more sequence data are available for immunoglobulins than for any other system of proteins, even including the hemoglobins. The almost infinite variability in immunoglobulin sequence that has been demonstrated accounts for antibody diversity and is thought to reflect antibody specificity.

4. Regions of variable and constant sequence

Each immunoglobulin polypeptide chain is divided into a region of variable amino acid sequence indicated schematically by the jagged line in fig. 1 and a region of constant amino acid sequence indicated schematically by the solid line. These are often called the v region and the c region, respectively. The variable region is always at the NH_2-terminal end; it comprises the first half of the light chain and the first quarter of the heavy chain or about 110-120 residues in each case. The variable region has thus far differed in amino acid sequence for each light chain examined whereas the constant region has an identical sequence except for one or two substitutions that may represent genetic differences. The heavy chain structure is based on analogous principles though less is known because of the greater length of heavy chains – about 450 residues. As a result, the emphasis on the structural study of immunoglobulins has turned to the sequencing of the γ and μ heavy chains.

Fig. 2 summarizes schematically much that is known about the general structure of immunoglobulins. A Y-shaped molecule with twofold symmetry is suggested by electron micrographs of regular aggregates of certain antibody-

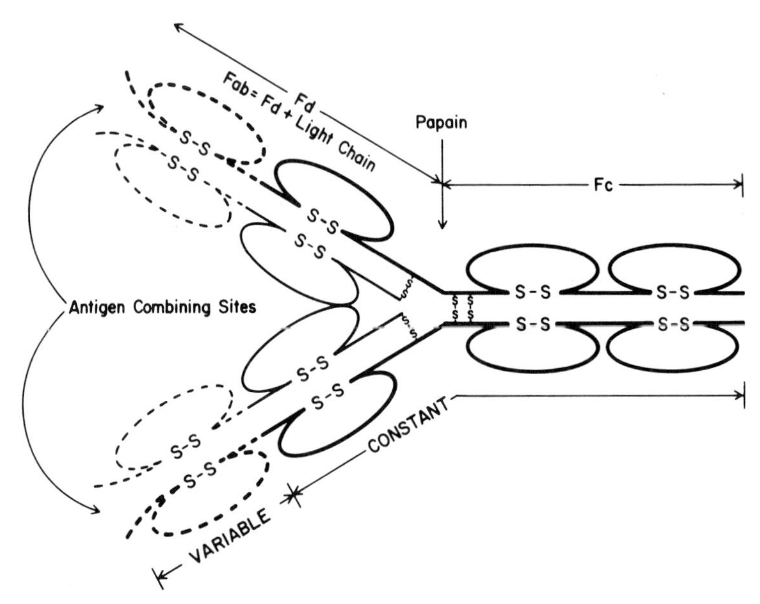

Fig. 2. Schematic diagram of the linear polypeptide chain structure of immunoglobulins. Two identical half molecules are symmetrically arranged; each contains a light chain (the light line) and a heavy chain (the darker line). The interchain and intrachain disulfide bridges are located in the correct position for human γG1 globulin and are similar for other immunoglobulins. The variable region indicated by the dash line occupies the first half of the light chain but perhaps only the first quarter of the heavy chain. Fc and Fab are fragments produced by limited proteolysis with papain. Electron micrographs and other data suggest a forked structure with a flexible region near the hinge peptide bridging Fc and Fd in the heavy chain. From Putnam and Köhler [3].

hapten complexes [7]. This fits with the presence of two antigen-combining sites in each antibody molecule, each of which is defined by an Fab unit. The Fab unit contains a whole light chain and the NH_2-terminal half of the heavy chain, the portion called Fd. By limited digestion with proteolytic enzymes the antibody molecule may be cleaved to yield two Fab pieces each of which is univalent with respect to antigen, and also an Fc piece. The latter is a dimer of the carboxyl half of the heavy chain. The Fc piece has some biological properties but no combining capacity for antigen. Preliminary X-ray diffraction studies have confirmed the pseudosymmetry of the Fc piece, which has most or all of the carbohydrate.

Light chains of the same antigenic type from the same species have an identical sequence for the carboxyl half of the chains comprising about 107 amino acid residues but vary greatly in amino acid sequence in the NH_2-terminal half, also containing about 107 residues. In fig. 2 the region of constant sequence (the c unit) is indicated by a straight line and the variable region (the v unit) by a dash line. By variability we mean that individual Bence-Jones proteins of the same antigenic kind differ in from 10 to 60 residues in the NH_2-terminal half. The single variation in the COOH-terminal half of human κ chains is inherited, whereas the explanation of the diversity of sequence in the NH_2-terminal half is still unknown. Normal light chains appear to consist of a hundred, a thousand, or perhaps 10,000 different sequences in the NH_2-terminal part but in the COOH-terminal half are identical with Bence-Jones proteins of the same antigenic type (either κ, or λ). Thus far, each Bence-Jones protein differs in amino acid sequence from every other one but is identical to the light chain of the homogeneous myeloma globulin or macroglobulin produced by the same patient.

The general principles of structure of heavy chains are similar to those of light chains except that heavy chains are about twice as long, and as is indicated by the dash line in fig. 2, probably only the first quarter of heavy chains is variable and the remaining three-fourths is constant. Thus, the variable portions of both light and heavy chains are about of equal size, approximately 110 residues.

The disulfide bonds of immunoglobulins are of two general types, intrachain and interchain. The large intrachain disulfide loops each contain about 60 residues. There are two of these in light chains and four in heavy chains. Their location imparts both symmetry and complementarity of structure. The second class of disulfide bond is in the interchain bridge of which there are three subclasses – the heavy-light bond bridging heavy and light chains, the interchain bonds bridging two heavy chains on the same molecule and the intermolecular disulfide bond. The latter is present in most of the μ and α chains and links μA and γM monomers into polymers. A similar COOH-terminal half-cystine is present on both κ and λ light chains but is lacking in γ heavy chains.

Though these results on the sequence of thousands of residues in light and heavy chains are stated in general terms, we can make one important conclusion and one prediction. Because all the variability in sequence of both heavy and light chains is located on the Fab piece which also has the combining site, we can conclude that the capacity for variability in sequence gives rise to antibody diversity and we can predict that it is related to antibody specificity.

5. Primary structure of light chains

The basic principles of the structure of immunoglobulin polypeptide chains have been derived from complete or nearly complete amino acid sequence analysis of seven human κ chains [8–13], seven human λ chains [14–17], two mouse κ chains [13] and partial sequence data on many fragments of the κ and λ chains of man and other species [18]. With an average of 214 residues per light chain the total number of residues determined approaches 4000. Sequence analysis is now complete for two human γ chains [19,20] and half a

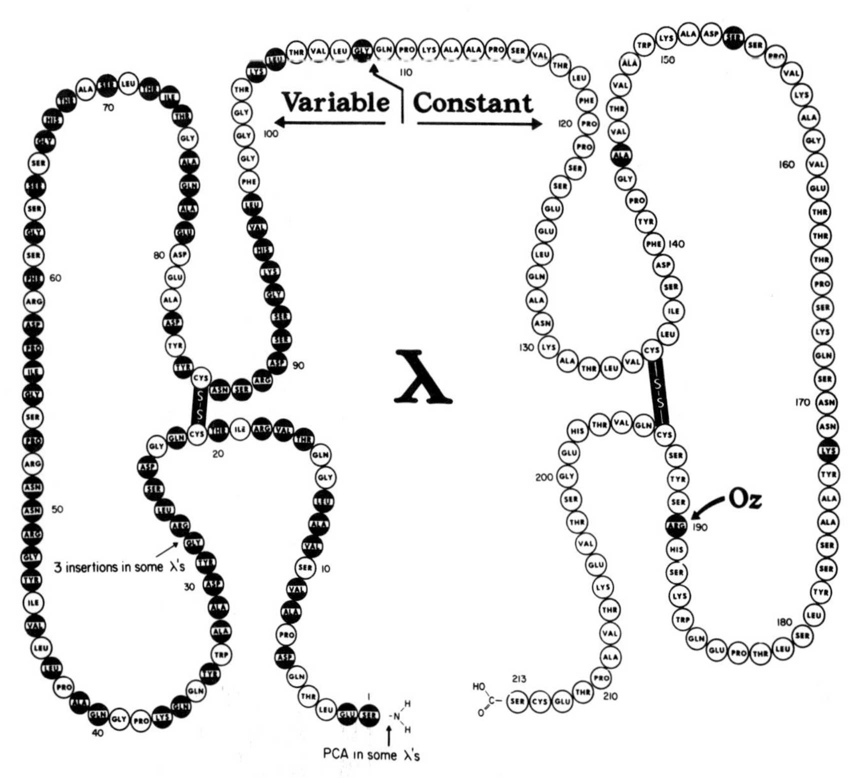

Fig. 3. Amino acid sequence of the human λ Bence-Jones protein Sh [23]. The black circles in the variable half mark the many loci where different amino acids have been found in other human λ chains. In the constant half, residue 190 is arginine in λ chains of the Oz^+ serological type and lysine in the Oz^- type. Interchanges in the constant part may occasionally occur at the three other positions marked by black circles. Data taken from Putnam et al. [14], Langer et al. [15], Milstein et al. [16], Ponstingl and Hilschmann [17] and Hood et al. [18].

rabbit γ chain [21] and is well advanced for a human μ chain [22]. The results are in general agreement with those for light chains as summarized below. The essential finding is that the light chains of a given antigenic type and species have an identical amino acid sequence in the approximate COOH-terminal half of the chain (the c unit) except for single amino acid substitutions associated with allotype or isotype. In contrast, the NH_2-terminal half (the v unit) is subject to multiple differences that are not correlated with allotypic or isotypic markers. The same holds true for heavy chains except that the v unit though being of similar length as in light chains corresponds to only the first quarter of the heavy chain.

The four characteristic features of light chain structure are illustrated in fig. 3, which gives the amino acid sequence of the human λ Bence-Jones protein Sh, the first light chain for which the entire sequence was reported [23]. The figure identifies the 77 positions that differ in other human λ chains and the 138 positions that thus far have appeared to be invariant. A very similar figure has been published for human κ chains [3].

The four unique characteristics of both κ and λ chains are:

1) A nearly exact partition into a variable region and a constant region each approximating 110 residues in the NH_2- and COOH-terminal portions of the chain, respectively.
2) A pseudosymmetrical structure determined by two homologous intrachain disulfide bonds, one in each half, each joining about 60 residues into a loop.
3) Variability in length of the chain owing to the insertion or deletion of up to six residues at a few restricted sites, notably around residues 27-30.
4) A COOH-terminal residue that participates in the disulfide bridge to heavy chains. This half-cystine is penultimate in λ chains and last in κ chains.

Thus far, no two Bence-Jones proteins of the same antigenic type from the same species have been found to have the same amino acid sequence. The differences are multiple. Unlike the single amino acid substitutions that occur in the abnormal hemoglobins, κ chains fall into three subtypes, κ_I, κ_{II}, and κ_{III}. The number of differences in sequence is as few as 16 within the subtypes but as many as 50 between specimens from different subtypes. All of the variable loci are in the NH_2-terminal half of the κ chain except for the allelic substitution of leucine for valine at position 191 which is associated with the hereditary Inv marker.

Unlike κ chains, λ chains can not readily be classified into a few related subtypes. When the NH_2-terminal residue is considered, as well as the length of the chain and the presence of apparent deletions, a subgrouping at first

Table 1.
Characteristic sequence differences in λ chains.
(Gaps in the variable region and interchanges in the constant region)

Bence-Jones Protein	Variable region								Constant region					Number of residues	Reference
	1	1	27a	27b	27c	94	95	96	108	144	153	172	190		
Ha	PCA	Ser	Gly	Thr	Gly	Ser	Ala	Val	Arg	Ala	Ser	Lys	Arg	217	[14]
Bo	PCA	Ser	Val	Gly	Asx	Asx	Phe	——	Arg	Ala	Ser	Lys	Arg	216	[14]
Vil	His	Ser	Val	Gly	Gly	Ser	——	Val	Gly	Ala	Ser	Lys	Arg	216	[17]
New	PCA	Ser	Ile	——	Gly	Asn	Ala	Val	Gly	Ala	Ser	Lys	Arg	216	[15]
Sh	——	Ser	——	——	——	Lys	His	Val	Gly	Ala	Ser	Lys	Arg	213	[14]
Kern	——	Tyr	——	——	——	——	——	Ala	Ser	Ala	*Gly*	Lys	Arg	211	[15]
X	——	Tyr	——	——	——	——	——	Val	Ser	Ala	Ser	Lys	Arg	211	[16]
Mz										*Val*	Ser	*Asn*	Arg		[16]
Oz													*Lys*		[24]

seems to occur. This is illustrated in table 1 which shows the characteristic sequence differences in human λ chains, both the gaps in the variable region and the interchanges in the constant region. Deletions may occur at the NH_2-terminal position (here numbered -1 because Sh is the reference protein), at 27a, b, and c, and also from 94 to 96. However, interchanges in the constant region are apparently not under allelic control and are unrelated to the sequence in the variable half.

Ponstingl and Hilschmann [17] have concluded that there are subgroup specific positions in the variable region of λ chains and that some deletions are subgroup specific. Accordingly, they have proposed four subclasses as follows: I (Ha and New), II (Bo and Vil), III (Sh), IV (Kern and X). Although there is some evidence for this classification, it is known that proteins even of the same subgroup differ by 30 residues in the variable part. This leads to the conclusion that λ chains vary greatly in sequence and cannot readily be classified into subgroups analogous to those of κ chains.

The remarkable variation in amino acid sequence of individual κ and λ Bence-Jones proteins though unique to the immunoglobulins is not an abnormal phenomenon but merely reflects the great heterogeneity in sequence of the NH_2-terminal region of normal light chains. Cumulative sequence data are now available on the first 20 residues of more than 30 human κ Bence-Jones proteins. Among these, at least 20 different sequences have been recorded for this small fragment of the κ chain. These differ one from another by as many as nine and by as few as one or two residues. However, at each variable position the chief alternative residues that may be present in different κ Bence-Jones proteins mirror the amino acids found in the mixed sequence of normal human κ light chains by Niall and Edman [25]. For example, at position 15, different human κ Bence-Jones proteins have had valine, leucine or proline, and in the mixed sequence of normal human κ chains all three amino acids are present. This is the most convincing evidence that Bence-Jones proteins are like the normal light chains in primary structure.

Because there is a strong tendency for conservation of certain residues, a point of diminishing returns is being reached in the search for more variable positions by the complete sequencing of κ and λ Bence-Jones proteins. For example, we previously showed that there were 34 apparently invariant positions in the NH_2-terminal 108 residues of the six κ chains for which complete or nearly complete sequence data were then available: Ag [8], Roy [9], Mil [13], Cum [9], Eu [11], and Tew [12]. Since then we have completed the sequence of the κ light chain of the γM globulin Ou [10], and no new variant positions have been found. As shown later, this light chain is of the κ_I type and does differ significantly in sequence even from other κ_I chains.

Recently, Suter et al. [9] gave the complete sequence of the variable part of the κ chain Ti, which is the first of subgroup κ_{III} to be sequenced. Despite the difference in subgroup, only one additional variable position was added. Yet, in its variable part this κ_{III} chain differed in sequence by about 40% from all other human κ chains that had been sequenced.

The same conclusion holds for λ light chains. The first three (Sh, Ha and Bo) were sequenced in our laboratory [14]. Among these only 54 positions were identical in the first 109 residues, though the remaining 105 were completely identical. After three more λ Bence-Jones proteins were sequenced in other laboratories (Kern [15], New [15], and X [16]), only 38 of the first 109 positions were unsubstituted. However, only one new variable position was added when the complete sequence of a seventh λ Bence-Jones protein (Vil) was reported by Ponstingl and Hilschmann [17].

6. Homology in the light chains of different species

The structural features described above for human κ and λ light chains undoubtedly hold in general for the light chains of all vertebrate species. This conclusion is based on two complementary sets of findings: 1) partial sequence analysis of short carboxyl-terminal and amino-terminal peptides of the normal light chains of many species [18,26] and 2) nearly complete sequence analysis of short carboxyl-terminal and amino-terminal peptides of the normal light chains of many species [18,26] and 2) nearly complete sequence analysis of several mouse κ Bence-Jones proteins [13]. The data on light chains various workers, as tabulated by Hood and Talmage [26], show heterogeneity in the NH_2-terminus of the light chains of all species examined. The COOH-terminus is required for the identification of the κ and λ types in other species. This has to be done by homology to primary structure of the human κ and λ chains; for the division into κ and λ types was initially based on serological reaction with antibodies to human κ and λ Bence-Jones proteins, which give little or no cross reaction with light chains from non-primates. The work of Franek et al. [27] has demonstrated that the disulfide bridge peptides likewise are homologous in the light chains of man and animals.

Sequence analysis of several mouse κ Bence-Jones proteins [13] has indicated a common evolutionary origin of the light chains of man and the mouse. The results are illustrated schematically in fig. 4 which compares the degree of sequence identity of the λ chains of man with that of the κ chains of man and the mouse. Human κ and λ chains have only about 40% identity in primary structure compared to about 60% identity between the κ chains of these

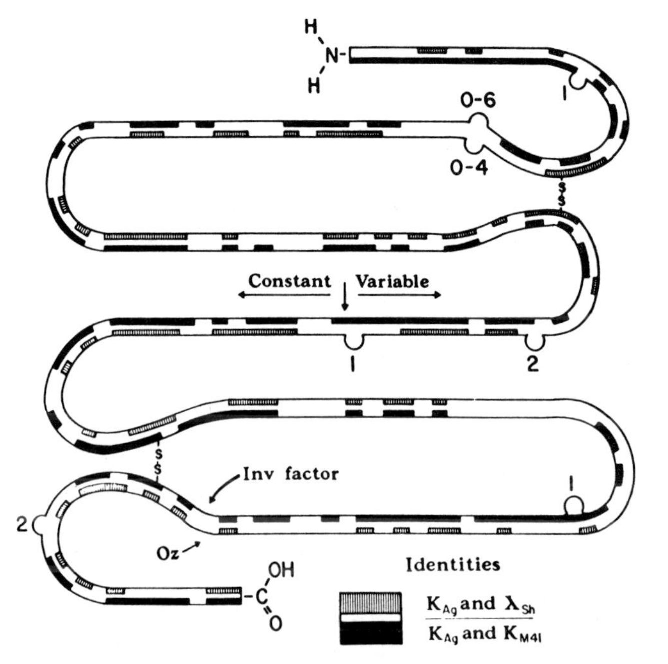

Fig. 4. Homology in amino acid sequence of human κ and λ light chains and mouse κ light chains. The sequences compared are the human κ Bence-Jones protein Ag (κ_{Ag}), the human λ Bence-Jones protein Sh (λ_{Sh}), and the mouse κ Bence-Jones protein M_{41} (κ_{M41}). The blocks mark identical residues, the smallest block being equivalent to one residue. The small loops identified by Arabic numbers are the locus of additional amino acids in one chain or another. In the variable region the structural homology holds only for the two proteins being compared (Ag and Sh shown by the blocks with vertical lines) (Ag and M_{41} shown by the solid blocks); however, in the constant region it holds for all human κ and λ light chains and mouse κ chains except for point substitutions such as occur at the Oz and Inv loci.

two species. Thus, the interspecies homology of light chains of the same antigenic type is greater than the intraspecies homology of light chains of different type. This is quite analogous to the finding that the α chains of the hemoglobins of different species are more alike in amino acid sequence than are the α and β chains of the same species. Just as workers in the hemoglobin field have concluded that these results indicate that the genes for α and β chains diverged early in evolutionary history, we have concluded that the genes for κ and λ chains diverged long before the species differentiation of man and the mouse [28].

7. Evolutionary relationships of light and heavy chains

Similar considerations apply to the light and heavy chains for which structural relationships bespeak a common evolutionary origin regardless of class, subclass, allotype, or antigenic type. These relationships are indicated by the schematic illustrations of the polypeptide chain formulas of fig. 1 and the more detailed drawing in fig. 2. The latter shows that each complete tetrachain immunoglobulin is composed of 12 subunits: one v unit and one c unit in each light chain and one v unit and three c units in each heavy chain. Each

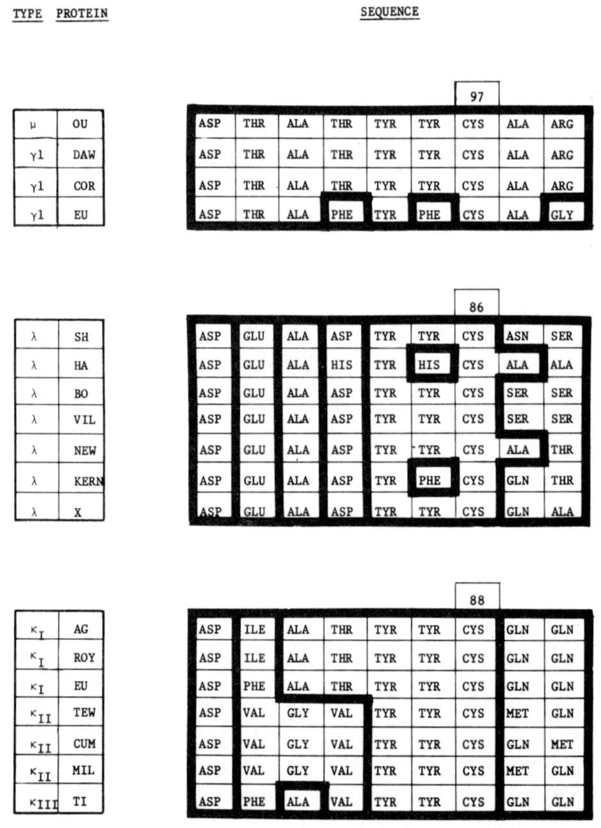

TYPE	PROTEIN	SEQUENCE						97		
μ	OU	ASP	THR	ALA	THR	TYR	TYR	CYS	ALA	ARG
γ1	DAW	ASP	THR	ALA	THR	TYR	TYR	CYS	ALA	ARG
γ1	COR	ASP	THR	ALA	THR	TYR	TYR	CYS	ALA	ARG
γ1	EU	ASP	THR	ALA	PHE	TYR	PHE	CYS	ALA	GLY

TYPE	PROTEIN							86		
λ	SH	ASP	GLU	ALA	ASP	TYR	TYR	CYS	ASN	SER
λ	HA	ASP	GLU	ALA	HIS	TYR	HIS	CYS	ALA	ALA
λ	BO	ASP	GLU	ALA	ASP	TYR	TYR	CYS	SER	SER
λ	VIL	ASP	GLU	ALA	ASP	TYR	TYR	CYS	SER	SER
λ	NEW	ASP	GLU	ALA	ASP	TYR	TYR	CYS	ALA	THR
λ	KERN	ASP	GLU	ALA	ASP	TYR	PHE	CYS	GLN	THR
λ	X	ASP	GLU	ALA	ASP	TYR	TYR	CYS	GLN	ALA

TYPE	PROTEIN							88		
κ_I	AG	ASP	ILE	ALA	THR	TYR	TYR	CYS	GLN	GLN
κ_I	ROY	ASP	ILE	ALA	THR	TYR	TYR	CYS	GLN	GLN
κ_I	EU	ASP	PHE	ALA	THR	TYR	TYR	CYS	GLN	GLN
κ_{II}	TEW	ASP	VAL	GLY	VAL	TYR	TYR	CYS	MET	GLN
κ_{II}	CUM	ASP	VAL	GLY	VAL	TYR	TYR	CYS	GLN	MET
κ_{II}	MIL	ASP	VAL	GLY	VAL	TYR	TYR	CYS	MET	GLN
κ_{III}	TI	ASP	PHE	ALA	VAL	TYR	TYR	CYS	GLN	GLN

Fig. 5. Homology in amino acid sequence around the second half-cystine residue of the first intrachain bridge in μ and $\gamma 1$ heavy chains (Cys-97) and in λ (Cys-86) and κ (Cys-88) light chains of man. Residues that differ from the μ chain sequence are boxed for emphasis. Data obtained from the following references: μ heavy chain [22], $\gamma 1$ heavy chains [19,20], λ light chains [14–17,23], κ light chains [2,8–13].

of these subunits contains some 110 to 120 amino acid residues including one intrachain disulfide loop joining about 60 residues. For this reason and because of the internal homology of primary structure within chains and between chains, it is believed that the primordial immunoglobulin gene coded for a primitive chain of approximately 110 residues. By contiguous duplication of the primordial gene a new gene twice as long could be formed which would code for a light chain of some 220 residues. The process could be repeated to form the gene for a primitive heavy chain. Tandem duplications of these genes followed by independent mutations could give rise to the κ and λ types of light chains and the γ, α, and μ heavy chains.

Support for the above hypotheses can be found in the similarity in amino acid sequence of homologous segments of human light and heavy chains such as is illustrated in fig. 5. In this case, the sequence around Cys-97 in μ and γ heavy chains is placed in opposition to the corresponding residue in the κ and λ light chains. The half-cystine compared in all these chains is the one at the COOH-terminus of the first intrachain disulfide bridge.

8. Primary structure of heavy chains

In the past year great emphasis has been given to the determination of heavy chain sequence in human immunoglobulins. The results have confirmed that heavy chains, like light chains, are divided into a variable NH_2-terminal

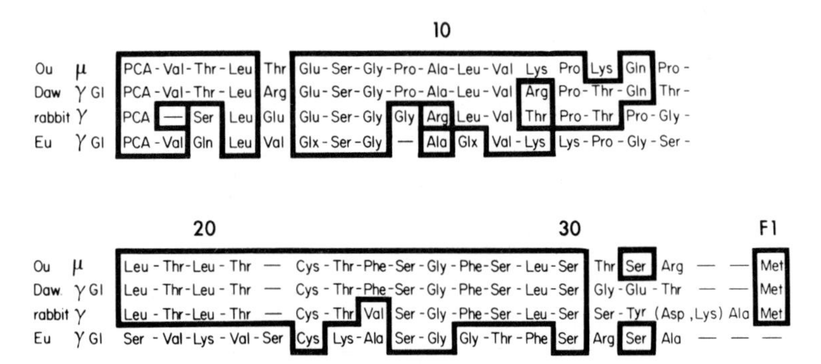

Fig. 6. Comparison of the first third of the NH_2-terminal variable sequence of the human μ chain Ou [22], the human γ1 chain Daw [19], a subtype of rabbit γ chain [30], and the human γ1 chain Eu [20]. Identical residues in any two or more of the proteins are included in boxes. Gaps have been introduced in the sequences to secure the maximum number of identiti.s. Fl refers to the first CNBr fragment of the first three chains listed.

region and a constant COOH-terminal region. However, the variable region of heavy chains though of approximately the same length as that of light chains, comprises only the first quarter of the chain and the constant region is three times as long as in light chains. The conclusions are based on the almost simultaneous reports of two $\gamma 1$ heavy chain sequences for proteins Daw and Cor by Press and Hogg [19] and of the complete sequence of the $\gamma 1$ chain of protein Eu [20]. Earlier, Wikler et al. [22] from our laboratory had reported the NH_2-terminal 105 residues of the variable sequence of the human μ chain Ou. The surprising result of these studies, illustrated in fig. 6 was the finding that the variable region of human μ and γ chains is not characteristic of the class of the chain, whereas the sequence of the constant region is. The variable regions of the Ou μ chain and the Daw $\gamma 1$ chain have a 73% identity in amino acid sequence provided a few gaps are introduced to achieve maximum homology [3]. However, partial sequence of the μ chain constant region already reported from our laboratory [10,22], as well as a great deal of our unpublished data [29] indicate that the μ and γ constant regions have no more than 40% identity in amino acid sequence. Conversely, the heavy chains od Daw and Eu, which are both of the $\gamma 1$ subclass, have only a 33% identity in their variable regions comprising the first 116 residues, whereas, they have only one known amino acid difference in their constant regions comprising the last 330 amino acid residues [19,20].

9. The concept of the pool of variable genes for heavy chains

One evidence that there are subgroups of μ chains and that these are analogous to subgroups of γ chains is given in fig. 6. This shows the sequence of a CNBr fragment comprising the first 35 amino acids of the human μ chain Ou, the human $\gamma 1$ chain Daw and of a subgroup of the normal rabbit γ chain in comparison with the corresponding segment of the human $\gamma 1$ chain Eu. It is evident that the two human $\gamma 1$ chains are more unlike in their variable sequences than are the three chains representing different classes and different species, i.e. the human μ chain Ou, the human $\gamma 1$ chain Daw, and the rabbit γ.

Sequence data on other human μ chains from our laboratory [10] and on other human γ chains from other laboratories [31–33] strongly indicate that the v regions of human μ and γ chains can be classified into four subgroups that are independent of the class of the c region. This has led us [10] to propose that the variable and constant regions of heavy chains are coded for by two sets of separate genes. This hypothesis assumes a large pool of variable genes for the v regions of heavy chains. Each v gene would code for a segment of from 110 to 110 to 120 amino acids which would be joined to the c region of either a μ chain or a γ chain. This would permit economy in the number of genes required to code for heavy chains. However, such a mechanism apparently does not operate for light chains, for the v regions of κ and λ chains are as characteristic of the light chain type as are the c regions.

The concept of a separate pool of v genes raises the question of whether the variable regions on the light and heavy chains of the same immunoglobulin molecules have some unique structural relationship. For example, Burnet [34] earlier proposed that the v regions of the light and heavy chains on the same molecule might be identical.

Because of the special role of the variable part in determining antibody specificity, we have investigated the structural relationships of the light and heavy chains on several macroglobulins. For example, we recently reported the complete sequence of a light chain from the human γM globulin Ou [10], for which the μ chain variable sequence had earlier been given [22]. The sequence of the light chain was of the κ_I type and differed from the κ_I reference light chain Ag by only 18 positions in the variable regions of the two chains. The light chain of the γG1-globulin Eu likewise is of the κ_I class and is equally related in primary structure to the Ou light chain. Yet, as shown in fig. 6 the heavy chains of these two immunoglobulins are of different classes and are quite unlike in sequence. We have other observations that confirm that there is no closer similarity in sequence between the variable regions of the light and heavy chains of the same immunoglobulin molecule than there is between the light and heavy chains of different molecules. This is not so surprising as it may seem for one would expect sequences of the light and heavy chain segments of the antigen-combining site to be complementary rather than identical.

In considering the evolutionary origin of immunoglobulins one question that arises is the degree of structural similarity among the constant regions of the μ, γ, and α chains. There is indeed a close relationship in the COOH-terminal sequences of the γ chains of different species. This is illustrated in fig. 7 giving the sequence of the COOH-terminal octadecapeptide of human γ chains of different isotypes and allotypes and of the γ chains of several animal species. The resemblance is remarkable; of the 19 positions shown, ten are identical in all ten chains. The order of difference is about the same between allotypes as between species. In a more extended comparison [20] it has been shown that there is a 65% identity in primary structure in the 220 residues of the carboxyl half of the heavy chain of human γG1 globulin and rabbit γ globulin.

This close similarity in primary structure of the c regions of chains of different isotypes, allotypes and species does not hold for the c regions of heavy

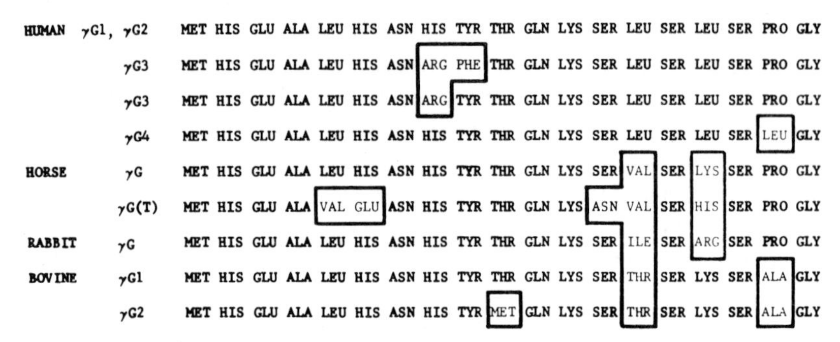

Fig. 7. Differences in the COOH terminal sequences of human γ chains of differing isotypes and allotypes and the γ chains of other species. Data from Porter [30], Press and Piggott [35], Prahl [36], Weir et al. [37], Hill et al. [21], and Milstein and Feinstein [38].

chains of different classes within the same species. The sequence of the COOH-terminal octapeptide of the human μ chain is -Ser-Asp-Thr-Ala-Gly-Thr-Cys-Tyr [22]. This bears no relationship to any of the COOH-terminal sequences of human γ chain isotypes depicted in fig. 7. On the other hand, the sequence of the COOH-terminal tripeptide of the human α chain is identical to that of the human μ chain given above.

We are now undertaking further study to determine whether the c regions of the μ and α chains are much more closely related in structure than are the c regions of μ and γ chains. At present, we are well on the way to completing the first sequence of a μ chain and have data on 140 different thermolysin peptides and about 50 tryptic peptides [29]. At no point in the c region of the μ chain have we found an identity of more than about 40% with any portion of the γ chain. This suggests that the genes for c regions of μ and γ chains diverged early in evolutionary history – earlier, in fact, than the species differentiation of the ancestors of man and the rabbit.

10. **Conclusion**

It is impossible to review the principles of immunoglobulin structure and outline perspectives for the future without recognizing the explosive growth of sequence data in the past five years and without acknowledging some concern about its assimilation and uncontrolled proliferation. There is no early end in sight. In fact, in 1970 the total sequence information on immunoglobulins will double, and perhaps it may quadruple in 1971 or 1972. With automatic protein sequenators now operating in at least five laboratories devoted to immunoglobulins the theoretical capacity for generating new sequence data has been increased by about 400 residues per week. There will probably soon be as much sequence data available on immunoglobulins as on all other proteins combined, except perhaps cytochromes and hemoglobins. This is a far cry from 1950 when Porter [39] published the first pentapeptide sequence of any immunoglobulin, or 1953 when I reported the first NH_2-terminal analyses of Bence-Jones proteins and myeloma globulins [40], or even early 1965 when Titani and Putnam [41] gave the first sequence data on tryptic peptides of light chains.

Where is this vast accumulation of structural data leading us? This question is particularly apt in view of Dr. Porter's prediction that the amino acid sequences of most antibodies will prove to be heterogeneous. There are, of course, many specific questions remaining to be answered of the kind I have proposed today. As one example, we would like to know whether the variable sequence is the same on the μ and γ heavy chains of the γM and γG globulins produced successively in the primary and secondary immune responses to a single antigen. Similarly, we may ask whether in this case the light chains have an identical or very similar sequence for the γM and γG antibodies against the same antigen. Many similar question remain to be answered that will help clarify the mechanism of antibody formation.

The great proliferation of sequence data on immunoglobulins in the past five years has hardly given us pause to reflect. Each new theory of antibody variability that is spawned in response to a fresh burst of sequence data has a half-life about equal to the time from submission to publication. This does not mean the data are superfluous or the theories valueless. Few fields of science have seen such a rapid acceleration of knowledge or such a flux of provocative hypotheses. Despite Dr. Porter's pessimism about the significance of the sequence variability as contributing to the specificity of the antigen-combining site we are dealing with a fundamental phenomenon involving the biosynthesis, structure, and genetic control of a special class of proteins. The explanation of the genetic origin of immunoglobulin structure is important in its own right as well as in relation to antibody specificity. The elucidation of this remarkable phenomenon will contribute to broader problems in biology such as the understanding of the sequence of molecular events in cellular differentiation and in gene repression and expression.

Acknowledgement

I am indebted to Drs. Heinz Köhler, Akira Shimizu and Claudine Paul for discussion of portions of this manuscript and of unpublished data on macroglobulin sequence. This work was supported by NIH grant CA-08497 from the National Cancer Institute and by a grant (GB-18483) from the National Science Foundation.

References

[1] Nomenclature for Human Immunoglobulins, Bull. World Health Org. 30 (1964) 447.
[2] F.W. Putnam, Science 163 (1969) 633.
[3] F.W. Putnam and H. Köhler, Naturwissenschaften 56 (1969) 439.
[4] J.L. Fahey, E.C. Franklin, H.G. Kunkel, E.F. Osserman and W.D. Terry, J. Immunol. 99 (1967) 465.
[5] J.B. Natvig, H.G. Kunkel and T. Gedde-Dahl Jr., Nobel Symp. 3 (1967) 313.
[6] F.W. Putnam, Biochim. Biophys. Acta 63 (1962) 539.
[7] R.C. Valentine and N.M. Green, J. Mol. Biol. 27 (1967) 615.
[8] F.W. Putnam, K. Titani and E.J. Whitley Jr., Proc. Roy. Soc. London Ser. B 166 (1966) 124.
[9] L. Suter, H.V. Barnikol, S. Watanabe and N. Hilschmann, Z. Physiol. Chem. 350 (1969) 275.
[10] H. Köhler, A. Shimizu, C. Paul and F.W. Putnam, Science, in press.
[11] P.D. Gottlieb, B.A. Cunningham, M.J. Waxdal, W.H. Konigsberg and G.M. Edelmen, Proc. Natl. Acad. Sci. U.S. 61 (1968) 168.
[12] E.J. Whitley Jr. and F.W. Putnam, unpublished results.
[13] W.J. Dreyer, W.R. Gray and L. Hood, Cold Spring Harbor Symp. Quant. Biol. 32 (1967) 353.
[14] F.W. Putnam, T. Shinoda, K. Titani and M. Wikler, Science 157 (1967) 1050.
[15] B. Langer, M. Steinmetz-Kayne and N. Hilschmann, Z. Physiol. Chem. 349 (1968) 945.
[16] C. Milstein, J.B. Clegg and J.M. Jarvis, Biochem. J. 110 (1968) 631.
[17] H. Ponstingl and N. Hilschmann, Z. Physiol. Chem. 350 (1969) 1148.
[18] L. Hood, W.R. Gray, B.G. Sanders and W.J. Dreyer, Cold Spring Harbor Symp. Quant. Biol. 32 (1967) 133.
[19] E.M. Press and N.M. Hogg, Nature 223 (1969) 807.
[20] G.M. Edelman, B.A. Cunningham, W.E. Gall, P.D. Gottlieb, U. Rutishauser and M.J. Waxdal, Proc. Natl. Acad. Sci. U.S. 53 (1969) 78.
[21] R.L. Hill, H.E. Lebovitz, R.E. Fellows Jr. and R. Delaney, Nobel Symp. 3 (1967) 109.
[22] M. Wikler, H. Köhler, T. Shinoda and F.W. Putnam, Science 163 (1969) 75.
[23] M. Wikler, K. Titani, T. Shinoda and F.W. Putnam, J. Biol. Chem. 242 (1967) 1668.
[24] E. Appella and D. Ein, Proc. Natl. Acad. Sci. U.S. 57 (1967) 1449.

[25] H.D. Niall and P. Edman, Nature 216 (1967) 262.
[26] L. Hood and D. Talmage, Science 168 (1970) 325.
[27] F. Franěk, B. Keil, J. Novotný and F. Šorm, European J. Biochem. 3 (1968) 422.
[28] F.W. Putnam, K. Titani, M. Wikler and T. Shinoda, Cold Spring Harbor Symp. Quant. Biol. 32 (1967) 9.
[29] A. Shimizu, H. Köhler, C. Paul and F.W. Putnam, unpublished results.
[30] R.R. Porter, Biochem. J. 105 (1967) 417.
[31] J.R.L. Pink and C. Milstein, FEBS Symposium 15 (1969) 177.
[32] B.A. Cunningham, M.N. Pflumm, U. Rutishauser and G.M. Edelman, Proc. Natl. Acad. Sci. U.S. 64 (1969) 997.
[33] B. Frangione and C. Milstein, Nature 224 (1969) 597.
[34] M. Burnet, Nature 210 (1966) 1308.
[35] E.M. Press and P.J. Piggott, Cold Spring Harbor Symp. Quant. Biol. 32 (1967) 45.
[36] J.W. Prahl, Biochem. J. 105 (1967) 1019.
[37] R.C. Weir, R.R. Porter and D. Givol, Nature 212 (1966) 205.
[38] C.P. Milstein and A. Feinstein, Biochem. J. 107 (1968) 559.
[39] R.R. Porter, Biochem. J. 46 (1950) 473.
[40] F.W. Putnam, J. Am. Chem. Soc. 75 (1953) 2785.
[41] K. Titani and F.W. Putnam, Science 147 (1965) 1304.

ON THE COMBINING SITES OF ANTIBODIES TO DEFINED DETERMINANTS

M. SELA, I. SCHECHTER, B. SCHECHTER
and A. CONWAY-JACOBS
*Department of Chemical Immunology,
The Weizmann Institute of Science, Rehovot, Israel*

Abstract: Sela, M., Schechter, I., Schechter, B., and Conway-Jacobs, A. On the Combining Sites of Antibodies to Defined Determinants. *Miami Winter Symposia* 2, pp. 382–396. North-Holland Publishing Company, Amsterdam.

The reaction of a specificity determinant on the antigen with the combining site on the antibody represents a uniquely specific pattern of recognition on a molecular level. In protein antigens it is possible to distinguish between 'sequential' determinants and conformation-dependent determinants. Previous studies on the size and nature of sequential determinants, making use of alanine peptides attached to proteins, led to the conclusion that the size of the combining region of the antibodies is such as to accommodate 3–4 alanine residues, and that the region of the antigenic determinant furthest removed from the protein carrier is of paramount importance in determining the specificities of the antibodies formed. The same conclusions were reached when, instead of preparing peptidyl proteins by polymerization techniques, we have used as immunogens proteins to which peptides of defined length and structure have been attached. The protein participates in the antigenic determinant only when the hapten attached is smaller than a tetrapeptide. A comparison of the combining sites of IgG and IgM antibodies showed that they are similar in size. Antibodies with specificity directed to D-alanine peptides may be fractionated by elution from an immunoadsorbent with di-D-alanine, followed by tetra-D-alanine. The main difference between the two fractions was that the dialanine reacted much better with the first fraction than with the second. With both fractions pentapeptides reacted not significantly better than tetrapeptides.

Spatial folding of proteins plays an important role in determining their antigenic specificity. In order to elucidate the role of the conformation of the antigen in immunogenicity and in antigenic specificity, we have investigated the immune response to synthetic polypeptides of defined sequence and conformation. The same tripeptide, L-Tyr-L-Ala-L-Glu, may be either polymerized to yield an α-helical molecule (under physiological conditions) or attached to a branched poly-DL-alanine. The first polymer served successfully as an example of "conformational", and the second of "sequential" determinants. There is almost no cross-reaction between the two systems. The tripeptide TyrAlaGlu and related peptides are efficient inhibitors of the antigen-antibody reaction in the case of the branched polymer, but not in the case of the helical polymer.

Antisera obtained by immunization with the ordered sequence helical polymer may also contain antibodies which do not react with the immunizing antigen but are able to react with a random copolymer of Tyr, Ala and Glu, implying that such antibodies are formed against determinants derived *in vivo* from the helical copolymer. Antibodies to the polymer of the tripeptide L-Pro-Gly-L-Pro, which has the characteristic triple helix of collagen, cross-react with fish collagen, rat collagen and even guinea pig skin collagen. These results support the notion that antigenic determinants of protein are controlled to a large extent by the higher order structure of proteins, and thus that the antigenic determinants are recognized largely while the immunogenic macromolecule is still intact.

1. Introduction

The world of immunological phenomena may serve as a wonderful example for biological organization, as such organization – in the three-dimensional space as well as in time – is typical for all its stages. This is valid for the recognition of the antigenic determinant, for the cooperation of cells in the induction of immune response, for antibody synthesis and the purely cellular immunological reaction, for the secondary response which implicates the "immunological memory", for the induction and persistance of immunological tolerance, and – last but not least – for the tremendously specific interaction of the determinant group on the antigen with the combining site on the antibody. This interaction represents a uniquely specific pattern of recognition on a molecular level, and it involves two macromolecules.

In protein and polypeptide antigens it is possible to distinguish between "sequential" determinants and conformation-dependent determinants [1]. We would like to call a sequential determinant one due to an amino acid sequence in a random coil form, and antibodies to such a determinant are expected to react with a peptide of identical, or similar, sequence. On the other hand, a conformational determinant results from the steric conformation of the antigenic macromolecule, and leads to antibodies which would not necessarily react with peptides derived from that area of the molecule. It seems that antibodies to native proteins are directed mostly against conformational rather than sequential determinants.

2. Antibodies of polyalanyl specificity

Antibodies with specificity directed toward peptides, obtained upon immunization with proteins to which such peptides are covalently attached, may serve as good examples of antibodies to sequential determinants. The size and chemical nature of such peptide determinants may be investigated and conclusions drawn about the complimentary combining sites of antibodies, by means of techniques so successfully used by Kabat for oligosaccharide determinants [2]. Several studies have appeared in recent years in which anti-polypeptide antibodies were characterized by the extent of inhibition with peptides of increasing size [3–6]. From a detailed study of antibodies with specificity directed either to poly-L-alanyl or to poly-D-alanyl determinants, obtained by immunization with polyalanyl proteins prepared by reacting the proteins with the respective *N*-carboxyalanine anhydrides, we concluded that the size of the specific combining region of anti-polyalanyl antibodies is such as to accommodate a maximum of 3–4 alanine residues [6]. We also drew from this study the conclusion that the region furthest removed from the protein carrier, i.e. in our case the amino termini of the peptides attached, is of major importance in defining the specificities of the antibodies formed.

More recently, instead of preparing peptidyl proteins by polymerization techniques, we have used as immunogens – for a more precise evaluation of the results – proteins to which peptides of defined length and structure have been attached [7]. Peptides of the structures $(DAla)_n$ - Gly(n = 1,2,3, 4) were coupled to ribonuclease. The resulting conjugates induced peptide-specific antibodies, detected by their precipitin reaction with rabbit serum albumin to which the same peptides were attached. These precipitin reactions were inhibited with peptides of general structures: $(DAla)_n$; $(DAla)_n$-Gly; and $(DAla)_n$-Gly-ϵ-aminocaproic acid. In all cases penta- or hexapeptides did not inhibit significantly better than the appropriate tetrapeptides, indicating that the combining sites of antibodies accommodate four amino acid residues. The lysine residue in the protein carrier participates in the antigenic determinant only when the peptide attached is smaller than a tetramer. Also in this case it was clear that the most exposed part of the antigenic determinant (the *N*-terminal amino acid residue of the peptide attached) plays an immunodominant role.

IgG antibodies differ from IgM [8] antibodies in their molecular size, electrical charge, antigenic properties and number of combining sites. It was, therefore, of interest to investigate the efficiency of inhibition by means of D-alanine peptides of increasing size, of the reaction of the IgM anti-poly-D-alanyl antibodies with a poly-D-alanyl protein conjugate and with poly-D-alanyl bacteriophage T4, and to compare it with the IgG antibodies derived from the same animals [9]. The inhibitions were carried out either with serum fractions or with antibodies isolated specifically on immunoadsorbents

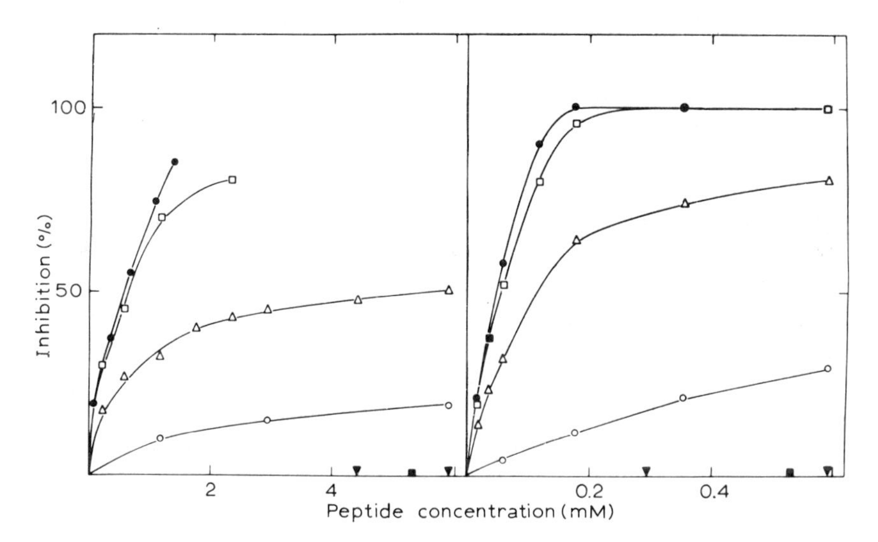

Fig. 1. Inhibition by alanine peptides of the precipitates obtained by reacting poly-D-alanyl human serum albumin with IgM (left) and IgG (right) anti-poly-D-alanyl antibodies isolated from a goat antiserum to poly-D-alanyl ribonuclease: ▾, D-alanine; ○, D-alanyl-D-alanine; △, tri-D-alanine; □, tetra-D-alanine; ●, penta-D-alanine; ■, tetra-L-alanine. From Haimovich et al. [9].

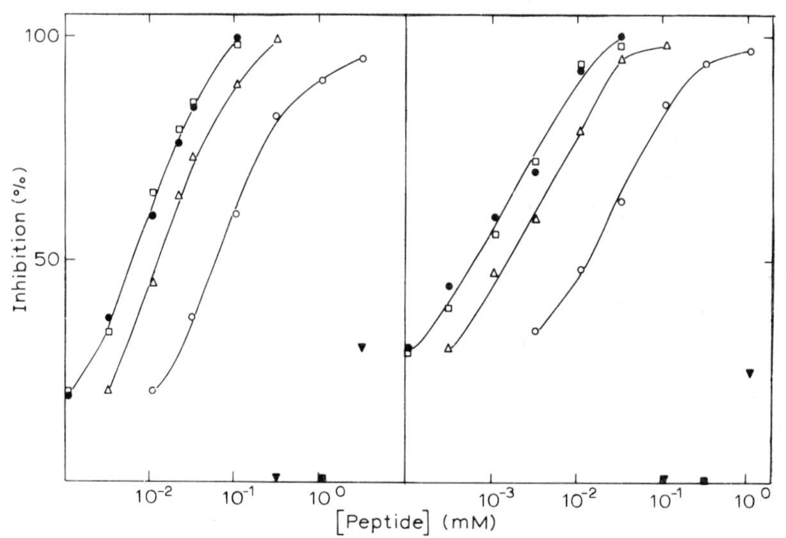

Fig. 2. Inhibition by alanine peptides of the inactivation of poly-D-alanyl bacteriophage T4 by IgM (left) and IgG (right) fractions separated from a pooled rabbit antiserum to poly-D-alanyl ribonuclease: ▾, D-alanine; ○, D-alanyl-D-alanine; △, tri-D-alanine; □, tetra-D-alanine; ●, penta-D-alanine; ■, tetra-L-alanine. From Haimovich et al. [9].

and separated by gel filtration. Both goat and rabbit antisera were used. As seen in fig. 1, the precipitin reaction of purified IgM anti-poly-D-alanyl antibodies with poly-D-alanyl human serum albumin is not inhibited significantly better by penta-D-alanyl than by tetra-D-alanine, whereas tri-D-alanine is a less efficient inhibitor. Thus, the behavior of IgG and IgM antibodies is very similar.

The inactivation of poly-D-alanyl bacteriophage T4 by IgM and IgG fractions of rabbit antiserum to poly-D-alanyl ribonuclease may be inhibited by D-alanine peptides, as seen in fig. 2. Also in this case, the efficiency of inhibition as a function of the size of the inhibitory peptide increased up to the tetrapeptide for both classes of antibodies. Thus, the antigen-combining sites of both IgG and IgM antibodies are similar in size and capable of accommodating a peptide composed of four alanine residues. It appears that the two immunoglobulin classes differ in their strength of interaction with the haptens, since significant differences are observed in the concentrations of peptides necessary to obtain the same extent of inhibition (fig. 1 and 2).

3. Fractionation of antibodies with peptides

Individual sera may show differences in their specificity as evaluated from inhibition studies and in some cases the upper limit for optimal inhibition slightly differs [6,10–12]. Kabat and his colleagues [13–15] fractionated human anti-dextran antibodies by specific adsorption on Sephadex (insolubilized dextran) and successive elution with oligosaccharides of the isomaltose series of increasing size. From inhibition studies they concluded that the isolated antibody fractions differed in the size of the combining sites. Antibodies of blood group A specificity were also fractionated, leading to similar conclusions [16].

We have now, in analogy, fractionated the initial population of antibodies specific to peptide chains composed of D-alanine [17,18]. The antibodies were obtained from rabbits immunized with poly-D-alanyl ribonuclease and with $(DAla)_3$-Gly-ribonuclease. Specific fractionation of the antibodies was achieved by adsorption on the water-insoluble poly-D-alanyl-rabbit serum albumin-cellulose conjugate, followed by stepwise extraction with di-D-alanine and tetra-D-alanine.

Similarly to the unfractionated antibodies, the separated fractions reacted with a pentapeptide not significantly better than with a tetrapeptide, whereas the inhibition reactions making use of tripeptide led to the conclusion

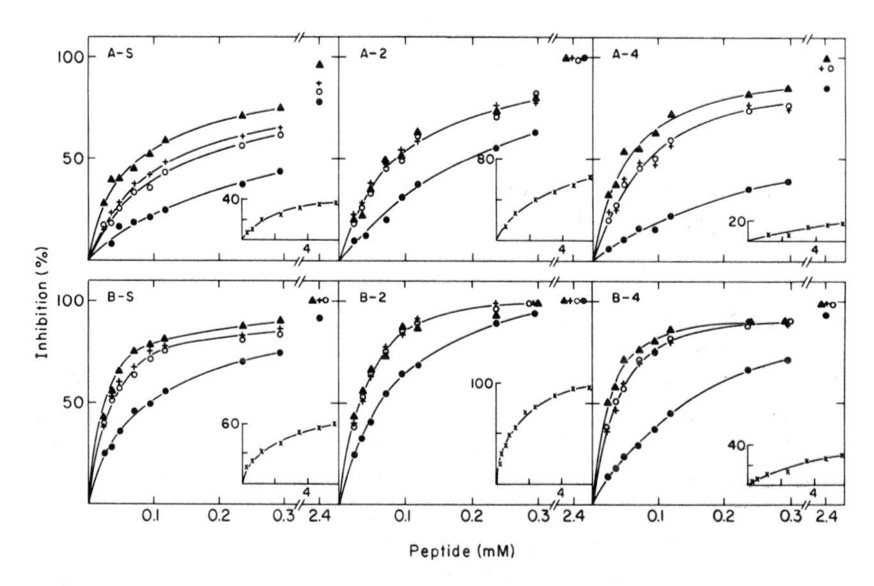

Fig. 3. Inhibition of the precipitation of anti-$(DAla)_3$-Gly-ribonuclease by $(DAla)_3$-Gly-rabbit serum albumin (animal A) and by poly-D-alanyl rabbit serum albumin (animal B). Whole sera (A–S, B–S), and two purified antibody fractions, obtained by elution with $(DAla)_2$ (A–2, B–2) and with $(DAla)_4$ (A–4, B–4) were used. The inhibitors are: X, D-alanyl-D-alanine; •, tri-D-alanine; ○, tetra-D-alanine; +, tri-D-alanyl-glycine; ▲, tri-D-alanyl-glycyl-ϵ-aminocaproic acid. From Schechter et al. [18].

that in all cases the tetrapeptide was a better inhibitor than the tripeptide. Two examples are illustrated in fig. 3. The tripeptide was in all experiments a much better inhibitor than the dipeptide. The inhibition studies thus indicate that all the antibody preparations investigated contain antibodies with combining sites accommodating a tetrapeptide.

The main difference between the antibody fraction eluted with the dipeptide and the antibody fraction eluted with the tetrapeptide is in their susceptibility to inhibition with D-alanyl-D-alanine. The dipeptide inhibited the precipitin reaction with antibodies eluted by means of the dipeptide at least ten times more efficiently than the reaction with antibodies eluted by means of the tetrapeptide (fig. 3). The tripeptide was also more inhibitory in the system involving antibodies eluted with the dipeptide than in the system involving antibodies eluted with the tetrapeptide.

It seems thus possible that all the antibodies formed possess combining sites complementary to a tetrapeptide, but that they are heterogeneous in terms of their capacity to bind the dipeptide. Certainly some antibodies complementary to a tetrapeptide could be eluted from the immunoadsorbent with a dipeptide, whereas other antibodies – even though complementary to a peptide of the same size – were not eluted with the dipeptide.

4. Sequential and conformational determinants composed of tyrosine, alanine and glutamic acid.

Spatial folding of proteins plays an important role in determining their antigenic specificity, as apparent from the poor reaction, or total lack of cross-reaction, between denatured proteins and antibodies to the same proteins in their native form. Thus, for example, antibodies to native bovine pancreatic ribonuclease do not react at all with the performic acid-oxidized ribonuclease, which is a randomly coiled chain devoid of disulfide bridges nor do antibodies to performic acid-oxidized ribonuclease react with the native enzyme [19]. Similarly, antibodies prepared in goat against rabbit immunoglobulin G do not react with the rabbit IgG after all its disulfide bridges were opened by reduction with 2-mercaptoethanol [20]. This is undoubtedly due to changes within the conformation of the protein molecule, resulting in loss of the original antigenic determinants.

Studies of synthetic antigens were helpful in the elucidation of many points concerning the molecular basis of antigenicity [21,22]. We have, therefore, asked ourselves whether we could use the same peptide sequence to prepare synthetic antigens possessing either a 'sequential' or a conformation-dependent antigenic determinant. We have tried to answer this question by building two types of polymers based on the tripeptide TyrAlaGlu [1]. In one case we attached the peptide to the amino termini of poly-DL-alanine side chains in multi-poly-DL-alanyl-poly-L-lysine (fig. 4, left). In the other case the peptide was polymerized to yield a periodic polymer (fig. 4, right). The periodic polymer of TyrAlaGlu, to be denoted $(TAG)_n$, exists under physiological conditions of pH and ionic strength in an α-helical form [23,24], whereas the TyrAlaGlu sequence in the branched polymer, to be denoted (TAG)-A--L, is representative of a 'sequential' determinant [1].

Both polymers were immunogenic in rabbits, but the antibodies against one polymer did not cross-precipitate with the other one (figs. 5 and 6). For comparison we immunized rabbits also with a random copolymer of L-tyrosine, L-alanine and L-glutamic acid (in a residue molar ratio of 1:1:1), obtained by copolymerization of the *N*-carboxyanhydrides of L-tyrosine, L-alanine and the γ-benzyl ester of glutamic acid, followed by debenzylation in anhydrous hydrogen bromide. This polymer, denoted $(TAG)_r$, cross-reacts partially with

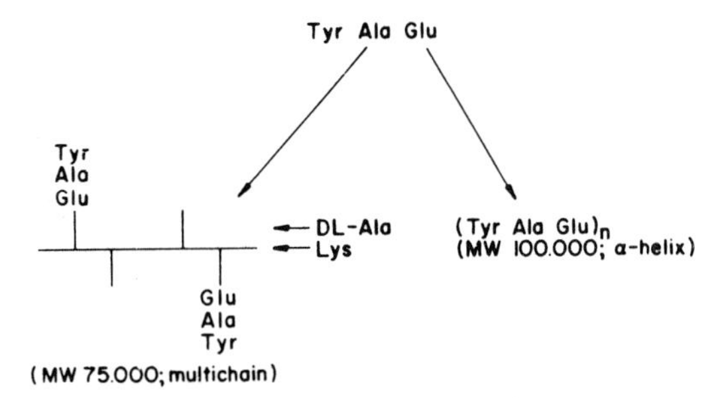

Fig. 4. Schematic representation of a synthetic branched polymer in which peptides of sequence TyrAlaGlu are attached to the amino termini of polymeric side-chains in multi-poly-DL-alanyl-poly-L-lysine (left), and of a periodic polymer of the tripeptide TyrAlaGlu (right). From Sela [22].

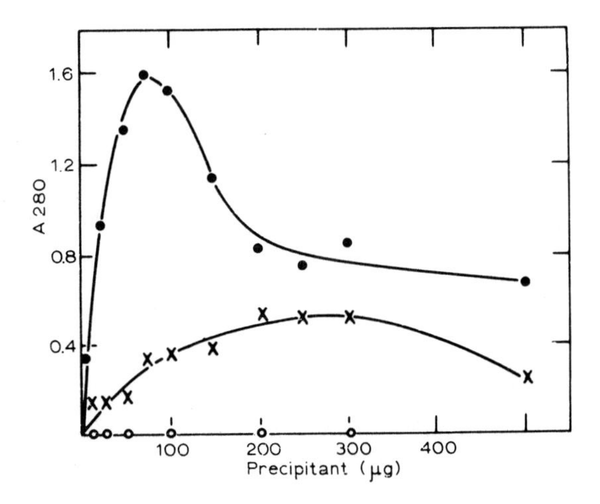

Fig. 5. Absorbancy at 280 nm of solutions in 0.1 N sodium hydroxide of precipitates obtained by the addition of: •, $(TAG)_n$; X, $(TAG)_r$; and ○, (TAG)-A–L, to a rabbit anti-$(TAG)_n$ serum (0.5 ml).

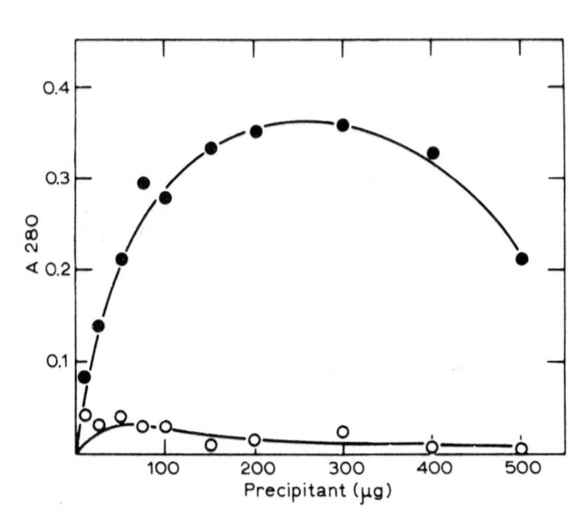

Fig. 6. Absorbancy at 280 nm of solutions in 0.1 N sodium hydroxide of precipitates obtained by the addition of: •, (TAG), and ○, $(TAG)_n$, to a solution of isolated rabbit antibodies against (TAG) (1.0 ml, containing 0.65 mg protein).

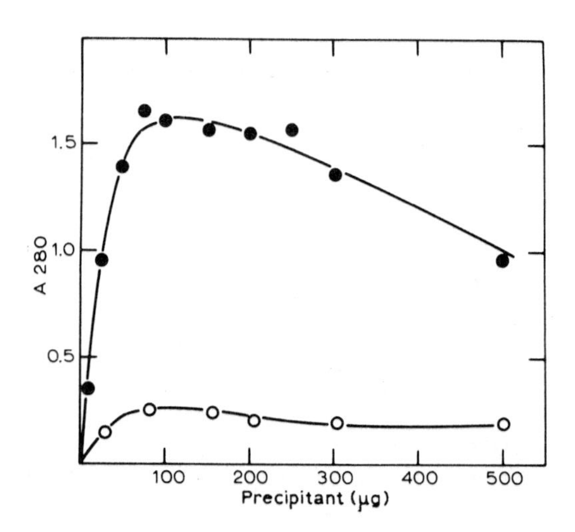

Fig. 7. Absorbancy at 280 nm of solutions in 0.1 N sodium hydroxide of precipitates obtained by the addition of: •, $(TAG)_r$; and ○, $(TAG)_n$, to a rabbit anti-$(TAG)_r$ serum (0.5 ml).

antibodies against $(TAG)_n$ (fig. 5). Its homologous and heterologous precipitin reactions are given in fig. 7.

The lack of cross-precipitation between the (TAG) and the $(TAG)_n$ systems would suggest that the two polymers do not possess common antigenic determinants. Indeed, the tripeptide TyrAlaGlu inhibits efficiently the (TAG) (fig. 8), but not the $(TAG)_n$ system (fig. 9). It may be thus concluded that the determinants on (TAG) are sequential and consist mainly of TyrAlaGlu, whereas the antigenic determinants on $(TAG)_n$ are conformation-dependent and result from the particular juxtaposition of the amino acid side chains in the conformation of the helical macromolecule.

Efforts to inhibit the reaction between $(TAG)_n$ and antibodies to $(TAG)_n$ by means of peptides of increasing size are depicted in fig. 9. The nonapeptide $(TyrAlaGlu)_3$, is already inhibitory, whereas peptide fractions, with average degrees of polymerization $(TyrAlaGlu)_7$ and $(TyrAlaGlu)_9$ are very efficient inhibitors of the precipitin reaction. The nonapeptide is not yet helical in solution [25], but the reaction with the combining site of the antibody might lead to a decrease in entropy compatible with its conversion into the helical form.

Partial cross-reactions were observed between the periodic ordered polymer, $(TAG)_n$, and the random copolymer, $(TAG)_r$, systems (figs. 5 and 7), with great variations between individual animals. Thus, 6–17% of anti-$(TAG)_r$ antibodies cross-precipitated with $(TAG)_n$, whereas 13–40% of anti-$(TAG)_n$ antibodies cross-precipitated with $(TAG)_r$. In order to understand better the nature of these cross-reacting antibodies, we have investigated their reaction with a mixture of peptides obtained upon digestion of the random copolymer $(TAG)_r$ with pronase.

$(TAG)_r$ (150 mg) was digested with pronase (2 mg) in 10 ml phosphate-buffered saline, pH 7.4. After 45 min the mixture was applied to a Sephadex G-15 column equilibrated with 0.05 M ammonium carbonate, pH 8.8. Six peaks were obtained. The material under each peak was subjected to high voltage electrophoresis, amino acid analysis and Edman degradation. Each peak was a mixture of 3–5 peptides. The peptides in peaks 2 to 6 were less than five amino acids long, and these peptide mixtures were used for inhibition studies.

The peptides inhibited efficiently the precipitin reaction between $(TAG)_r$ and anti-$(TAG)_r$, but did not inhibit at all the reaction between $(TAG)_n$ and anti-$(TAG)_n$. This suggested that essentially all of the antigenic determinants on $(TAG)_r$ are 'sequential'. The peptides also completely inhibited the partial cross-reactions between $(TAG)_n$ and anti-$(TAG)_r$ and between $(TAG)_r$ and

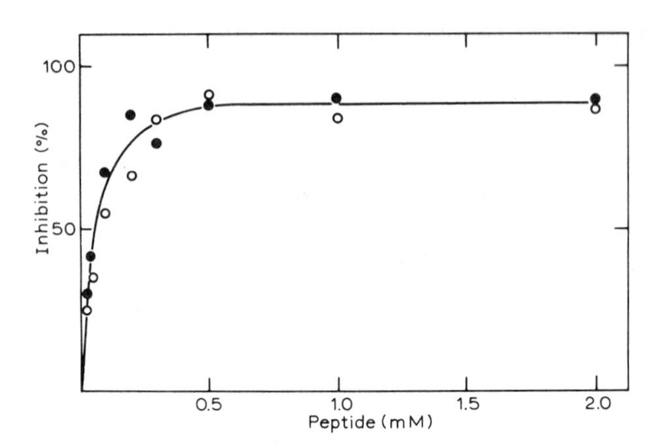

Fig. 8. Inhibition of the homologous reaction between (TAG) and antibodies isolated from its antiserum, by the peptides: •, TyrAlaGlu; and ○, $(TyrAlaGlu)_2$.

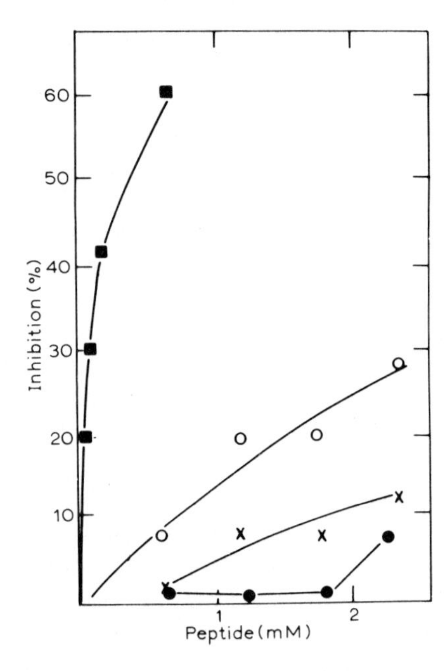

Fig. 9. Inhibition of the homologous reaction between $(TAG)_n$ and antibodies isolated from its antiserum, by the peptides: •, TyrAlaGlu; X, $(TyrAlaGlu)_2$; ○, $(TyrAlaGlu)_3$; and ■, $(TyrAlaGlu)_7$ and $(TyrAlaGlu)_9$. The last two peptides were obtained by polymerization techniques, and they represent average values.

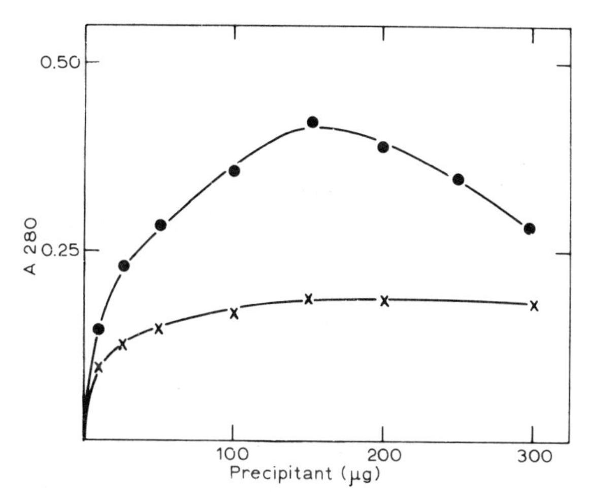

Fig. 10. Precipitin reaction of those anti-$(TAG)_n$ antibodies which were adsorbed on a $(TAG)_r$-immunoadsorbent (0.5 ml., containing 0.26 mg protein) with: •, $(TAG)_r$; and X, $(TAG)_n$.

anti-$(TAG)_n$. To explain the cross-precipitation between anti-$(TAG)_r$ and $(TAG)_n$ we may assume that there are a few non-helical segments in $(TAG)_n$, capable of cross-reacting with antibodies against determinants on $(TAG)_r$ possessing a similar sequence.

As mentioned above, the peptides derived upon digestion of $(TAG)_r$ with pronase were capable of inhibiting the precipitin reaction of anti-$(TAG)_n$ antibodies with $(TAG)_r$ but not at all of anti-$(TAG)_n$ with $(TAG)_n$ itself. This would imply that the antibody fraction which is formed upon immunization with $(TAG)_n$ and which cross-reacts with $(TAG)_r$, is directed against 'sequential' determinants. Indeed, when an immunoadsorbent prepared from $(TAG)_r$ and bromoacetylcellulose [26] was used to isolate specifically the cross-reacting anti-$(TAG)_n$ antibodies, we observed – as seen in fig. 10 – that these antibodies precipitated to a much larger extent with the heterologous $(TAG)_r$ than with the homologous $(TAG)_n$. It is not clear whether these antibodies were formed against some non-helical fragments on the intact $(TAG)_n$ molecules, or whether they were produced against some *in vivo* degradation products.

5. Other studies on the role of conformation

Another periodic polymer of a tripeptide which has been shown to possess conformation-dependent antigenic determinants is $(ProGlyPro)_n$ [1,27]. This polymer, known to have the characteristic triple helix of collagen [28,29], led both in guinea pigs and in rabbits to antibodies which cross-reacted with fish collagen, rat collagen and even guinea pig collagen. This was the first instance that antibodies to a synthetic antigen reacted with a natural protein, and clearly the reason for this reaction is in their higher order structure.

In *Ascaris* cuticle collagen disulfide bridges play a crucial role in the preservation of its higher order structure [30]. Indeed, we have recently observed [31] that the periodic polymer $(ProGlyPro)_n$ cross-reacts with antibodies prepared in rabbits against the native *Ascaris* cuticle collagen [32], but gave no cross-reaction with antibodies obtained after immunization with the reduced and carboxymethylated *Ascaris* cuticle collagen [30].

As the determinants of globular proteins are mostly conformation-dependent , their elucidation clearly requires a detailed knowledge both of the amino acid sequence and of the three-dimensional structure of the protein. We have shown recently that antibodies reacting with a natural protein, the hen egg-white lysozyme, may be obtained upon immunization with a synthetic antigen conjugate [33].

A peptide containing the sequence 60–83 of lysozyme, and denoted the 'loop' peptide as it still contains one disulfide bridge, was attached to multichain poly-DL-alanine (multi-poly-DL-alanyl- -poly-L-lysine). The resulting synthetic conjugate elicited in rabbits and goats the formation of antibodies with specificity directed against a unique region in native lysozyme. This was shown by the capacity of lysozyme to inhibit the homologous antigen-antibody reaction as well as by isolation of the antibodies capable of reacting with lysozyme on an immunoadsorbent prepared from lysozyme and bromoacetylcellulose [26]. A fraction of the antibodies obtained upon immunization with lysozyme has a similar specificity and these antibodies could be isolated with an immunoadsorbent prepared from bromoacetylcellulose and the 'loop' peptide. In this case the capacity of the isolated antibodies to react with lysozyme is efficiently inhibited by the 'loop' peptide, but not by the open-chain peptide derived from it after reduction and carboxymethylation. Thus, it seems that the antibodies to the 'loop' are directed against a conformation-dependent determinant.

6. Concluding remarks

Use of defined synthetic polypeptide antigens and of polypeptidyl proteins has permitted a more clear understanding of the nature and size of antigenic determinants and of the respective combining sites of antibodies. Examples, illustrated here, of 'sequential' and conformation-dependent determinants throw light on the nature of the specific determinants of globular and fibrillar proteins.

If one considers the results described above, concerned with the role of steric conformation in the antigenic specificity of proteins and of synthetic polypeptides, it is unavoidable to draw the conclusion that no significant splitting by proteolytic enzymes may occur between the moment an immunogen is administered and the moment it is being recognized at the biosynthetic site. Such proteolysis would have to result in the destruction of the conformation of most protein determinants. It seems, therefore, that if proteolytic destruction of the antigen plays a role in controlling antibody formation, it would have to occur *after* the determinant has been 'recognized' at the site of biosynthesis. Studies on the role of the optical configuration in determining the antigenicity of synthetic polymers are in agreement with this interpretation [22]. Namely, the destruction of the immunogenic macromolecule, or at least of its antigenic determinants, may be necessary to prevent the overwhelming of the immune system with excess antigen which may obstruct antibody synthesis. It should nevertheless be stressed that the *recognition* of the antigenic determinant occurs while the immunogenic molecule is still intact.

References

[1] M. Sela, B. Schechter, I. Schechter and F. Borek, Cold Spring Harbor Symp. Quant. Biol. 32 (1967) 537.
[2] E.A. Kabat, J. Immunol. 97 (1966) 1.
[3] R. Arnon, M. Sela, A. Yaron and H.A. Sober, Biochemistry 4 (1965) 948.
[4] H. Van Vunakis, J. Kaplan, H. Lehrer and L. Levine, Immunochemistry 3 (1966) 393.
[5] H.J. Sage, G.F. Deutsch, G.D. Fasman and L. Levine, Immunochemistry 1 (1964) 133.
[6] I. Schechter, B. Schechter and M. Sela, Biochim. Biophys. Acta 127 (1966) 438.
[7] B. Schechter, I. Schechter and M. Sela, J. Biol. Chem. 245 (1970) 1438.
[8] Bull. World Health Organ. 30 (1964) 447.
[9] J. Haimovich, I. Schechter and M. Sela, European J. Biochem. 7 (1969) 537.
[10] E.A. Kabat, J. Immunol. 84 (1960) 82.
[11] E.A. Kabat, Structural Concepts in Immunology and Immunochemistry, (Holt,

Rinehart and Winston, New York, 1968).
[12] J.W. Goodman, D.E. Nitecki and J.M. Stoltenberg, Biochemistry 7 (1968) 706.
[13] S.F. Schlossman and E.A. Kabat, J. Exptl. Med. 116 (1962) 535.
[14] J. Gelzer and E.A. Kabat, J. Exptl. Med. 119 (1964) 983.
[15] J. Gelzer and E.A. Kabat, Immunochemistry 1 (1964) 303.
[16] C. Moreno and E.A. Kabat, J. Exptl. Med. 129 (1969) 871.
[17] B. Schechter, I. Schechter and M. Sela, Israel J. Med. Sci. 5 (1969) 436.
[18] B. Schechter, I. Schechter and M. Sela, Immunochemistry, in press.
[19] R.K. Brown, R. Delaney, L. Levine and H. Van Vunakis, J. Biol. Chem. 234 (1959) 2043.
[20] M.H. Freedman and M. Sela, J. Biol. Chem. 241 (1966) 2383.
[21] M. Sela, Advan. Immunol. 5 (1966) 29.
[22] M. Sela, Science 166 (1969) 1365.
[23] J. Ramachandran, Abstr. 7th Internat. Congr. Biochem. Tokyo, 1967, J-144, p.982.
[24] J. Ramachandran, A. Berger and E. Katchalski, private communication.
[25] J. Ramachandran, private communication.
[26] J.B. Robbins, J. Haimovich and M. Sela, Immunochemistry 4 (1967) 11.
[27] F. Borek, J. Kurtz and M. Sela, Biochim. Biophys. Acta 188 (1969) 314.
[28] W. Traub and A. Yonath, J. Mol. Biol. 16 (1966) 404.
[29] J. Engel, J. Kurtz, E. Katchalski and A. Berger, J. Mol. Biol. 17 (1966) 255.
[30] O.W. McBride and W.F. Harrington, Biochemistry 6 (1967) 1484.
[31] A. Maoz, S. Fuchs and M. Sela, unpublished data.
[32] J. Josse and W.F. Harrington, J. Mol. Biol. 9 (1964) 269.
[33] R. Arnon and M. Sela, Proc. Natl. Acad. Sci. U.S. 62 (1969) 163.

MYELOMA PROTEINS WITH ANTIBODY-LIKE ACTIVITY IN MICE

M. POTTER

U.S. Dept. Health, Education, and Welfare, National Instituees of Health, National Cancer Institute, Bethesda, Md. 20014, U.S.A.

Abstract: M.Potter. Myeloma Proteins with Antibody-like Activity in Mice. *Miami Winter* Symposia 2, pp. 397–408. North-Holland Publishing Co., Amsterdam, 1970.

In a series of 3 consecutively transplanted plasmacytomas induced by mineral oils or 2,6,10,14-tetramethylpentadecane in the inbred BALB/c mouse, a heavy chain gene was expressed in 94 of the tumors. In 66 of the 94 the IgA chain was active indicating a marked trend to differentiate the IgA heavy chain genes. Among the 66 IgA myeloma 14 were found that reacted proteins with antigens in a manner characteristic of antigen-antibody interactions. Many of the antigens were polysaccharides which have been isolated from the normal intestinal flora of the BALB/c mouse gut. Different myeloma proteins often react with the same antigen. Serologic and chemical studies were presented to show that the IgA myeloma proteins that react with the same antigen are structurally different. The results thus far suggest that the precursors of the neoplastic plasma cells were active in forming antibody at the time when they underwent neoplastic transformation.

1. Introduction

Chemically homogeneous immunoglobulins (myeloma proteins) produced by some of the plasma cell tumors of mice precipitate or agglutinate with specific macromolecular antigens and bind haptens in solution [1–8]. These proteins behave like an individual molecular species of immunoglobulin within a population of antibody molecules. It has been known for some time that some human M-proteins, i.e. myeloma proteins, Waldenström macroglobulins or M-proteins produced in non-neoplastic conditions also are biologically active [9] (for further references see review by Metzger [10]). A particularly promising feature of the myeloma proteins in mice is that highly specific chemically defined haptens have been identified for many of the active proteins. These proteins along with their specific haptens will make it possible to examine the chemistry of the antigen-antibody interaction, in much the same way as enzyme-substrate interactions have been characterized. From a comparative study of a series of homogeneous immunoglobulins, it should become possible to correlate amino acid sequence variations with the folding of the immunoglobulin polypeptide chains and the formation of a three dimensional combining site. With this knowledge it may even be possible to make specific antibodies in the test tube.

Realization of these goals depends upon finding suitable active proteins. Thus far the number of proteins available is limited. The limitation is due in part to our lack of understanding of the biological basis of plasma cell development and differentation, and plasma cell tumor formation. One of the key questions is whether neoplastic plasma cells are derived from cells participating in immune responses.

Plasma cell tumors are induced in the inbred BALB/c strain of mice by the intraperitoneal injection of solid plastics [11], mineral oils [12] or chemically defined branched chain saturated hydrocarbons such as 2,6,10,14-tetramethylpentadecane or pristane [13].

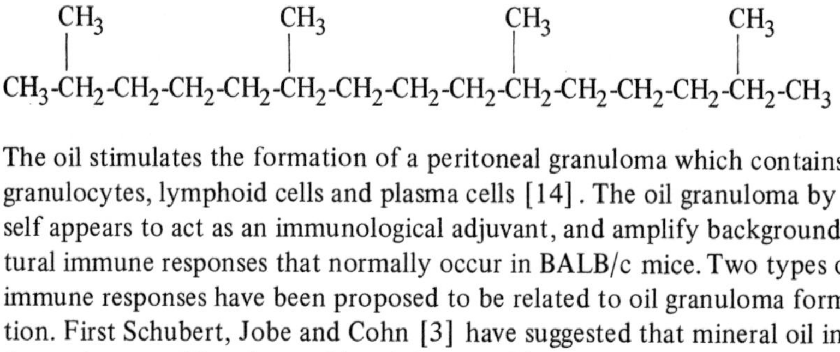

$$\begin{array}{c} \quad CH_3 \qquad\qquad\qquad CH_3 \qquad\qquad\qquad CH_3 \qquad\qquad\qquad CH_3 \\ CH_3\text{-}\overset{|}{C}H_2\text{-}CH_2\text{-}CH_2\text{-}CH_2\text{-}\overset{|}{C}H_2\text{-}CH_2\text{-}CH_2\text{-}CH_2\text{-}\overset{|}{C}H_2\text{-}CH_2\text{-}CH_2\text{-}CH_2\text{-}\overset{|}{C}H_2\text{-}CH_3 \end{array}$$

The oil stimulates the formation of a peritoneal granuloma which contains granulocytes, lymphoid cells and plasma cells [14]. The oil granuloma by itself appears to act as an immunological adjuvant, and amplify background natural immune responses that normally occur in BALB/c mice. Two types of immune responses have been proposed to be related to oil granuloma formation. First Schubert, Jobe and Cohn [3] have suggested that mineral oil induces tissue proliferation and breakdown and hence generates antigens (antigens of autogeneous origin) to which antibodies are formed locally. A second hypothesis is that the oil granuloma amplifies one of the common natural immune responses, the formation of antibodies to the microflora of the gastrointestinal tract. These are not mutually exclusive.

2. **Antigens of autogenous origin**

In Waldenström's macroglobulinemia, cold agglutinin syndrome, multiple myeloma in man, a number of biologically active proteins have been described that bind antigens of autogenous origin, e.g. γG-immunoglobulin, I, a red blood cell antigen-system, streptolysin-o, lipoprotein, etc. [10]. In the mouse the evidence for antigen of autogenous origin is indirect. A common type of activity found in mouse myeloma proteins is the ability to bind antigens containing nitrophenyl derivatives. These are artificial antigens made by derivatizing proteins with dinitrophenyl (DNP) or trinitrophenyl (TNP) groups. Two IgA proteins with relatively high affinity for DNP groups have been described, the MOPC 315 protein which has an intrinsic association constant K_A for ϵ-DNP-L-lysine of 1.6×10^7 moles^{-1} [2] and the MOPC 460 protein which has a K_A of 1.2×10^5 moles^{-1} [6]. Since the mice in which these tumors arose were not immunized with DNP containing antigens, why should these myeloma proteins be active against DNP? One suggestion is that the original normal antibody forming precursor cells were responding to antigen other than DNP, and their ability to bind nitrophenyl groups is a fortuitous cross-reaction [3].

For example, Schubert, Jobe and Cohn [3] screened a large series of myeloma proteins for their ability to precipitate DNP, TNP antigens and in addition a number of other derivatized proteins including 5-acetyl uracil (5-AU) and purine-6-oyl substituted proteins. They found 10–15% of their IgA anti-DNP, TNP proteins precipitated with 5-AU or purine-6-oyl antigens. We have found proteins with similar activity in our collection (fig. 1, table 1). Since the purine and pyrimidine ligands are chemically different from DNP they propose that the DNP activity could represent a cross reactivity to purine or pyrimidine containing antigens. Essentially the selecting antigen with which the antigen sensitive cell reacts is not DNP or TNP but is denatured nucleic acid derived from the oil granuloma reaction. They further support this notion by demonstrating that mouse mueloma proteins specifically bind DNA (Schubert, Roman and Cohn, 1970).

Eisen and colleagues have also found numerous cross reactions with the 315 anti-DNP protein [25], including 5-AU, caffeine, pyridoxal. The intrinsic association constants for these ligands are 100 to 1000 times lower than that for DNP. They also found, however, that MOPC 315 binds menadione 2-methyl 1, 4-naphthoquinone with a very high intrinsic association constant. A possible explanation offered by these workers is that the active site of 315 contains tryptophan which forms molecular charge transfer complexes with a number of chemically different ligands, e.g. TNP, menadione.

3. Antigens of microbial origin

The basic facts supporting the hypothesis that plasma cell tumors are de-

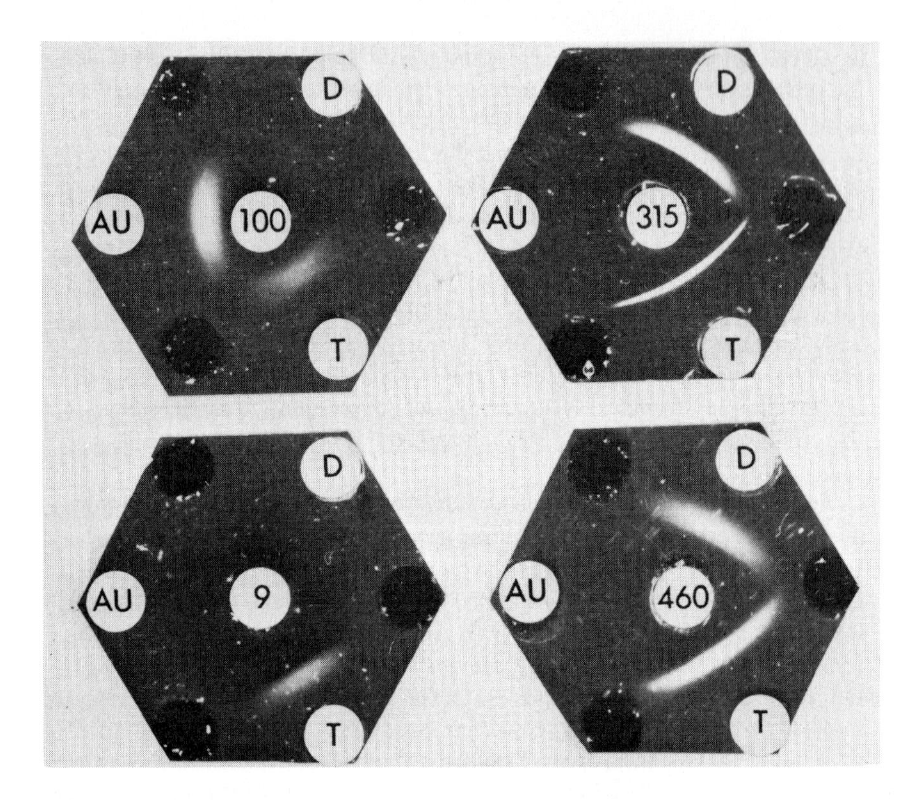

Fig. 1. Ouchterlony reactions in which sera from mice bearing the IgA myeloma protein producing plasmacytomas, TEPC 100, MOPC 315, MOPC 460 and HOPC 9 were reacted with D (= DNP-bovine serum albumin), T (=TNP_{28}-horse serum albumin) and AU (= 5-acetyl uracil, human serum albumin). Note the difference in intensity and sharpness of the precipitin lines. Sharp, intense lines appear to be associated with higher binding affinities. The 5-AU-HuSA apparently differs from the 5-AU-BSA prepared by Schubert et al. [3] by not being precipitated by MOPC 315. I thank Dr. Herman Eisen, Washington University, St. Louis for the D and T antigens and Dr. Sam Beiser, College of Physicians and Surgeons for the AU antigen.

rived from cells participating in immune responses to antigens of gut flora origin is now reviewed.

3.1. *Selective differentiation of IgA class immunoglobulins in neoplastic plasma cells*

In the mouse there are five major heavy chain class genes that code for common region polypeptides of the IgM, IgA, IgF (γl), IgG (γ2a) and IgH

Table 1
Classification of immunoglobulins produced by 111 consecutively transplanted BALB/c plasmacytomas induced by mineral oil.

Immunoglobulin	No.	No.	No. active proteins
IgM	0	0	0
IgA	66	66	14
IgF	10		
IgG	7	28	0
IgH	11		
Kappa only	6		
Lambda only	2	17	0
None	9		

(γ2b) heavy chains. Plasma cells making IgA class immunoglobulins are most frequently involved in neoplastic transformation following intraperitoneal mineral oil [1,11,15]. Among 111 consecutive successfully transplanted plasma cell tumors, we found 94 that synthesize and secrete complete immunoglobulin molecules, of these 66 (69%) were IgA (table 1).

Table 2
Binding activity of 14 myeloma proteins (table 1)

Protein	Antigen	Hapten specificity	Ref.
MOPC 460	DNP-BSA	Dinitrophenyl	[6]
MOPC 329,378,471 TEPC 17*	TNP-BSA	Trinitrophenyl	[2]
MOPC 384	*Proteus mirabilis* sp. 2 LPS	α-Methyl D-galactoside	[8]
MOPC 467	*Pasteurella pneumotropica* LPS	-	[8]
MOPC 406	*Salmonella weslaco* LPS	*N*-Acetyl D-mannosamine	[8]
MOPC 332	*Escherichia coli*? type LPS	-	[8]
MOPC 299,511 HOPC 8*,TEPC 15*	*Lactobacillus* antigen and *Pneumococcus* C polysaccharide	Phosphoryl choline	[5,26, 27]

LPS = lipopolysaccharide
MOPC (abbreviated M) = mineral oil plasma cell
HOPC (abbreviated H) = 7-*N*-hexyloctadecane induced plasma cell tumor
TEPC (abbreviated T) = 2,6,10,14-tetramethylpentadecane induced tumor
*TEPC and HOPC tumors induced by Dr. Paul Anderson, NCI.

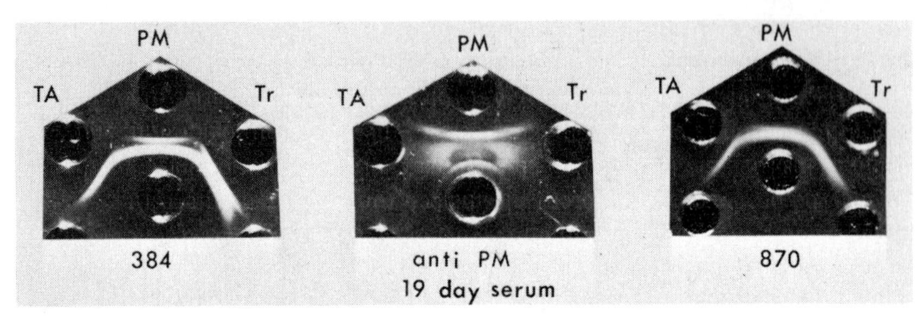

Fig. 2. Ouchterlony reactions in which sera from mice bearing the plasmacytomas, MOPC 384 and MOPC 870 were reacted with lipopolysaccharides (isolated by the phenol method of Westphal) from TA (=*Salmonella tel aviv*), PM (= *Proteus mirabilis* species 2, isolated from BALB/c mice in our colony) and Tr (= *Salmonella tranoroa*). Note similar reactions for the 3 antigens. *Salmonella tel aviv* and *tranoroa* are in different *O*-specificity serogroups. The double lines may be due to variations in molecular weight of antigen molecules. In the center reaction a 19 day serum from BALB/c mice immunized with formalinized *Proteus mirabilis* species 2 organisms was reacted with the same 3 antigens. Note this antiserum was specific for the *Proteus mirabilis* lipopolysaccharide. The 384 and 870 myeloma proteins thus have extensive cross reactions.

Thus far most active proteins are IgA type immunoglobulins (tables 1 and 2). By contrast in four series of multiple myelomas in man [16–19] including 794 cases, a complete immunoglobulin molecule was produced in 666 cases and of these only 30% were of the IgA class.

In the normal animal, IgA forming cells are found predominantly in connective tissues lining the gastrointestinal and respiratory tracts [20]. Although the complete sequence of immune events is not known, it is thought that IgA producing cells make antibody that reacts with antigens which enter the organism through the respiratory and gastrointestinal tracts [21,22]. The localization of normal plasma cells in the connective tissues of the intestine may result from migration of circulating IgA producing cells into these tissues or by direct development of the IgA differentiation in cells stimulated by these antigens.

Mineral oil induced plasma cell tumors do not arise, however, in the lamina propria of the gut but rather in the peritoneal oil granuloma itself [14], which is separated from the lamina propria by the muscularis layers. To link the IgA producing plasmacytomas to the normal lamina propria IgA producing plasma cell, migration of antigen selected cells into the oil granuloma must be postulated. An alternative explanation is that the oil granuloma specifically triggers differentiation of the IgA genes.

3.2. *Low incidence of induced plasma cell tumors in germfree mice.*

The largest potential source of natural antigens in vertebrates are the normal microflora of the gastrointestinal tract. Germfree mice lack bacteria and are also deficient in antibody producing cells in the lamina propria of the gastrointestinal tract. McIntire and Princler [15] attempted to induce plasma cell tumors by intraperitoneal injection of mineral oil in germfree BALB/c mice and found a very low incidence of plasma cell tumors. They also successfully immunized the mice during the induction period to several protein antigens. Thus, germfree mice which were immunologically competent but lacked the continuous immunologic stimulus provided by the bacteria and microorganisms in the gastrointestinal tract, were much less susceptible to plasma cell tumor formation. This finding suggests the continued stimulation by the gastrointestinal microbial antigen is an important factor in plasma cell tumor formation.

3.3. *Some of the antigens identified by active myeloma proteins are produced by bacteria residing normally in the BALB/c intestinal tract.*

Active myeloma proteins have been detected by a screening procedure in which myeloma proteins are reacted with antigens which were available from other studies for example, hapten substituted proteins and the polysaccharides, lipopolysaccharides, teichoic acids, used in the serologic classification of bacteria. These were generously supplied by individuals who had prepared them for other purposes. For example, Dr. Otto Lüderitz of the Max-Planck-Institut gave us 40 samples of *Salmonella* lipopolysaccharides. When we screened myeloma proteins for their ability to precipitate with these polysaccharides, we found proteins that specifically identified the *Salmonella* lipopolysaccharides including such seemingly exotic species as *S. tel aviv, S. adelaide, S. weslaco,* etc. At first these results seemed to be irrelevant, but when we began isolating lipopolysaccharides from gram negative enterobacterial species residing in the gastrointestinal tracts of BALB/c mice in our colony, we found that the 2 different IgA myeloma proteins which bound the *S. tel aviv* lipopolysaccharide, for example, also precipitated with a lipopolysaccharide produced by a species of *Proteus mirabilis* (#2). Further two other proteins that precipitated the lipopolysaccharide of *S. adelaide* also precipitated equally well a lipopolysaccharide from *Pasteurella pneumotropica* [8]. This organism which was also isolated from the gastrointestinal tracts of BALB/c mice is a common cause of chronic respiratory disease in mice.

A third example of a relevant antigen will be discussed in more detail. In 1967 Dr. Melvin Cohn of the Salk Institute [1,4] discovered two mouse IgA myeloma proteins, S63 and S107, that precipitated with the *Pneumococcal*

INHIBITION OF PRECIPITATION

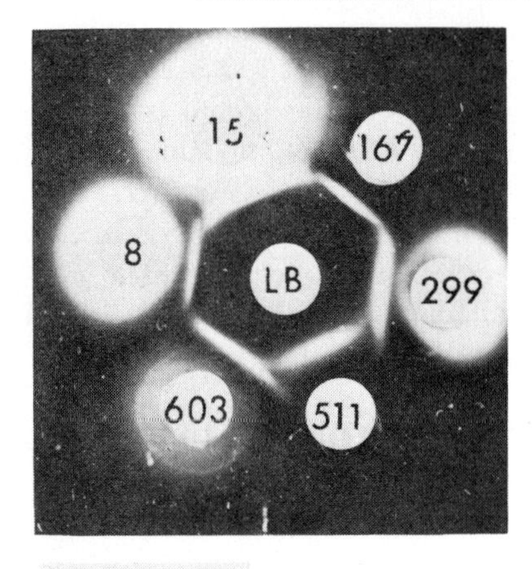

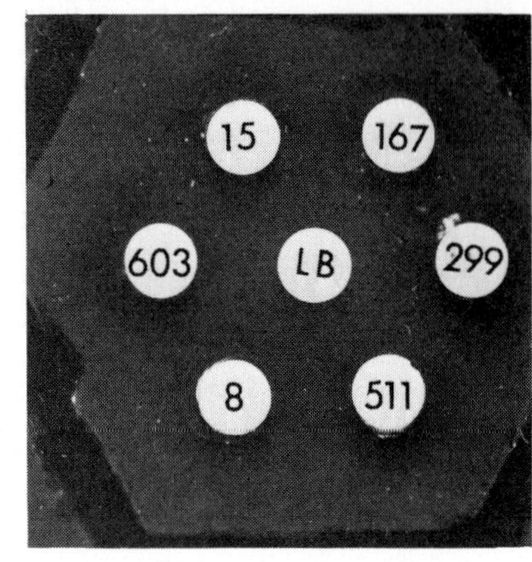

Fig. 3. *Left*: Precipitin reactions with 6 different IgA myeloma proteins with LB (*Lactobacillus* antigen). Sera from mice bearing tumors were used, although purified proteins give similar reactions; 8 (= HOPC 8), 15 (= TEPC 15), 167 (= MOPC 167), 299 (= MOPC 299), 511 (=MOPC 511) and 603 (= McPC 603). Note the spurring of the 511 reaction with the others. The prominent halos around 8, 15 and 299 are frequently observed and are as yet unexplained. The LB antigen was isolated from Na-deoxycholate treated organisms by precipitation in 80% ethyl alcohol. *Right*: The reaction on the right contained the same materials except the wells were prefilled 10^{-3} M phosphoryl choline. The hexagon was isolated by a trench that prevents diffusion of the phosphoryl choline. Note all precipitations were inhibited.

C polysaccharide. Subsequently we have found 6 more IgA myeloma proteins, HOPC8, TEPC 15, MOPC 167, MOPC 299, MOPC 511 and McPC 603 that precipitated this polysaccharide. Four of these occurred in the 111 consecutive cases described in table 1. We have recently found an antigen in *Lactobacilli* [23] which is also precipitated by all of the proteins that precipitate the *Pneumococcus* C polysaccharide. It seems significant that *Lactobacilli* are among the most common residents of the gastrointestinal tracts of mice [24].

The *Lactobacillus* antigen was isolated from two species of *Lactobacillus acidophilus* cultured from BALB/c mice in our colony [23] by Mr. Charles K. Mills of the American Type Culture Collection, by extraction of pressure disrupted cells with trichloroacetic acid. The antigen is intracellular and not on the bacterial surface [24].

These findings provide direct evidence that antigens produced by the normal bacterial flora, are specifically identified by some myeloma proteins.

3.4. *More than one IgA myeloma protein binds the same relevant antigen.*

When an antigen is found which one IgA myeloma binds, it is usually possible to find proteins produced by other plasma cell tumors which bind that antigen. Two or more proteins have been found thus far that react with dinitrophenyl [1,3,6] or 5-acetyl-uracil substituted proteins [3,25], *Proteus mirabilis* sp. 2 antigen, or the *Pasteurella pneumotropica* antigen [8]. The largest series found thus far which contains 8 proteins is the anti-*Lactobacillus* or *Pneumococcus* C polysaccharide group. Dr. Myron Leon of St. Luke's Hospital, Cleveland, Ohio discovered that the immunodominant group on the *Pneumococcus* C polysaccharide recognized by the IgA myeloma proteins, S63, S107, T15, M167, M299 and M603 was phosphoryl choline by demonstrating that low concentrations of phosphoryl choline would inhibit precipitation of the myeloma proteins with *Pneumococcus* C polysaccharide [27]. We have also been able to inhibit the precipitation of the *Lactobacillus* antigen with phosphorylcholine (fig. 1). Thus the myeloma proteins produced by 8 independently induced plasma cell tumors react with the same antigenic determinant. This appears to represent a remarkable frequency of proteins with a similar type of activity.

It would seem reasonable that 'natural' immune responses would be similar in mice of the same inbred strain and colony as these mice may often carry very similar microflora during their lives. This supposition provides an explanation for why more than one tumor produces a protein that identifies the same antigen.

3.5. *Similarities and differences in the structure of IgA proteins that bind the same antigen.*

Five of the 6 IgA proteins that bind phosphoryl choline containing antigens H8, T15, M167, M299 and M603 have kappa light chains and one, M511, has a lambda light chain.

The kappa chains of H8, T15, M167 and M603 were compared by the tryptic peptide map technique and partial amino acid sequence analysis. This work was done in collaboration with Dr. Leroy Hood and Mr. David McKean [28]. The *N*-terminal amino acid sequences were determined on isolated light chains in a Spinco Amino Acid Sequencer. The PTH-residues were analyzed by gas chromatography and confirmed (with the exception of serine and threonine) by high voltage paper electrophoresis. The sequences are shown in table 3. While the 4 chains have similarities, e.g. by all beginning with the sequences Asp-Ile-Val differences were found within the first 15 residues in the comparison of T15, M167 and M603. Thus far H8 resembles T15.

Table 3
Amino terminal sequences of kappa chains isolated from IgA myeloma proteins that bind antigens containing phosphoryl choline

H8	Asp-Ile-Val-Met-Thr-Gln-Ser -Pro-Thr-Phe-Leu-Ala-Val-Thr-Ala
T15	Asp-Ile-Val-Met-Thr-Gln-Ser -Pro-Thr-Phe-Leu-Ala-Val-Thr-Ala
M167	Asp-Ile-Val-Ile -Thr-Gln-Asx-Glu-Leu-Ser-Asp-Pro-Val-Thr-Ser
M603	Asp-Ile-Val-Met-Thr-Glx-Ser -Pro-Ser -Ser-Leu-Ser-Val-Ser -Ala

Previously, we [5] reported multiple tryptic peptide differences in M167 and M603, indicating that these chains vary in multiple regions in the variable sequence.

Another sensitive method for detecting similarities and differences among myeloma proteins is by the use of homologous myeloma specific antisera [28]. We have previously described antigenic differences in pairs of proteins that identify the same antigen [8]. These findings have been extended to include the large group of proteins that bind phosphoryl choline containing antigens. In collaboration with Miss Rose Lieberman of the NIAID [27] homologous antisera were prepared to the BALB/c H8, T15, M167, M299, M511 and M603 myeloma proteins in strain A/He or Al mice. Antisera prepared to M167, M511 and M603 were specific and precipitated only the immunizing myeloma protein and no other IgA proteins in our collection. A remarkable result was the inability to prepare a specific antiserum for H8, T15 or M299. Antisera prepared to anyone of these three proteins precipitated the other two proteins but not M167, M511 or M603. In addition, anti-H8, anti-T15 and anti-M299 precipitated S63 and S107 myeloma proteins kindly sent to us by Dr. Melvin Cohn of the Salk Institute. The anti-H8, T15 and M299 antisera did not precipitate 60 other IgA myeloma proteins in our collection. This result suggests that H8, T15, M299 and the S63 and S107 proteins may be nearly identical or identical proteins. Cohn et al. [4] previously reported that S63 and S107 had the same individual specificity.

The tumors producing these proteins were induced at different times and in different laboratories and are not contaminants. The MOPC2 99 tumor, the oldest in our collection, was induced in 1966. This tumor has stopped producing large quantities of myeloma protein. The HOPC 8 tumor was induced with 7 *N*-hexyloctadecane and the TEPC 15 tumor was induced with 2,6,10, 14-tetramethylpentadecane [13] by Dr. Paul Anderson of the NCI. We do not carry S63 or S107 in our laboratory.

The chemical and serological studies show that when more than one protein is found which binds the same antigen, that structural differences in the proteins are often demonstrable. In conventionally induced immune responses, antibodies of differing specificity and affinity appear in response to the same antigenic determinant. Although the proteins produced by plasma cell tumors induced in different mice of the same inbred strain are being compared artificially here, there is a general correspondence to what might be expected of an immune response in a single individual to a single antigenic determinant. Differences in affinity and specificity of myeloma proteins that bind the same antigen have been identified [2,6,7,8,26].

An additional point of considerable interest is the possibility that some proteins produced that react with the same antigen may be structurally alike. If this evidence can be substantiated by chemical data, then the first evidence of strict control of antibody variability will have been established.

3.6. *Discussion*

The evidence has been summarized here to support the hypothesis that the neoplastic plasma cells that arise following intraperitoneal injection of mineral oils are derived from cells participating in immune responses. That is, the normal precursor cells were fully differentiated to make specific immunoglobulin and had further interacted with antigen causing maturation to active immunoglobulin synthesis and secretion. The nature of the natural immune response favored here was the normal response to antigens that challenge the organism from the gastrointestinal tract. The proposal of Schubert, Jobe and Cohn [3] that an autoimmune response generated by the oil itself was not excluded, and requires further consideration. The particular facts that were cited to support the gut-flora concept were: 1) The preferential development of IgA immunoglobulins – a class of immunoglobulins that is normally associated anatomically with the gastrointestinal tract; 2) The low incidence of induced plasma cell tumors in germfree mice which lack normal gut-flora; 3) The isolation of antigens from common gastrointestinal organisms with which some myeloma proteins specifically identify and 4) The independent development of more than one myeloma protein that identifies the same relevant antigen (the large anti-phosphoryl choline series was the most striking example).

The finding that closely related proteins or even the same myeloma protein may be produced by different plasma cell tumors was an interesting finding. This suggests further that the neoplastic transformation has a predilection to develop in cells that make specific types of immunoglobulin, and that antigen itself may play a prominent role in the complex series of events in plasma cell tumor development.

Acknowledgements

The author wishes to gratefully acknowledge the very able technical assistance of Miss Elizabeth Bridges and Mr. Alvado Campbell in various phases of this work. I also thank Dr. Leroy Hood, Mr. David McKean and Miss Rose Lieberman for their permission to quote their data sited in the text before publication.

References

[1] M. Cohn, Cold Spring Harbor Symp. Quant. Biol. 32 (1967) 211.
[2] H.N. Eisen, E.S. Simms and M. Potter, Biochemistry 7 (1968) 4126.
[3] D. Schubert, A. Jobe and M. Cohn, Nature 220 (1968) 882.
[4] M. Cohn, G. Notani and S.J. Rice, Immunochemistry 6 (1969) 111.
[5] M. Potter and M.A. Leon, Science 162 (1968) 369.
[6] B.M. Jaffe, H.N. Eisen, E.S. Simms and M. Potter, J. Immunol. 103 (1969) 872.
[7] H. Metzger and M. Potter, Science 162 (1968) 1398.
[8] M. Potter, Federation Proc. (1970) in press.
[9] J. Waldenström, S. Winblad, J. Hällén and S. Liungman, Acta Med. Scand. 176 (1964) 619.
[10] H. Metzger, Am.J. Med. (1970) in press.
[11] R.M. Merwin and L.W. Redmon, J. Natl. Cancer Inst. 31 (1963) 997.
[12] M. Potter and C.R. Boyce, Nature 193 (1962) 1086.
[13] P.N. Anderson and M. Potter, Nature 222 (1969) 994.
[14] M. Potter and R.C. McCardle, J. Natl. Cancer Inst. 33 (1964) 497.
[15] K.R. McIntire and G.L. Princler, Immunology 17 (1969) 481.
[16] R. Bachmann, Scand. J. Clin. Lab. Invest. 18 (1966) 273.
[17] J.R. Hobbs, Brit. Med. J. 3 (1967) 699.
[18] J. Hällén, Acta Med. Scand. Suppl. 426 (1966).
[19] E.F. Osserman, R.A. Rifkind, K. Takatsuki and D.P. Lawlor, Ann. N.Y. Acad. Sci. 113 (1964) 627.
[20] P.A. Crabbé, A.O. Carbonara and J.F. Heremans, Lab. Invest. 14 (1965) 235.
[21] P.A. Crabbé, D.R. Nash, H. Bazin, H. Eyssen and J.F. Heremans, J. Exptl. Med. (1969) 723.
[22] G.W. Kriebel, Jr., S.C. Kraft and R.M. Rothberg, J. Immunol. 103 (1969) 1268.
[23] C.K. Mills and M. Potter, unpublished observations.
[24] R.W. Schaedler and R.J. Dubos, J. Exptl. Med. 115 (1962) 1149.
[25] H.N. Eisen, B. Underdown, M. Michaelides and E.S. Simms, Federation Proc. 29 (1970) 437.
[26] M.A. Leon and N.M. Young, Federation Proc. 29 (1970) 437.
[27] M.Potter, R. Lieberman, L. Hood and D. McKean, Federation Proc. 29 (1970).
[28] M. Potter and R. Lieberman, Cold Spring Harbor Symp. Quant. Biol. 32 (1967) 187.

NATURAL ANTIBODIES IN PRIMITIVE VERTEBRATES. THE SHARKS

M.M. SIGEL, E.W. VOSS, Jr., S. RUDIKOFF,
W. LICHTER and J.A. JENSEN
University of Miami School of Medicine, Miami, Florida, USA,
University of Illinois, Urbana, Ill., USA
and Lerner Marine Laboratory, Bimini, Bahamas

Abstract: Sigel, M.M., Voss, E.W., Jr., Rudikoff, S., Lichter, W., and Jensen, J.A. Natural Antibodies in Primitive Vertebrates. The Sharks. *Miami Winter Symposia* 2, pp. 409–428. North-Holland Publishing Company, Amsterdam, 1970.

The serum of unimmunized nurse sharks reacts with a great variety of antigens. It neutralizes influenza and Rous sarcoma viruses, kills gram-negative and gram-positive bacteria, agglutinates and lyses erythrocytes of many species, is growth inhibitory and cytotoxic for cells and precipitates DNP. The lytic and neutralizing reactions are complement-dependent and require fixation of shark C′1. Guinea-pig complement components C′4, 2, 3, etc. can be substituted for shark complement only after the fixation of shark C′1. The "antibody" contained in the IgM fraction is present in low amounts at birth but increases fairly rapidly in the young shark. The natural "antibody" differs from antibody engendered by artificial immunization in that it cannot be adsorbed from the whole serum on to a solid DNP absorbent. Interestingly enough, after precipitation with DNP the "antibody" can be adsorbed. Furthermore the amount of the natural "antibody" does not increase with continued immunization. Most of the "antibody" activity has been associated so far with 19 S macroglobulin. Equilibrium dialysis gave an average intrinsic association constant, K_0, of approximately 10^4 l/mole.

1. Introduction

Studies in this laboratory [1–4] and in others [5,6] have revealed that sharks respond with humoral antibody formation to a variety of antigens including a hapten [7] and that these antibodies are present in two forms of IgM. Fig. 1 depicts the profile of shark anti-BSA serum chromatographed on Sephadex G-200. The antibodies measured by passive hemagglutination of BSA-coated red cells are contained both in the 19 S and 7 S fractions. The

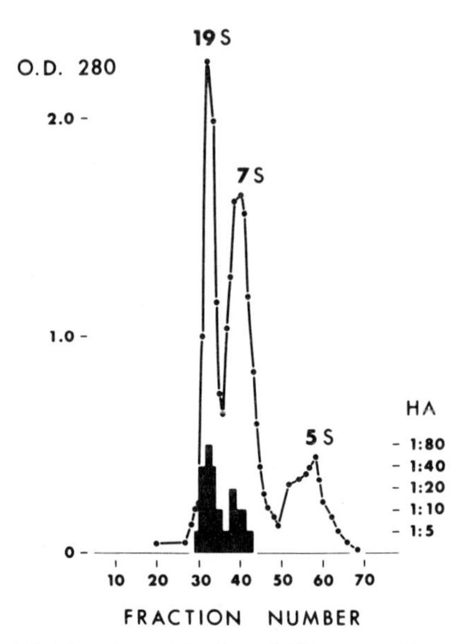

Fig. 1. Sephadex G-200 filtration of shark anti-BSA serum. Shaded area represents antibody.

5 S component represents transferrin [8]. The 7 S protein has many properties of the 19 S protein and is considered to be a monomeric form of IgM.

This paper relates our findings on naturally occurring antibodies of the nurse shark. The serum of normal unimmunized nurse sharks (*Ginglymostoma cirratum*) reacts with an unusually large and diverse array of antigens. It was both surprising and intriguing to find that these ocean inhabiting animals possessed such a wide range of activity against antigens which are not known to be present in their environment. The possibility that the shark's immune mechanism is degenerate is not very likely because immunization with known antigens usually leads to specific responses. It is, of course, possible that in addition to having acquired many components of adaptive immunity which came to characterize the higher animals the shark has retained some of the elements of the primitive mechanisms of invertebrates. Thus, the wide spectrum of reactivity of shark serum could result from the presence of some such primitive substance. While this possibility has not been discounted the data in the present paper reveal that at least some of the natural reactivity resides in true immunoglobulin (IgM) and that the amount of reactive substances increases with age. There is a third possibility, namely, that the shark immune recognition system may react to an antigenic determinant which is shared by a large variety of antigens and that the polyspecific reactivity of the shark serum is merely a reflection of such a recognition. This hypothetical common antigenic determinant may have escaped previous detection because the immune mechanisms of conventionally used animals may fail to recognize it.

As important as these considerations are they will not be developed in this paper because of lack of knowledge. This presentation is addressing itself exclusively to the reactions observed and to the findings bearing on the nature of the shark serum protein associated with them. The nurse shark is a surviving member of an ancient suborder and its natural antibody and immunoglobulins are therefore of vital importance in the over-all studies of immune response, especially the development, function and structure of IgM immunoglobulin which phylogenetically and ontogenetically is regarded as a primordial protein.

2. Materials and methods

In the initial study both lemon and nurse sharks were used. The subsequent investigations were restricted to the nurse shark.

The methods of holding, anesthetizing and bleeding of sharks were previously described [2]. "Adult" refers to sharks weighing 2½ to 25 kg; "immature" refers to sharks weighing less than ½ kg.

In view of the fact that freezing of serum destroys complement, the various tests were performed within a few days after serum collection and the sera were kept at 4°.

2.1. *Hemolytic and hemagglutination assays.*

The preparation of diluents, nurse shark sera, erythrocyte suspensions and nurse shark antibody sensitized sheep erythrocytes has been described previously [9]. Measurements of hemolytic and hemagglutinating activity of shark sera were done by microtitration using cell suspensions of 1×10^8 cells/ml. The reaction mixtures per microtiter well contained four or five 0.025 ml volumes. One volume was the sample dilution, one volume the cell suspension, leaving two or three volumes for other reagents or diluent. The 50% lytic endpoint was estimated after centrifugation of the "V" type microtiter plates, the 2+ agglutination endpoint was read after the "U" type microtiter plates were left undisturbed at room temperature until the cells had completely settled–reading was facilitated by tilting the plates and observing the behaviour of the cell button.

2.2. *Measurement of agglutinating antibodies*

One volume serum dilutions in 0.02 M EDTA saline were incubated with 2 volumes of EDTA saline and 1 volume of erythrocytes in gelatine-veronal-buffer (GVB++) for 10 min at 30° on a microplate vibrator. The plates were then left to settle at room temperature until readable patterns developed (3–5 hr).

2.3. *Measurements of hemolytic activity*

2.3.1. Lysis by whole shark serum (combined activity of antibody and complement)

One volume of serum dilutions in GVB++ was incubated with 3 volumes of GVB++ and 1 volume of erythrocyte suspensions (in GVB++) for 60 min at 30° on a microplate vibrator. The plates were then centrifuged and read.

2.3.2. Lysis by shark complement

Same as above except that the cell suspension consisted of sheep erythrocytes that were optimally sensitized with natural shark antibody.

2.4. *Inhibition of growth of cells*

One-day stationary cultures of KB or primary human amnion cells containing 1 ml of medium (Eagle's minimal essential (EME) with 10% calf serum) were used. To these cultures were added 0.1 ml amounts of nurse shark serum. The cultures were then incubated for an additional 3 days at 37° and the extent of cell growth was measured by the method of Oyama and Eagle [10] comparing the amount of protein at this time with the base line protein in the one-day cultures. The endpoint (ED_{50}) was the highest dilution of serum causing 50% inhibition of growth.

2.5. *Cytotoxicity test*

The cytotoxic effect of shark serum for KB cells was determined by viable count using trypan blue. The use of trypan blue is based on the ability of viable cells to exclude the dye and therefore cells with dye (stained blue) are considered dead. One ml of tissue culture medium containing 400,000 cell/ml was mixed with 0.5 ml of shark serum in various dilutions. The mixture was allowed to incubate at 36° for 15 min. After incubation 0.05 ml of 0.2% trypan blue was added to 0.5 ml of the mixture and after approximately 5 min total and viable cell counts were made on a hemocytometer. The percentage of unstained cells represented the viable population. At the same time the remaining 1 ml of the mixture was diluted 1:10 and a set of tissue cultures was initiated in stationary tubes which were then incubated for 3 days at 37° as above.

2.6. *Killing of E. coli B*

Nutrient agar slants were inoculated with *E. coli* B. After incubation at 37° for 18-20 hr, the growth was suspended in a small amount of nutrient broth and then diluted to about 1×10^7 organisms/ml. To 1.8 ml aliquots of the bacterial suspension was added 0.2 ml of 2-fold dilutions of shark serum. The mixtures were held at 37° for 30 min and then immediately diluted for plating to determine the percent reduction in living organisms.

2.7. *Neutralization of influenza virus*

Tenfold dilutions of PR8 influenza virus were mixed with equal amounts of nurse shark serum (in 1 or more dilutions), incubated at room temperature for 30 min and injected intra-allantoically into 10-day embryonated eggs. The same dilutions of virus, not treated with shark serum, were inoculated to establish the virus infectivity titer. The allantoic fluids were harvested 48 hr later and presence of virus was determined by the hemagglutination reaction. The neutralization index represents the reduction of infectivity by the serum, (virus infectivity titer in saline/virus infectivity titer in serum).

2.8. *Neutralization of Rous sarcoma virus (RSV)*

RSV in a concentration of approximately 30 tumor-producing doses was mixed with equal amounts of adult nurse shark serum in several dilutions. The mixtures were allowed to stand at room temperature for 30 min and injected into the wing web tissue of 3 to 10 day old chicks. The birds were observed for tumor formation for a period of 2 wks.

2.9. *Mixed agglutination*

Chicken or pigeon RBC (1×10^8) were sensitized with varying dilutions of normal serum or purified immunoglobulin. After incubation for 60 min at 30° and overnight in the cold, the cells were washed 3 times in GVB++ and resuspended to 1×10^8. The sensitized cells were then mixed with an equal volume of unsensitized RBC from a variety of species, also standarized to 1×10^8. The mixtures were incubated for 30 min at 30° and 30 min at 0°. A sample was removed and examined microscopically for mixed agglutination.

2.10 *Immunoelectrophoresis*

Immunoelectrophoresis [11] was performed on microscope slides using 1% agarose (Mann Research, N.Y., N.Y.) in barbital buffer ($m\mu = 0.04$, pH 8.6). Approximately 2.5 V/cm was applied to slides for 45–60 min at room temperature.

2.11. *Centrifugation in sucrose gradient*

Centrifuge studies were carried out in 10–40% sucrose gradients. Gradients were centrifuged for 17 hr at 35,000 rpm in a Spinco Model L centrifuge equipped with an SW39 rotor. Tubes were punctured from the bottom and 3 drop fractions collected.

2.12. *Polyacrylamide gel disc electrophoresis*

Disc gel electrophoresis was performed at pH 9.5 according to the method of Ornstein [12] and Davis [13]. One hundred to two hundred mg of protein was run in each tube at 4 mA/tube for two hr at room temperature.

2.13. *Preparation of DNP-bovine serum albumin (BSA)*

The method of Eisen [14] was used for preparation of dinitrophenylated antigen. BSA (Pentex, Inc., Kankakee, Ill.) was dissolved in water (10 mg/ml) and an equal weight of K_2CO_3 and 2,4-dinitrobenzene-sulfonic acid, sodium salt (Eastman Organic Chemicals), was added. The reactions were allowed to proceed for 24 hr at 37° in the dark. The mixture was passed through Dowex 1 X 8 (J.T. Baker Chemical Co.) equilibrated in water to remove dinitrophenolate and dinitrobenzenesulfonate. The amount of protein was determined by dry weight of the product and the number of DNP molecules by absorbance at 360 nm.

2.14. *Immunoadsorption*

DNP immunoadsorbent was prepared by the method of Robbins et al. [15] as modified by Gallagher and Voss [16] using DNP-human serum albumin (HSA) (23 moles per mole of protein) which was linked covalently to bromoacetyl cellulose. Shark serum was mixed with the immunoadsorbent and the antibody was allowed to adsorb for 1 hr at room temperature. Elution was accomplished by means of 0.1 M DNP-glycine in 0.05 M phosphate buffer pH 8.0. The antibody was separated from the DNP on a Dowex 1 X 8 column (no DEAE) equilibrated with the same buffer.

2.15. *Purification of natural antibody*

Equal amounts (usually 10 ml) of normal shark serum and DNP-BSA (4 mg/ml in 0.05 M phosphate buffer, pH 8) were mixed and first incubated for 60 min at 37° and then overnight at 4°. The precipitate was centrifuged and washed thrice in cold 0.05 M phosphate buffer. The precipitate was dissolved in 0.1 M 2,4-DNP, pH 8. The antibody was separated from the antigen by column chromatography. For this purpose we used a column composed of two layers; the bottom layer (5 X 2 cm) consisted of Dowex 1 X 8 in an

0.1 M phosphate buffer, pH 8; the top layer (10 × 2 cm) consisted of DEAE cellulose equilibrated in the same buffer. The protein eluted with the same buffer was monitored at 280 nm.

2.16. *Equilibrium dialysis*

Two chambered containers separated by dialysis membrane were used. To one was added the antibody and to the other ^{3}H-DNP–L-lysine (ligand). The chambers were equilibrated for 24 hr at 5°. Following equilibration 25 μl samples from each chamber were placed in 8 ml of Bray's scintillation fluid. Samples were counted either in a Nuclear Chicago or a Beckman LS-100 liquid scintillation spectrometer.

3. **Results and discussion**

3.1. *Reactions with red blood cells*

The sera of unimmunized sharks were found to agglutinate sheep red blood cells. They were also able to agglutinate human and chicken erythrocytes. Unheated shark serum produced lysis of these cells. Spurred by these findings

Table 1
Hemagglutination and hemolysis by sera from adult and immature nurse sharks.

Source of RBC	Hemagglutination		Hemolysis	
	Adult	Immature	Adult	Immature
Dog	256	16	64	72
Human (Group O, Rh$^+$)	1000	2000	64	24
Hamster	32	2	32	72
Rat	750	32	48	48
Guinea-pig	64	32	128	48
Mouse	512	256	n.d.	n.d.
Chicken	2000	16	32	96
Sheep	64	2	128	12
Cameroon goat	64	2	128	12
Nubian goat	8	Neg	128	9
Abyssinian ass	512	4	64	24
Llama	32	Neg	32	6
Tapir	750	2	32	6
Patas monkey	512	256	256	48
Rabbit	2000	2000	756	24
Pig	2000	6	128	96

we proceeded to test erythrocytes from 13 additional species. Table 1 shows that normal shark serum agglutinated cells of all species tested but the potency of the reaction differed. There was no obvious relation to the presence of Forssman antigen as the Forssman negative rabbit red cells were agglutinated to a higher titer than were Forssman positive red cells of the guinea-pig and the sheep. Human red cells (not only group O but group A or B) and pig cells invariably reacted to higher titers than did sheep cells. The sheep cells titer in table 1 is 64 but titers as high as 512 were obtained in some shark sera. With four exceptions, serum of the immature shark had low agglutinating activity, considerably lower than the serum of the adult sharks. The exceptions were human, mouse, monkey and rabbit erythrocytes. At present we cannot explain the differential behaviour of serum from immature sharks. This question is being pursued further but as detailed below the immature shark serum was also deficient in its reaction with bacterial and viral antigens.

The natural hemagglutinin was isolated by means of DEAE chromatography and gel filtration. It had the properties of 19 S IgM. Fig. 2 shows that the hemagglutinin peak was located in the rapidly sedimenting fraction corresponding to 19 S. More precise studies utilizing the ultracentrifuge will be published separately.

The observation that adult sera contained greater activity than the sera of immature sharks suggested that perhaps the activity was due to acquisition of antibodies from exposure to antigens in the environment, including food. To

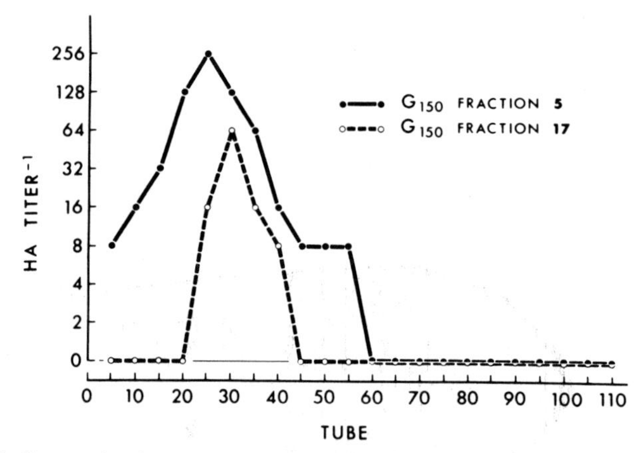

Fig. 2. Sedimentation in sucrose gradient of fractions from first and second peaks obtained from Sephadex G-150. Both fractions contained rapidly sedimenting hemagglutinin, the major part of which is in fraction 5. The activity in fraction 17 presumably represents trailing of the first peak.

test this possibility a small experiment was performed in which some baby sharks were maintained on a diet of lamb while others were kept on a standard diet of fish food. There were no differences in the anti-sheep cell activities of the respective sera.

Hemolytic activity effected by shark complement will be a subject of a separate report. For the purpose of this presentation, suffice it to say that in some instances the lytic activity was higher than the agglutinating activity whereas in others the reverse was true. This presumably reflects a difference in fragility of the cells. It must be kept in mind that in the lytic reaction, the amount of complement is a limiting factor and in no instance were lytic titers greater than 640–960 observed. When sheep RBC were first sensitized by heated shark serum (heated to destroy C'1) and subsequently mixed with the complete shark complement system the lytic titer was doubled. Lysis of red cells by shark serum depended on the presence of natural antibodies and an intact complement system. Guinea-pig complement components could lyse red cells sensitized by shark antibody only if the first complement component (C'1) was of shark origin. In other words, shark anti-erythrocyte antibody was incompatible with guinea-pig C'1, but cell-antibody-shark-C'1 complexes could activate the remaining guinea-pig complement components.

3.2. *The effect of shark serum on cells in tissue culture*

Normal nurse shark serum exerted a growth inhibitory effect in cultures of many kinds of cells. Human malignant and normal cells, hamster cells transformed by adenovirus, mouse L cells and fish cells were susceptible to the ac-

Table 2

Effect of shark serum on KB and human primary amnion cell cultures.

Sample	KB cells	ED_{50}
Mature shark serum		460*
Immature shark serum		180
△ Mature shark serum		<50
△ Mature shark serum supplemented with native immature shark serum (1:300)**		2400
	Amnion cells	
Mature shark serum		690
△ Mature shark serum		<50

* Reciprocal of final dilution.

** Immature shark serum 1:300 had no inhibitory effect by itself.

tion of the serum and there was no specific selection. Table 2 shows some typical results obtained with KB (human cancer) cells and human primary amnion cells. Once again it can be seen that adult serum had a significantly higher activity than did serum from immature sharks. This reaction was also complement dependent and could be abolished by heating of serum at 56° for 30 min. Even though immature shark sera contained lesser concentrations of cell sensitizing activity they contained complement at levels on a par with those encountered in adult sera. Table 2 also shows that complement from baby shark sera could restore the activity of heated adult serum. It should also be noted that in the restoration experiment the inhibitory titer of the heated serum was greatly increased, suggesting that complement was a limiting factor in this reaction.

Unheated adult shark serum was cytotoxic to a variety of cells as measured by the dye exclusion test (table 3).

Absorption of the serum with one kind of cells lowered the inhibitory activity against these and other cells. It is as yet impossible to draw conclusions about the specificity of this reaction because a definitive experiment to resolve this point requires the use of shark C'1 free of antibody. We are currently attempting to purify and fractionate shark complement.

3.3. *Neutralization of viruses*

In our previous studies [1,2] normal shark sera were found to contain little or no hemagglutination inhibiting activity against the influenza virus. In contrast, such sera could significantly neutralize the infectivity of PR8 in-

Table 3
Cytotoxic and growth inhibitory effect of adult shark serum.

	Extent of growth*	Percent of live cells
Control	100	100
Untreated shark serum (1:45)	25	6
Untreated shark serum (1:135)	35	10
Untreated shark serum (1:405)	55	83
Untreated shark serum (1:1215)	95	98
Heat treated shark serum 56° X 30 min (1:45)	95	97
Heat treated shark serum 56° X 30 min (1:405)	95	100

* As measured by protein determination after 3 days incubation.

Table 4
Effect of nurse shark serum on influenza virus type A, PR8.

PR8 virus exposed to	Neutralization index
Adult serum (1:5)	220
△ Adult serum (1:5)	7.9
Immature serum (1:5)	11
Immature serum (1:2.5) + Δadult (1:2.5)	91

△ Heated at 56°, 30 min.

fluenza virus. As represented in table 4 adult shark serum was much more effective than the serum of immature sharks. The activity was heat-sensitive suggesting participation of complement and the activity of heated adult serum could be restored by the addition of unheated baby shark serum which by itself had no significant activity. Similarly, unheated, but not heated, shark serum neutralized the tumor inducing capacity of Rous sarcoma virus as is demonstrated in table 5.

3.4. *Effect of shark serum on bacteria*

Fig. 3 reveals that normal nurse shark serum killed *E. coli*. Adult serum was more effective than the serum of immature sharks and heating of serum once again abolished the effect. The reaction was not specific for *E. coli* as shark serum was also able to kill *S. aureus*. Since shark serum contains a relatively high concentration of urea (about 2%) the question arose whether the antibacterial and antiviral effects were caused by this substance. Control experiments were therefore performed and it was found that urea had little or no effect in the dilutions used. Moreover, urea was still present in the heated sera which were without effect.

Table 5
Effect of adult nurse shark serum on RSV tumor formation in chicks.

RSV 10^{-5} treated with	Final serum dilution	% Animals with tumors	Av. tumor size (mm)
Unheated serum	1:2	37.5	7.3
Heated serum	1:2	87.5	22.9
Unheated serum	1:6	22.2	25.0
Heated serum	1:6	74.4	25.6
Saline	–	75.0	28.0

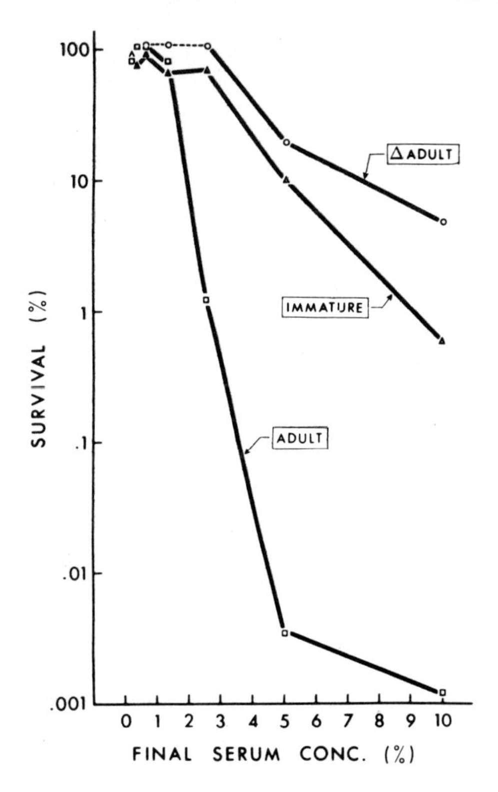

Fig. 3. The killing of *E. coli* B by normal shark serum. △ = heated 56°, 30 min.

All of these findings bespoke the fact that nurse sharks contain in their serum an unusually wide range of activity against a large variety of antigens. The most important breakthrough in these studies occurred when it was found that normal shark serum would also react with DNP causing precipitation of this hapten on a protein carrier. This discovery was important for three reasons: 1) It added an additional antigen to the catalogue of substances with which shark serum reacts, 2) it suggested an analogy between shark serum and myeloma protein, but 3) most importantly, it provided a point of departure for quantitative and more definitive immunochemical studies.

3.5. *Response of shark to DNP immunization*

Before discussing our findings on the natural antibody reacting with DNP it would be helpful to draw the course of the immune response of a shark to immunization with this hapten. Fig. 4 profiles the immune response of one

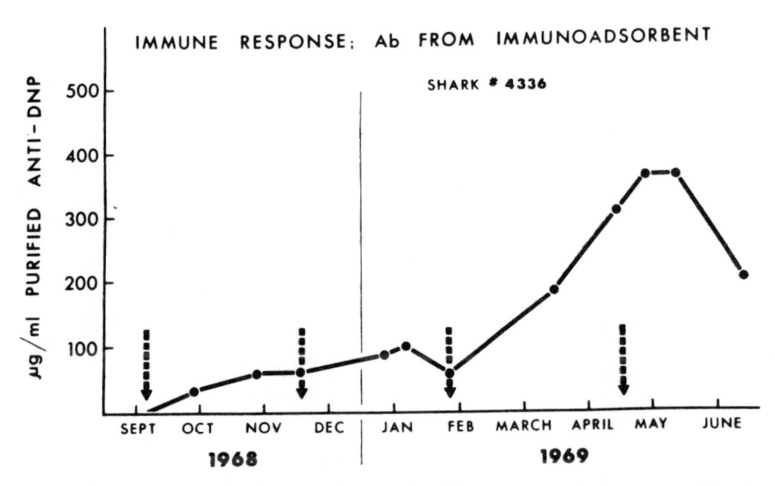

Fig. 4. Antibody response in shark immunized with DNP hemocyanin as measured by the amount of antibody. Antibody purified by immunoadsorption. Arrows indicate times of immunization.

shark to repeated immunizations with DNP hemocyanin. In this part of the experiment the antigen was administered in September and November of 1968 and in February and April of 1969. The antibody was measured in terms of micrograms of anti-DNP antibody purified by immunoadsorption. Not only was there a progressive increase in the amount of antibody but there was a recurrent shifting of the type of antibody from 19 S to 7 S. Antigen restimulation stimulated 7 S production, whose levels decreased progressively with time. Simultaneous with a decrease in 7 S was a concomitant increase in the levels of 19 S. The nature of the antibody was determined by molecular sieve, chromatography and disc electrophoresis in acrylamide gel. Fig. 5 shows that at the time of injection on 1/25/69 the antibody was virtually all macroglobulin, about 7 weeks later it was associated predominantly with the 7 S immunoglobulin. A month later both 19 S and 7 S antibody was present in relatively large amounts. We wish to emphasize the two basic characteristics of the antibody produced by immunization to DNP: 1) increase in the amount with time and with immunization and 2) the 19 S to 7 S shift. Furthermore, this antibody could be isolated by means of adsorption to DNP-BSA bromoacetyl cellulose. Elution was effected by means of DNP-glycine. The curve in fig. 4 was based on antibody purified by this procedure. A description of purification and binding properties of immune anti-DNP antibody has been published [7].

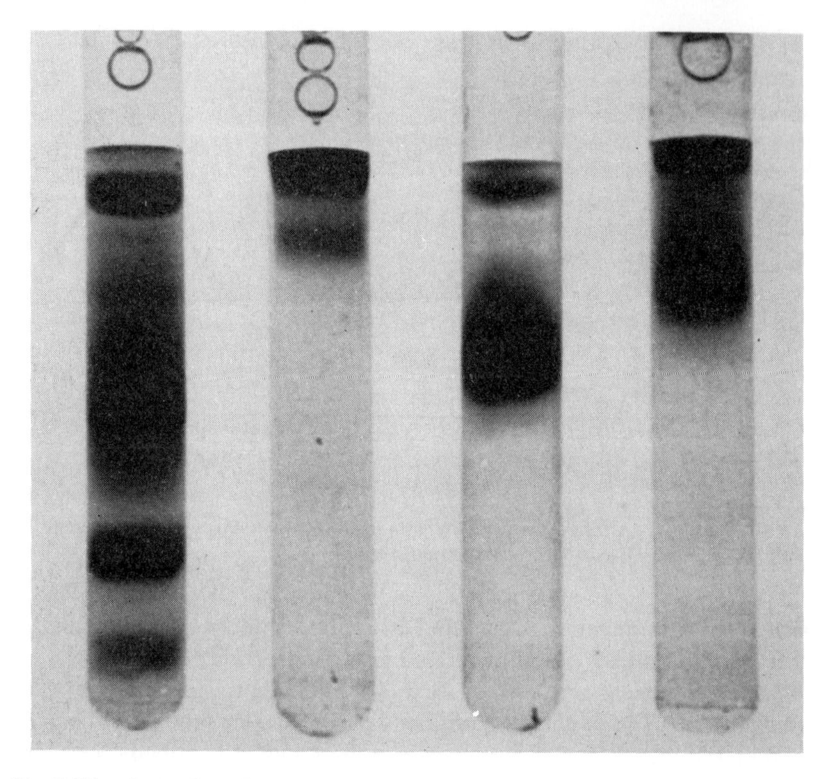

Fig. 5. Disc electrophoresis on serum and purified antibody from shark 4336 (see fig. 4). Left to right: whole serum, antibodies from 1-25-69, antibodies from 3-16-69 and antibodies from 4-16-69.

3.6. *Natural DNP "antibody"*

As stated previously, precipitation was obtained when shark serum was mixed with DNP on a protein carrier such as BSA. This reaction was exploited for the isolation of the precipitating substance using DNP-BSA as described under Materials and methods. This substance had an electrophoretic mobility of IgM immunoglobulin (fig. 6) and could be precipitated in agar by rabbit anti-shark IgM. Studies to date have indicated that the DNP precipitating activity resides mainly in the 19 S molecule.

The fact that the natural occurring precipitating substance was found in the IgM component suggests that this substance was an antibody but certain properties of this substance distinguish it from the antibody produced by immunization. First of all, this "antibody" was not removed by the immuno-

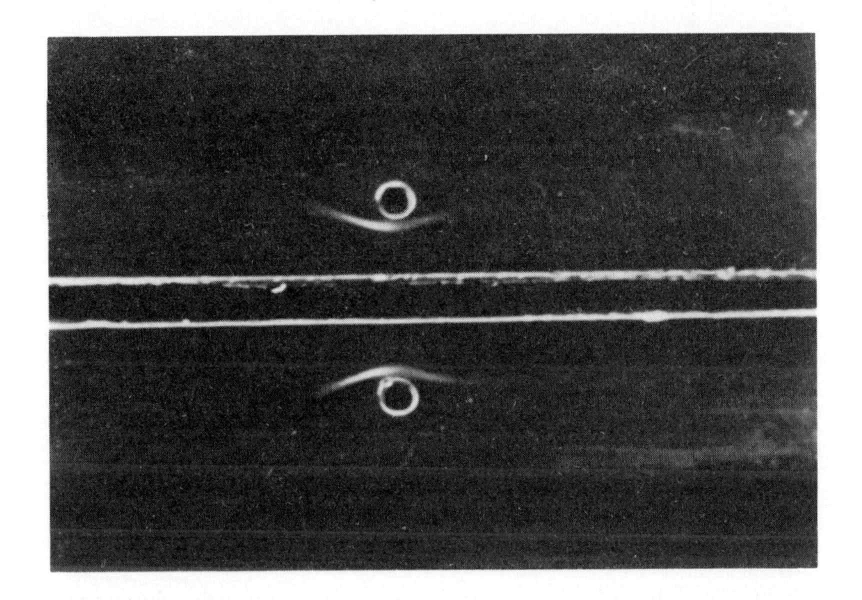

Fig. 6. Immunoelectrophoresis of natural "antibody" obtained by precipitation with DNP-BSA. Rabbit anti-shark serum in trough.

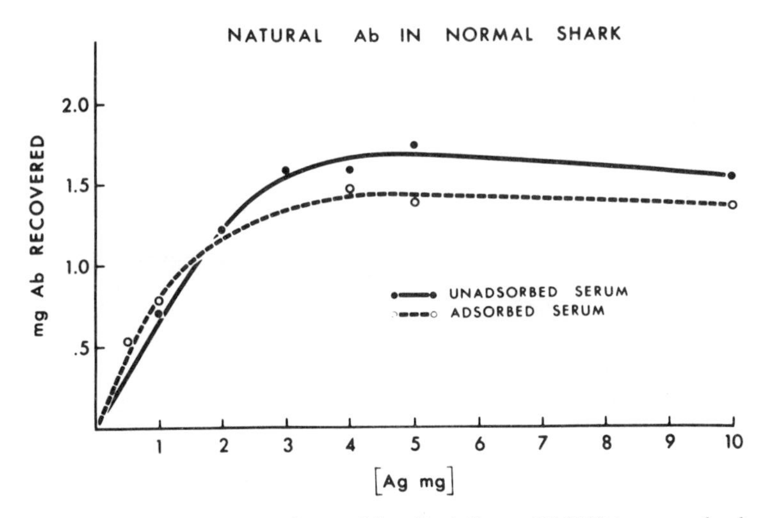

Fig. 7. Attempted adsorption of natural "antibody" on solid DNP immunoadsorbent. No adsorption occurred from the serum.

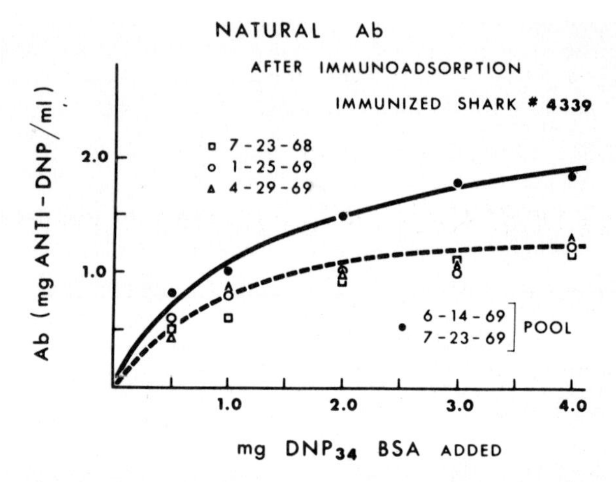

Fig. 8. Quantitative precipitation of natural "antibody" from 3 sera of shark 4339. No significant increase in antibody level with time and with continued immunization.

adsorbent. Fig. 7, based on quantitative precipitation, shows that approximately the same amount of antibody was present before and after adsorption of serum on the DNP-BSA bromoacetyl cellulose immunoadsorbent. It should

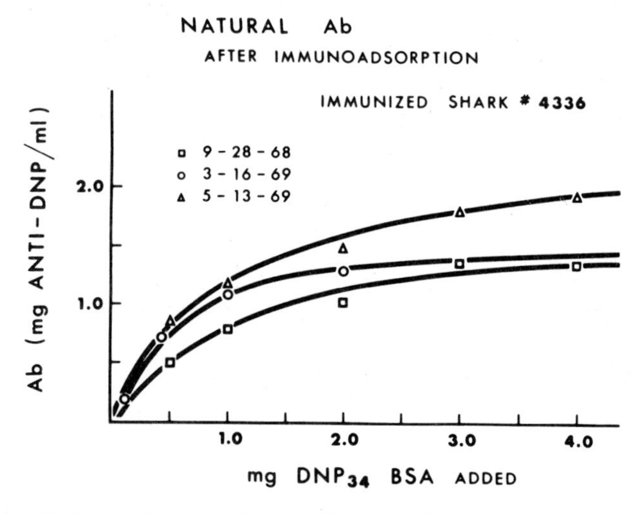

Fig. 9. Quantitative precipitation of natural "antibody" from 3 sera of shark 4336. No significant increase in antibody level with time and with continued immunization.

be pointed out, however, that natural "antibody" obtained from immunoprecipitation could be adsorbed on to this immunoadsorbent. The reason for this is unknown. Secondly, as shown in figs. 8 and 9 the amount of the natural "antibody" did not increase with time or with immunization. The ability of this "antibody" to bind ligand was demonstrated by equilibrium dialysis experiments which gave an average intrinsic association constant, K_0, of approximately 10^4 l/mole. Thus, the affinity of the natural "antibody" is approximately 1/10 that of the antibody raised in the shark by immunization with DNP [7].

We are currently conducting experiments to determine whether the wide spectrum of reactivity is associated with separate "antibody" molecules or is associated with multiple specific reactivities on single molecules. We have approached this question from several directions, one being mixed hemagglutination reaction as shown in figs. 10, 11 and 12. The results in figs. 10 and 11 indicate that chicken cells sensitized by natural "antibody" can agglutinate unsensitized sheep cells. This finding can be interpreted as reflecting either the presence of multiple reactive specificities on the same antibody molecule or the presence of common antigens in the two types of cells. Other experi-

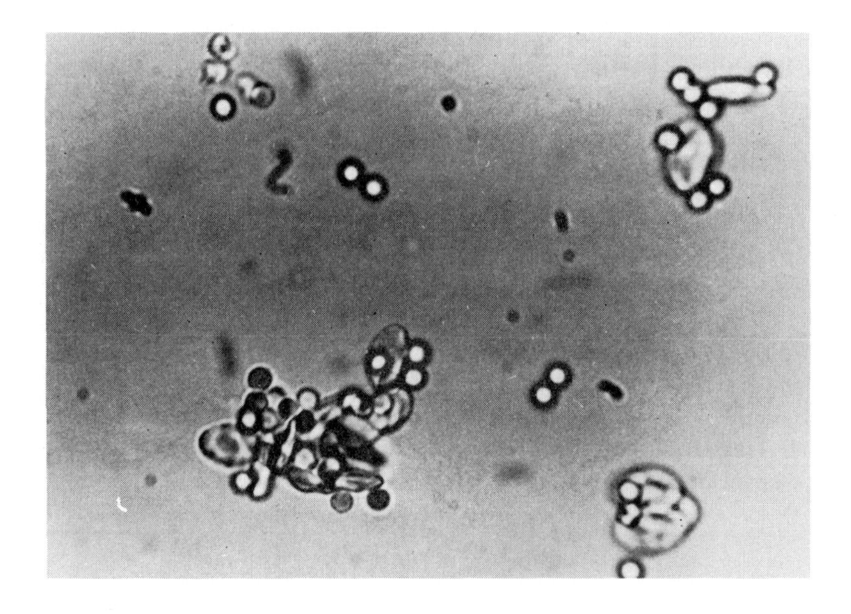

Fig. 10. Positive mixed agglutination, chicken and sheep RBC.

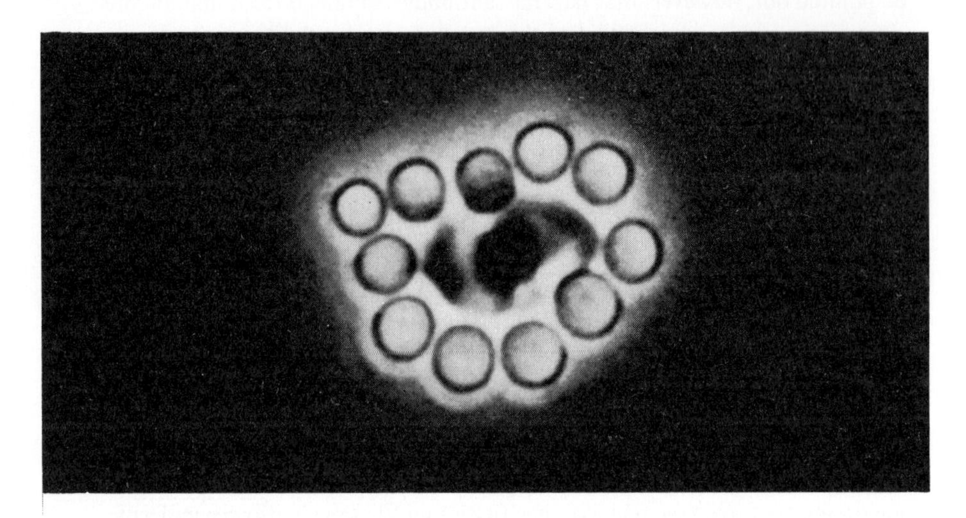

Fig. 11. Single chicken red blood cell with adherent sheep RBC in a positive mixed agglutination.

ments are necessary to distinguish between these two possibilities and these will be reported elsewhere.

Acknowledgements

This investigation was supported by U.S. Public Health Service Research Grant Nos. AI-05758 and AI-08288 from the National Institute of Allergy and Infectious Diseases. Part of the equipment used was purchased with funds provided by the American Cancer Society, Florida Division, Inc.

We are very grateful to Dr. Warren J. Russell for providing us with the data on the killing of *E. coli* by normal shark serum.

It is a pleasure to acknowledge the highly competent technical assistance by Mr. Mantley Dorsey, Jr., Mr. Larry Wellham, Mrs. Julia A. Williams and Mr. Samuel Watkins, Jr.

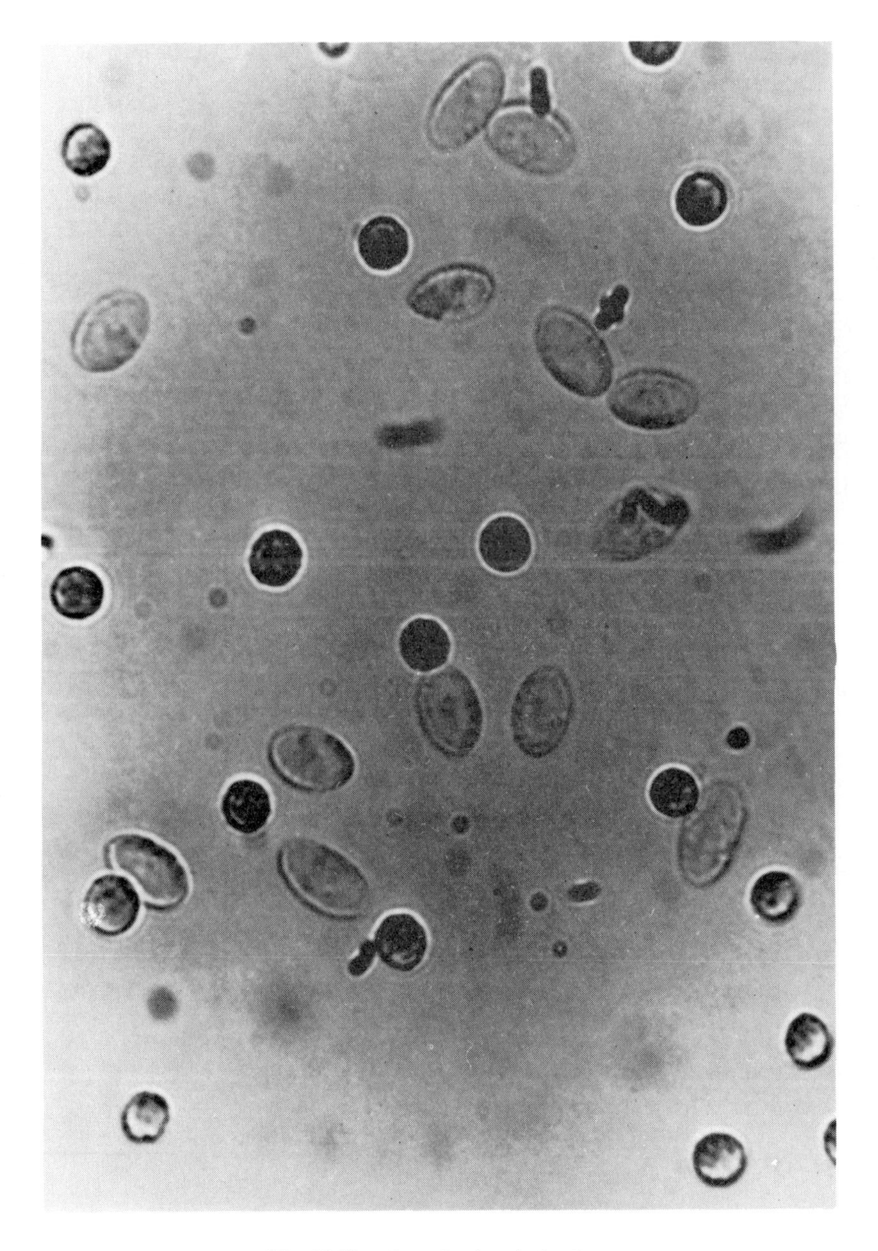

Fig. 12 Negative mixed agglutination.

References

[1] L.W. Clem and M.M. Sigel, Federation Proc. 22 (1963) 1138.
[2] M.M. Sigel and L.W. Clem, Ann. N.Y. Acad. Sci. 126 (1965) 662.
[3] L.W. Clem and P.A. Small, J. Exptl. Med. 125 (1967) 893.
[4] L.W. Clem, F. De Boutaud and M.M. Sigel, J. Immunol. 99 (1967) 1226.
[5] R.A. Good and B.W. Papermaster, Advan. Immunol. 4 (1964) 1.
[6] J. Marchalonis and G.M. Edelman, J. Exptl. Med. 122 (1965) 601.
[7] E.W. Voss, W.J. Russell and M.M. Sigel, Biochemistry 8 (1969) 4866.
[8] F. De Boutaud, L.W. Clem and M.M. Sigel, to be published.
[9] J.A. Jensen, J. Exptl. Med. 130 (1969) 217.
[10] V.I. Oyama and H. Eagle, Proc. Soc. Exptl. Biol. Med. 91 (1956) 305.
[11] A.J. Scheidegger, Arch. Allergy Appl. Immunol. 7 (1955) 103.
[12] L. Ornstein, Ann. N.Y. Acad. Sci. 121 (1964) 321.
[13] B.J. Davis, Ann. N.Y. Acad. Sci. 121 (1964) 404.
[14] H.N. Eisen, Methods Med. Res. 10 (1964) 94, 115.
[15] J.B. Robbins, J. Haemovich and M. Sela, Immunochemistry 4 (1967) 11.
[16] J.S. Gallagher and E.W. Voss, Immunochemistry 6 (1969) 199.

CONCEPTUAL AND THERAPEUTIC IMPLICATIONS OF STRUCTURAL STUDIES ON NUCLEOLAR AND RIBOSOMAL RNA OF HEPATOMAS

H. BUSCH, Y.C. CHOI, W. SPOHN and J. WIKMAN
Department of Pharmacology, Baylor College of Medicine, Houston, Texas 77025, USA

Abstract: Busch, H., Choi, Y.C., Spohn, W., and Wikman, J. Conceptual and Therapeutic Implications of Structural Studies on Nucleolar and Ribosomal RNA of Hepatomas. *Miami Winter Symposia* 2, pp. 429–446. North-Holland Publishing Co., Amsterdam, 1970.

The remarkable advances in the methods for isolation of nuclei of cancer cells and their constituents such as nucleoli, chromatin and chromosomes along with the evolution of techniques for purification and structural analysis of histones, acidic nuclear proteins, high molecular weight RNA and low molecular weight RNA of the nucleus has provided a remarkable opportunity for workers in cancer research to comparatively evaluate the nuclear products of cancer cells and other cells. The differences initially reported from this laboratory between RNA of cancer cells and other cells have now been documented by a variety of techniques including ^{32}P nucleotide analysis, oligonucleotide frequency analysis, hybridization and isolation of RNA fragments of low molecular weight. In addition, significant quantitative differences have been found in acidic nucleolar proteins of hepatomas and normal liver with particular respect to mobility of appropriately solubilized proteins no polyacrylamide disc gels. Moreover, in recent studies some differences have been found in banding patterns of low molecular weight RNA in nucleoli of tumors and normal liver. These findings suggest that the constellation of genes referred to previously as "cancer operon" may be responsible for producing very different readouts in tumors and other tissues. The crucial needs for the future are the improvement of methods for separation and characterization of the low molecular weight RNAs and proteins of the nucleolus and the acidic proteins of the nucleus and nucleolus. Rapid advances in these techniques are already emerging. In addition critical information is now being obtained on the substructures of the large nucleolar RNA molecules which may point the way to definitive cancer chemotherapy.

1. Introduction

The major goal of basic research on the Biochemistry of Cancer is the determination of exploitable differences between cancer cells and other cells [1].

Fundamentally, the initiating lesion in oncogenesis seems to be one of sensitivity to a wide variety of chemical agents, physical agents, some infectious agents and other less well defined causal factors. From the point of view of the processes involved in the continuation of the cancer trait in neoplastic cells the following may be of particular importance: (a) specific "cancer gene sets" or a "cancer polyoperon" [1] which may produce inhibitors of normally operative inhibitory genes or (b) in cancer cells there may be a deletion of control or repressor genes; at present this seems to be a less likely factor in oncogenesis.

As part of the activation of "master" or "operator" cancer genes, phenotypically functional gene sets seem to be inactivated in hepatomas, insulinomas and other tumors except for the gene sets in the growth process [2]. The phenomenon of continued expression of the "cancer gene set" in cells which contain all of the normal genes may resemble the phenomenon of "paramutation" in plants [3]. The key feature of the concept is that it eliminates the need for mutations which would have to be of high frequency, or lysogeny, which would bring new DNA into cells.

Inherent in this concept is the idea that all cells contain "cancer gene sets" which can evolve the "cancer phenotype" [1]. Clearly some structural gene members of the cancer gene set that are operative for normal functions in other cells may be utilized by a single master "cancer gene" in ways that have survival values for the cancer cells but are lethal to the host. For example, cells of a number of types multiply at high rates. However, in most instances their multiplication reflects a homeostatically critical mechanism, such as production of white cells, red cells, gastrointestinal epithelium, etc. In the case of cancer cells, rapid growth simply results in the heaping up of a huge amount of cells. In addition, many types of cells, particularly white cells, migrate from place to place in the organism. However, this behavior is not associated with excessive cell production. On the other hand, in the case of cancer cells, metastasis involves cell migration coupled with a high level of cell production in unusual sites. Because of the invasive character of cancer cells, this combined behavior produces properties that are uncommon to other cell types. There is local tissue destruction and infringement on normal function of other tissues. Thus, it would appear that the "operator" or master genes involved in the development of the cancer phenotype may utilize normal gene sets that are available for growth, invasiveness, metastasis, etc. The ultimate effect is the destruction of the host.

1.1. *The nucleolus in cancer cells*

The aberrations of the nucleolus in cancer cells have been recently reviewed

[4,5]. In addition to morphologic aberrations which have been reported almost 40 years ago, the nucleolus has been shown to have an unusual ^{32}P nucleotide composition in a series of studies in this laboratory [6]. These differences in ^{32}P nucleotide composition have led to the series of studies dealt with in the following section which support the concept that aberrant synthesis of nucleolar RNA occurs in cancer cells and further, that the information forthcoming from this data may have important chemotherapeutic potentialities.

2. Nucleic acid structure

The approach to the structure of nucleic acids of very large size has been two-fold: (a) analysis of elementary constituents such as nucleotide composition or oligonucleotide composition and (b) degradation reactions leading to structural analysis of the types recently reported by the Cambridge and Geneva groups. In a series of studies reported from this laboratory the ^{32}P nucleotide compositions of nucleolar 45 S RNA were found to be significantly different from those of nontumor tissues [5–7] in that the adenylic acid content of the nucleolar RNA of the tumors was low (13–15%) and the cytidylic acid content was high (29–34%). In the liver the corresponding values were 20 and 24%, respectively and in the kidney, 14 and 21%, respectively. Parenthetically, it should be noted that these values suggest that there is some

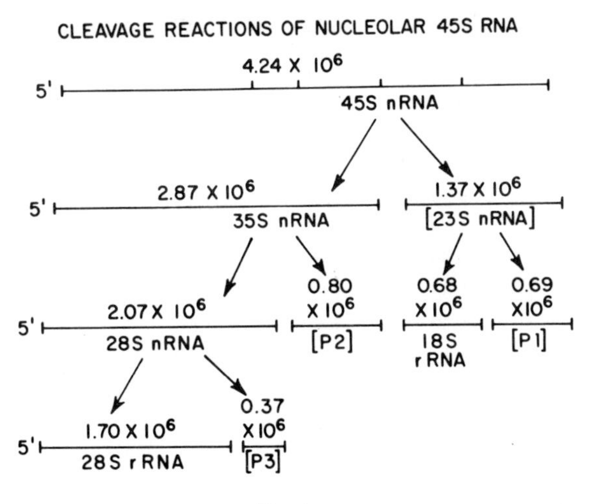

Fig. 1.

specialization in nucleolar gene readouts. In any event these high molecular weight RNAs are known to undergo specific cleavages (fig. 1) to form ribosomal RNAs and other polynucleotides (P1, P2, P3).

2.1. *Oligonucleotide frequency studies*

Analysis of the oligonucleotide frequencies is made by determining the distribution of labeled ^{32}P in a series of mono- to tetranucleotides derived by pancreatic RNAase digestion. This technique was evolved by Rushizky and Knight [8] who "fingerprinted" the oligonucleotides using a combined electrophoretic and chromatographic procedure. Initial use of this procedure in our laboratory provided the data shown in table 1 [9]. These studies indicate that in the whole nuclear RNA thc various sedimentation classes of RNA had marked differences in their content of C, U, AC, GC and AGC. Of particular

Table 1
Percentage of radioactivity as a proportion of total radioactivity recovered in oligonucleotides obtained by ribonuclease digestion of ^{32}P-labeled nuclear RNA of normal liver [9].

Oligonucleotide	% Radioactivity				
	18 S RNA	28 S RNA	35 S RNA	45 S RNA	55 S RNA
C	21.07	25.36	25.15	24.30	20.63
U	18.50	15.93	15.66	18.62	19.13
Ψ	0.57	0.40	0.51	0.38	0.32
AC	6.43	5.42	6.20	6.20	7.10
AU	4.83	3.60	3.73	3.65	5.31
GC	10.31	12.00	10.66	11.15	9.61
GU	5.82	5.72	6.07	5.86	6.05
A_2C	2.56	2.40	2.43	2.55	2.59
(AG)C	6.52	8.62	7.23	7.81	6.10
A_2U	2.78	2.25	2.31	2.79	2.50
(AG)U	3.32	2.42	2.65	2.50	2.90
G_2C+A_3U	3.36	3.02	3.52	2.57	3.37
G_2U	1.85	2.16	2.02	1.48	1.77
A_2C	1.37	1.57	1.66	1.60	1.50
$(A_2G)C$	4.13	3.16	4.40	3.54	4.25
$(AG_2)C$	2.78	2.62	2.20	1.74	2.67
$(A_2G)U$	2.24	2.14	2.40	1.98	2.07
$(AG_2)U$	1.00	0.93	1.81	0.86	1.25

Each number represents an average of 3 experiments with different samples of RNA.

Table 2
Percentage of radioactivity as a proportion of total radioactivity recovered in oligonucleotides obtained by ribonuclease digestion of ^{32}P-labeled nucleolar RNA of normal liver [9].

Oligonucleotides	% Radioactivity		
	28 S RNA	35 S RNA	45 S RNA
C	27.4	28.1	27.3
U	13.8	13.1	12.9
Ψ	0.97	0.48	1.37
AC	4.30	4.30	4.70
AU	2.49	2.66	2.93
GC	13.2	14.2	13.4
GU	6.00	5.91	5.82
A_2C	2.01	2.46	2.84
(AG)C	6.38	6.63	5.66
A_2U	1.80	2.25	1.55
(AG)U	3.84	3.00	3.73
G_2C+A_3U	4.54	4.96	4.81
G_2U	2.42	2.10	2.55
A_3C	1.19	1.39	1.50
$(A_2G)C$	2.64	2.72	2.92
$(AG_2)C$	3.10	2.17	2.03
$(A_2G)U$	2.68	2.14	2.81
$(AG_2)U$	1.22	1.23	1.93

Each number represents an average of 4 experiments with different samples of RNA.

interest was the finding that the 28 S and 35 S nuclear RNAs had an oligonucleotide frequency similar to that of nucleolar RNA (table 2). In the nucleolar RNA the C content was higher and the U content was lower than that of the whole nuclear RNA. In addition, the content of AC and AU was lower and that of the GC was substantially higher than that of the whole nuclear RNA. Less notable changes were found in the tri- and tetranucleotides. When a comparison was made of the oligonucleotide frequencies of the 45 S RNAs of the Novikoff hepatoma (table 3), marked differences were found, i.e., in the nucleolar RNA there was a greater C, a lower U, a greater GC and GU and a lower AC and AU than was found in the 45 S RNA of the chromatin fraction [5]. In the trinucleotides there was a lower A_2C and AGU and a higher G_2C+A_3U than in the chromatin fraction. Similarly, there was a lower A_3C, A_2GC, AG_2C, A_2GU and AG_2U than in the chromatin fraction. It is particular-

Table 3
Comparative distribution of oligonucleotides in 45 S RNA of the chromatin, nuclear residue and nucleolar fractions of the Novikoff hepatoma [5].

	45 S Nucleolar RNA	45 S Residue RNA	45 S Chromatin RNA
C	27.90	23.67	21.50
U	14.77	17.17	18.85
Pseudo U	0.60	0.37	0.60
AC	4.47	4.85	6.85
AU	2.97	5.82	6.80
GC	16.25	11.62	9.25
GU	12.47	9.27	8.60
A_2C	1.24	1.92	2.52
A_2U	1.93	1.70	2.33
(AG)C	4.98	4.85	5.23
(AG)U	3.45	4.32	4.72
G_2U	2.25	2.36	2.17
G_2C+A_3U	3.21	3.87	2.23
A_3C	0.36	1.09	2.00
$(A_2G)C$	1.58	2.25	2.45
$(AG_2)C$	0.86	1.66	2.20
$(A_2G)U$	1.35	1.57	1.92
$(AG_2)U$	0.80	1.32	1.32

ly notable that the sum of AC and AU is 7.4% in the nucleolar RNA and 13.5% in the chromatin RNA. Conversely, the sum of GC+GU is 29% and only 18% in the chromatin RNA.

The values for the residue RNA were frequently, but not always, intermediate between the values of the nucleolar and the chromatin fractions, i.e., the values for G_2U and G_2C+A_3U were higher than those of the nucleolar and chromatin fractions. The results indicate that there are RNA species present in the residue fraction that are either specifically localized in this fraction or are in higher concentrations in the residue fraction.

2.2. *Comparative oligonucleotide frequencies of tumors and other tissues*

Previous studies from our laboratory [5–7] have indicated that the ^{32}P nucleotide composition of 45 S nucleolar RNA of tumors is characterized by a low content of adenylic acid and a high content of cytidylic acid. These results were confirmed for nucleolar 35 S and 28 S RNA as shown in table 4.

Table 4
Nucleotide composition of rapidly sedimenting nucleolar RNAs from normal rat liver, Novikoff hepatoma, and Morris hepatoma 9618A after 60 min labeling with orthophosphate-^{32}P [7].

RNA	AMP	UMP	GMP	CMP	$\frac{A+U}{G+C}$
Liver					
28 S	22.4	20.4	33.3	23.8	0.75
35 S	21.0	19.5	35.5	23.8	0.68
45 S	21.1	18.4	35.5	25.2	0.65
Novikoff hepatoma					
28 S	13.8	21.0	33.6	31.7	0.53
35 S	13.7	20.9	33.5	31.3	0.53
45 S	13.5	20.6	33.6	31.7	0.52
Morris hepatoma 9618A					
28 S	15.8	18.4	35.3	30.3	0.52
35 S	16.6	18.7	34.7	30.1	0.54
45 S	15.7	19.4	33.2	31.6	0.54

Each rat received 2 mCi of orthophosphate-^{32}P i.p. 60 min before it was sacrified. The values are percentages of total radioactivity in the four necleotides. The values are averages of 3 experiments for each tissue. AMP, adenylic acid; UMP, uridylic acid; GMP, guanylic acid; CMP, cytidylic acid; A + U/G + C, ratio of adenylic acid and uridylic acid to guanylic acid and cytidylic acid.

To determine whether these differences resulted from differences of pool sizes of precursor molecules and general disparities throughout the molecule, studies were undertaken of the distribution of isotope of oligonucleotides as shown in table 5. With this technique, little difference was found in the labeling of C and U in the normal rat liver, Novikoff hepatoma and the highly differentiated Morris hepatoma 9618A. However, marked and statistically significant differences were found in the labeling of GC and GU. In both instances, the values for the hepatomas were greater than those of normal rat liver, i.e., the isotope in GC was 12.5% in normal liver, 15.5% in Morris hepatoma 9618A and 18.7% in the Novikoff hepatoma; the isotope in GU was 4.5% in normal liver, 9.5% in the Novikoff hepatoma and 6.4% in Morris hepatoma 9618A. Evidence that these differences did not result from either pool sizes or random differences was obtained by analysis of the isotope content of AC and AU. In both the Novikoff hepatoma and Morris hepatoma 9618A the isotope content of AC was the same as that of normal rat liver. The isotope content of AU was essentially the same as that of the rat liver in Novikoff hepatoma.

Table 5
Radioactivity distribution in mono- and oligonucleotides of nucleolar RNAs after 60 min labeling with orthophosphate-^{32}P [7].

(oligo) nucleotides[a]	Normal rat liver			Novikoff hepatoma			Morris hepatoma 9618A		
	28 S[b]	35 S	45 S	28 S	35 S	45 S	28 S	35 S	45 S
C	28.8	28.5	27.6	28.1	26.6	26.1	28.8	28.6	28.0
U	13.8	14.6	15.9	12.8	12.9	15.0	14.7	14.2	14.5
Pseudo U	1.7	1.3	0.9	0.5	0.5	0.5	0.8	0.6	0.6
Mono-nucleotides	44.3	44.4	44.4	41.4	40.0	41.6	44.3	43.4	43.1
AC	5.1	5.0	5.0	5.5	5.2	4.9	5.2	5.3	5.1
GC	12.2	12.3	12.9	15.1	15.5	15,6	18.3	18.5	19.0
AU	2.8	2.7	2.8	3.1	2.4	2.8	1.9	1.9	1.8
GU	4.4	4.8	4.3	9.7	9.1	9.8	6.3	6.4	6.4
Di-nucleotides	24.5	24.8	25.0	33.4	32.2	33.2	31.7	32.1	32.4
AAC	3.1	3.2	3.1	2.2	2.1	1.8	2.0	2.0	1.8
(AG)C	7.4	8.1	8.3	6.4	6.5	5.5	7.5	7.2	7.3
GGC/A_3U	2.3	2.5	1.9	3.0	3.5	3.2	2.5	2.2	2.4
AAU	2.2	2.0	2.4	1.3	2.2	1.6	1.5	1.7	1.5
(AG)U	3.2	2.9	3.1	3.0	3.7	3.2	2.2	2.7	2.5
GGU	1.1	1.8	1.4	3.0	2.2	3.1	1.0	1.0	0.9
Tri-nucleotides	19.3	20.5	20.2	18.9	20.2	18.4	16.7	16.9	16.4
A_3C	1.7	1.6	1.2	1.7	1.2	0.8	1.1	1.4	1.1
(A_2G)C	4.9	4.8	4.5	2.5	1.7	1.7	2.6	2.6	3.0
(AG_2)C	2.5	1.5	1.4	1.8	2.5	1.6	1.4	1.1	1.1
(A_2G)U	1.6	1.4	2.4	1.6	1.6	2.4	1.4	1.5	2.3
(AG_2)U	1.0	0.8	1.2	0.7	0.9	1.2	0.6	0.8	0.6
Tetra-nucleotides	11.7	10.7	10.7	8.3	8.0	7.7	7.1	7.4	8.1

The values for each sequence are the percentages of the total radioactivity from each RNA found in 18 eluted spots.
[a]Abbreviations are those used by Rushizky and Knight [8].

The primary labeling pool of AC and GC is CTP and the primary labeling pool of AU and GU is UTP. Accordingly, the marked differences in the results for AC and GC and AU and GU reflect the relative amounts of these dinucleotides in the long RNA molecules rather than simply pool effects.

The values for the trinucleotides (table 5) were not markedly different. However, in the tumors there was a decrease in the tetranucleotide A_2GC, and the overall percent of isotope in the tetranucleotides was less than that of the normal rat liver.

2.3. *Substructures of high molecular weight RNA*

The chief goal of the initial research on high molecular weight RNA in our laboratory is that of determination of marker segments of these very large molecules. Partial hydrolysis of 28 S, 35 S and 45 S nucleolar and ribosomal RNA has been undertaken. Among the types of digestions being made are alkaline hydrolysis, pancreatic RNAase digestion and T_1 RNAase digestion. Recently, a very large fragment (the B_3 fragment) composing 13% of the 28 S rRNA molecule and containing 600 nucleotides was isolated from ribosomal and nucleolar 28 S RNA [10]*. Another related fragment, the B_4 fragment,

* The B_3 fragment was derived by digestion of the RNA of the 28 S and other RNAs with 20 U of T_1 RNAase per mg of RNA (37°) for 30 min [10].

Table 6
^{32}P-Nucleotide composition of B bands obtained by gel electrophoresis [10].

Novikoff ribosomal 28 S	AMP	UMP	GMP	CMP	(AMP + UMP)/(GMP + CMP)
28 S					
^{32}P	16.0	17.3	36.8	30.0	0.50
Ultraviolet	16.0	19.0	36.0	29.0	0.54
Digestion A (8 units of T_1 RNAase per mg of RNA)					
B_{3a+b} (^{32}P)	6.6	15.0	39.9	38.5	0.27
Digestion C (20 units of T_1 RNAase per mg of RNA), 4°					
B_{3a+b} (^{32}P)	4.7	14.1	40.2	41.0	0.23
Digestion D (20 units of T_1 RNAase per mg of RNA), 37°					
B_{3a} (^{32}P)	4.6	13.8	40.6	40.7	0.23
B_4 (^{32}P)	5.7	12.0	39.6	42.6	0.22

which is somewhat smaller, was also isolated. In table 6, the ^{32}P nucleotide compositions of the B_3 and B_4 fragment indicate that they are very rich in guanylic and cytidylic acids, i.e., these together compose more than 80% of the total residues in these fragments. The resistance of the B_3 fragment to T_1 RNAase digestion indicates that it has a great deal of hydrogen bonding or secondary structure. The high melting temperature (T_m) of this fragment (75°) supports this idea.

The oligonucleotide frequency analysis of the B_3 and B_4 fragments is pre-

Table 7

Distribution of ^{32}P in oligonucleotides after pancreatic RNAase digestion of B_{3a} and B_4 band from ribosomal and nuclcolar 28 S RNA extracted from polyacrylamide gels [10].

(Oligo)nucleotide	Total isotope recovered in oligonucleotides (Digestion D (20 units of T_1/mg RNA, 37°, 30 min))			
	B_{3a} Band		B_4 Band	
	Ribosomal	Nucleolar	Ribosomal	Nucleolar
	%		%	
C	45.8	46.1	45.6	45.0
U	13.3	13.0	11.1	11.6
Pseudo U	0.4	0.4	0.2	0.2
Mononucleotides	59.5	59.5	56.9	56.8
AC	2.5	2.1	3.2	3.3
AU	0.9	1.0	0.7	0.6
GC	17.4	18.1	19.8	19.7
GU	4.4	4.0	4.8	4.9
Dinucleotides	25.2	25.2	28.5	28.5
AAC	1.0	1.1	1.5	1.7
AAU	0.7	0.9	1.1	1.0
(AG)C	2.2	2.6	2.6	2.7
(AG)U	0.5	0.3	0.8	0.8
GGC/A_3U	5.0	5.0	4.8	4.6
GGU	1.5	1.3	0.8	0.8
Trinucleotides	10.9	11.2	11.6	11.6
AAAC	0.6	0.5	0.5	0.4
(AAG)C	0.7	0.6	0.6	0.6
(AAG)U	0.7	0.6	0.8	0.7
(AGG)C	1.3	1.5	1.8	1.8
(AGG)U	0.4	0.3	0.5	0.5
Tetranucleotides	3.7	3.5	4.2	4.0

Table 8
Distribution of ^{32}P in oligonucleotides after pancreatic digestion of the B_3 fragment.

(Oligo)nucleotides	B_3 Band of ribosomal 28 S RNA	
	Novikoff	Rat liver
	(% Total isotope in recovered oligonucleotides)	
C	45.8	44.7
U	13.3	12.0
Pseudo U	0.4	0.6
Mononucleotides	59.5	57.1
AC	2.5	3.4
AU	0.9	1.7
GC	17.4	16.0
GU	4.4	4.1
Dinucleotides	25.2	25.2
AAC	1.0	1.9
AAU	0.7	1.5
(AG)C	2.2	2.4
(AG)U	0.5	0.9
GGC/A_3U	5.0	4.8
GGU	1.5	1.3
Trinucleotides	10.9	12.4
AAAC	0.6	0.9
(AAG)C	0.7	0.9
(AAG)U	0.7	1.1
(AGG)C	1.3	1.7
(AGG)U	0.4	0.6
Tetranucleotides	3.7	5.2
Residues at origin	32.0	32.0

From J. Wikman, E. Howard and H. Busch, Cancer Res. 30 (1970) 773.
Each value represents an average of 4 determinations from separate experiments.

sented in table 7. This analysis is meaningful because of the large size of the fragment, i.e., approximately 600 nucleotides. These results show that both the B_3 and B_4 fragments are esentially the same in ribosomal and nucleolar RNAs.

Table 8 shows that the ^{32}P nucleotide compositions of the B_3 fragment of the nucleolar and ribosomal RNAs are not significantly different in the Novikoff hepatoma and the rat liver. These results illustrate the nonrandomness of the differences between the hepatoma and the normal liver and indicate that

Table 9
Distribution of alkali resistant dinucleotides in segments of nucleolar 45 S RNA.

5′-Terminal	Alkali stable dinucleotides*	Mononucleotides	Alkali stable dinucleotides	Mononucleotides	Alkali stable dinucleotides	Mononucleotides
	3 AmpAp		0 AmpAp		3 AmpAp	
	6 AmpUp		1 AmpUp		4 AmpUp	
	12 AmpGp		3 AmpGp		4 AmpGp	
CmpUp(80%)	+ GmpAp		+ GmpAp		+ GmpAp	
	6 AmpCp		1 AmpCp		3 AmpCp	
	2 UmpAp		0 UmpAp		2 UmpAp	
Ap(10%)	1 UmpUp		0 UmpUp		1 UmpUp	
	4 UmpGp		1 UmpGp		1 UmpGp	
	5 UmpCp	5,700 Np	0 UmpCp	2,100 Np	2 UmpCp	5,000 Np
	+ CmpUp		+ CmpUp		+ CmpUp	
	6 GmpUp		2 GmpUp		2 GmpUp	
	10 GmpGp		3 GmpGp		4 GmpGp	
Cp(10%)	1 GmpCp		1 GmpCp		1 GmpCp	
	2 CmpAp		1 CmpAp		2 CmpAp	
	4 CmpGp		1 CmpGp		1 CmpGp	
	4 CmpCp		2 CmpCp		1 CmpCp	
	Total 66		Total 16		Total 31	

← Nucleolar 28 S RNA →

← Nucleolar 35 S RNA →

← Nucleolar 45 S RNA →

* The alkali stable trinucleotide, UmpUmpCp, is not included.
Y. Choi and H. Busch, J. Biol. Chem. 245 (1970) 1954.

Fig. 2.

these differences are in regions other than the highly hydrogen bonded B_3 fragment.

2.4 *Markers in nucleolar and ribosomal RNA*

In studies on the structural elements of the nucleolar 45 S RNA, the alkali resistant 5′-terminals (mainly pCmpUp), an alkali resistant trinucleotide, (UmpUmpCp) and a series of dinucleotides have been found. Their localization in the nuclear 45 S RNA and its subproducts are indicated in table 9. Further studies on partial cleavages of very high radioactive nucleolar RNA are in progress and approximately 30 sub-bands have been isolated (fig. 2). At present it appears that the B_3 band is an electrophoretically homogeneous species.

GUGU - More recent studies on oligonucleotides have indicated that in addition to high content of GU and GC in the tumors, there is an increased amount of the tetranucleotide GUGU. This finding was obtained in experiments in which high molecular weight RNA was hydrolyzed with T_1 RNAase, an enzyme which specifically cleaves RNA at the G-3′-P residue. Oligonucleotide frequency analysis has demonstrated that the dinucleotide, UG, is present in higher concentrations in RNA of the nucleoli of tumor cells than in non-tumor cells. These high values suggest that the trinucleotide UGU is unusually frequent in the tumor. Because the trinucleotide, UGU, is derived by T_1 RNAase hydrolysis, the UG must be adjacent to another G, i.e., the trinucleotide UGU is adjacent to G. Thus, the tetranucleotide GUGU is an oligonucleotide present in high concentrations in the tumors (fig. 3).

2.5. *Positions in the molecule*

In any given RNA, a specific trinucleotide or tetranucleotide can be present in only a limited number of positions and in a limited frequency. For example, in the 45 S RNA molecule containing 12,000 nucleotides, approximately 1/3 are G residues, or a total of 4000 G residues. In addition, approximately 20% of the residues are U residues, which means that there are approximately 2400 U residues. The possibility that these are linked as GUGU is only 1 in

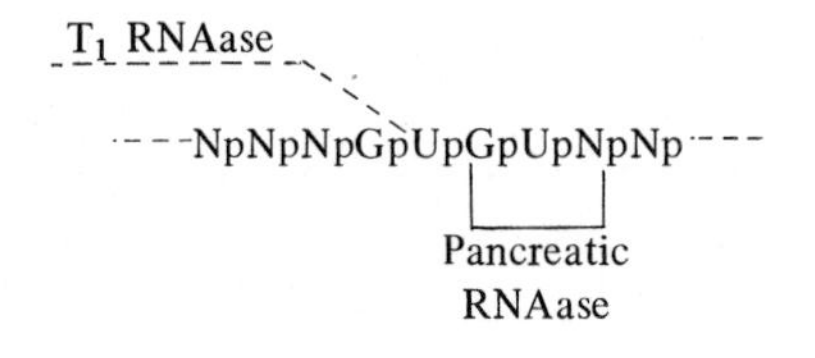

Fig. 3. Cleavages by which the GUGU sequence was identified.

256 for any tetranucleotide or in other words, of the total possible number of tetranucleotides, only 1 in 256 would produce this specific sequence.

On the basis of information available at the present time, it is not possible to specify whether GUGU is a specific sequence of the tumor nucleolar RNA or reflects the presence of a tumor specific larger sequence. However, the evidence for the presence of this sequence in high concentration in tumor nucleolar RNA has therapeutic potentialities which will be discussed below.

2.6. *Other abnormalities of nucleolar constituents in cancer cells*

A search for other possible aberrations was instituted as part of a more extensive evaluation of the RNA and protein of the nucleoli. In this laboratory studies on *low molecular weight RNA* are proceeding with considerable rapidity, i.e., a significant number of new species of low molecular weight RNA have been identified [11–16]. Not all of these low molecular weight RNAs have been highly purified, but two, 4.5 S RNA_I and 4.5 S RNA_{III}, have been isolated as individual molecular species in high states of purity [15–17]. In the nucleolus there are a number of low molecular weight RNAs including 4 S RNA, 5 S RNA, U3 RNA and 8 S RNA. In recent studies of Prestayko et al. [18] it has been possible to isolate 4 sub-bands of U3 RNA in tumor cells. However, in normal liver cells, only 3 such bands have been found. It seems possible that the extra U3 band may be a messenger RNA that may have the function of coding for some specific nucleolar protein.

It has only recently become possible to isolate the nucleolar proteins in any satisfactory state of purity [19]. Systems employing high concentrations of urea have now been utilized for polyacrylamide gel electrophoresis and chromatography on Sephadex. These sytems have permitted the demonstration that nucleolar proteins of neoplastic cells differ markedly in electrophoretic behavior from those of nontumor cells. Particularly, a number of protein bands, referred to as bands 8 and 9 are present in high concentrations in tumor nucleoli and are apparently absent from nucleoli of other cells. Although none of these proteins have been adequately characterized to this point, an effort to effect these characterizations is now in progress in this laboratory. If some of these proteins are present in high concentrations in the tumor cells, it seems possible that their presence reflects synthetic activities of messenger RNAs that are either present in low concentrations or are nonfunctional in other cell types, i.e., the U3 RNAs noted above may be involved in the synthetic reactions that are producing these proteins

3. Discussion

Although these findings are of interest because of the importance of nucleolar function from both a quantitative and qualitative point of view, these new findings may be critical as an inroad into the general problem of cancer chemotherapy. Although the relationship of these differences to the concept of the cancer gene cannot be completely defined at the present time, it does seem possible that the abnormal nucleolar gene readouts may reflect the activity of structural genes of the "cancer gene set" or possibly even the "master" or "operator" genes of the cancer gene sets. It is not possible to specify at the present time whether fragments of the newly synthesized nucleolar RNA may be coded for by "special" cancer genes or the whole readouts may reflect the activity of special cancer genes. In any event, the demonstration of different frequencies of special populations of oligonucleotides in tumor cells and other cells and the possibility that newly synthesized high molecular weight nucleolar RNA may contain unique sequences suggests that genes coding for such sequences may be targets for selective destruction. Of particular importance is the provisional demonstration that a given tetranucleotide, GUGU, is present in higher frequencies in tumors than in other tissues.

It now seems possible to make a preliminary test of the possibility that oligonucleotide therapy may be directed against specific targets, although the tetranucleotide GUGU may not be completely adequate as a specific target. In recent months, critical evaluations of potential strategies have been made, and they include the following:

1. Synthesis of oligonucleotide analogs of GUGU by chemical means
   (a) incorporation of base analogs such as thioguanine or fluorocytosine;
   (b) incorporation of sugar analogs such as arabinosides or deoxyribosides in place of ribose;
   (c) substituted phosphates may be utilized such as methyl phosphates or ethyl phosphates;
   (d) elimination of the sugar and phosphates by substitution with methylene groups.
2. Analogs may be prepared of the tetranucleotide coding for GUGU (dCdAdCdA)
   (a) appropriate analogs may include 6 MP or fluorocytosine;
   (b) alternating sugars of the deoxyribose and ribose variety might be employed.
3. Oligonucleotides may be obtained from poly U, poly UG or other hetero- or homopolymers by cleavage into low molecular weight oligonucleotides ranging from mono- to tetradecanucleotides.

(a) substituents can be added to these molecules directly, i.e., sulphur derivatives of uracil;
(b) oxidations of the 3′-terminal by periodate and appropriate analogs may be incorporated on the aldehyde groups.

All of these potential operations are based upon the possibility that some form of carrier can be developed to bring these molecules into the cell. Such a carrier might be a lipid spherule or a protein which could be taken up easily by phagocytosis or by pinocytosis. One type of protein carrier might be albumin [1]. In addition, low or high molecular weight lipids may be employed. It is necessary that these carriers and their associated products enter the cell with the analogs intact.

The analytical approach continues actively. The important techniques developed at Cambridge and in Belgium and Switzerland for structural analysis of high molecular weight RNAs can now be applied to these RNAs of neoplastic cells because of the development of methods which permit the isolation in good yield of highly radioactive RNA from tumors [20]. Very extensive structural information on the high molecular weight RNAs will become available within the next 12 to 18 months. It seems possible that deca- and dodecanucleotides can be isolated from these RNAs of tumors and other cells and that appropriate comparisons will permit synthesis of structurally specific elements. At that time another important possibility for chemotherapy can be envisioned; namely, the insertion into tumor cells of normal regulatory RNA fragments. Some of these may exert a specific regulatory effect on nucleolar function and if they can reach the proper positions within the tumor nucleoli, they may serve as regulatory macromolecules within tumor nucleoli. If so, chemotherapy may be achieved through mechanisms that simply involve replacement rather than destruction of specific DNA templates of the nucleolus.

The continuing development of analytical information about the nucleolus of neoplastic cells has provided a demonstration of low molecular weight RNA species in these nucleoli and a demonstration of new types of proteins which may be coded for by some of the lower molecular weight RNA products. It is generally accepted that the high molecular weight nucleolar RNA serves as a source of ribosomal RNA and other products which need to be defined. It is essential that new information be developed about the functional roles of the various types of nucleolar RNAs and in particular, mechanisms which control the synthetic reactions involved. At least at this point, it is feasible to explore the uptake of such analogs into the cell and into the nucleoli.

References

[1] H. Busch, An Introduction to the Biochemistry of the Cancer Cell (New York, Academic Press, 1962).
[2] G. Weber and M. Lea, The Molecular Correlation Concept, in: Methods in Cancer Research, Vol. 2 (New York, Academic Press, 1967) pp. 523–578.
[3] A. Brink, Genetic Repression of R Action in Maize, in: The Role of Chromosomes in Development (New York, Academic Press, 1964).
[4] H. Busch, P. Byvoet and K. Smetana, Cancer Res. 23 (1963) 313.
[5] H. Busch and K. Smetana, The Nucleolus (New York, Academic Press, 1970).
[6] H. Busch, J.L. Hodnett, H.P. Morris, R. Neogy, K. Smetana and T. Unuma, Cancer Res. 28 (1968) 672.
[7] E. Yadzi, T.S. Ro-Choi, J. Wikman, Y.C. Choi and H. Busch, Cancer Res. 29 (1969) 1755.
[8] G.W. Rushizky and C.A. Knight, Proc. Natl. Acad. Sci. U.S. 46 (1960) 947.
[9] F. Marks, E.J. Hidvegi, E. Yazdi and H. Busch, Biochim. Biophys. Acta 195 (1969) 340.
[10] J. Wikman, E. Howard and H. Busch, J. Biol. Chem. 244 (1969) 5471.
[11] T. Nakamura, A. Prestayko and H. Busch, J. Biol. Chem. 243 (1968) 1368.
[12] Y. Moriyama, J.L. Hodnett, A.W. Prestayko and H. Busch, J. Mol. Biol. 39 (1969) 335.
[13] J.L. Hodnett and H. Busch, J. Biol. Chem. 243 (1968) 6334.
[14] A.Prestayko and H.Bush, Biochim. Biophys. Acta 169 (1968) 327.
[15] T.S.Ro-Choi, Y.Moriyama, Y.C.Choi and H.Bussch, J. Biol. Chem. 254 (1970) 1970.
[16] A.W. Prestayko, M. Tonato and H. Busch, J. Mol. Biol. 47 (1970) 505.
[17] S.M. El-Khatib, T.S. Ro-Choi, Y.C. Choi and H. Busch, Federation Proc. 29 (1970).
[18] A.W. Prestayko, M. Tonato and H. Busch, Proc. Am. Assoc. Cancer Res. 10 (1970).
[19] J. Dworak, M. Knecht, R. McParland and H. Busch, Federation Proc. 29 (1970)
[20] C.M. Mauritzen and Y.C. Choi, in preparation.

CONTROL OF DNA SYNTHESIS IN MAMMALIAN CELLS*

R. BASERGA

Department of Pathology and Fels Research Institute,
Temple University School of Medicine, Philadelphia, Pa. 19140, USA

Abstract: Baserga, R. Control of DNA Synthesis in Mammalian Cells. *Miami Winter Symposia* 2, pp. 447–461. North-Holland Publishing Company, Amsterdam, 1970.

Mammalian cell populations can be divided, in relation to DNA synthesis and mitosis, into three groups or subpopulations: 1) continuously dividing cells that continuously move around the cell cycle; 2) G_0 cells, that is, cells that leave the cell cycle but can be induced to synthesize DNA and divide by an appropriate stimulus, and 3) non-dividing cells that have permanently left the cell cycle and are destined to die without dividing again. In most cases a cell that has taken the decision to replicate its genetic material is a cell that has taken the decision to divide. Therefore, a study of the factors that control the initiation of DNA synthesis in mammalian cells is of importance in understanding the regulation of cell proliferation. Of the two models described above (continuously dividing cells and G_0 cells) we have chosen G_0 cells to illustrate the biochemical events that precede and presumably control the onset of DNA synthesis in mammalian cells. There are several models of stimulated DNA synthesis both *in vivo* and *in vitro* but the model we have selected to study the sequence of biochemical events that occur between the applications of the stimulus and the burst of mitoses is the isoproterenol-stimulated salivary gland of mice.

When mice or rats are given a single injection of isoproterenol (IPR), a synthetic catecholamine, the salivary glands respond, after a lag period of about 30 hr, with a burst of cellular proliferation. The proliferative response is limited to the salivary glands, and among the major salivary glands of rodents the parotid responds with the greatest stimulation, with as many as 80 percent of the parotid cells being stimulated to divide by a single injection of IPR. The advantages offered by this model are: 1) cell proliferation is induced by a single administration of a chromatographically pure chemical compound; 2) the metabolic changes occurring between time zero and the burst of mitoses can be timed with great accuracy, and 3) the fraction of cells stimulated to divide is considerable. The two questions we are asking ourselves in this system are: 1) what is the sequence of metabolic events that occur between the application of the stimulus and the burst of mitoses? and 2) what is the initial event that triggers the entire sequence of changes leading to cell division?

* The work described in this article has been supported by United States Public Health Service Research Grant CA-08373 from the National Cancer Institute, and DE-02678 from the National Institute for Dental Research.

A careful analysis of the various biochemical events that occur between the application of the stimulus and the burst of mitoses permits us to conclude that cell proliferation in mammalian cells is controlled by a segment of the genome itself and that the onset of mitoses is preceded by a series of steps which include various rounds of protein and RNA synthesis and culminate in the replication of DNA that precedes mitosis.

1. Introduction

The cells of the adult animal can be divided, in respect to DNA synthesis and cell division, into three populations: 1) continuously dividing cells, that keep moving around the cell cycle from one mitosis to the next. A good example of this kind of population are the epithelial cells lining the crypts of the small intestine; 2) non-dividing cells, such as polymorphonuclear leukocytes and keratinizing cells of the epidermis, which have permanently left the cell cycle and are destined to die without dividing again; and 3) quiescent cells, that have left the cell cycle and ordinarily do not synthesize DNA or divide but can be stimulated to do so by an appropriate stimulus. The regenerating liver cells after partial hepatectomy constitute the best illustration of this third type of cell.

Although DNA synthesis and cell division can be dissociated, in the great majority of cases they are tightly coupled; that is, in most instances, a cell that is synthesizing DNA is a cell that has taken the decision to divide. Since mitosis is almost invariably preceded by replication of DNA, a study of the factors that control the initiation of DNA synthesis is of critical importance in investigating the control of cell proliferation in mammalian cells. In both continuously dividing cells and quiescent cells that can be stimulated to divide by an appropriate stimulus, the onset of DNA synthesis is preceded by a series of metabolic events. However, this discussion is limited to the biochemical changes that precede the onset of DNA synthesis in quiescent cells stimulated to divide, and specifically to the metabolic events that occur in the salivary glands of rodents after stimulation with isoproterenol. For the functional and structural changes preceding the initiation of DNA synthesis in continuously dividing cells, and for other models of stimulated DNA synthesis and cell division, the reader is referred to a recent book on the biochemistry of cell division [1]. In this paper I will only attempt to summarize our knowledge of the various biochemical events occurring in the salivary glands of rodents between the application of the stimulus, isoproterenol, and the burst of mitoses that occurs in the salivary gland cells 30 hr later.

The isoproterenol-stimulated salivary gland was first introduced as a model of induced cellular proliferation by Barka [2,3] who showed that a single injection of isoproterenol caused, after a lag period of about 20 hr, a marked increase in DNA synthesis and mitoses in rats. Isoproterenol is a commercially available synthetic catecholamine whose structure is almost identical to that of epinephrine, except that the methyl group at the end of the side chain of epinephrine has been replaced by a bulkier isopropyl group. When injected intraperitoneally into either rats or mice it causes a marked hyperplasia as well as a hypertrophy of the cells of the submandibular and parotid glands [2,4,5]. However, after a single injection of isoproterenol, the response of the salivary glands is essentially limited to cellular proliferation. In a typical experiment, mice are injected intraperitoneally with 0.8 μmole per g of body weight of isoproterenol. Thirty hr later both the parotid and the submandibular glands contain numerous mitoses, which are exceedingly infrequent in non-stimulated salivary glands of rodents. It is this burst of mitoses that is the object of the present discussion. The problem can be stated as follows (fig. 1): a single injection of isoproterenol at time zero causes a large number of mitoses to appear in the salivary glands of mice between 30 and 36 hours later. The questions we are asking ourselves are: 1) what is the sequence of biochemical changes that occur between the application of the stimulus and the burst of mitoses? and 2) what is the initial event that triggers the entire sequence of changes leading to cell division? These two questions are the key to the whole problem of the regulatory mechanisms that control cell division in mammalian cells both *in vitro* and *in vivo* [6,7]. We have selected this model for our studies because of some unique features it offers, namely: 1) DNA synthesis and cell division are induced by a single administration of a chromatographically pure chemical compound; 2) the metabolic events occurring between time zero and the burst of mitoses can be timed with great accuracy; 3) the stimulation, as we shall see later, can be irreversibly inhibited; and 4) the number of cells in the salivary glands that are stimulated to synthesize DNA and divide is quite high, as we shall also see later.

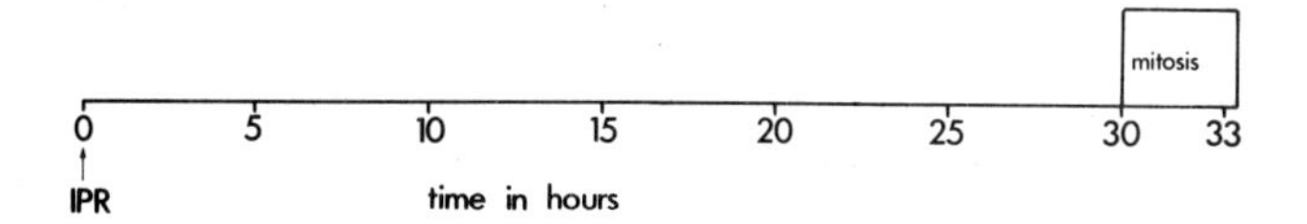

Fig. 1. The problem to be considered in the present discussion, namely, what is the sequence of biochemical events occurring between the application of the stimulus, isoproterenol, at zero time and the burst of mitoses 30 hr later.

In trying to answer the first question (what is the sequence of metabolic events occurring between the administration of isoproterenol and the burst of mitoses?) we shall proceed backward, that is, we shall start from mitosis and from it retrace our steps backward to the early chemical events that from a remote distance control the mitotic division of cells.

2. Late pre-replicative phase

The first step in our homeward journey is simple. Several years ago, Nygaard and Rusch [8] and Van Potter and coworkers [9,10] showed that cell division in the regenerating liver is preceded by DNA synthesis, that is, by the exact replication of the genetic material so that each daughter cell may receive the euploid amount of DNA. This is true in most models of stimulated cell proliferation [6] and also in the isoproterenol-stimulated salivary gland. Fig. 2 shows the percentage of cells in DNA synthesis and the mitotic index of mouse salivary glands at various times after the administration of isoproterenol. The number of cells in DNA synthesis was determined by injecting thymidine-^{3}H 30 min before killing and by autoradiography of the fixed salivary glands. The wave of mitoses is preceded by a wave of DNA synthesis. Of

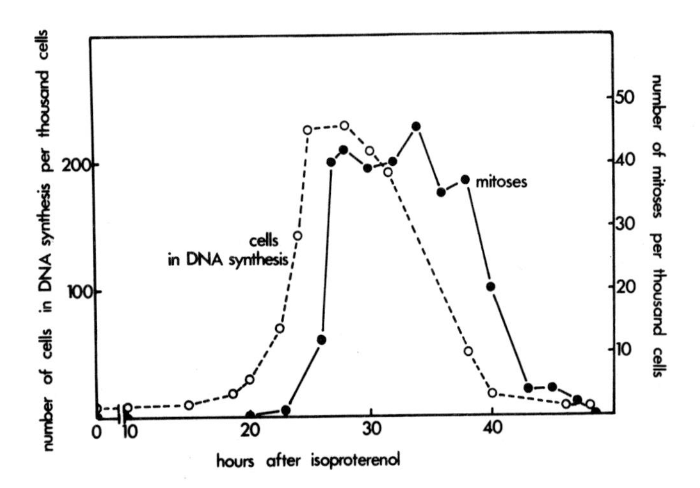

Fig. 2. DNA synthesis and mitosis in salivary gland cells after a single injection of isoproterenol into Fels A mice. All animals were killed 30 min after an injection of thymidine-^{3}H and the number of cells in DNA synthesis was determined by autoradiography.

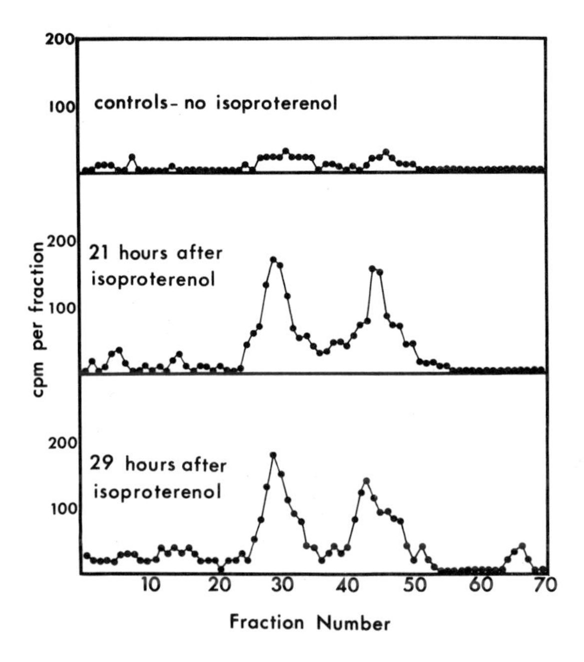

Fig. 3. Histone synthesis in the isoproterenol-stimulated salivary gland of mice. The animals, except the controls, were injected with 0.8 μmole/g of body weight of isoproterenol. One hour before killing they were injected with lysine-^{3}H. Histones were isolated and electrophoresed according to the method of Borun et al. [12]. On the abscissa is the fraction number of the polyacrylamide gel. On the ordinate are the counts per minute in the respective fraction. Note the marked increase in the amount of lysine-^{3}H incorporated into proteins in the two histone peaks at 21 and 29 hr after the administration of isoproterenol.

the three major salivary glands of rodents, the parotid is the most responsive, the submandibular gives an intermediate response, and the sublingual is practically insensitive to the stimulation by isoproterenol [11]. In the parotid, from 60-80% of the acinar cells are stimulated to enter DNA synthesis by a single administration of isoproterenol [11].

At the same time that DNA is replicated, histone synthesis also reaches a maximum and, in fact, according to a number of authors [12] histone synthesis occurs only during the S phase. In collaboration with Ted Borun (unpublished results), I have studied the synthesis of histones in the isoproterenol-stimulated salivary gland. Figs. 3 and 4 summarize our results. The mice were injected with isoproterenol and at various time afterwards with lysine-^{3}H. The histones were extracted from the isolated, purified nuclei and electrophoresed

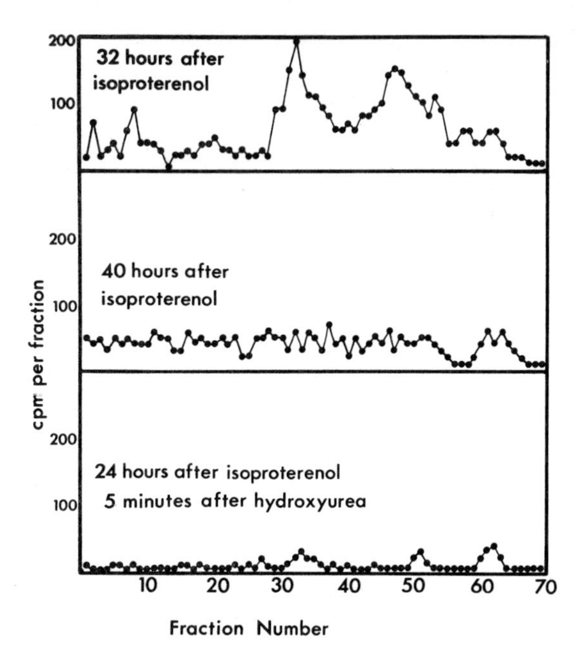

Fig. 4. Same experimental conditions as in fig. 3. The histone synthesis is still increased at 32 hr after isoproterenol but it has returned to control levels at 40 hr when DNA synthesis has also returned to control values. A single injection of hydroxyurea, 5 min before the injection of lysine-^{3}H, causes a complete suppression of DNA synthesis as well as histone synthesis in the stimulated salivary gland. Fels A mice were used in all these experiments.

on polyacrylamide gel. Fig. 3 shows that in the control animals, not injected with isoproterenol, the incorporation of lysine-^{3}H into the two histone peaks is very modest. However, at 21, 29 and 32 hr after isoproterenol, when the synthesis of DNA is at its peak, the two histone peaks are also considerably increased, denoting a very active incorporation of lysine-^{3}H into salivary gland histones. The incorporation of lysine-^{3}H returns to control levels by 40 hr after isoproterenol when DNA synthesis has also subsided (fig. 4). We were also able to demonstrate that the administration of hydroxyurea, an inhibitor of DNA synthesis, promptly inhibits the synthesis of histones in the isoproterenol-stimulated salivary gland even at the peak of DNA synthesis, that is, 24 hr after injection of isoproterenol. These results confirm on broad lines the results previously obtained by Borun et al. [12] in HeLa cells that histone synthesis replicates *pari passu* with DNA synthesis.

Four enzymes tightly coupled with DNA synthesis have been studied in the

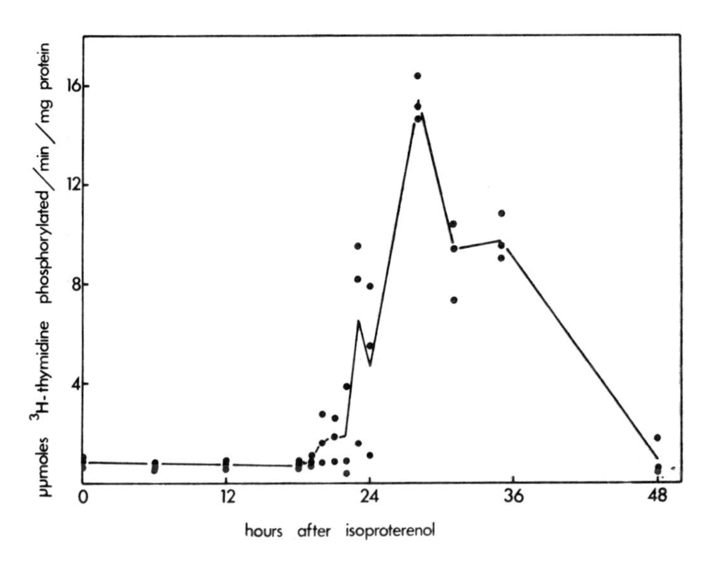

Fig. 5. Thymidine kinase activity in the isoproterenol-stimulated salivary gland of Fels A mice. The animals injected with isoproterenol were killed at the intervals indicated on the abscissa. The salivary glands were homogenized and the 105,000 g supernatant was used for thymidine kinase determination (see text).

isoproterenol-stimulated salivary gland. Fig. 5 illustrates an experiment in which one of these enzymes was studied. The animals were injected with 0.8 μmole/g of body weight of isoproterenol and thymidine kinase activity was determined in the 105,000 *g* supernatant of salivary gland homogenate at various intervals after the administration of isoproterenol. Thymidine kinase activity, determined by the method of Breitman [13], follows very much the curve of DNA synthesis, beginning to increase at about 20 hr after isoproterenol, and reaching a peak at about 28–30 hr, while by 48 hr it has returned essentially to control levels.

Three other enzymes were studied in a similar way: DNA polymerase by Barka [3] and thymidylate synthetase and thymidylate kinase by Pegoraro and Baserga [14]. The results of these studies are shown in fig. 6. The activities of thymidylate synthetase and DNA polymerase increase at about 18 hr after isoproterenol, that is, roughly 2 hr before the onset of DNA synthesis, whereas the activities of thymidine kinase and thymidylate kinase begin to increase at 20 hr, at the very time when DNA synthesis begins to increase.

At about the time that the activities of thymidylate synthetase and DNA polymerase begin to rise, the glycogen stores of the salivary glands undergo a

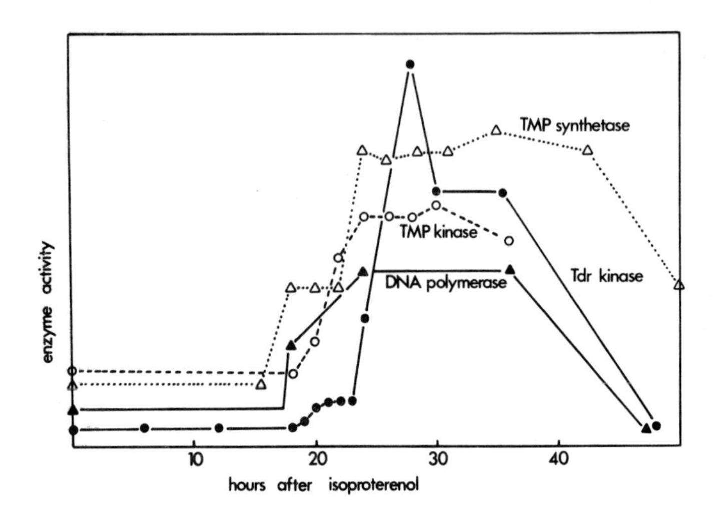

Fig. 6. Activities of certain enzymes tightly coupled with DNA synthesis in the isoproterenol-stimulated salivary gland of Fels A mice. The abscissa gives the time after a single administration of isoproterenol. On the ordinate is enzyme activity on an arbitrary scale. The values for DNA polymerase have been taken from the paper by Barka [3], and for the other three enzymes from the paper by Pegoraro and Baserga [14].

rapid transformation. The concentration of glycogen in salivary gland tissues, which had increased to a level five times the control values, decreases very rapidly [15]. Similar changes in glycogen concentration have been reported in two other models of stimulated DNA synthesis, namely, the phytohemagglutinin-stimulated lymphocytes [16] and the alloxan-diabetic rat liver stimulated to synthesize DNA by insulin [17]. The significance of these changes is still obscure, but our data indicate that they are specific for the salivary gland and are closely correlated with the onset of DNA synthesis.

Let us now return to three of the enzymes mentioned above as tightly coupled with the process of DNA replication, namely, thymidine kinase, thymidylate kinase and thymidylate synthetase. Pegoraro and Baserga [14] have determined the time of appearance of the templates coding for these enzymes by injecting actinomycin D at varying intervals after isoproterenol and then measuring enzymatic activity at 28 hr, which is the peak of activity for the three enzymes. When actinomycin D was injected at 18 hr, or later, after the administration of isoproterenol, enzyme activity at 28 hr was the same as in isoproterenol-stimulated salivary glands of animals not injected with actinomycin D. On the contrary, if actinomycin D was injected before the 12th hr

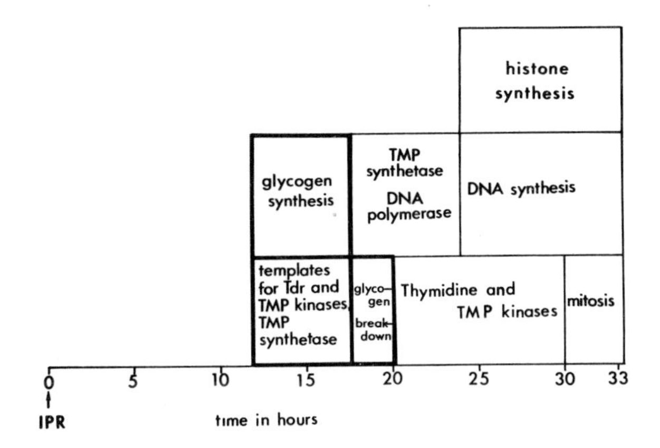

Fig. 7. A summary of the biochemical events occurring in the late prereplicative phase and during DNA synthesis in the isoproterenol-stimulated salivary gland.

after isoproterenol, enzyme activity at 28 hr remained at the levels of control mice not injected with isoproterenol. Between 12 and 18 hr the effect of actinomycin D was intermediate. The results indicated that the templates coding for these three enzymes were synthesized between 12 and 18 hr after isoproterenol. The templates, which it is reasonable to assume are RNA templates, are not only made during a discrete period of the prereplicative phase, but are made distinctly in advance of the enzymes. The findings up to this point are summarized in fig. 7, an attempt to make the *tabula rasa* between zero time and mitosis less frightening.

3. Early pre-replicative phase

Sasaki et al. [18] have shown that isoproterenol causes two bursts of protein synthesis in salivary glands. These two bursts occur at 1 and 8 hr after isoproterenol. The increased protein synthesis is limited to the free ribosomes fraction, whereas no increase in protein synthesis can be detected in the membrane-bound ribosomes fraction. This is of particular interest since several investigators have recently demonstrated that free ribosomes synthesize proteins necessary for cell growth, while membrane-bound ribosomes, which constitute the majority of salivary gland ribosomes, synthesize export proteins [19–21]. The increase in protein synthesis in the free ribosomes fraction could also be demonstrated in a cell-free amino acid incorporating system. It was noted, howe er, that the increase at 8 hr was inhibited by the previous administration of actinomycin D, but not the increase at one hour. The relevance of these processes to cell division was further confirmed by the finding that when inhibitors of protein synthesis (such as cycloheximide and puromycin) were given one hour after isoproterenol, the subsequent stimulation of DNA synthesis, 20 hr later, was effectively and irreversibly suppressed, although these compounds inhibited protein synthesis in the salivary glands of mice for only a very short time. Sasaki et al. [18] concluded from their investigations that at 1 and 8 hr after isoproterenol, a protein (or class of proteins) is synthesized in the salivary glands which is necessary for the onset of DNA synthesis 20 hr later, and that when the synthesis of this or these proteins is inhibited, the whole process of stimulated DNA synthesis comes to a standstill. The fact that cycloheximide, which has only a fleeting effect on protein synthesis in the salivary glands of mice, effectively suppresses isoproterenol-stimulated DNA synthesis when given one hour after isoproterenol, also indicates that the templates coding for these proteins must be labile.

I mentioned above that when actinomycin D is given 30 min before isoproterenol there is no inhibition of the burst of protein synthesis occurring at one hour in the free ribosomes fraction. This suggests that no template synthesis is necessary for the increased protein synthesis occurring at this time. However, there is no question that the whole process of isoproterenol-stimulated DNA synthesis is very sensitive, at later times, to even small doses of actinomycin D [22]. This inhibitory effect of actinomycin D suggests that other RNA templates distinct from the ones previously mentioned in connection with thymidine kinase and other DNA enzymes must be made between 1 and 8 hr after isoproterenol. I must emphasize the fact that the existence of these templates is purely based on the circumstantial evidence derived from the inhibitory effect of actinomycin D. At variance with other models of stimulated DNA synthesis, in which a brisk stimulation of RNA synthesis in the early pre-replicative phase has often been described [6], no increase in RNA synthesis can be detected in the isoproterenol-stimulated salivary glands of mice [22]. Some changes do occur in the rate of incorporation of uridine-^{3}H into salivary gland RNA after administration of isoproterenol, but these changes can be explained by changes in the size of the uridine triphosphate precursor pool [23]. In this respect it is of interest to note that 7 hr after isoproterenol there is also a decrease in the activity of the enzyme uridylate kinase [24]. The decrease in uridylate kinase activity is again specific for the salivary gland and seems to be relevant to the onset of DNA synthesis several hours later.

Another early effect of isoproterenol is a marked stimulation of salivary gland secretion which can be shown both ultrastructurally [25] and functionally, as evidenced by a marked decrease in *α-amylase act*ivity [26,27], DNAase activity [14], and concentration of calcium and magnesium [28]. α-Amylase and DNAase activity decrease within 2 hr after administration of isoproterenol to a minimum which is about 10 percent of control values. After the 10th hr the activities of these two enzymes begin to increase and by 24 hr both enzymes have returned to control levels. The resynthesis of α-amylase, which begins 10 hr after isoproterenol, is totally unaffected even by large doses of actinomycin D [27], a finding which indicates that the previously described effects of actinomycin D on DNA synthesis and enzyme induction are selective and not due to a generalized toxic effect. Similar results were obtained with DNAase, whose resynthesis is not inhibited by actinomycin D given at various times between 2 and 18 hr after isoproterenol [14].

An important question in our model of isoproterenol-stimulated DNA synthesis is whether or not secretion plays an important role in the stimulation of DNA synthesis and cell division. This is not only of theoretical importance but also has some practical aspects in deciding the priority of future investigations. On the one hand, physaelemin, a polypeptide that is a hundred thousand times more potent than isoproterenol in stimulating secretion, has no effect on DNA synthesis in salivary glands [29]. On the other hand, we have not been able yet to stimulate DNA synthesis in the salivary glands without first producing a marked stimulation of secretion, as we shall see a little later. Relevant to the problem of secretion is the recent finding by Malamud [30] that adenyl cyclase activity in the mouse salivary gland increases quickly, reaching a maximum within 2.5 min after the administration of isoproterenol.

This brings us almost inadvertently to the second question we asked ourselves at the beginning, namely, what is the initial event that triggers the entire sequence of changes leading to cell division. An answer to this question would be highly desirable since it is the key to the control of cellular proliferation and, consequently, of both normal and abnormal growth. Needless to say, we do not know the answer to this question but we think that a gallant attempt ought to be made at least to place the unknown answer within finite boundaries. In the first place, the evidence indicates that the stimulating effect of isoproterenol on DNA synthesis is exerted directly at the level of the salivary gland cells and is not mediated through other organs or tissues. This statement is based on the findings that isoproterenol-stimulated DNA synthesis is not inhibited by adrenalectomy [31], hypophysectomy [32], thyroidectomy [33], or by extirpation of the superior cervical ganglion [34,35]. In addition, we have recently found that when 0.1 mg of isoproterenol, an amount which is ineffective in stimulating DNA synthesis when injected intraperitoneally, is injected directly through a cannula into the salivary gland, there is a marked stimulation of DNA synthesis in both the homo- and the contralateral salivary glands (unpublished results). Also recently, Kreider [36] has removed the last doubts by achieving stimulation of DNA synthesis by isoproterenol in an organ culture. If the effect of isoproterenol is exerted directly at the level of the salivary gland cells, then it is of interest to determine the fate of the injected isoproterenol. The catabolism of isoproterenol in the intact animal is, fotunately, quite simple. Isoproterenol is either methylated on the 3-hydroxyl group, or glucuronized on the 4-hydroxyl group of the ring [37]. These compounds can be easily separated by column chromatography [38] and studies in our laboratory have conclusively demonstrated that: 1) while in other organs isoproterenol is rapidly catabolized, unchanged isoproterenol accumulates in the salivary gland in the first 30 min after injection, and 2) the unchanged isoproterenol molecule, and not any of its catabolites, is the compound actually responsible for stimulating DNA synthesis and cell division [39].

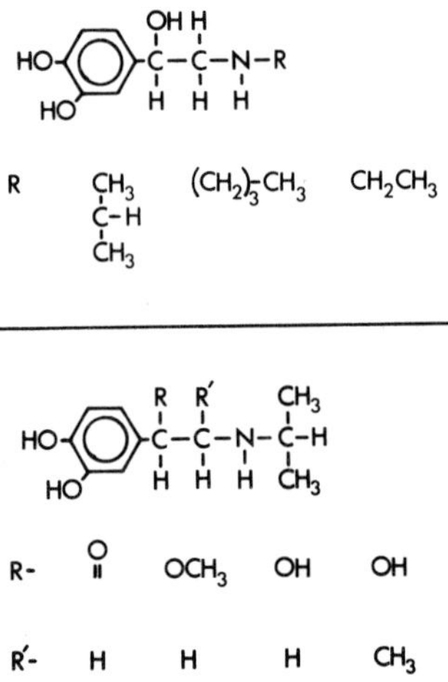

HO–(ring)(HO)–C(OH)(H)–C(H)(H)–N(H)–R

R	$CH(CH_3)_2$	$(CH_2)_3CH_3$	CH_2CH_3

HO–(ring)(HO)–C(R)(H)–C(R′)(H)–N(H)–C(CH_3)(H)(CH_3)

R-	=O	OCH_3	OH	OH
R′-	H	H	H	CH_3

Fig. 8. Analogs of isoproterenol. The upper part of the figure shows substitutions of the end group of the side chain. When R is isopropyl, the structure represented in the upper part of the figure is isoproterenol. When R is replaced by an ethyl group the compound is inactive in stimulating DNA synthesis. Bulkier groups (R in the middle) give derivatives that are still active in stimulating DNA synthesis.

The lower part of the figure shows substitutions on the side chain, exlusive of the end group. When R is OH and R′ is H, the structure given in the lower part of the figure is isoproterenol. Various substitutions at the R or R′ positions have no effect, or very little effect, on the ability of the analog to stimulate DNA synthesis in the mouse salivary gland.

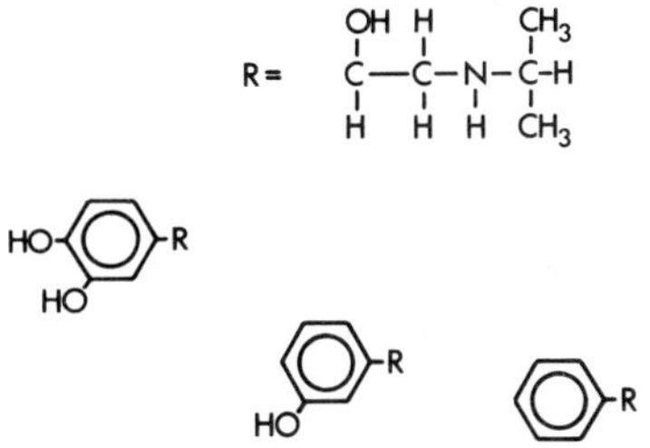

Fig. 9. Analogs of isoproterenol. The side chain is the same for all three compounds. When the phenyl ring has two hydroxyl groups, the compound is isoproterenol itself. Removal of one hydroxyl group has no effect on the ability of the analog to stimulate DNA synthesis, but when both hydroxyl groups are removed the compound is totally inactive in stimulating DNA synthesis in mouse salivary glands.

We have then investigated which group in the isoproterenol molecule is necessary for the compound to be active in stimulating DNA synthesis and cell division. This was done in collaboration with Drs. Kirby and Swern at Temple University, and a number of isoproterenol analogs were studied in order to determine which among them were capable of stimulating DNA synthesis and secretion in the salivary glands [40]. The results can be summarized as follows (figs. 8 and 9): 1) substitutions of various kinds on the alpha and beta carbons of the side chain produce compounds that are still quite active in stimulating DNA synthesis; 2) the isoproterenol customarily used in our experiments is the racemic form, *dl*-isoproterenol. When the racemic compound is resolved into the *d*- and *l*-isoproterenol, both derivatives are active in stimulating DNA synthesis; 3) when the isopropyl group at the end of the side chain is replaced by a less bulky group (as, for instance, an ethyl group), activity is markedly decreased, whereas larger end groups have no effect on the ability to stimulate DNA synthesis; 4) the most marked effects was obtained when the hydroxyl groups on the phenyl ring were eliminated or replaced. When both hydroxyl groups were missing, the compound was totally inactive in stimulating DNA synthesis, although it could still stimulate salivary gland secretion, a finding which assures us that the compound was taken up by cells. On the contrary, when only the 4-hydroxyl or the 3-hydroxyl groups are removed, the molecule is not only still active but it is even more active than the parent compound; 5) we have been able to find compounds that will produce salivary gland secretion without stimulating DNA synthesis, but thus far we have not been able to synthesize a compound that will produce DNA synthesis without causing salivary gland secretion.

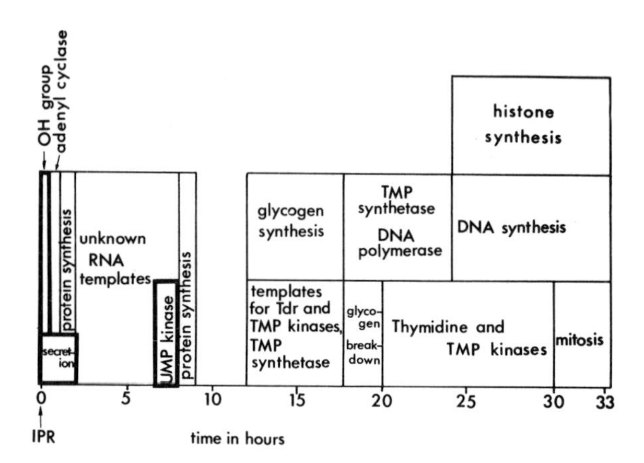

Fig. 10. A summary of our knowledge of the sequence of biochemical events occurring in the salivary glands of mice and rats between the application of the stimulus (an intraperitoneal injection of isoproterenol) and the burst of mitoses 30 hr later.

This is as far as our story goes. Our knowledge of the various biochemical events occurring between the application of the stimulus, isoproterenol, and the burst of mitoses 30 hr later, is summarized in fig. 10. In less than an hour we have then moved backward 30 hr from mitosis to the hydroxyl group that interacts with the salivary gland cells and presumably stimulates the sequence of events leading to the onset of DNA synthesis and subsequently cell division. To some of the readers I may now look a squid, which is defined in the dictionary as an animal that rapidly moves backward emitting at the same time large quantities of ink. One must admit, however, that the blank space between zero time and mitosis now looks less frightening, and that despite the numerous gaps, some kind of intelligent order begins to appear in the sequence of metabolic events that lead from the interaction of the hydroxyl group with the cell to the division of the cell itself.

References

[1] Biochemistry of Cell Division, ed. R. Baserga (Charles Thomas, Springfield, Ill, 1969).
[2] T. Barka, Exptl. Cell. Res. 37 (1965) 662.
[3] T. Barka, Exptl. Cell. Res. 39 (1965) 355.
[4] K. Brown-Grant, Nature 191 (1961) 1076.
[5] H. Selye, R. Veilleux and M. Cantin, Science 133 (1961) 44.
[6] R. Baserga, Cell Tissue Kinetics 1 (1968) 167.
[7] R. Baserga, in: Human Tumor Cell Kinetics, ed. S. Perry, Nat. Cancer Inst. Monogr. 30 (1969) 1.
[8] O. Nygaard and H.P. Rusch, Cancer Res. 15 (1955) 240.
[9] F.J. Bollum and V.R. Potter, J. Biol. Chem. 233 (1958) 478.
[10] L.I. Hecht and V.R. Potter, Cancer Res. 16 (1956) 988.
[11] R. Baserga, in: Proliferation and Spread of Neoplastic Cells. M.D. Anderson Hospital and Tumor Institute, Texas Medical Center, Houston, 1968, p. 261.
[12] T.W. Borun, M.D. Scharff and E. Robbins, Proc. Natl. Acad. Sci. U.S. 58 (1967) 1977.
[13] T.R. Breitman, Biochim. Biophys. Acta. 67 (1963) 153.
[14] L. Pegoraro and R. Baserga, Invest. 22 (1970) 266.
[15] D. Malamud and R. Baserga, Exptl. Cell. Res. 50 (1968) 581.
[16] D. Quaglino, D.C. Cowling and F.G.J. Hayhoe, Brit. J. Haematol. 10 (1964) 417.
[17] D.F. Steiner and J. King, J. Biol. Chem. 239 (164) 1292.
[18] T. Sasaki, G. Litwack and R. Baserga, J. Biol. Chem. 244 (1969) 4831.
[19] S.J. Hicks, J.W. Drysdale and H.N. Munro, Science 164 (1969) 584.
[20] C.M. Redman, J. Biol. Chem. 244 (1969) 4308.
[21] M.C. Ganoza and C.A. Williams, Proc. Natl. Acad. Sci. U.S. 63 (1969) 1370.
[22] R. Baserga and S. Heffler, Exptl. Cell. Res. 46 (1967) 571.
[23] D. Malamud and R. Baserga, Biochim. Biophys. Acta 196 (1969) 258.
[24] D. Malamud and R. Baserga, Science 162 (1968) 373.
[25] J.V. Simson, Z. Zellforsch. 101 (1969) 175.
[26] P. Byrt, Nature 212 (1966) 1212.
[27] J.P. Whitlock Jr., R. Kaufman and R. Baserga, Cancer Res. 28 (1968) 2211.
[28] R.H. Dreisbach, Proc. Soc. Exptl. Biol. Med. 116 (1964) 953.
[29] G. Bertaccini, G. De Caro and R.Cheli, J. Pharm. Pharmacol. 18 (1966) 312.
[30] D. Malamud, Biochem. Biophys. Res. Commun. 35 (1969) 754.
[31] D. Malamud and R. Baserga, Life Sci. 6 (1967) 1765.
[32] J.J. Argonz, Acta Physiol. Latino Am. 12 (1962) 231.
[33] G. A. Bray, Endocrinology 79 (1966) 554.
[34] H. Wells, Am. J. Physiol. 207 (1964) 313.
[35] T. Barka, Exptl. Cell. Res. 47 (1967) 564.
[36] J. Kreider, Cancer Res., in press.
[37] G. Hertting, Biochem. Pharmacol. 13 (1964) 1119.
[38] I.J. Kopin, J. Axelrod and E. Gordon, J. Biol. Chem. 236 (1961) 2109.
[39] R. Baserga, T. Sasaki and J.P. Whitlock Jr., in: Biochemistry of Cell Division, ed. R. Baserga (Charles Thomas, Springfield, Ill., 1969) p. 77.
[40] K.C. Kirby Jr., D. Swern and R. Baserga, Mol. Pharmacol. 5 (1969) 572.

RESPIRATION, GLYCOLYSIS AND ENZYME ALTERATIONS IN LIVER NEOPLASMS

S. WEINHOUSE

Fels Research Institute, Temple University School of Medicine, Philadelphia, Pennsylvania 19140, USA

Abstract: Weinhouse, S. Respiration, Glycolysis and Enzyme Alterations in Liver Neoplasms. *Miami Winter Symposia* 2, pp. 462–480. North-Holland Publishing Co., Amsterdam, 1970.

Intensive study of the Morris hepatomas in recent years has established that tumors, even of a single cell type, can display a wide diversity in differentiation, paralleled by marked differences in metabolism and enzyme pattern. Well differentiated Morris hepatomas differ sharply from most experimentally induced rodent tumors in that they have a low glycolytic capability, a high respiration, and utilize fatty acids as metabolic fuel. These well differentiated tumors are low in glucokinase activity, but are comparable with liver in the activity of certain enzymes of fatty acid metabolism.

To ascertain relationships between the growth rate and state of differentiation of liver tumors and their degree of retention or loss of hepatic funtional activities, a series of hepatic "marker" enzymes have been studied in a spectrum of Morris tumors. Highly differentiated liver tumors, which grow very slowly, have the full complement of four hexokinase isoenzymes including high levels of glucokinase and fructokinase, have essentially only the liver type fructose-1-phosphate aldolase B, have a preponderance of liver type pyruvate kinase, have four adenylate kinase isozymes, with liver type distribution; and have a moderate-to-low LDH activity with a preponderance of M type subunits. Less highly differentiated tumors, which make up the bulk of Morris hepatomas, and which grow slowly to moderately, retain generally the liver type isozymes. However, the rapidly growing, poorly differentiated tumors have nearly or complemetely lost all of the liver "marker" isozymes, and in some instances these have been replaced by high activities of isozymes normally low or absent in liver. The poorly differentiated hepatomas are characterized by high levels of hexokinase isozyme III, which is very low in liver; virtually complete replacement of aldolase B by the non-hepatic aldolase A; and a very high level of muscle pyruvate kinase accompanied by an isoenzyme absent in liver or muscle. These results are compatible in general with Weber's concept of enzymatic correlation with tumor growth rate.

The high hexokinase and pyruvate kinase levels of the poorly differentiated tumors readily explain their high glycolytic activity on the basis of competition for ADP between the glycolytic and respiratory systems, coupled with rapid regeneration of ADP via hexokinase. Underlying the differentiation of hepatic tumors there appears to be a switch in genome readout involving the suppression of enzymes involved in hepatic function and the unmasking of other enzymes normally suppressed in the tissue of origin.

The neoplastic transformation of cells most certainly involves some direct or indirect alteration of the genome; however, the nature of this alteration is still elusive, despite intensive effort to discover differences in DNA composition between normal and cancer tissue. In contrast, the transformed cells which make up the neoplasms exhibit a wide multiplicity and diversity of metabolic and enzymic behavior, and thus far it has not been possible to determine which of the many deviations from the tissue of origin are causally implicated, and which ones are resultants of the inevitable progression that follows upon the successive replications of the tumor cell. Up until comparatively recent years it was generally held that the cancer cell is dedifferentiated, and this view was supported by many studies which showed that neoplasia results in loss of the funtional activity associated with normal tissues. Functionally active, differentiated cells have distinctive enzyme patterns which make each tissue unique, whereas the rapidly proliferating cells composing the neoplasms were observed to have largely or completely lost the unique synthetic or metabolic capabilities of their tissue of origin. On the basis of a comprehensive comparison of a variety of experimental neoplasms, Greenstein [1] enunciated his well-known generalization, which held that in their enzyme makeup tumors as a class resembled one another more closely than they resembled their tissues of origin. The deletion of certain classes of liver proteins in hepatic tumors resulting from feeding of chemical carcinogens is the basis for the so-called "deletion hypothesis", which proposes that the neoplastic transformation is initiated by removal of specific proteins by reactions with the carcinogen [2]. These findings, which point to the neoplastic cell as a unique cell type, with a characteristic enzymatic makeup which underlies its biological behavior, is in accord with the Warburg theory [3], according to which the cancer cell has lost its respiratory capability and survives by developing a mechanism for energy transduction through anaerobic utilization of glucose. Without dwelling on the long-standing, acrimonious controversy which accompanied the Warburg hypothesis, it will suffice to say that although the respiratory incapability of tumors proved to be an untenable hypothesis [4], it was generally accepted that high aerobic and anaerobic glycolysis is a common feature of all neoplastic cells.

With the development of the series of hepatic tumors of the rat known as the Morris hepatomas* [5], by Harold P. Morris in 1960, two important con-

* All of the work on the Morris hepatomas was done with the active collaboration of H.P.Morris, and the following collaborators participated in the work from the author's laboratory: D.L.DiPietro, C.Sharma, R.Manjeshwar, J.Shatton, R.C.Adelman, F.A. Farina, C.H.Lo, W.Criss, G.Litwack, L.Bloch-Frankenthal, J.Langan and K. Ohe.

cepts arose which were in decided opposition to the above-mentioned conception of the tumor cell as a unique cell type. They showed: (1) that tumors of a single cell of origin could display a wide diversity in degree of differentiation, paralleled by variations in growth rate and in the retention and loss of hepatic function; and (2) that a high aerobic glycolysis is not necessary for tumor growth. I have chosen to discuss work from our laboratory, conducted by Jennie Shatton and a series of pre- and postdoctoral students** during the past 8 or 9 years, on the biochemical characterization of these tumors. These tumors display a wide spectrum in degree of differentiation as revealed by histologic examination*** [5], but we have found it convenient to divide these into two groups; namely, well differentiated and poorly differentiated. In addition, Dr. Morris has developed a few highly differentiated tumors, and data on these will also be included. A summary of their properties is in table 1. Closely correlated with the degree of differentiation is the growth rate, which is very low in the few highly differentiated tumors, extending to nearly a year for a transplant generation†. Growth rates are considerably faster, at 2 to 6 months for well differentiated tumors, and are extremely rapid, at 2 weeks or less for the poorly differentiated hepatomas [5]. The highly differentiated tumors have the normal liver chromosome number and karyotype, the well differentiated differ slightly, and the poorly differentiated deviate markedly in chromosome karyotype and number, from those of rat liver [6]. Respira-

** Work carried out by the author and his associates was aided by grants CA-07174, CA-10439, AM-05487, and CA-10916 from the NIH; and P-202 from the American Cancer Society.

*** We are indebted to Dr. David Meranze for histologic examination of these hepatic tumors.

† Growth rate is expressed as the average time between inoculation and transplantation.

Table 1
General properties of Morris hepatomas [5,6,39].

Property	Degree of differentiation		
	High	Well	Poor
Growth rate	Very low	Low	Rapid
Chromosome number	Normal	Nearly normal	Abnormal
Chromosome karyotype	Normal	Nearly normal	Abnormal
Respiration	High	Moderate	Moderately low
Glycolysis	Low	Low	High
Enzyme pattern	Liver-like	Some deletions	Many deletions

tion decreases moderately with loss of differentiation, but the striking feature of these tumors is the low or negligible glycolysis in the well differentiated, in contrast with the usual high level of glycolysis in the poorly differentiated tumors [7,8]. These well differentiated tumors, with their low glycolytic activity, make it clear that high lactic acid production is not an absolute requirement for tumor survival; these low glycolyzing tumors can grow, albeit slowly, and they can also metastasize, and can ultimately kill their hosts. Our curiosity as to the reasons for this grossly improper behavior of these tumors prompted us to carry out a collaborative program with Dr. Morris to explore its enzymatic basis. Although we have not as yet a definitive answer as to why these tumors have a low glycolysis, the study has provided some interesting clues on regulatory factors of glycolysis as well as some thought-provoking data on the molecular basis of the state of tumor differentiation.

Our first efforts were naturally directed to examining the activity of the initial step in glucose utilization, namely, its phosphorylation to glucose-6-phosphate. This reaction is catalyzed by a glucose-ATP phosphotransferase which exists in multiple forms in rat liver, as well as in liver of other species [6]. Fig. 1 shows the activity of this enzyme in rat liver. The term, hexokinase, is employed to cover actually three isozymes, which have similar properties, and which are present in liver at a low total activity which does not change

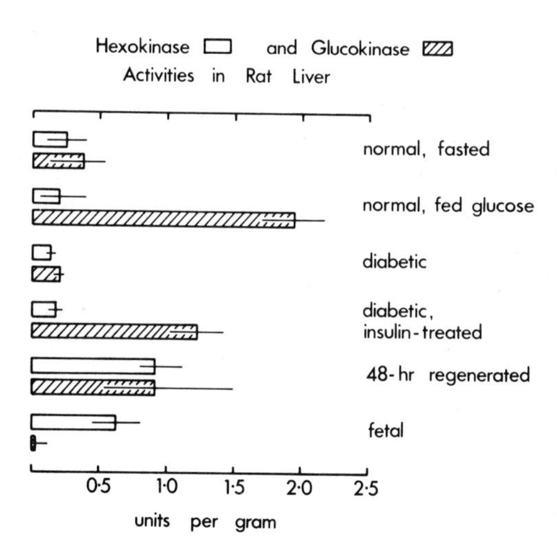

Fig. 1. Hexokinase and glucokinase activities in normal rat liver under various conditions. Values in this and subsequent tables and figures are given in units (μmoles per min) per gram tissue [14].

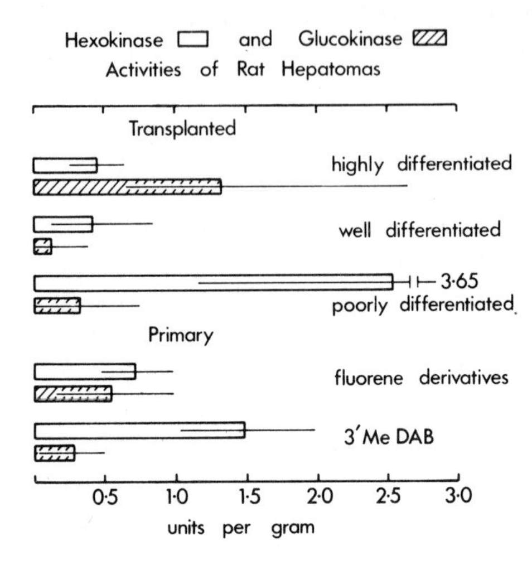

Fig. 2. Hexokinase and glucokinase activities in hepatic tumors [16].

with diet or hormonal conditions [9–12]. On the other hand, the fourth isozyme, called glucokinase, is highly responsive to dietary and hormonal conditions, being low in fasted normal animals, and high in carbohydrate-fed animals [9, 10]. It is also insulin-dependent, being extremely low in diabetes and is restored by insulin injection [13, 14]. This isozyme has an important physiological function in hepatic glucose utilization. It has a very high K_m for glucose, and it is this property that is responsible for the fact that the liver takes up glucose only when the blood glucose concentration is high. During liver regeneration after partial hepatectomy, both hexokinase and glucokinase are high, and it is striking to note that fetal liver has essentially only hexokinase and little or no glucokinase; the latter appears at normal levels only 16 to 21 days after birth [15]. Fig. 2 shows how the glucose-ATP phosphotransferase pattern changes in liver neoplasms. In a few highly differentiated, ver slow-growing hepatomas the isozyme pattern was very similar to that of normal liver, with high glucokinase and low-to-moderate hexokinase. In a large number of well differentiated tumors, studied over a wide range of transplant generations, the glucokinase dropped to very low levels, with essentially no change in hexokinase [16,17]. This is a very striking observation in view of the fact that tumors in general have rather high hexokinase levels. Indeed, when broken cell preparations of such tumors were incubated aerobically, glucose utilization and lactic acid production were very low; and the rate-limiting role of the low glucose phosphotransferase activities was verified by greatly increased glycolysis if such preparations were fortified with exogenous crystalline yeast hexokinase [18]. In contrast, a large number of poorly differentiated hepatocarcinomas exhibited high hexokinase activity with little or no glucokinase, and similar preparations of these tumors invariably had a high glycolytic activity [18]. Thus, loss of differentiation in hepatic tumors, with greatly increased growth rate, is accompanied by virtual loss of a glucose phosphotransferase which is physiologically functional and under regulation by diet and hormones, being replaced by isozymes which are ordinarily low in normal liver. Recent study of these phosphotransferase isozymes in the Novikoff and 3924A hepatomas revealed that all three hexokinase isozymes share in the marked rise, but that predominant activity is present in isozyme III [17].

Aldolase. Aldolase also exists in multiple forms which, like lactate dehydrogenase [19] are tetramers of subunits with different primary structures distinguishable by immunological or kinetic criteria [20]. In addition to aldolase A, which is the sole form of muscle aldolase, and aldolase B, which is the major form in liver, a new form, termed aldolase C has been found only in brain, where it exists largely as an A-C hybrid with a large preponderance of A [20, 21]. The three forms are most conveniently detected by electrophoresis. The relative quantities of A and B subunits in liver and hepatomas can be assayed kinetically by the ratio of activity toward fructose-1,6-diphosphate and fructose-1-phosphate. The ratio for the A form is about 50, whereas that for the B form is exactly 1, so that intermediate values for the ratio indicate the presence of both forms. In fig. 3 the FDP and F-1-P values are plotted in bar graphs for normal liver and for hepatomas of varying degree of differentiation

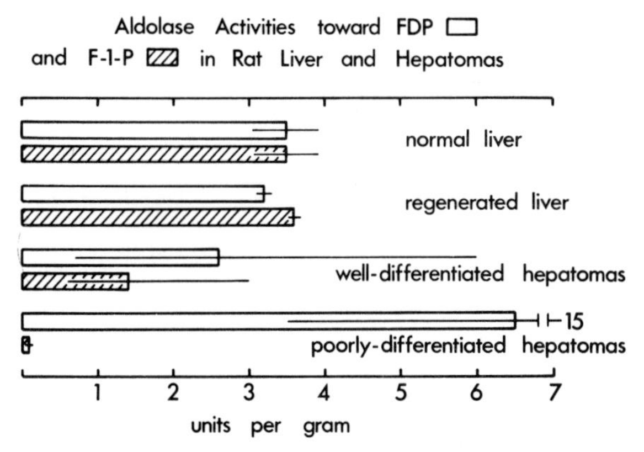

Fig. 3. Activities of aldolase towards fructose diphosphate and fructose-1-phosphate in liver and hepatic tumors [22].

[22]. In normal liver and in regenerating liver the aldolase is nearly or entirely in the B form, with equal activities toward both substrates. In the well differentiated, slow growing tumors, the activity is somewhat greater with FDP than F-1-P, but the ratio of approximately 2 indicates that aldolase B is preponderant. However, the high FDP/F-1-P ratio in the poorly differentiated tumors leaves no doubt that the B form normally present in liver has been essentially completely replaced by the A form, which is low or absent from liver. A similar observation has been made by Nordmann and Schapira [23]. An interesting example of the persistence of the C form in brain tumors has been provided recently by Sugimura [24].

Pyruvate kinase. A crucial enzyme in the glycolytic pathway is the transphosphorylase which catalyzes the transfer of phosphate from phosphoenolpyruvate to ADP. This enzyme, like the glucose-ATP phosphotransferase, exists in multiple forms. Two major forms are found in liver; an A form, similar in kinetic properties to the muscle enzyme, and a B form, which is found as the predominant form in normal liver [25,26]. The latter is highly responsive to carbohydrate in the diet and has a number of distinctive kinetic properties which differentiate it sharply from the former [25,26]. It is also low in diabetic rats, and its activity is restored by insulin treatment [27]. Thus it shares in common with other liver "marker" enzymes specific functions that give the liver cells their unique metabolic significance. Fig. 4 shows how the activities of the A and B forms vary with the degree of tumor differentiation [28]. A single, highly differentiated tumor, the 9618A, had the same isozyme

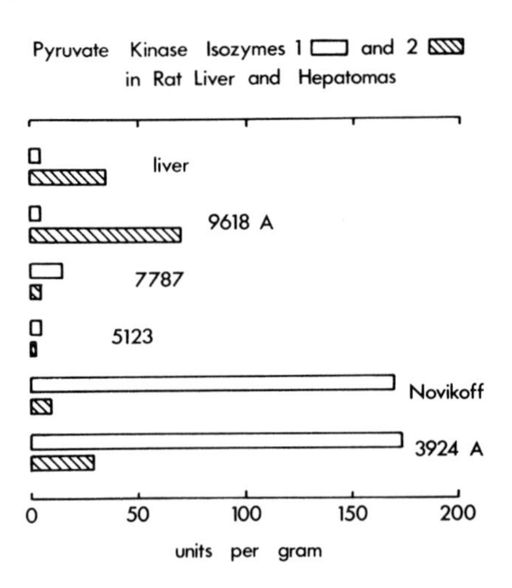

Fig. 4. Pyruvate kinase isozymes in liver and hepatic tumors [28].

pattern as liver, with predominance of the B type. However, a sharp distinction was observed between the well differentiated and the poorly differentiated tumors. The former had very low levels of both isozymes, whereas the rapidly growing, poorly differentiated Novikoff and 3924A tumors had extremely high levels of an isozyme which has the properties of the A type.
A question which arises from the present results is whether the muscle type hexokinases, or the high FDP/F-1-P ratio aldolase, or the A type pyruvate kinase, which appear in the poorly differentiated tumors, are true muscle type isozymes. By starch gel electrophoresis and by column chromatography on DEAE-cellulose, the respective, I, II, and III hexokinase isozymes of the tumor are identical with the ones in liver [17]. In a recent report, Gracy et al. [29] purified the aldolases from rat muscle and the ascitic form of the Novikoff hepatoma. Both enzymes had the same molecular weight of 158,000, each dissociated into inactive 40,000 M.W. subunits, and liberated tyrosine from the carboxyl terminus on treatment with carboxypeptidase. The muscle and tumor enzymes, whether native or carboxypeptidase-treated, were identical in kinetic properties, in amino acid composition, and in "fingerprint" pattern after tryptic digestion. Thus the aldolase isozyme that replaces the liver form in the poorly differentiated tumor is the true muscle form. The situation may be different in regard to the hepatoma pyruvate kinase. Tanaka et al. [25] reported that normal liver has four pyruvate kinase isozymes detectable by starch gel electrophoresis. Taylor, Morris and Weber [30] partially purified the major isozymes from rat liver, muscle, and the 3924A, a poorly differentiated hepatoma, and observed that the muscle and tumor isozymes differed in their electrophoretic migration on starch gel, in their stability, and in their susceptibility to SH reagents. More recently, Criss in our laboratory [31] separated the multiple forms of this enzyme by means of electrofocussing and obtained results depicted in fig. 5. Liver and a highly differentiated hepatoma, the 9618A, exhibited similar patterns of 4 isozymes, one of which was identical with that in skeletal muscle. In a well differentiated hepatoma, the 9633, the same 4 isozymes appeared, but were accompanied by a fifth form not detectable in normal liver; and in the poorly differentiated Novikoff hepatoma, the first three were very low, whereas the preponderant activity was in the muscle type and the new type. According to Dr. Criss, the new form is probably identical with the above-mentioned tumor enzyme purified by Taylor et al. [30].

Adenylate kinase. An ATP-AMP phosphotransferase, responsive to diet and insulin, was reported recently by Adelman et al. [32] to be present in liver cytosol. This enzyme is high in fasting rat liver and is lowered markedly by glucose ingestion; it is also very high in fed diabetic rat liver and is lowered

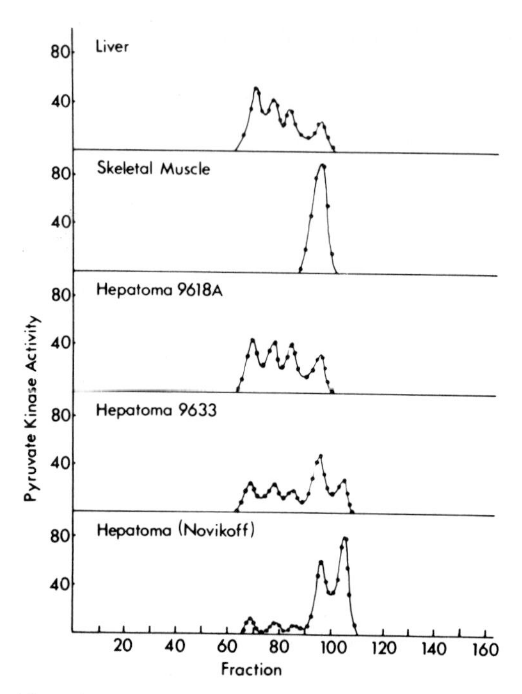

Fig. 5. Pyruvate kinase isozyme patterns for liver and hepatic tumors, using pH isoelectric focussing [31].

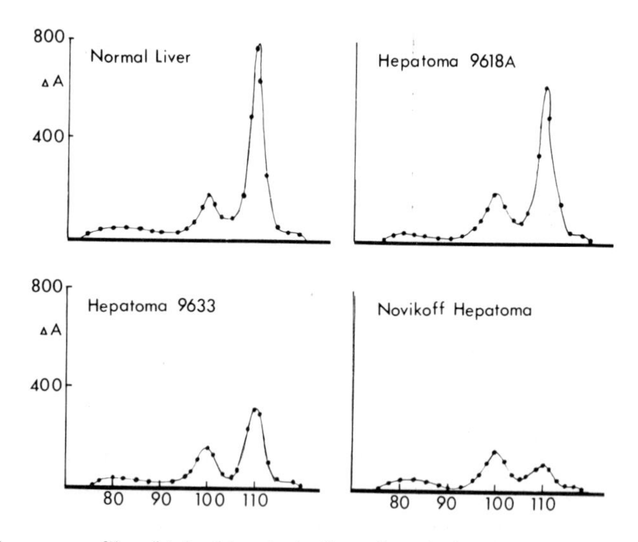

Fig. 6. Enzyme profiles obtained by electrofocussing of adenylate kinase from liver and hepatomas.

Table 2
Adenylate kinase isozyme activity in the liver and tumors from normal, fasted, and glucose refed rats [33].

Tissue	Transplant generation	Tumor growth rate (months)	Isozyme activity				
			I	II	III	IV	Total
Normal rat liver			5±2	28±4	121±9	2±1	156±12
Fasted liver			4±1	50±5	220±10	3±1	275±11
Refed liver			2±1	23±3	80±8	1±1	105±10
9618A	4	6	3±2	25±5	105±8	1±1	134±10
9633	7	4.5	3±2	28±3	47±7	2±1	78±8
Novikoff hepatoma	(709 to 711)	0.3	2±2	26±3	14±4	2±1	45±6

markedly by insulin treatment. Since this enzyme catalyzes the equilibrium among the three adenine nucleotides, AMP, ADP, and ATP, substances that are substrates or modulators of many enzymes of great functional significance in liver, its importance can hardly be overestimated. Isoelectric focussing was employed by Criss et al. [33] to separate the liver enzyme into 4 distinct forms, the predominant one of which is the adaptive enzyme. Fig. 6 shows that although all 4 forms are retained in the hepatomas, the activity of the adaptive form which is unique to liver drops sharply with loss of differentiation. Numerical data on the activities of these isozymes are displayed in table 2. Again we see another example of the loss of a liver "marker" enzyme with loss of differentiation of hepatic tumors.

Pyruvate kinase in relation to glycolysis and respiration. The high aerobic and anaerobic glycolysis of tumor tissues has been a subject of controversy that burned brightly for many years and is still flickering. The great German biochemist, Otto Warburg, first observed this phenomenon in 1922 and propounded his well-known theory of respiratory impairment in cancer cells [3]. He observed that both normal and neoplastic tissue slices utilized glucose and produced lactic acid when incubated *in vitro* in a medium of physiological pH and ionic environment. He made the further significant observation that this process was lowered quantitatively when the incubation was conducted in oxygen rather than in nitrogen. He named this process the Pasteur Effect, since Pasteur had noted some 60 years earlier that when yeast cells were placed in an aerobic environment they ceased to make alcohol, that is, fermentation stopped. Although the Pasteur Effect has not yet been explained satisfactorily it has been a focal point for much discussion and experimentation on regulatory mechanisms of carbohydrate metabolism.

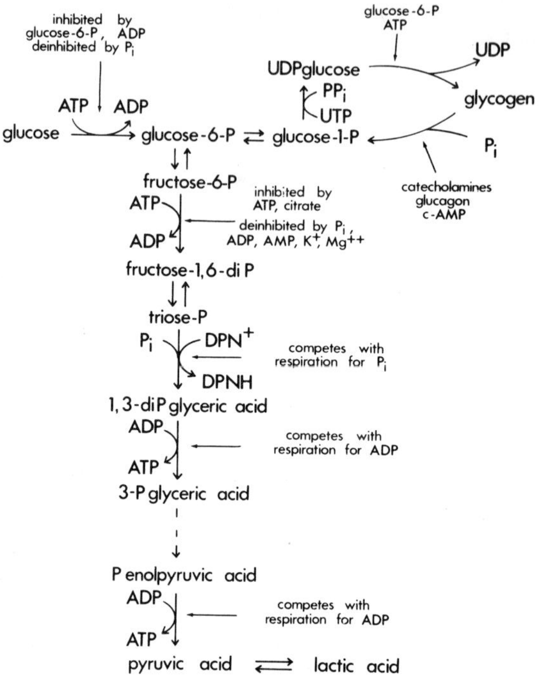

Fig. 7. Glycolytic pathway, indicating possible control sites (see [38]).

According to Warburg, the Pasteur Effect operates in tumors but is quantitatively deranged in that it does not inhibit glycolysis as efficiently as it does in normal tissues. A generation ago Johnson [34] and Lynen [35] independently suggested that the Pasteur Effect may be mediated by competition for substrates between the transphosphorylating enzymes of glycolysis and the respiratory ADP and P_i acceptor systems for oxidative phosphorylation. The three enzyme sites, as shown in fig. 7, are at triose phosphate dehydrogenase, where P_i is taken up into organic combination; and at phosphoglycerate kinase and pyruvate kinase, where ADP is converted to ATP. The Johnson and Lynen theory languished for many years without definitive evidence *pro* or *con*; in recent years attention has shifted to phosphofructokinase and hexokinase, two irresversible steps at earlier stages in the glycolytic sequence. Although phosphofructokinase is likely to be a control site for the Pasteur Effect in muscle, pyruvate kinase seemed to us to be a very good possibility in the Morris hepatomas, since there was an excellent correlation between low pyruvate kinase and the low glycolysis of the well differentiated tumors on the

Table 3
Effect of fructose-1,6-diphosphate, (FDP) on glycolytic and respiratory phosphorylation (condensed from [38]).

Tissue	Substrate	2-DG Uptake	O_2 Uptake	Lactate	P resp	P glyc
Liver	None	19.7	11.5	0.9	19.7	-
	FDP	25.3	10.0	9.1	7.1	18.2
Hepatoma						
5123D	None	18.3	9.5	0	18.3	-
	FDP	23.7	10.0	3.0	17.7	6.0
3683	None	2.6	0.8	0.6	2.6	-
	FDP	28.4	3.5	13.0	2.4	26.0

one hand, and the high pyruvate kinase and high glycolysis of the poorly differentiated tumors on the other. Without going into details of the experimental procedures, which are described [36], we were able to show that the addition of fructose diphosphate to whole, respiring homogenates of the well differentiated tumors, resulted in only a low production of lactate, and a low yield of glycolytic ATP, with no decrease in the high respiratory ATP production. However, when FDP was added to respiring homogenates of the poorly differentiated tumors, glycolysis was very high, accompanied by a high glycolytic ATP production; and the otherwise low respiratory phosphorylation was reduced to zero (table 3). By intermixing experiments in which particulate fractions of one tumor type were mixed with supernatant fractions of the other type (table 4) we observed that the glycolytic activity of the supernatant fraction did indeed determine the degree of glycolytic phosphorylation. When FDP was added to a mixture of the poorly differentiated tumor super-

Table 4
Effect of intermixing supernatants and particles of well and poorly differentiated tumors in glycolytic and respiratory phosphorylation with FDP as substrate (condensed from [38]).

Supernatant	Particles	2-DG Uptake	O_2 Uptake	Lactate	P resp	P glyc
5123	5123	18.6	9.6	1.8	15.0	3.6
3683	3683	36.0	4.2	17.6	0.8	35.2
5123	3683	27.0	4.0	13.5	~ 0	27.0
3683	5123	36.1	9.8	8.2	19.7	16.4

natant and the well differentiated tumor respiratory system, the respiratory activity was characteristic of that of the well differentiated tumor homogenates, while the glycolytic activity and resultant glycolytic phosphorylation were markedly decreased below that of the poorly differentiated tumor homogenate. When the mixing was reversed, that is, when the well differentiated tumor supernatant was mixed with the poorly differentiated tumor particles, glycolytic activity was greatly increased and accounted for a larger proportion of the total phosphorylation than that exhibited by the well differentiated tumor homogenate. Evidently, a powerful glycolytic system can carry on glycolytic phosphorylation at the expense of respiratory phosphorylation, and a normally low glycolytic system can increase markedly when the competition by respiration is removed. Although it would be premature to propose that pyruvate kinase is the site of the Pasteur Effect in all tumors, such a possibility suggests itself on the basis of the behavior of this model system.

Oxidation of fatty acids. The survival and growth of hepatomas that have low glycolytic activity requires consideration of their source of metabolic fuel, and studies conducted in our laboratory point to fatty acids. As shown in table 5 [37] when ^{14}C-labelled palmitate or butyrate was incubated with respiring whole homogenates of well differentiated tumors, the conversion to $^{14}CO_2$, indicative of complete oxidation, was in the same range as in the same preparations of liver. However, their oxidation in homogenates or poorly differentiated tumors was negligible. This figure also shows the same relationship for β-hydroxybutyrate dehydrogenase activity [38], and recent work of Langan [5] indicates that the levels of fatty acid activating enzyme (acyl-CoA synthetase) and mitochondrial ketone body production also are high in well differentiated and low or absent in poorly differentiated tumors. It would appear from this work that the degree of glucose utilization bears an inverse

Table 5

Fatty acid metabolism in liver and hepatomas [37]. ^{14}C fatty acids were used for oxidation studies. Hydroxybutyrate dehydrogenase was assayed according to Ohe et al. [38].

	Butyrate Oxidized (%)	Palmitate Oxidized (%)	β-OH Butyrate dehydrogenase (units/g)
Liver	79±10	30±3	16.1 (13.0–22.9)
Well differentiated tumors	48± 7	24±3	2.55 (0–7.5)
Poorly differentiated tumors	1	0	0.3 (0–0.6)

relationship to fatty acid utilization; and that with loss of differentiation there is a switch from fatty acids to glucose, accompanied by loss of enzyme activities involved in fatty acid catabolism.

Discussion

Time will permit only a brief and superficial discussion of the manner in which these tumors have been employed to promote a deeper understanding of the neoplastic process. The literature is vast and growing and has been summarized in various reviews [5,39–42]. I will reserve for discussion only those aspects that related closely to the foregoing metabolic and enzymatic data.

Glycolysis in relation to enzyme alteration. The Morris hepatomas illustrate a growing recognition that the neoplastic transformation of a single cell type, the parenchymal liver cell, may give rise to an exceedingly diverse family of tumors. At the one extreme are the few highly differentiated tumors with high respiration and low glycolysis, and with a full complement of enzymes of fatty acid catabolism, making fatty acids their predominant metabolic fuel. At the opposite extreme are the virtually undifferentiated tumors with high glycolysis and low-to-moderate respiration; whose absence of enzymatic equipment for fatty acid catabolism points to glycolysis as their major source of metabolic energy. From the foregoing studies of our own and other laboratories it seems clear that high aerobic glycolysis is a resultant of progression rather than initiation of neoplasia and the availability of tumors that exhibit a wide range of glycolytic activity has made it possible to draw some inferences concerning the enzymatic background of these variations in glycolysis. Since metabolic control is probably exerted at many steps in the glycolytic sequence [43], it is an oversimplification to designate any one enzyme as the site of glycolytic control. Nevertheless, two enzymes stand out for consideration; these are glucose-ATP phosphotransferase and pyruvate kinase, the first and last irreversible steps in the conversion of glucose to lactic acid. Low glycolysis in the well differentiated hepatomas appears to be due to low hexokinase and glucokinase activities. In the poorly differentiated liver tumors, however, the high hexokinase activity would permit a high rate of glucose utilization which would be abetted by the extraordinarily high activity of pyruvate kinase. Tumors whose glycolytic activity has been found to be high are invariably those which, through repeated transplantation, have undergone progression to a virtually undifferentiated state. It is conceivable that the high aerobic glycolysis of tumors as well as the Pasteur Effect, is attributable generally to successful competition for ADP by a high pyruvate kinase activity when associated with a low or moderate respiratory system. Before this speculative hypothesis is accepted, however, more information will be required, not only on the levels of pyruvate kinase activity in a large number of tumors, but also on the properties and relative activities of the isoenzyme and on the signal mechanisms by which variable fluxes of intermediates through this reaction are fed back to the initial step of glucose phosphorylation.

Isozyme pattern in relation to tissue differentiation. The four enzymes discussed in this paper illustrate the complexity and diversity of phenotypic expression in liver neoplasia. With the glucose-ATP phosphotransferases and the pyruvate kinases a common pattern occurs. The most highly differentiated tumors exhibit isozyme patterns essentially identical with that of liver. With partial loss of differentiation there is a marked decrease in the typically hepatic isozyme, but no marked change in the other types. With continued loss of differentiation, a process accompanied by morphologic, ultrastructural, and chromosomal alterations [6] as well as in growth rate, there is a large increase in non-hepatic isozymes to levels much greater than their total activity in liver. With aldolase, loss of differentiation does not lead to changes in total activity, but there is an essentially complete switch from the hepatic type to the non-hepatic type. With adenylate kinase the liver enzyme drops sharply, but the other isozymes do not increase with loss of differentiation. These results make it clear that mere comparisons of total enzyme activity between a tumor and its tissue of origin, or among different tumors, are inadequate to detect profound phenotypic alterations which may accompany the neoplastic condition. Moreover, the continued presence of liver marker isozymes in well differentiated hepatomas shows that this alteration or loss in a neoplasm need not relate causally to the initiation of neoplasia, but rather represents the underlying molecular mechanism of tumor progression. This conception does not necessarily discount the significance of such alterations; indeed they are probably crucial to the growth and development of the tumor. The loss of those enzymes concerned with functional activities of the tissue of origin would conceivably stimulate cell replication by removal of competing energy-requiring processes. The further replacement of isozymes geared to function by other isozymes geared to rapid and efficient utilization of metabolic fuel would possibly impart to such cells advantages of growth rate that would insure their survival despite the possible existence of host defense mechanisms. The end result of this sequential series of enzyme alterations would be an anaplastic, rapid growing tumor which, by loss of those distinguishing morphologic and biochemical characteristics of the tissue of origin, would resemble only other tumors similarly derived. One can therefore understand why, according to the so-called Greenstein generalization, the many times transplanted tumors resemble each other more closely than they resemble their tissue of origin [1].

It is notheworthy that with the loss of regulation of cell division in the hepatoma there is a loss or suppression of a number of enzymes whose activity in liver is under regulation by hormones; and it is of further significance that as regulatory control is further lost by decreased differentiation the new enzymes that appear are not of the adaptive type. A corollary question now emerges, whether those adaptive enzymes that are retained in the well differentiated tumors are still responsive. Our own studies have shown that neither glucokinase, pyruvate kinase nor adenylate kinase in well differentiated liver tumors were responsive to dietary or hormonal manipulation. Furthermore, these enzymes could not be made to appear by treatments that lead to enhancement of their activity in normal liver. However, a more recent study [44] indicates that hormones such as cortisol and glucagon may affect the activities in certain hepatomas of such enzymes as tyrosine-α-ketoglutarate aminotransferase, serine dehydratase, glucose-6-phosphatase, and glucose-6-phosphate dehydrogenase.

How are the activities of these isozymes switched off and on, and how are such changes related to the process of dedifferentiation? In the absence of information on the regulation of protein synthesis in cells of higher organisms, one can only resort to analogy with these processes in bacteria. On this basis we envision that the various isozymes or their polypeptide subunits are coded by specific operator genes, and their synthesis is controlled by the operation of specific repressors. The loss of differentiation would require first a repression of those genes that code for the liver type isozymes, and as dedifferentiation proceeds there is a derepression of other genes that code for enzymes that are repressed in the normal liver. Both processes do not occur simultaneously, and it would appear that repression of the liver-specific isozymes has to occur before derepression of the non-hepatic isozyme, since the non-hepatic isozymes have never appeared in high activity in the same tumors having high activity of the liver type isozyme. However, there is as yet no clear-cut evidence that differences in the respective activities are due to differences in rates of protein synthesis; and even if so, it is not certain that the transcriptional phase rather than the translation phase is involved.

It is also possible that the enzyme alterations result from a stepwise, perhaps random, deletion of genes during successive cell division, a likely explanation for the loss of many functional activities and alterations of structure that accompany tumor progression. This mechanism would be compatible with the appearance of non-hepatic enzymes if we assume that regulatory genes are lost that normally suppress their synthesis.

Resemblance of poorly differentiated tumors to fetal liver. In their isozyme patterns the poorly differentiated tumors closely resemble the fetal liver.

Since the embryonal liver cell only develops the liver type isozymes at or near parturition, it is perhaps not too surprising that tumors resemble fetal liver, just as poorly differentiated tumors resemble one another. However, there may be more to this similarity than meets the eye, since Abelev [45] has recently found the production of fetal globulins in chemically induced liver tumors, and Gold and his coworkers [46] found a fetal antigen in certain gastrointestinal tumors. These findings support the view that tumor dedifferentiation may, in its basic biochemical mechanism, be a reversal of normal embryonic development and may account for the often noted resemblance of tumors to embryonic tissue. In these few examples we observe an acquisition of characteristics which again may be attributable to the unmasking of operator genes normally repressed in the adult liver.

There is a growing body of evidence to indicate that neoplasia may be associated with bizarre aberrations of protein synthesis. Time will permit the mention of only a few specific illustrations. Some of the best of these are found in the clinical literature. Lipsett [47] has cited over 100 cases of Cushing's syndrome associated with a variety of clinical non-pituitary neoplasms, principally bronchogenic carcinoma. This was apparently due to secretion of substances having the hormonal activity of ACTH. Severe hypoglycemia has been frequently reported as an accompaniment of various tumors of non-endocrine origin, and recently Miyabo et al. [48] found evidence of immunoreactive insulin in a gastric tumor associated with hypoglycemia. Goodall [49] and Eliel [50] have also reviewed the clinical literature which points to production and secretion of hormones from a variety of tumors of non-endocrine origin. Evidently the gene readout mechanism can be distorted in certain neoplasms so that genes normally completely repressed in the tissue of origin are unmasked for transcription and translation.

When the Morris hepatomas first appeared they were hailed as the hope for an eventual understanding of the molecular basis of cancer formation. According to Potter [42] the availability of liver tumors that are undoubtedly malignant, yet differ in few biochemical parameters from the normal, differentiated liver cell of origin, opens the possibility for identifying the crucial first step of the chemically induced neoplastic transformation. Since this process is a lengthy one, and probably involves several stages of preneoplasia [41] it is probably too much to hope that it would be explainable on the basis of one or a few enzyme abberations, At any rate, if this hope has not yet been realized, nonetheless the intensive biochemical studies of these tumors are providing a much clearer picture of the underlying molecular alterations which accompany the spectrum of variations in growth rate and degree of differentiation which characterize neoplasia.

References

[1] J.P. Greenstein, Biochemistry of Cancer, 2nd ed. (Academic Press, New York,1954).
[2] J.A. Miller and E.C. Miller, Advan. i. Cancer Res. 1 (1953) 339.
[3] O. Warburg, The Metabolism of Tumors (Arnold Constable, London, 1930).
[4] S. Weinhouse, Oxidative Metabolism of Tumors 3 (1955) 269.
[5] H.P. Morris, Advan. Cancer Res. 9 (1965) 227.
[6] P.C. Nowell, H.P. Morris and V.R. Potter, Cancer Res. 27 (1967) 1561.
[7] G. Weber, G. Banerjee and H.P. Morris, Cancer Res. 21 (1961) 933.
[8] A.C. Aisenberg and H.P. Morris, Nature 191 (1961) 1314.
[9] D.L. DiPietro, C. Sharma and S. Weinhouse, Biochemistry 1 (1962) 455.
[10] E. Vinuela, M. Salas and A. Sols, J. Biol. Chem. 238 (1963) 1175.
[11] L. Grossbard and R.T. Schimke, J. Biol. Chem. 241 (1966) 3546.
[12] H.M. Katzen and R.T. Schimke, Proc. Natl. Acad. Sci. U.S. 54 (1965) 1218.
[13] M. Salas, E. Vinuela and A. Sols, J. Biol. Chem. 238 (1963) 3535.
[14] C. Sharma, R. Manjeshwar and S. Weinhouse, J. Biol. Chem. 238 (1963) 3840.
[15] F.J. Ballard and I.T. Oliver, Biochem. J. 90 (1964) 261.
[16] R.M. Sharma, C. Sharma, A.J. Donnelly, H.P. Morris and S. Weinhouse, Cancer Res. 25 (1965) 193.
[17] J.B. Shatton, H.P. Morris and S. Weinhouse, Cancer Res. 29 (1969) 1161.
[18] J.C. Elwood, Y.C.Lin, V.J.Cristofalo, S.Weinhouse and H.P.Morris, Cancer Res. 23 (1963) 906.
[19] T.P. Fondy and N.O. Kaplan, Ann. N.Y. Acad. Sci. 119 (1965) 888.
[20] E. Penhoet, T. Rajkumar and W.J. Rutter, Proc. Natl. Acad. Sci. U.S. 56 (1966) 1275.
[21] E.E. Penhoet, M. Kochman and W.J. Rutter, Biochemistry 8 (1969) 4391.
[22] R.C. Adelman, H.P. Morris and S. Weinhouse, Cancer Res. 27 (1967) 2408.
[23] Y. Nordmann and F. Schapira, European J. Cancer 3 (1967) 247.
[24] T. Sugimura, S. Sato, S. Kawabe and N. Suzuki, Nature 222 (1969) 1070.
[25] T. Tanaka, Y. Harano, F. Sue and H. Morimura, J. Biochem. (Tokyo) 62 (1967) 71.
[26] E. Bailey and P.E. Walker, Biochem. J. 111 (1969) 359.
[27] G. Weber, N.B. Stamm and E.A. Fischer, Science 149 (1965) 65.
[28] F.A.Farina, R.C.Adelman, C.H.Lo, H.P.Morris and S.Weinhouse, Cancer Res. 28 (1968) 1897.
[29] R.W. Gracy, A.G. Lacko, L.W. Brox, R.C. Adelman and B.L. Horecker, Arch. Biochem. Biophys. 136 (1970) 480.
[30] C.B. Taylor, H.P. Morris and G.Weber, Life Sci. 8 (1969) 635.
[31] W.E. Criss, Biochem. Biophys. Res. Commun. 35 (1969) 901.
[32] R.C. Adelman, C.H. Lo and S. Weinhouse, J. Biol. Chem. 243 (1968) 2538.
[33] W.E.Criss, G.Litwack, H.P.Morris and S.Weinhouse, Cancer Res. 30 (1970) 370.
[34] M.J. Johnson, Science 94 (1941) 200.
[35] F. Lynen, Ann. Chem. 546 (1941) 120.
[36] C.H. Lo, V.J. Cristofalo, H.P. Morris and S. Weinhouse, Cancer Res. 28 (1968) 1.
[37] L. Bloch-Frankenthal, J. Langan, H.P. Morris and S. Weinhouse, Cancer Res. 25 (1965) 732.
[38] K. Ohe, H.P. Morris and S. Weinhouse, Cancer Res. 27 (1967) 1360.
[39] G. Weber, Advan. i. Cancer Res. 6 (1961) 403.

[40] V.R. Potter, in: Cellular Control Mechanisms and Cancer, eds. O. Muhlbock and P. Emmelot (Elsevier, Amsterdam, 1964) pp. 190–210.
[41] E. Farber, Cancer Res. 28 (1968) 1210.
[42] V.R. Potter, Cancer Res. 21 (1961) 1331.
[43] P.A. Srere, Biochem. Med. 3 (1969) 61.
[44] M. Watanabe, V.R. Potter, H.C.Pitot and H.P.Morris, Cancer Res. 29 (1969) 2085.
[45] G.I. Abelev, S.V. Assecritova, N.A. Kraevsky, S.D. Perocla and N.J. Perovodchikova, Internat. J. Cancer 2 (1967) 551.
[46] S. von Kleist and P. Burtin, Cancer Res. 29 (1969) 1961.
[47] M.B. Lipsett, Ann. Int. Med. 61 (1964) 733.
[48] S. Miyabo, T. Fujimura and M.Murakarni, Diabetes 17 (1968) 286.
[49] C.M. Goodall, Intern. J. Cancer 4 (1969) 1.
[50] L.P. Eliel, Cancer Bull. 20 (1968) 37.

MICROFLUORIMETRIC STUDY OF INTRACELLULAR ENZYME KINETICS IN SINGLE CELLS

E. KOHEN, C. KOHEN and B. THORELL

Department of Pathology, Karolinska Institute, Stockholm, Sweden, and Papanicolaou Cancer Research Institute, Miami, Florida, USA

Abstract: Kohen, E., Kohen, C. and Thorell, B. Microfluorimetric Study of Intracellular Enzyme Kinetics in Single Cells. *Miami Winter Symposia* 2, pp. 481–515. North-Holland Publishing Company, Amsterdam, 1970.

The transient changes in NAD^+ reduction which result from the microelectrophoretic addition of substrate in stepwise increasing amounts can be followed continuously in single living ascites cancer cells (EL2 cells) by means of a beam-splitter supplemented Chance-Legallais microfluorimeter for fluorescence recordings synchronous with cell manipulations. Microfluorimetry provides a basis for the direct verification in the living cell and its localized compartments of various hypotheses derived from conventional biochemical studies, concerning metabolic control and enzymic behaviour. So far, the most illustrative examples of the technique have been obtained by adding glycolytic intermediates to the cytoplasm of EL2 cells and following the fluorescence changes in the extramitochondrial compartment (cytoplasmic or nuclear region). Upon addition of glycolytic substrate (e.g. glucose-6-phosphate), a cycle of NAD^+ reduction and reoxidation is observed. The associated fluorescence curve resembles closely the kinetic curves of formation and disappearance of enzyme-substrate complexes, as described in transient state relationships. With glycolytic substrates, each fluorescence cycle represents at least the summation of activities at the level of glyceraldehyde phosphate dehydrogenase (NAD^+-reducing) and lactic dehydrogenase (NADH-reoxidizing). In initial experiments, the plot of the integrated area of the fluoresence curve (a measure of accumulated NADH) against the concentration of glycolytic substrate is sigmoidal and favours a higher-order relationship with regard to substrate in the optimum concentration range. The substrate levels required for half-maximal or maximal reduction of glycolytic NAD^+ and the rate of substrate utilization are altered by adenine nucleotides and P_i as well as growth in presence of drugs (e.g. amytal) or hormones (e.g. triiodothyronine). Preliminary experiments are suggestive of alterations in glycolytic activity when attempts are made to modify or block the mitochondrial-extramitochondrial exchange (e.g. with atractylate, an inhibitor of adenylate translocase). Furthermore, there is some preliminary evidence that the delicate balance between various metabolite modulated enzymes is affected not only by allosteric activators or inhibitors, but also by physical effects. The glycolytic flux estimated from the half-time of the transient changes in NAD^+ reduction following addition of substrate, can be increased by several times simply by raising the temperature at which the EL2 cells are incubated in the microfluorimeter. The maximal increase in flux at higher temperature (e.g. 37°) occurs in conditions where at room temperature NADH reoxidation by lactic dehydrogenase lags behind NAD^+ reduction by glyceraldehyde phosphate dehydrogenase: e.g. with glucose-6-phosphate in presence of P_i or lactate. Simultaneously the point of maximal NAD^+ reduction is shifted towards greater substrate levels at higher temperature. Thus at 37° the glycolytic flux may proceed with less hindrance through the various bottlenecks along the glycolytic chain and NADH is more effectively reoxidized by lactic dehydrogenase. As the temperature dependency of the glycolytic flux is critically altered by the metabolites added or the conditions under which the EL2 is grown, the microfluorimeter offers promising prospects for the study of metabolic regulation and competition between various enzyme systems in the complex environment of the cell.

1. Introduction

In approaching the problem of enzyme reactions in the living cell it may be a proper introduction to recall a principle from the physics of elementary particles, the Heisenberg uncertainty principle [1], which dwells upon the impossibility to determine simultaneously the velocity and position of a particle, thus implying in more generalized terms that each scientific observation involves a disturbance of the object studied. A common analogy to the Heisenberg principle has been the long professed belief that the study of life implies the destruction of life, as it is still true for most of the biochemical studies on cell extracts and homogenates. However the introduction of optical methods, among others microspectrophotometry [2–5] and microfluorimetry [6–9] has made it possible to observe chemical reactions in localized compartments of the living cell, without interfering with the integrity of the whole, the "integrating unit" that is the cell.

The microfluorimetric technique [6–9] allows the determination of 10^{-18} mole fluorochrome in cell regions down to 3 μm in diameter [6, 10, 11]. In such studies the information yielded by a fluorochrome can be greater if the compound exhibits a certain number of characteristic properties: reduced pyridine nucleotides are particularly suitable as natural fluorochromes, because of the relatively high efficiency of their blue fluorescence [11–13] (emission peak at 454 nm) in the "protein-bound" state and their relatively low sensitivity to damage by excitation at 366 nm; also their ubiquitous distribution throughout the cell and their association with a large number of dehydrogenases of respiration and glycolytic pathways, places them in a "strategic position" for an extensive attack on the control of energy metabolism. Therefore, by looking at the fluorescence of a single group of compounds (pyridine nucleotides) it is possible to follow the activity of a large number of enzymes and of various intracellular compartments, provided the proper substrates (or combinations thereof) are selected each time. Furthermore by the use of selective inhibitors [16] it is to some extent possible to switch off the functioning of a selected pathway or compartment [17–20], so that the fluorimetric observations can be specifically focused on a mitochondrial or extramitochondrial target, as well as a selected enzyme system.

2. Materials and methods

The Chance-Legallais microfluorimeter [6] developed in the years 1958-59 consists essentially of a fluorescence microscope supplemented with a photomultiplier located at the aperture normally used for a camera, and from there on a cathode follower, a push-pull arrangement and a switch circuit controlled by reed operated contacts. The microfluorimeter allows differential measurements between the fluorescence of an object located at a specific site in the microscope field (measuring aperture) and that of a nearby free, nonfluorescent space (reference aperture). The positions of the reference and measuring apertures coincide with the projections in the microscope field of the extreme excursions of a circular hole in a vibrating diaphragm placed in front of the photocathode. The vibration of the diaphragm and the operation of the contacts controlling the switch circuit are synchronized with the sinusoidal fluctuations in light intensity of the mercury arc used for fluorescence excitation, so that the diaphragm is at the extremes of its travel at the peaks of the light intensity pulses. The portions of the photocurrent selected by the switch are used to charge a condenser, but because of the push-pull arrangement the signals coming from two consecutive positions will charge the condenser in reverse, so that its potential represents the difference between the signals coming from the measuring and reference apertures.

The first biological problem encountered in the application of the microfluorimetric technique has been the damaging influence of the ultraviolet light used for fluorescence excitation. This can be prevented by the use of narrow-band blocking filters which select from the light source, a water-cooled highpower mercury arc, the emission band required for fluorescence excitation (the 366 nm band) but eliminate the more harmful shorter wavelengths [6,7]. Also the cells are incubated in a modified Krebs-Ringer medium supplemented with albumin and compounds rich in thiol groups (such as glutathione and cysteine) which exhibit a radiation-protective effect [9, 17, 21].

To study intracellular enzyme kinetics it has been necessary to impose metabolic transients on the living cell (fig. 1). Initially this purpose was

Fig. 1. Front view of the equipment used for the study of intracellulai enzyme kinetics: a beam splitter and heated stage supplemented Chance-Legallais microfluorimeter with a micromanipulation assembly for microelectrophoresis.

achieved by cell perfusion [22] with substrates and drugs. However a detailed study of intracellular enzymes implies the ability to alter the internal environment of the cell. This in turn requires the introduction of metabolic intermediates, activators or inhibitors directly into the cytoplasm of the living cell. Because of the pitfalls related to microinjection [23,24], preference was given to microelectrophoresis [25–28]. In practice for the compounds of intermediate metabolism which have been so far investigated, the required intracellular concentrations [29] are such that the intensity of the microelectrophoretic current can be kept safely below the levels required for major upheavals in the ion balance of the living cell. At such current intensities the Joule effect also has turned out to be quite negligible.

Thus, provided the above safeguards are respected, the living cells tolerate well [28] the microfluorimetric and microelectrophoretic procedures required for the evaluation of intracellular enzyme activity. Unlike in the test tube, in the living cell such activities are closely associated with the architecture of intracellular enzyme compartments [30,31], which controls the distribution of endogenous activators and cofactors. At an early stage in these studies, by a simple perfusion method [17,22] it has been possible to discriminate between the main two metabolic compartments, insofar as pyridine nucleotides are concerned. Upon perfusion with amytal [32,33] or rotenone [33,34], inhibitors of electron transport at the level of NADH-cytochrome *b* reductase, there is a selective reduction of mitochondrial pyridine nucleotides, while subsequent perfusion with glucose leads to the reduction of extramitochondrial (cytoplasmic or nuclear) pyridine nucleotides. Later the independent responses of the two main pyridine nucleotide compartments were further confirmed [35,36] by microelectrophoretic treatment with Krebs cycle or glycolytic intermediates, the former interacting with mitochondrial pyridine nucleotides, the latter with the extramitochondrial [10,37,38].

Although microfluorimetric studies are in principle applicable to a large variety of animal or even vegetal cells, in practice cells of suitable morphology are selected, since the independent study of various intracellular organelles requires as little as possible overlapping of these organelles in the field of observation. Initially insect spermatids were used [7,9] for these studies because they exhibited distinct nuclear, cytoplasmic and mitochondrial regions, but the difficulties in immobilizing spermatids (a precondition for cell perfusion and microelectrophoresis) as well as their low responsiveness to extramitochondrial substrates led to the preference for mammalian cells.

So far the most illustrative responses (fig. 2) have been observed in ascites cell cultures (EL2 cells) treated with glycolytic substrates [17,28,35,36,39–41]. It was in these cells that the clear cut metabolic differences between the mitochondrial and extramitochondrial pyridine nucleotides were first established by microfluorimetry. Further studies [40–42] have revealed discrete differences in the extramitochondrial compartment between the nucleus and cytoplasm in terms of pyridine nucleotide reduction and reoxidation.

The study of intracellular enzyme kinetics from the very first instant of substrate addition demands the uninterrupted observation of the cellular site where a response is expected. A beam-splitter supplemented Chance-Legallais microfluorimeter [43–45] (see fig. 1) allows the continuous recording of the NADH or NADPH-dependent fluorescence changes synchronously with cell manipulations performed under visible light.

Provided a certain substrate threshold is exceeded each microelectrophoretic addition of glycolytic intermediates is followed by a cycle of NAD^+

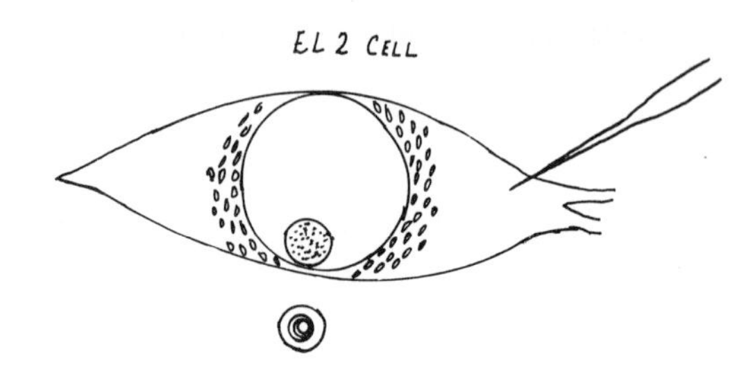

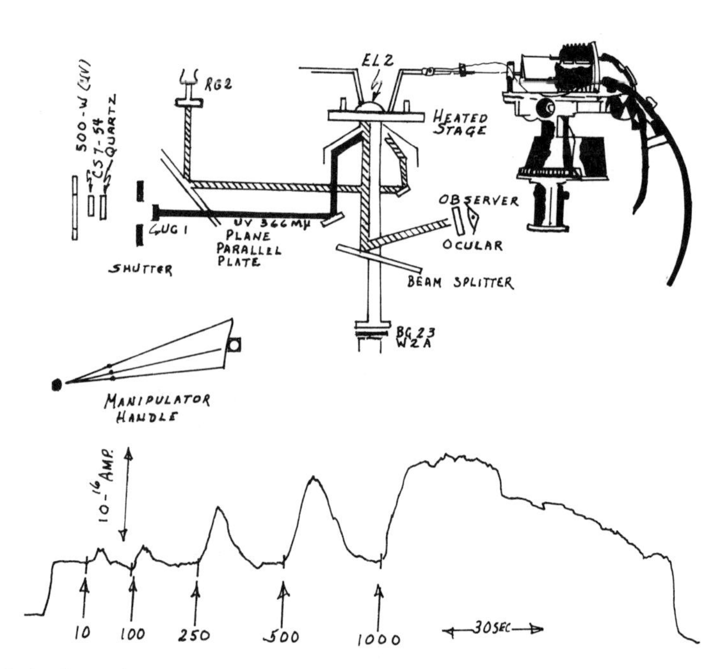

Fig. 2. A schematic representation of the EL2 cell, the microfluorimetry-microelectrophoresis system and a fluorescence curve recorded from the EL2 upon repetitive additions of substrates in stepwise increasing amounts.
a) The mitochondrial, nuclear and cytoplasmic regions of the EL2 cell, the position of the micropipette and the two apertures of the differential microfluorimeter (measuring and reference) are indicated in the upper drawing.
b) The center drawing shows the beam splitter system which allows the simultaneous illumination of the microscope field by red light (for cell visualization and manipulations) and ultraviolet light (for fluorescence excitation). While the red light is reflected towards the ocular, the blue light emitted by fluorescent cell structures is transmitted towards the photocell (CS7-54, UG1, RG2, BG23, W2A glass and gelatin blocking filters).
c) The lower drawing shows the fluorescence changes in the extramitochondrial region of an EL2 cell in response to repeated additions of Glc-6-P in gradually increasing amounts. 10 corresponds to a microelectrophoretic current of 9×10^{-11} A. The baseline with both apertures of the differential microfluorimeter on free space is seen at both ends of the tracing. The initial level of fluorescence prior to the first addition of substrate is seen on the left of the first arrow. The magnitude of the primary photocurrent is indicated in 10^{-16} A. (Expt. EK69/11A, cell 2).

reduction and reoxidation (the so-called fluorescence pulse or fluorescence cycle, cf. fig. 2). The fluorescence pulse [45] corresponds to the imposition of a metabolic transient and resembles the kinetic curve of formation and disappearance of the enzyme-substrate complex, as described in transient-state relationships [46]. Under the most simplified assumptions the fluorescence curves recorded with glycolytic substrates represent the algebraic sum of forward and backward reactions at the levels of glyceraldehyde phosphate dehydrogenase and lactic dehydrogenase [47, 48].

Each fluorescence cycle (fig. 2) exhibits three characteristic phases [45,49]: a rapidly rising ascending phase (NAD^+ reduction), a steady state at a plateau level and a descending phase (NADH reoxidation); at about halftime there is a change in the slope of the descending branch and a slowing down in the rate of NADH reoxidation. Such fluorescence cycles can be analyzed in terms of peak amplitude ((p_1)max), halftime of NAD^+ reduction and reoxidation ($t_{½\,off}$), rise time, duration of steady state at plateau level, integrated area of the fluorescence pulse ($\int_0^\infty p_1 \, dt$), etc. (fig. 2).

From the amount of added substrate and the halftime of a NAD^+ reduction and reoxidation cycle it is possible to estimate the rate of substrate utilization (fig. 3) according to a simple formula derived empirically from differential analyzer solutions for the kinetics of the Michaelis complex [46].

$$k_3 E = \frac{S}{t_{½\,off}}$$

(k_3 = velocity constant of the decomposition of the enzyme-substrate complex, E = enzyme concentration, S = substrate concentration)

The initial concentration S of substrate (e.g. glucose-6-phosphate=Glc-6-P) reached in the EL2 cell corresponds to the microelectrophoretically ejected amount in mole/cell volume in liters [49]. For a 1-sec microelectrophoresis at 10^{-8} A, the number of elementary charges displaced should be roughly 6×10^{18} (number of electrons per A) $\times 10^{-8} = 6 \times 10^{10}$.

Dividing by the Avogadro number and the number of charges per Glc-6-P molecule, this corresponds to approximately $1/6 \times 10^{-13}$ mole in the case of a fully ionized molecule. Since there will be some uncertainty as to the degree of ionization, the ion population at the microelectrode tip and the relative electrophoretic mobility of Glc-6-P ions as compared to other ions (H^+, Na^+, etc.), for practical purposes the above figure is rounded off to 10^{-14} mole Glc-6-P for a 1-sec duration of a 10^{-8} A current. This value when divided by an estimated cell volume of 5-10 pl gives an initial intracellular concentration of 1-2 mM (which agrees well with the physiological levels of glycolytic intermediates [29] in ascites cells treated with saturating amounts of glucose). A rough estimate of the average rate of substrate utilization is then possible. In these experiments, substrate amounts and rates of utilization are expressed in A/pl for an ellipsoid model exhibiting a short axis of 5 μm, with perpendicularly to this axis and along the greatest axis of the ellipsoid, an elliptical cross section of area equal to the cross section of the cell in the microscope field. Thus, for the actual A/pl values in a particular cell, the above values have to be corrected according to the cell thickness and geometry, as the cells can be 2-4 times thicker than the ellipsoid model.

Generally three modes of substrate [49,50] administration were used for each experimental series: 1) *dose-response curves*, adding consecutively to the same cell, substrate amounts from below threshold levels (required for a barely detectable fluorescence response) to supraoptimal levels (over amounts required for maximal response); *maximal or submaximal doses at once*; 3) *continuous* delivery of suboptimal substrate amounts. Microelectrophoretic additions lasted 1 sec for the two former and around 30 sec for the last method.

For "dose-response curves" the microelectrophoretic current was started at about 10^{-11} A [49,50]. Glc-6-P was used first alone and then in combination with adenine nucleotides, as well as other substrates, cofactors, activators or inhibitors [49,50] (see fig. 3). The rationale for using activators of inhibitors of glycolysis in the same micropipette has been described previously [16, 36,41]. The micropipette was generally filled with a 1-2 M solution of Glc-6-P. In glycolytic combinations [49,50] : Glc-6-P/adenine nucleotides = 8, Glc-6-P/P_i = 2, Glc-6-P/oxamate = 1 to 5, Glc-6-P/succinate = 1, Glc-6-P/atractylate = 50 to 160.

3. Results

3.1. *The kinetics of the fluorescence response to glucose and phosphorylated intermediates*

The initial phase of the fluorescence changes which follow the addition of glycolytic substrate exhibits considerable analogies, regardless of whether microelectrophoresis or perfusion is applied [45,49]. In all cases the fluorescence curve presents an ascending branch (see Methods, figs. 2 and 3) which starts to rise almost immediately after substrate addition (microelectrophoresis of phosphorylated intermediates) or shortly thereafter (5-7 sec later for glucose perfusion). Finally a new steady state is reached. It is from there on that the fluorescence curves start to follow very divergent courses. Between the treatment by microelectrophoresis and the addition of glucose to the medium, there is an essential methodological difference which influences con-

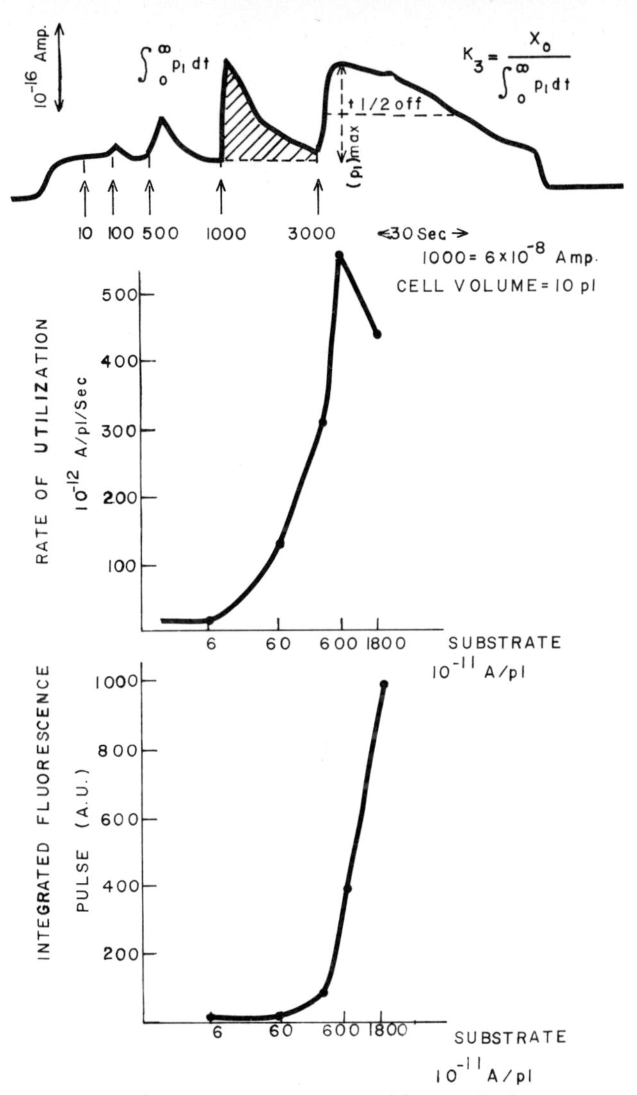

Fig. 3. The analysis of a fluorescence curve recorded from an EL2 cell upon repetitive additions of glycolytic substrate in stepwise increasing amounts (Expt. EK69/8AI, cell H5).
a) The upper drawing shows the fluorescence curve. The conditions are as shown in fig. 1c, except for 1000 corresponding to a microelectrophoretic current of 6×10^{-8} A. The integrated area of a fluorescence pulse $\int_0^\infty p_1\,dt$, the half-time of NAD^+ reduction and reoxidation ($t_{½off}$) and the amplitude (p_1) max are indicated. The formula on the right upper corner corresponds to the velocity constant k_3 of the decomposition of the enzyme-substrate complex (X_0 = substrate concentration).
b) According to a simple relationship found empirically from differential analyzer solutions for the kinetics of the Michaelis complex, it is possible to estimate roughly the rate of glycolytic substrate utilization (center drawing) from the relation $k_3E = \frac{X_0}{t_{½off}}$ (E = enzyme concentration).
c) The lower drawing shows the relationship between the integrated area of the fluorescence pulse $\int_0^\infty p_1\,dt$ (in arbitrary units) and the microelectrophoretic current (as a measure of the added Glc-6-P).

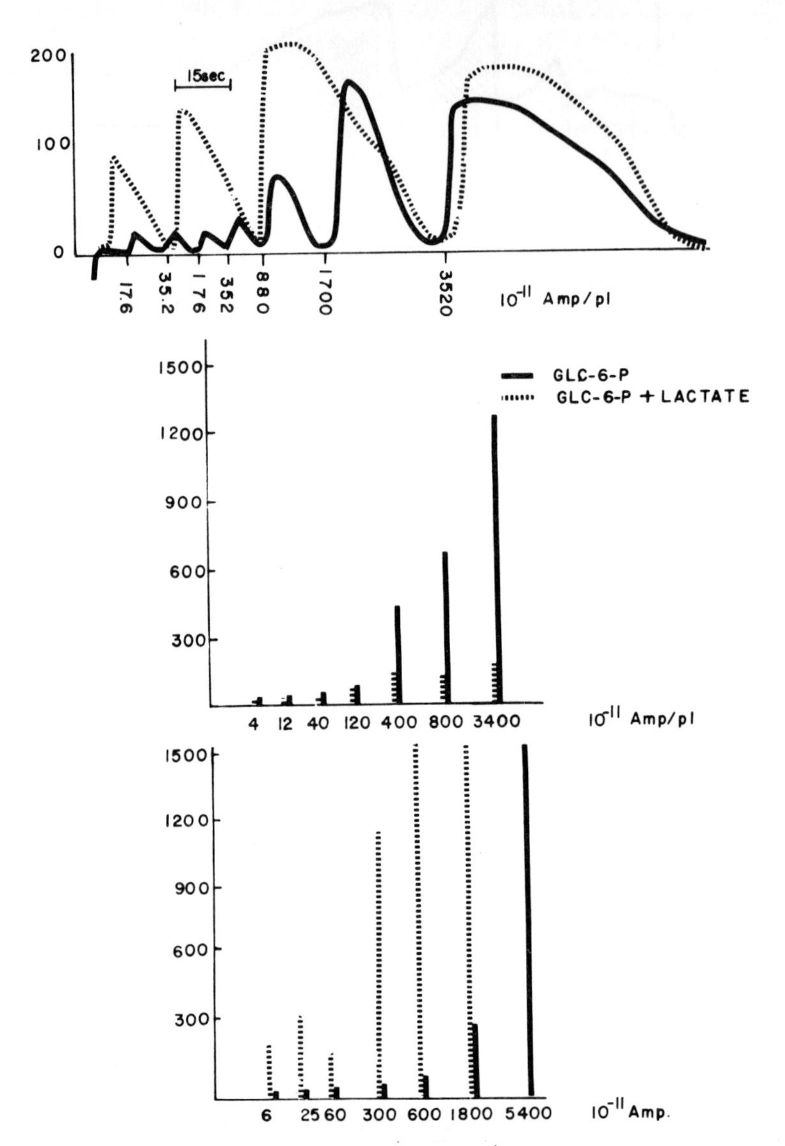

Fig. 4. The influence of lactate on the kinetics of NAD^+ reduction in an EL2 cell treated with stepwise increasing amounts of Glc-6-P.

a) The upper drawing shows the fluorescence curves for Glc-6-P alone or in combination with lactate. Conditions as shown in fig. 1c, but the curves are summed from at least five cells (ordinate in arbitrary units).

b) The rate of Glc-6-P utilization calculated for the curves in the upper drawing (center drawing: ordinate shows rate in 10^{-12} A/pl/sec, abscissa substrate in 10^{-11} A/pl).

c) The integrated area of the fluorescence pulse $\int_0^\infty p_1 \, dt$ for the same curves (lower drawing: ordinate shows area in arbitrary units, abscissa substrate in 10^{-11} A/pl).

siderably the time course of the fluorescence changes. Upon microelectrophoretic addition of substrate, a transient metabolic change is being produced and because (or as long as) the metabolite is introduced in a pulsed way and in limited amount there will be a half-time for its utilization (as expressed by the transient aspect of the NAD^+ reduction). However upon the addition of glucose to the medium, the cells which are located at the bottom of an open chamber are placed in presence of a practically inexhaustible supply of substrate. In this case rather than a complete cycle of NAD^+ reduction and reoxidation, there should be a relatively permanent transition from one steady state to another. In microelectrophoresis, the same situation can be duplicated by adding supraoptimal amount of substrate which cannot be rapidly metabolized.

When dose-response curves are made (figs. 2,3 and 4), starting from threshold levels of substrate barely sufficient to elicit a detectable fluorescence response, gradually a concentration range is reached for which the rate of glycolytic NAD^+ reduction rises the most rapidly. This occurs at an intermediate concentration range below the substrate level required for maximal NAD^+ reduction. Thus for a doubling of substrate concentration in this region, there can be a 4–16 times increase in the integrated area of the fluorescence cycle ($\int_0^\infty p_1 \, dt$). When $\int_0^\infty p_1 \, dt$ is plotted against substrate concentration (figs. 3,4

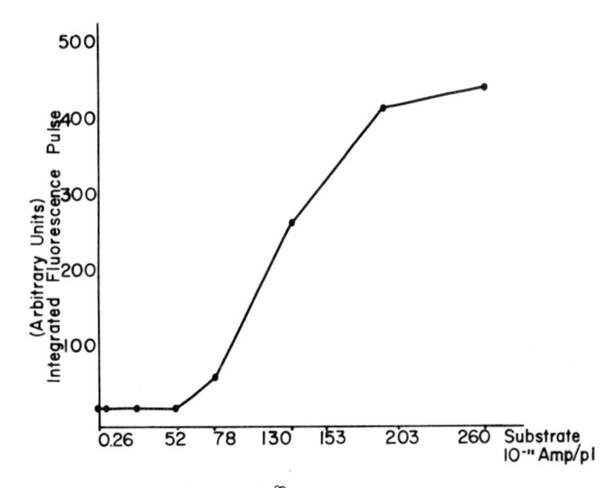

Fig. 5. A characteristic sigmoid plot of $\int_0^\infty p_1 \, dt$ against substrate concentration, showing stabilization at an almost plateau level once substrate saturation is reached. Further addition of substrate would result in substrate inhibition, cf. fig. 5. (Expt. W68-11, cell 15).

and 5), the ascending branch of the resulting sigmoid curve corresponds to the critical concentration range for which NAD^+ reduction rises the most rapidly. Quite often, as indicated above, the slope of the ascending branch favours a higher order relationship with regard to substrate in the optimum concentration range (fig. 5). The position of the ascending branch in the sigmoid is characteristic for each experimental series under standard conditions; it can be displaced laterally or its slope can be affected by treatment with drugs or inhibitors, as well as by changing the conditions under which the cells are grown. In fact EL2 cells maintained under similar conditions exhibit almost identical dose-response curves. Thus both the dose-response curve and its analysis provide a key to the study of intracellular enzyme kinetics in localized cell regions.

In several instances the above described second or higher order relationship between substrate concentration and pulse area tends to return to a first order relationship before a maximum is reached (fig. 5). Then, with subsequent larger doses of substrate, there can be no further increase in the amplitude of the fluorescence curve. However, the integrated area of the fluorescence cycle can continue to increase through considerable prolongation of the NAD^+ reduction-reoxidation half-time. Somewhere between the optimum and maximal concentration ranges, the rate of substrate utilization tends to level off and finally decreases, if saturation is reached (fig. 5).

As compared to the "dose-response curves" (figs. 2,3 and 4), the EL2 cells respond quite differently to the sudden addition of a maximal (or submaximal, in the optimum concentration range) level of substrate [49]. In this case a much higher NAD^+ reduction and a considerably larger NADH accumulation can be attained. The half-time of the NAD^+ reduction and reoxidation is now shorter than it would be for a comparable level of substrate in a dose-response curve. This in itself suggests that large additions at once are more effectively utilized (higher degradation rate) than if the substrate levels are raised gradually. It is true, however, that such drastic overshoots in NAD^+ reduction cannot be sustained for a long time and the cell ceases to respond following a certain number of similar additions.

As to the continuous delivery method of glycolytic substrate [49] by low or even very low microelectrophoretic current, it is somewhat analogous to the glucose-perfusion technique, whereby the cell membrane provides (unless hexokinase is implicated) the main hindrance to the penetration of substrate. In both cases the attainment of a steady state level at maximal NAD^+ reduction is considerably postponed. Since the cell is more or less continuously exposed to substrate, once attained this steady state is considerably or even indefinitely prolonged [49].

3.2. The influence of cofactors on metabolic transients

Generally speaking the EL2 cells grown under various conditions (e.g. with drugs or hormones) fit largely in two different metabolic categories. In the first (the untreated EL2 controls) they are quite self-sufficient in terms of endogeneous activators. It is not possible to change in these cells the rate of externally added substrate utilization by further additions of metabolic activators, e.g. adenine nucleotides, or P_i. A second category of EL2 cells, specially these grown under the influence of drugs (e.g. triiodothyronine) are metabolically sensitive to the addition of specific activators (fig. 6).

In the EL2 cells grown under standard conditions, the utilization of microelectrophoretically added glucose-6-phosphate, as calculated from the NAD^+

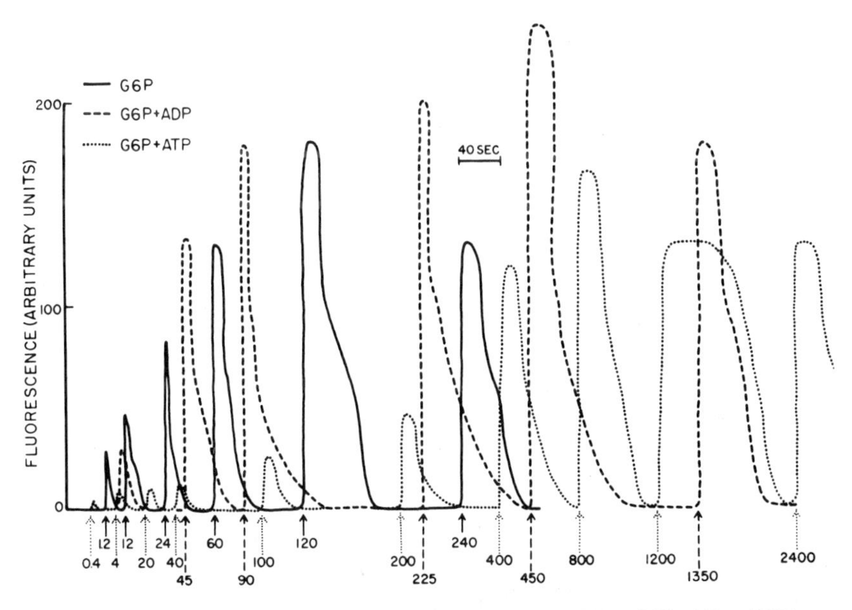

Fig. 6. Summed dose-response curves for Glc-6-P, Glc-6-P + ADP and Glc-6-P + ATP in T3-grown EL2 cells. Each curve corresponds to the summation of determinations in at least 5 cells. The fluorescence levels are indicated in the ordinate in arbitrary units. The numbers in the abscissa correspond to the added Glc-6-P in 10^{-11} A/pl. The level of fluorescence prior to the first addition of substrate is seen on the left of the first arrow and it is arbitrarily equated to 0. Each arrow corresponds to a microphoretic addition:

↑ = Glc-6-P, ⇡ (dotted) = Glc-6-P + ATP, ⇡ (dashed) = Glc-6-P + ADP reduction-reoxidation half-time, proceeds already at a maximal rate in the absence of externally added adenine nucleotides or P_i. In such cells the rates of Glc-6-P utilization exhibit considerable parallelism, upon treatment with this substrate alone or in combination with ADP and P_i. However ATP acts as an inhibitor and the rates of substrate utilization may be up to 4 times smaller in cells treated with Glc-6-P + ATP.

Little correlation is found between the maximal levels of NAD^+ reduction reached after substrate addition and the rates of Glc-6-P utilization *per se.* The maximum level of NAD^+ reduction is not a faithful indicator of the glycolytic rate in these cells. While the NAD^+ reduction is quite similar in cells given Glc-6-P alone or in combination with P_i, there is a certain decrease in the maximal amplitudes of the fluorescence pulses in the presence of ATP and ADP [49]. However the observed changes in the case of ATP and ADP do not have necessarily a similar meaning. While the pace of NAD^+ reduction at the glyceraldehyde phosphate dehydrogenase step is a direct expression of the glycolytic rate, reoxidation by lactic dehydrogenase also has to be taken into account. Thus a similar $NADH/NAD^+$ ratio can be obtained under diametrically opposed conditions: i.e. a low glycolytic activity or on the contrary a very high one, but in the latter case with efficient NADH reoxidation at lactic dehydrogenase preventing the accumulation of the reduced coenzyme. It is possible to discriminate between such opposite conditions by looking at the rate of substrate utilization (figs. 3 and 4), as calculated from the half-time of the NAD^+ reduction and reoxidation cycles. In the above example, with ADP the rate is comparable to Glc-6-P alone, but it is lower with ATP. The plots of substrate concentration against the integrated area of the fluorescence cycle ($\int_0^\infty p_1 \, dt$) provide themselves some indications about the glycolytic rate, from the position of the ascending branch in the resulting sigmoid curves (cf. preceding section, fig. 5): i.e. a displacement to the left means a decreased rate, since maximal NAD^+ reduction and substrate saturation will then occur with lesser substrate concentration. Similar sigmoid curves are observed for Glc-6-P alone or in combination with adenine nucleotides as well as P_i. However maximal NAD^+ reduction occurs with lower substrate levels in presence of adenine nucleotides (and to a lesser extent P_i). The maximum displacement to the left is seen with ATP [49].

As to the second category of EL2 cells, which are dependent upon the external addition of glycolytic activators (i.e. triiodothyronine-grown cells), their glycolytic NAD^+ can be vey highly reduced, even with Glc-6-P alone (fig. 6). However as compared to the normally grown EL2 cells (EL2 controls, which do not depend on external supply of activators), such maximal reduction of glycolytic NAD^+ is now attained in the triiodothyronine-grown cells with much lower levels of exogenous substrate. Under these conditions, substrate saturation is reached quite early with the resulting bottlenecks in the glycolytic flux; Since such bottlenecks seem to occur at various ADP or P_i-dependent enzymatic sites, it is quite understandable that these compounds are useful in relieving the metabolic block. It takes, in fact, the joint addition of ADP and P_i to bring in these metabolite dependent EL2 cells the rate of glycolytic substrate utilization to the levels recorded in the control EL2. It is noteworthy that in the triiodothyronine-grown EL2 cells the rate of Glc-6-P utilization is activated also by ATP, which acts as an inhibitor in the control. These apparently contradictory effects may be related to the compartmentalization of metabolites and their competitive trapping by various organelles [29,35,51–55]. What matters in fact is not the overall endogenous concentration of ATP or the average ATP level following microelectrophoretic addition, but rather the ATP concentration at the effective site where it is needed.

3.3. *The equilibrium between NAD^+ reduction at glyceraldehyde phosphate dehydrogenase and reoxidation at lactic dehydrogenase*

Each point in the fluorescence curve following the addition of Glc-6-P is the algebraic sum of the forward and backward reactions at the levels of these two enzymes, which explains (as discussed in the previous section) why similar fluorescence curves can be obtained under quite different conditions.

In these experiments, two methods were applied to influence the glyceraldehyde phosphate dehydrogenase-lactic dehydrogenase equilibrium: one through the use of a selective inhibitor of lactic dehydrogenase (oxamate) [56–58], the other by attempting to enhance backward reaction at lactic dehydrogenase through addition of lactate.

Both lactate and oxamate (figs. 4 and 7) make it easier to record the fluorescence changes following suboptimal or threshold doses of substrate. It seems that when the forward reaction is allowed to move freely at the lactic dehydrogenase, the fluorescence changes with small amounts of substrate are quite elusive, as NADH cannot accumulate adequately. In a way the blocking of this forward reaction or the stimulation of the reverse reaction help in the inscription of fluorescence changes which otherwise would have remained unnoticed. Thus it is a common feature with oxamate or lactate to record very narrow and quite tall fluorescence pulses, following substrate additions slightly over threshold. In these conditions, a prolongation of the NADH reoxidation half-time over the response time of the microfluorimeter facilitates the detection of the fluorescence changes.

With either lactate or oxamate, the rate of glycolytic substrate utilization

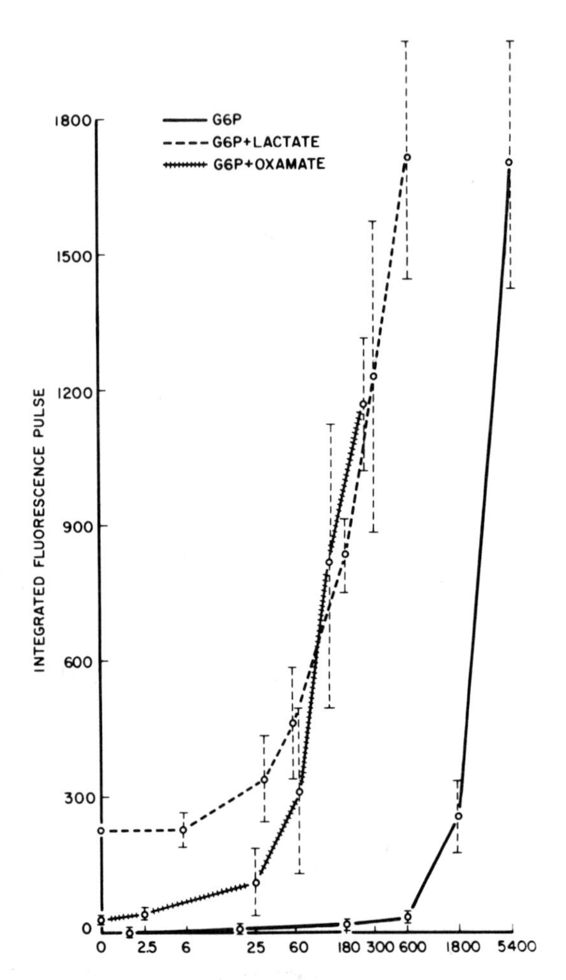

Fig. 7. The relationship between the integrated area of the fluorescence pulse $\int_0^\infty p_1\,\mathrm{d}t$ (in arbitrary units) and the microelectrophoretic current in EL2 cells treated with Glc-6-P alone and Glc-6-P + lactate or oxamate (conditions as in fig. 3c, but summed from at least five cells).

$$\mathrm{S.E.} = \pm\sqrt{\Sigma d^2/n\,(n-1)}.$$

is uniformly decreased (fig. 4) and maximal reduction of glycolytic NAD^+ is reached much earlier. Any hindrance of the progression of substrates along the glycolytic chain should result in a decreased rate of utilization. A similar situation occurs at the terminal end of the chain when the forward reaction at the lactic dehydrogenase is not allowed to move freely.

While the rate curves of Glc-6-P-utilization are quite similar in the presence of lactate or oxamate, all along the substrate concentration range, from threshold to maximal values, the sigmoid plots of NADH accumulation follow a similar course only up to suboptimal substrate levels (fig. 7). At larger substrate concentrations NAD^+ reduction in presence of lactate is equal to or higher than the levels observed with Glc-6-P alone. Unlike lactate, the maximal reduction of NAD^+ is somewhat lower in the oxamate-inhibited EL2 (30-50% below the level for Glc-6-P alone). It has been a common observation that in these EL2 cells, the inhibition of glycolytic metabolism is often accompanied by a drop in the level of NAD^+ reduction (cf. earlier sections).

3.4. *The influence of drugs and hormones or physical effects on intracellular enzyme kinetics*

Addition of drugs to the growth media of EL2 cells enables enzyme induction phenomena to be studied, provided the induced enzyme systems participate in reactions coupled with a change in the redox state of mitochondrial or extra-mitochondrial pyridine nucleotides. In cells treated with drugs, hormones or physical agents (e.g. X-rays, temperature) the actual level of glycolytic activity will be related to the algebraic sum of influences exerted upon it. Mitochondrial damage (as seen in radium or X-ray treated cells) or enzyme biosynthesis (e.g. in 1-15 days treatment with triiodothyronine) might enhance Glc-6-P–dependent NAD^+ reduction, while changes in the compartmentalization of NAD^+ splitting enzymes should have an adverse effect (e.g. in cells treated with cytostatics). In presence of amytal or rotenone, there is a characteristic prolongation of the NAD^+ reoxidation half-time in the Glc-6-P induced fluorescence cycles [59]. It is often noticed that whenever $t_{½\,off}$ is prolonged, there is a drop in the amplitude $((p_1)_{max})$ of the fluorescence cycle, as if the product $t_{½\,off} \times ((p_1)_{max})$ tends to stay constant [46]. There seems also to be a certain correlation [59] between the shape of the fluorescence curve and the ability of the EL2 cell to sustain the imposition of repeated metabolic transients. In cells which exhibit high-amplitude and short-lasting fluorescence cycles, such metabolic transients can be repeated from 6-15 times, while the low-amplitude and long–half-time responses are damped quite easily. It may be that the reoxidation of glycolytic NAD^+ is closely dependent upon mitochondrial-cytoplasmic exchanges, which cannot proceed adequately when electron transport is blocked in the mitochondria by amytal or rotenone. Under these conditions a compensatory stimulation of glycolysis in the cytoplasm can be expected. There is indeed a greater accumulation of NADH in the glycolytic compartment as suggested by a larger value of the $t_{½\,off} \times (p_1)_{max}$ product. However, at long range, a certain deficiency may

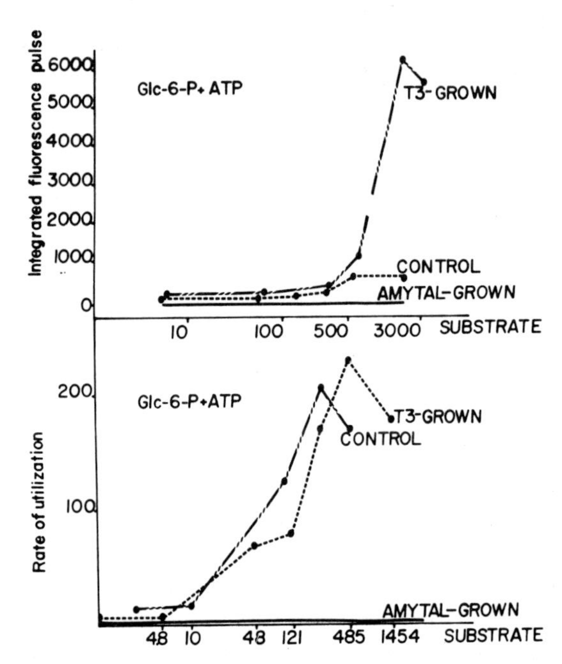

Fig. 8. NADH accumulation (as expressed by the integrated fluorescence pulse $\int^{\infty} p_1 \, dt$ in arbitrary units, see upper curve) and rate of Glc-6-P utilization (in 10^{-12} A/pl/sec, see lower curve) in triiodothyronine and amytal-grown EL2 cells versus the normally grown EL2 (controls). In all these experiments the glycolytic substrate was supplemented with ATP (ATP/Glc-6-P = 1/8). It is noteworthy that *with ATP present* there was an enhancement of Glc-6-P utilization and glycolytic NAD^+ reduction in the triiodothyronine-grown (compare with fig. 6), while the rate of utilization was decreased in the normally grown (as compared to Glc-6-P alone) and practically zero in the amytal-grown. T3-grown = = triiodothyronine-grown.

appear in the perpetuation of the glycolytic process, unless other outlets or pathways are found for the reoxidation of glycolytic NAD^+. It is noteworthy that EL2 cells maintained for short periods (e.g. 4 days) in the presence of amytal, continue to exhibit an amytal-inhibited metabolism (with low-amplitude, broad fluorescence pulses after Glc-6-P) even when transferred to an amytal-free medium. In cells maintained for longer periods in amytal (i.e. 2 weeks) there is a partial recovery of the fluorescence response, somewhat reminiscent of the tolerance which develops towards phenobarbital [60–65]. However the above described pattern of response (lowered $(p_1)_{max}$ and prolonged $t_{1/2\,off}$) continues to be observed in cells maintained for several months in amytal.

As to the responses in triiodothyronine-grown cells they are somewhat different in the ordinary EL2 and in giant EL2 cells produced by X-ray treatment [28]. In the radiation giants the most evident change is a rather large enhancement of the Glc-6-P-dependent NAD^+ reduction. In the non-irradiated EL2, following growth for two weeks in the presence of triiodothyronine, there is also a small increase in glycolytic NAD^+ reduction, but the most consistent change is the dependency of the glycolytic rate on external additions of adenine nucleotides and P_i (figs. 6 and 8). It is somewhat puzzling to find in the triiodothyronine-grown EL2 cells, a glycolytic flux much smaller than in the EL2 controls unless ATP, ADP or P_i is added or alternative pathways are involved. The thyroid hormone is known for its uncoupler properties [66–73] and usually when mitochondrial oxidative phosphorylation is uncoupled there is a compensatory increase in glycolytic activity, as suggested by dinitrophenol experiments [54]. It has been also reported that thyroxine should activate carbohydrate degradation via reoxidation of glycolytic NADH through the alpha-glycerophosphate shuttle [72,73]. However as indicated earlier, in cells treated with hormones the actual level of glycolytic activity will be related to the algebraic sum of influences exerted upon it. Thus, in triiodothyronine-grown cells there may be influences exerting an adverse effect upon the glycolytic rate. Triiodothyronine-grown cells are somewhat in the situation of oxamate-treated cells [50], as there seems to be a deficiency in the reoxidation at LDH of glycolytic NAD^+. LDH is inhibited *in vitro* by 10^{-5} M thyroxine [71]. It is possible that as in amytal-grown cells, the changes seen in long-term (around 2 weeks) cultures with triiodothyronine might be the result of biosynthetic, structural or organelle changes induced in these cells by the hormone, rather than the direct hormone effect *per se*. In liver cells prolonged exposure to thyroxine leads (according to preliminary observations) to an increased mitochondrial population [72], as if for compensating the uncoupler effect of the hormone. In the case of compensatory mitochondrial hyperactivity [74] it would be easier to understand the lowered glycolytic rate in the T3-grown (e.g. in terms of mitochondrial competition for glycolytic activators, such as ADP or P_i). It is possible to test this hypothesis, by introducing directly into the cytoplasm of triiodothyronine-grown cells such activators. In fact it is only in triiodothyronine-grown cells treated with ADP + P_i that the rate of glycolysis approximates the level seen in the EL2 controls grown in the absence of the hormone. Another indication that the observed metabolic changes might be due to the long-term consequences of growth in the presence of the hormone (adaptation) rather than to the hormone itself, comes from short-term (1-48 hr) studies with triiodothyronine; these experiments suggest to the contrary an acceleration of the glycolytic rate in the cytoplasm, but they are yet inconclusive.

Besides drugs, hormones, allosteric activators or inhibitors, there is also the possibility of action by physical factors on intracellular enzyme kinetics. Some of these effects, e.g. the metabolic alterations in X-ray-induced giants have been already discussed previously [16]. Another group of interesting effects, concerns the possible action of temperature on the delicate balance between various enzymatic steps. An illustrative example is provided by the temperature dependency [75–84] of the glycolytic flux in the EL2 cells. The

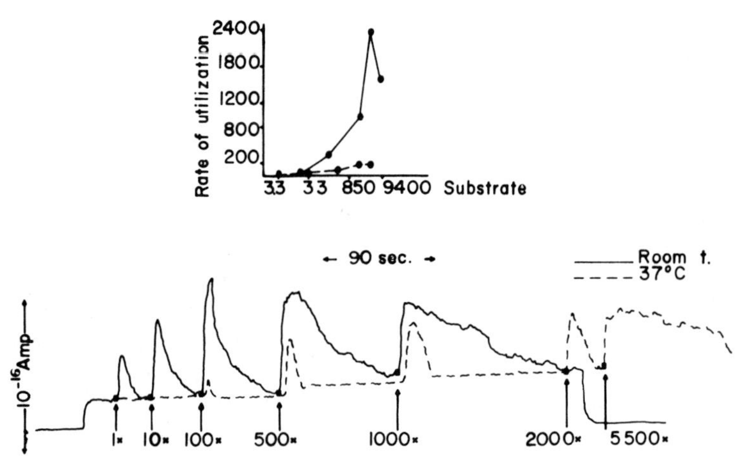

Fig. 9. Representative microfluorimetric recordings from EL2 cells treated with glycolytic substrate (Glc-6-P) at room temperature and 37°. Glc-6-P was supplemented with lactate (Glc-6-P/lactate = 3), a substrate which in these cells tends to prolong the time lag between NAD^+ reduction by glyceraldehyde phosphate dehydrogenase and reoxidation by LDH (especially at room temperature). 1x = 4.5 × 10^{-11} A. The rate of Glc--6-P utilization (10^{-12} A/pl/sec) as calculated from the half-time of each NAD^+ reduction and reoxidation cycle after the addition of substrate (10^{-11} A/pl) is indicated above the microfluorimetric recordings. It is noteworthy that the lower amplitudes of the fluorescence cycles at higher temperature are not indicative of a decreased Glc-6-P utilization, as the latter is considerably enhanced at 37°. Generally, there is little correlation *per se* between the rate of Glc-6-P utilization and the amplitude of Glc-6-P–induced fluorescence, since the latter corresponds only to the algebraic sum of NAD^+ reducing and reoxidizing enzyme activities (e.g. if NADH reoxidation can keep pace with NAD^+ reduction, the recorded fluorescence changes can be quite low despite a very active glycolytic flow). Thus the glycolytic flow proceeds with less hindrance at 37° when the reoxidation of extramitochondrial pyridine nucleotides can keep pace with the reducing equivalents released at the energy yielding steps. (Expt. EK69/8AI, cells 5αH3).

glycolytic flux estimated from the half-time of the transient changes in NAD^+ reduction can be increased by several times, simply by raising from 23° to 37° [85–88] the temperature at which the cells are incubated in the microfluorimeter (fig. 9). However the magnitude of this effect seems related to the pre-heating level of the glycolytic activity. In these EL2 cells so far there is a maximal level of the glycolytic flux, somewhere around 1200 to 1800 $\times 10^{-12}$ A/pl/sec weight, which cannot be exceeded regardless of further treatment with allosteric activators, heat, etc. (fig. 10). If the glycolytic rate in the extramitocondrial space is already close to this maximum at room temperature, there is little acceleration which can be expected from further heating. Thus, the maximal increase in glycolytic flux at higher temperature occurs in conditions where at room temperature NADH reoxidation (presumably by lactic dehydrogenase) lags behind NAD^+ reduction by glyceraldehyde phosphate dehydrogenase: either when glycolytic NAD^+ is being reduced too fast (i.e. in the presence of phosphate) or when NADH is reoxidized too slowly (i.e. in the presence of lactate). The above lag in NADH reoxidation is usually associated with a lowering of the glycolytic flux (fig. 4) at room temperature, which can then benefit from heating.

In the normally grown EL2 cells which show little dependence on an external supply of activating metabolites (see earlier sections), the glycolytic flow with Glc-6-P alone (or supplemented by adenine nucleotides) is already close to a maximum at room temperature and there is little benefit to be derived from heating. In such cases the ratio of the glycolytic flux at 37°/flux at 23° stays close to 1, but the same ratio doubles in the presence of phosphate. The most striking differences between 23° and 37° (up to a six-fold increase in flux) are significantly found in conditions where both the glycolytic flux and the reoxidation of extramitochondrial NADH are severely impaired by the presence of lactate. Thus at 37°, the glycolytic flux may proceed with less impediment through the various bottlenecks along the glycolytic chain and NADH is more effectively reoxidized by lactate dehydrogenase [85–90]. At higher temperature there is a drop in the ratio of NAD^+ reduced/Glc-6-P added, which may help in maintaining an active outpouring of reducing equivalents from the glycolytic chain. So far, as compared to normally grown ascites cells, such phenomena are not easily discernable in cells grown with triiodothyronine.

The above temperature studies in normally grown ascites cells provide possibly a glimpse to an actual race between two functionally antagonistic enzymes (glyceraldehyde phosphate dehydrogenase or lactic dehydrogenase) in the intact cell. Depending upon the temperature the equilibrium may be poised in favour of one or the other and the unhindered flow of glycolysis seems to re-

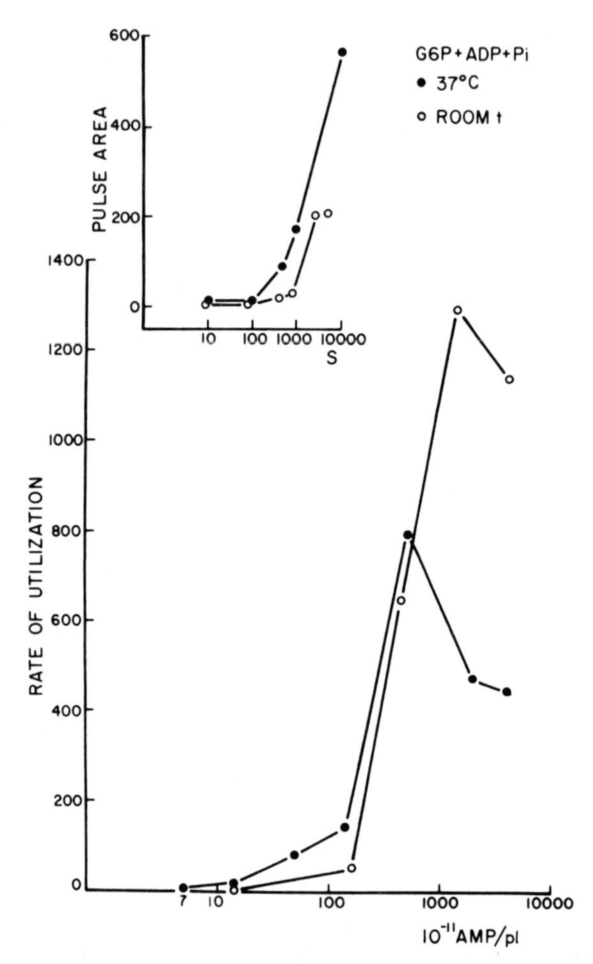

Fig. 10. The rate of Glc-6-P utilization at room temperature and 37° in triiodothyronine-grown EL2 cells treated with Glc-6-P + ADP + Pi. The rate of substrate utilization is indicated on the ordinate in 10^{-12} A/pl/sec. The values on the abscissa correspond to the added Glc-6-P in 10^{-11} A/pl. Each point represents the results of at least five determinations. The curves in the upper corner represent the corresponding sigmoid plots of $\int_0^\infty p_1 \, dt$ arbitrary units) against substrate concentration. ($10 = 6 \times 10^{10}$ A).

quire that the two enzymes keep relatively at pace with each other. Microfluorimetric determinations are facilitated by a certain lag in the activity of the reoxidizing enzyme, since such a lag will allow the assay of accumulating pyridine nucleotides, but the glycolytic rate will drop considerably if the NADH accumulation proceeds far out of pace with reoxidation. It is likely that substrates will move along the glycolytic sequence with less hindrance when the reoxidation of extramitochondrial pyridine nucleotides can keep pace with the reducing equivalents released at the energy yielding steps (figs. 4 and 9).

3.5. *Interactions between various metabolic compartments*

Although in the study of glycolytic metabolism the extramitochondrial space has been presented as a single functional unit, and computer representations have often tended to neglect the nuclear membrane, discrete differences are indeed found between nucleus and cytoplasm [40–42]. The glycolytic rates are approximately comparable and the levels of glycolytic NAD^+ reduction not unlikely in these two glycolytic sub-compartments; it seems more difficult to reoxidize nuclear pyridine nucleotides and the glycolytic response is somewhat harder to sustain in the cytoplasm upon repetitive additions of substrate. In these EL2 cells almost 50% of the glycolytic activity may be in the nucleus [40].

A promising area for exploration refers to the mitochondrial-extramitochondrial exchanges [35–37] (see fig. 11). The experiments with amytal, rotenone, uncouplers and triiodothyronine are already indicative of the possible interplays at this level. It is also noteworthy that in the presence of succinate, the glycolytic response proceeds at first adequately, but the rate of glycolysis drops if the cell is allowed to exhaust its endogenous activators (i.e. adenine nucleotides and P_i). However in the presence of succinate, a higher rate of glycolysis can be restored especially in the ATP-dependent triiodothyronine-grown EL2, if this nucleotide is added simultaneously.

In triiodothyronine-grown EL2 cells, the postulated lack of glycolytic activators (e.g. ADP or P_i) in the cytoplasm was tested by the simultaneous addition of these activators. Another way to test the hypothesis is by acting at the level of mitochondrial-extramitochondrial exchanges [91–97]. Although the mitochondrial content of total adenine nucleotides is reportedly constant, mitochondria tend to become enriched preferentially in ADP at the expense of the cytoplasm. This can be prevented by atractylate [92–97], an inhibitor of adenylate translocase. A preliminary attempt was made to introduce atractylate into the cytoplasm (and thus around the mitochondria) of EL2 cells, at a concentration capable of inhibiting the translocase. Under these circumstances, there was indeed an enhancement in the extramitochondrial utilization of glycolytic substrates.

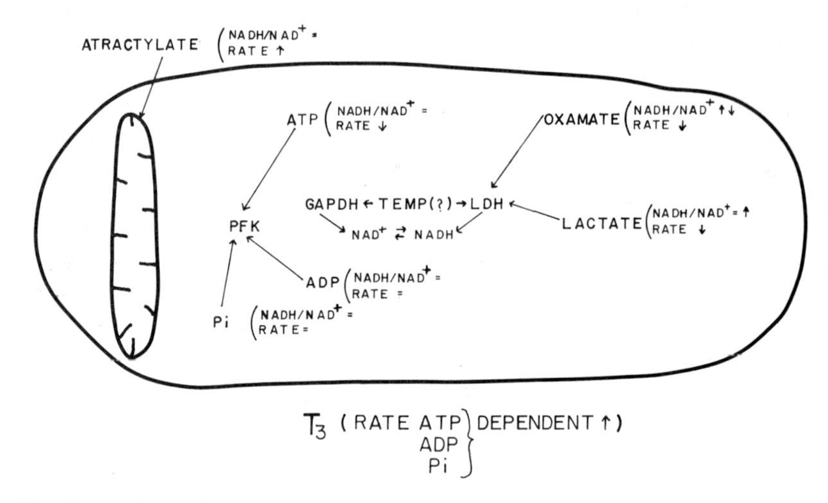

Fig. 11. A scheme of the main reactions studied so far in the living cell by microfluorimetry. In this oversimplified model of the EL2 cell, only the mitochondrial and extramitochondrial compartments are considered. It has been possible to alter the metabolic responses of the EL2 cell by actions at three levels: a) through activators or inhibitors of selected enzymes; b) by trying to modify mitochondrial-extramitochondrial exchanges; c) by growing cells in the presence of hormones or drugs. PFK = phosphofructokinase; GAPDH = glyceraldehyde phosphate dehydrogenase; LDH = lactic dehydrogenase; T3 = triiodothyronine. ↑ : increase, ↓ : decrease; = : no change.

4. Discussion

The reported experiments point to several avenues of exploration which are now open to microfluorimetry. The first point which had to be made concerns the definite possibility of achieving intracellular enzyme studies in localized compartments of the living cell. While such studies are slightly complicated by the sensitivity of the fluorochrome to more than one intracellular enzyme at the time, nonetheless a detailed study of the glycolytic response has been possible. Reduced to its most simple expression, the fluorescence response to glucose-6-phosphate represents the summation of the activities at the levels of glyceraldehyde phosphate dehydrogenase and lactic dehydrogenase. The apparent balance between these two enzymes can be poised on one side or the other by various actions: through temperature effects, overloading with substrate, inhibitors (e.g. oxamate) or factors favouring the reverse reaction (e.g. lactate). Thus, while dealing with at least two enzymes engaged in a perpetual race on which the NAD^+/NADH equilibrium in the glycolytic compartment is critically dependent, it is gradually becoming possible to determine the individual contributions of these two enzymes.

The cell is very far from being a "bag of enzymes" as some computer representations may have preferred to assume for the sake of simplicity. By an interplay of properly selected metabolites [28,49], substrates [16] or drug effects [50,59,98] it is becoming increasingly feasible to determine differences in the metabolic behaviour of compartments *in situ.* The experiments with amytal, rotenone, triiodothyronine, atractylate, etc. indicate that the living cell is a "metabolic complex" in which each part, organelle or compartment is in perpetual interaction and exchange with the others. In this field the mitochondrial-extramitochondrial interactions offer a natural avenue for exploration.

Besides such interactions between compartments, the rates of activity at the level of intracellular enzymes lend themselves to a parallel between *in vitro* and *in vivo* studies. In both EL2 controls and triiodothyronine-grown cells at the optimum concentration range of substrate, there is a high power relationship between the activity of glycolytic enzymes and substrate. Since each "dose-response curve" was determined by repeated additions in the same cell, such curves bear a degree of authenticity which cannot be duplicated by statistical studies of any sort. At this stage, it would be too farfetched to make any inference from the exponential rate to the subunit structure of intracellular enzymes. Although such rates would be compatible with allosteric enzymes [99,100] exhibiting multiple subunits [101,102], several alternative interpretations are possible (e.g. endogenous substrate thresholds, enzyme relaxation phenomena, etc.).

So far the main compartments of the living cell and the reactions which yield the highest fluorescence changes are within reach of the method. The reported experiments provide a preliminary view of the actual interplay between metabolic compartments or metabolite modulated enzyme systems (fig. 11). However, in the future it should be possible to make a more detailed study of still finer compartments with an improved time constant. The microfluorimetric technique which faces the metabolic control mechanisms, as they occur coherently within a "whole", offers certainly the prospect of a more complete understanding as compared to any reconstructed model of the living cell, made on the basis of biochemical analysis or computer simulation (table 1).

Table 1
Hypotheses or data from independent biochemical studies and their evaluation by the microfluorimetric- microelectrophoretic techniques

Hypotheses or data from independent biochemical studies	Microfluorimetric observations
NAD^+ compartmentalization between cytoplasm and mitochondria [7,17]	Simple visual observations and microfluorimetric determinations illustrate the functional separation of the two main pyridine nucleotide compartments: 1) a mitochondrial compartment associated with repiratory function, where the $NAD^+ \rightarrow NADH$ transition is favoured by anaerobiosis, inhibitors of electron transport at the NADH-flavoprotein level (e.g. amytal, rotenone) and Krebs cycle intermediates [17]; 2) an extramitochondrial compartment associated with glycolytic function where the $NAD^+ \rightarrow NADH$ transition is favoured by intermediates of glucose metabolism [17,28].
Correlation between extramitochondrial fluorescence, NAD oxidation-reduction and the rate of glycolysis [51]	The level of extramitochondrial fluorescence (around 450 nm) is closely associated with the reduction of extramitochondrial NAD^+. However there is no direct correlation with the glycolytic rate. Similar levels of NAD^+ reduction are observed in cells grown under various conditions (e.g. normally grown vs. triiodothyronine (T3)-grown). NAD^+ reduction levels provide a measure of NADH accumulation as a result of the time lag between NAD^+ reduction at glyceraldehyde phosphate dehydrogenase (GAPDH) and reoxidation at lactic dehydrogenase (LDH). Thus a low level of glycolytic NAD^+ reduction is compatible with a high rate of substrate utilization when reoxidation at LDH keeps pace with GAPDH activity (algebraic sum close to zero within the limits of detection).
The problem of eventual nucleo-cytoplasmic compartmentalization	No detectable compartmentalization despite discrete discrepancies between the nucleus and the cytoplasm (e.g. relatively higher level of nuclear NAD^+ reduction at anaerobiosis and with amytal or rotenone, also with various glycolytic intermediates

Hypotheses	Microfluorimetric observations
	[40–42]. In X-ray produced giant cells (ascites or liver), there is: rotenone-insensitivity of the nucleus vs. rotenone-enhancement of glycolytic NAD^+ reduction in the cytoplasm (no such discrepancy in normally grown cells). This might be due to membrane effects resulting from radiation damage [103, 104]. Also "partial physiological anaerobiosis" of the nucleus might be a possibility in giant cells with very large nuclei and high intracytoplasmic oxygen consumption [42].
The problem of intracellular compartmentalization with regard to substrates and intermediates [31,51,91,94]	No detectable permeability barriers to the intracellular diffusion of glycolytic or Krebs cycle substrates added to the cytoplasm.
Permeation of metabolites through the cell walls	1) Permeation of glucose through the cell wall: [105] rate of permeation is dependent upon the metabolic state of the cell (practically no transport in large numbers of ascites cells responding to microelectrophoretic additions of phosphorylated glycolytic intermediates). In cases where transport occurs, there is a 10-25 sec time-lag in reaching the maximal level of glycolytic NAD^+ reduction, as compared to substrate added microelectrophoretically. 2) Generally, there is no change in mitochondrial or extramitochondrial NAD^+ reduction after cell perfusion with phosphorylated glycolytic intermediates, Krebs cycle substrates, glutamate, etc. Thus the cell wall seems impermeable to these metabolites, except possibly for a weak penetration.
The problem of ATP compartmentalization between the mitochondria and cytoplasm [29,53–55]	1) No enhancement of glycolysis following addition of ATP to the cytoplasm in normally grown ascites cells, but marked enhancement in cells grown with triiodothyronine (T_3) 2) In presence of succinate, ATP enhancement of glycolysis. 3) However ATP experiments are complicated by uncontrollable ATP $\rightleftharpoons$ ADP interconversions.

Hypotheses	Microfluorimetric observations
The problem of interactions between the mitochondrial and extramitochondrial compartments [7,10,18,29,53–55]	1) Prolongation of glycolytic NAD^+ reoxidation half-time in cells perfused or grown up to several months with amytal (or rotenone). 2) Activation of cytoplasmic and nuclear glycolysis by mitochondrial uncouplers. 3) So far, in a variety of cells investigated, 5 discernable types of metabolic behavior depending upon the sensitivity of extramitochondrial glycolysis to mitochondrial inhibitors [39, 41]. 4) The mitochondrial oxidations of respiratory substrates are sensitive to simultaneous or prior treatment of these cells with extramitochondrial (glycolytic) substrates, and vice versa for the extramitochondrial oxidations. 5) Suggestion of mitochondrial-extramitochondrial competition for adenine nucleotides during glycolysis in presence of succinate. 6) Acceleration of glycolytic flux when adenylate translocase is blocked by atractylate to prevent trapping of cytoplasmic ADP by the mitochondria.
ATP/ADP. P_i control of mitochondrial and extramitochondrial NAD^+/NADH ratio [94]	1) In the mitochondria higher level of NAD^+ reduction with ATP, lower with ADP. 2) In the cytoplasm (or nucleus) where a reverse relationship is expected complexities depending upon conditions to which cells have been adapted: a) in normally grown ascites cells and amytal-grown cells *lower* NAD^+ reduction with ATP (no change or at times "paradoxical" decrease with ADP) b) in T_3-grown cells *higher* levels of NAD^+ reduction or NADH accumulation with ATP.
Control of glycolysis by "bottlenecks" along the glycolytic chain [52]	1) "Bottlenecks" along various steps of the glycolytic chain (PFK*, ALD, GAPDH, PGK, PK) with ADP, P_i or ATP rate-limiting depending upon conditions of cell growth. Suggestive evidence for multisite-multicomponent control from the activating effects

* *Abbreviations*: PFK: phosphofructokinase; ALD: aldolase; GAD: glyceraldehyde phosphate; GAPDH: glyceraldehyde phosphate dehydrogenase; PGK: phosphoglycerate kinase; PK: pyruvate kinase; F6P: fructose-6-phosphate; FDP: fructose-1,6-diphosphate; LDH: lactate dehydrogenase.

Hypotheses	Microfluorimetric observations
	of glycolytic mixtures (e.g. Glc-6-P + FDP + ADP) in cells where the response to Glc-6-P alone is inhibited. Also in many normally grown ascites cells no need for externally added activators, only substrate. 2) Competition with mitochondria for adenine nucleotides (see above). 3) Considerable slowing of glycolytic rate upon substrate saturation of LDH and in presence of LDH inhibitors or compounds interfering with the reoxidation of glycolytically reduced NAD^+ (e.g. oxamate, lactate, α-glycerophosphate).
Transitions from one metabolic pathway into another [20,31]	Irreversibility of F6P $\rightarrow$ FDP transition and lack of significant FDPase activity for the reverse transition suggested by the higher glycolytic NAD^+ reduction with Glc-6-P (or F6P) as compared to FDP (GAP). Glc-6-P and F6P are able to proceed along other pathways than the Embden Meyerhof sequence (e.g. pentose monophosphate shunt, glucose-6-phosphatase and glucose oxidase, etc.) to which FDP remains restricted.
Physical effects or drug effects on organelle structure [103,104]	1) Metabolic changes compatible with alterations of lipoprotein membrane structures (e.g. membranes of endoplasmic reticulum, mitochondria, lysosomes, etc.) in cells treated with X-rays or cytostatics. 2) Varying degrees of metabolic alterations depending on dosage [16,50,98]: a) with sublethal doses of radiation or alkylating agents massive inhibition of glycolysis and decrease of extramitochondrial NAD^+; b) with lesser doses giants exhibiting more or less active glycolysis. 3) Inverse relationship between peak level of glycolytic NAD^+ reduction and reoxidation half-times, suggestive of a difficulty in the reoxidation of glycolytic NAD^+ in cells with inhibited glycolytic rates.
Effects of T_3, enzyme activation and biosynthesis [67-74]	1) Great dependency of the glycolytic rate in T_3-grown cells on externally supplied ATP. 2) As compared to normally grown ascites cells, in the T_3-grown a more difficult reoxidation of glycolytic NADH which does not benefit from temperature (see below in

Hypotheses	Microfluorimetric observations
	this same table, effect of temperature on normally grown ascites cells): 2 of relative LDH deficiency or changes in the isozyme population. 3) In the T_3-grown cell, as compared to the normally grown, a lower sensitivity of the glycolytic rate to externally added LDH inhibitors (e.g. oxamate) possibly because of already present endogenous inhibition. 4) Shifts in the balance between various steps along glycolytic pathways as well as alterations in intracytoplasmic organelles affecting the compartmentalization of metabolites or mitochondrial-extramitochondrial competition for such metabolites.
Enzyme induction by drugs and tolerance [60–65]	Gradually increased tolerance to larger doses of amytal (e.g. prolonged survival in presence of 0.1 M amytal) and partial recovery of the metabolite response (with reference to the normally grown) in cells maintained up to 2 weeks in amytal.
Enzyme activity *in vivo* vs. *in vitro*, allostery ... [46,51,99–102]	1) Rates of glycolytic substrate utilization in ascites cells, in the upper range of values calculated by biochemical methods. 2) Exponential rate of increase in NADH accumulation and sigmoid kinetics favouring a higher order relationship with regard to substrate: suggestive of allosteric enzymes with multiple subunits, with a possible clue to the number of subunits (e.g. 2-4) from the exponent in the rate of increase (among other possible interpretations: endogenous substrate thresholds, enzyme relaxation phenomena).
Effects of modifiers on intracellular enzyme kinetic curves [99]	In normally grown ascites cells or the T_3-grown, lateral displacements of intracellular enzyme kinetic curves along the horizontal axis by addition of modifiers (e.g. adenine nucleotides, P_i, etc.), accompanied at times by changes in slope.
Temperature effect [75–84] on intracellular enzyme kinetic curves	Up to 6-fold increase in glycolytic flux at 37°/glycolytic flux at 23° in cells where NADH reoxidation by LDH lags behind NAD^+ reduction by GAPDH (e.g. with Glc-6-P in presence of P_i or lactate). Also at higher temperature a simultaneous shift in the point of maximal NAD^+ reduction towards greater substrate level. Thus, intracellular

Hyptheses	Microfluorimetric observations
	enzyme kinetic curves exhibit with temperature, changes in slope as well as displacement along the horizontal axis.
Correlation between the kinetic properties of isozymes, their intracellular distribution and compartmentalization (e.g. theory that lactate or pyruvate-inhibited LDH–1 is present in tissues with high degree of aerobiosis, while lactate or pyruvate-resistant LDH-5 is present in tissues with high degree of anaerobiosis and glycolysis. Cf. also observation that the LDH-1–LDH-5 differences are largely cancelled by raising the temperature from 25° to 37°, whereby lactate or pyruvate-inhibition of LDH-1 is relieved [85–88].	The inhibition of glycolytic NAD^+ reoxidation in presence of lactate at room temperature and its reactivation at 37°, suggest the occurrence of LDH-1 in cells.

Acknowledgements

The authors wish to express their thanks to Professor Britton Chance, The Johnson Foundation, University of Pennsylvania, for encouragement and fruitful exchange. These experiments were made possible by his generous loan of the Chance-Legallais microfluorimeter and related microelectrophoretic equipment. The authors are also grateful to research engineer Lennart Åkerman, and biomedical engineer Lars Nordberg, for cooperation, advice and stimulating discussions; to Mrs. Märta Thorell, Astrid Eklund, John Goodman, Mrs. Marilyn Foreman, Miss Robin Sulman, and Robert Steele for the drawings; to Miss Norma Orr, Mrs. Alvera Miller and Miss Sylvia Whitobsky for helping with the manuscript.

This work was supported by an Anna Fuller Fellowship, Swedish Medical Research Council Grants K68-12X-630-04, B69-12X-630-05, K69-12X-630-05K, B70-12X-630-06A, and Grant # P-518 from the American Cancer Society.

References

[1] W. Heisenberg, Philosophic Problems of Nuclear Science, translated by F.C. Hayes, (Pantheon, New York, 1952).
[2] T. Caspersson, Skand. Arch. Physiol. 73 (1936) Suppl. 8.
[3] B. Thorell, Faraday Soc. Discussions (1950) 432.
[4] B. Chance, R. Perry, L. Åkerman and B. Thorell, Rev. Sci. Inst. 30 (1959) 735.
[5] R. Perry, B.Thorell, L. Åkerman and B. Chance, Biochim. Biophys. Acta. 39 (1960) 24.
[6] B. Chance and V. Legallais, Rev. Sci. Inst. 30 (1959) 732.
[7] B. Chance and B. Thorell, J. Biol. Chem. 234 (1959) 3044.
[8] R. Perry, B. Thorell, L.Åkerman and B. Chance, Nature 184 (1959) 929.
[9] E. Kohen, Exptl. Cell. Res. 35 (1964) 26.
[10] B. Chance, E. Kohen, C. Kohen and V. Legallais, in: Advances in Enzyme Regulation, ed. G. Weber Vol. 5, (Pergamon Press, New York, 1967) p.3.
[11] E. Kohen, C. Kohen and B. Thorell, Biomedical Enging. 4 (1969) 554.
[12] P.O. Boyer and H. Thorell, Acta Chem. Scand. 10 (1956) 447.
[13] B. Chance and H. Baltscheffsky, J. Biol. Chem. 233 (1958) 736.
[14] H. Sund, in: Biological Oxidations, ed. T.P. Singer (Interscience, New York, London, Sydney, 1968) p. 603.
[15] H. Sund, ibid. pp. 641-705.
[16] E. Kohen, C. Kohen and B. Thorell, Exptl. Cell. Res. 49 (1968) 169.
[17] E. Kohen, Exptl. Cell. Res. 35 (1964) 303.
[18] D.E. Atkinson, Science 150 (1965) 851.
[19] J. Needham and H. Lehmann, Biochem. J. 31 (1937) 1913.
[20] W. J. Ray and G.A. Roscelli, J. Biol. Chem. 239 (1964) 1228.

[21] L. Rapkine, Biochem. J. 32 (1938) 1729.
[22] E. Kohen, Biochim. Biophys. Acta 75 (1963) 139.
[23] M.J. Kopac, Trans. N.Y. Acad. Sci. Sc. II 23 (1960) 200.
[24] M.J. Kopac, Phys. Tech. i. Biol. Res. 5 (1964) 191.
[25] D.R. Curtis, Phys. Tech. i. Biol. Res. 5 (1964) 144.
[26] E. Kohen and V.A. Legallais, Rev. Sci. Instr. 36 (1965) 1890.
[27] E. Kohen, V.A. Legallais and C. Kohen, Exptl. Cell Res. 41 (1966) 223.
[28] E. Kohen, C. Kohen and W. Jenkins, Exptl. Cell. Res. 44 (1966) 175.
[29] J.P.K. Maitra and B. Chance, Control of Energy Metabolism (Academic Press, New York, London, 1965) p. 157.
[30] L.J. Ree and D.J.Cox, Annual Rev. Biochem. 35 (1966) 57.
[31] V. Moses and K.K. Lonberg-Holm, J. Theoret. Biol. 10 (1966) 366.
[32] L. Ernester, H. Low and O. Lindberg, Acta Chem. Scand. 9 (1955) 200.
[33] J. Burgos and E.R. Redfearn, Biochim. Biophys. Acta 110 (1965) 475.
[34] L. Ernster, G. Dallner and G.F. Azzone, J. Biol. Chem. 238 (1963) 1124.
[35] E. Kohen and C. Kohen, Histochemie 7 (1966) 339.
[36] E. Kohen and C. Kohen, Histochemie 7 (1966) 348.
[37] N.S. Henderson, Ann. N.Y. Acad. Sci. 151, Art. 1 (1968) 429.
[38] N.O. Kaplan, Ann. N.Y. Acad. Sci. 151, Art 1 (1968) 382.
[39] E. Kohen, C. Kohen and B. Thorell, Histochemie 12 (1968) 95.
[40] E. Kohen, G. Siebert and C. Kohen, Histochemie 3 (1964) 477.
[41] E. Kohen, C. Kohen and B. Thorell, Histochemie 12 (1968) 107.
[42] E. Kohen, C. Kohen and B. Thorell, Histochemie 16 (1968) 170.
[43] E. Kohen, C. Kohen, L. Nordberg, B. Thorell and L. Åkerman, Digest of the 7th Internat. Conference on Medical and Biological Engineering, The Organizing Comittee for the 7th Intern. Conf. Med. Biol. Engng. (Stockholm, 1967) p. 274.
[44] E. Kohen, C. Kohen, B. Thorell and L. Åkerman, Biochim. Biophys. Acta 158 (1968) 185.
[45] E. Kohen, C. Kohen and B. Thorell, Biochim. Biophys. Acta. 167 (1968) 635.
[46] B. Chance, Modern Trends in Physiology and Biochemistry (Academic Press, New York, 1952) p. 25.
[47] L. Astrachan, S.P. Colowick and N.O. Kaplan, Biochim. Biophys. Acta 24 (1957) 141.
[48] S.F. Velick, J. Biol. Chem. 233 (1958) 1455.
[49] E. Kohen, C. Kohen and B. Thorell, Biochim. Biophys. Acta. 198 (1970) 1.
[50] E. Kohen, C. Kohen and B. Thorell, Exptl. Cell Res., 59 (1970) 307.
[51] E. Garfinkel and B. Hess, J. Biol. Chem. 239 (1964) 971.
[52] B.Hess, in: Control of Energy Metabolism, eds. B.Chance, R.Estabrook and J. Williamson (Academic Press, New York, London, 1965) p. 111.
[53] S. A; Neifakh, J.A. Avramov, V.S. Gaitskhoki, T.B. Kazakova, N.K. Monakhov, V.S. Repin, V.S. Turovski and I.M. Vassiletz, Biochim. Biophys. Acta 100 (1965) 329.
[54] I.Y. Lee and E.L. Coe, Biochim. Biophys. Acta 131 (1967) 441.
[55] C. Ritter, E. Kohen and B. Thorell, J. Histochem. Cytochem. 17 (1969) 188.
[56] W.B. Novoa, A.D. Winer, A.J. Glaid and G.W. Schwert, J. Biol. Chem. 234 (1959) 443.
[57] Papaconstantinou, J.and S.P. Colowick, J. Biol. Chem. 236 (1961) 278.
[58] J.C. Elwood, Cancer Res. 28 (1968) 2056.

[59] E. Kohen, C. Kohen and B. Thorell Z. Physiol. Chem. 350 (1969) 297.
[60] S. Orrenius, J. Cell. Biol. 25 (1965) 627.
[61] S. Orrenius, J. Cell. Biol. 26 (1965) 723.
[62] S. Orrenius, Studies on the Drug Hydroxylating Enzyme System of Rat Liver Microsomes (Almquist and Wiksels, Uppsala, 1965).
[63] S. Orrenius, J. Cell. Biol. 28 (1966) 181.
[64] G. Wahlstrom, Acta Pharmacol. Toxicol. 26 (1968) 64.
[65] G. Wahlstrom, Acta Pharmacol. Toxicol. 26 (1968) 92.
[66] J.R. Bronk, Biochim. Biophys. Acta 97 (1965) 9.
[67] L.I. Glick and J.R. Bronk, Biochim. Biophys. Acta 97 (1965) 16.
[68] L.I. Glick and J.R. Bronk, Biochim. Biophys. Acta 97 (1965) 23.
[69] Y.P. Lee and H.L. Laroy, J. Biol. Chem. 240 (1965) 1427.
[70] S.R. Tipton, M. Zirk and G.S. St. Amand, Federation Proc. 17 (1958) 163.
[71] J.F. Wolff and E.C. Wolff, Biochim. Biophys. Acta 26 (1957) 387.
[72] E.C. Wolff and J. Wolff, in: The Thyroid Gland, Vol. 1, eds. Pitt-Rivers and W.R.. Trotter (Butterworths, London, 1964) p. 237.
[73] S.C. Werner and J.A. Nauman, Annual Rev. Physiol. 30 (1968) 213.
[74] J.R. Tata, in: Recent Progress in Hormone Research, ed. G. Pincus, Vol. 18, (Academic Press, New York, 1964) p. 221.
[75] S. Arrhenius, Z. Physik. Chem. 4 (1889) 226.
[76] I.W. Sizer, J. Cell. Comp. Physiol. 10 (1937) 61.
[77] I.W. Sizer, J. Gen. Physiol. 21 (1938) 695.
[78] M. Eigen and G.G. Hammes, Advan. Enzymol. 25 (1963) 1.
[79] G. Cserlinski and F. Hommes, Biochim. Biophys. Acta 79 (1964) 46.
[80] G. Cserlinski and G. Schreck, J. Biol. Chem. 239 (1964) 913.
[81] G. Cserkinski, J. Theoret. Biol. 21 (1968) 387.
[82] G. Cserlinski, J. Theoret. Biol. 21 (1968) 398.
[83] G. Cserlinski, J. Theoret. Biol. 21 (1968) 408.
[84] G. Hammes, Advan. Protein Chem. 23 (1968) 1.
[85] M. Dawson, T.L. Goodfriend and N.O. Kaplan, Science 143 (1964) 929.
[86] C.R. Amarasingham and A. Uong, Ann. N.Y. Acad. Sci. 151 (1968) 424.
[87] N.O. Kaplan, Ann. N.Y. Acad. Sci. 151 (1968) 400.
[88] E.S. Vessel, Ann. N.Y. Acad. Sci. 151 (1968) 5.
[89] M.J. Wolin, Science 146 (1964) 775.
[90] P.J. Fritz, Science 150 (1965) 364.
[91] B. Chance and G. Hollunger, J. Biol. Chem. 236 (1961) 1534.
[92] A. Bruni and A.R. Contessa, Nature 191 (1961) 818.
[93] A. Bruni, S. Luciani and A.R. Contessa, Nature 201 (1964) 1219.
[94] M. Klingenberg, in: Control of Energy Metabolism, eds. B. Chance, R. Estabrook and J. Williamson (Academic Press, New York, London, 1965) p. 149.
[95] M. Klingenberg and E. Pfaff, in: Regulation of Metabolic Processes in Mitochondria, eds. J.M. Tager, S. Papa, E. Quagliariello and E.C. Slater (Elsevier, Amsterdam, 1966) p. 180.
[96] M. Klingenberg and E. Pfaff, in: Methods in Enzymology, Vol. 10, Oxidation and phosphorylation, eds. R.W. Estabrook and M.E. Pullman (Academic Press, New York, 1967) p. 680.
[97] M. Klingenberg and E. Pfaff, in: Metabolic Roles of Citrate, ed. T.W. Goodwin, (Academic Press, London, 1968) p. 105.

[98] E. Kohen, C. Kohen and B. Thorell, Acta Pharmacol. Toxicol. 26 (1968) 556.
[99] D.E. Atkinson, Ann. Rev. Biochem. 35 (1966) 85.
[100] J.E. Gander, Ann. N.Y. Acad. Sci. 138 (1967) 730.
[101] C.L. Market, Ann. N.Y. Acad. Sci. 151 (1968) 14.
[102] W.S. Allison, Ann. N.Y. Acad. Sci. 151 (1968) 180.
[103] J.M. Manteifel and M.N. Meisel, Federation Proc. Translation Supp. 25 (1966) T 981.
[104] H. Grunicke and H. Holtzer, in: The Cell Nucleus, Metabolism and Radiosensitivity, Proceedings of an International Symposium held in Rijswijk, 2. H., Netherlands, May 1966 (Taylor and Francis, London, 1966) p. 325.
[105] H.E. Morgan, J.R. Neely, J.P. Brineaux and C.R. Park, in: Control of Energy Metabolism, eds. B. Chance, R. Estabrook and J. Williamson (Academic Press, New York, London, 1965)p. 347.

PROTEIN CATABOLISM IN THE ISOLATED PERFUSED REGENERATING RAT LIVER. LOSS OF PROTEIN CATABOLIC RESPONSE TO GLUCAGON*

L.L. MILLER, L. MUTSCHLER NAISMITH and P.F. CLOUTIER

Department of Radiation Biology and Biophysics and Department of Biochemistry, School of Medicine and Dentistry, University of Rochester, Rochester, New York 14620, USA

Abstract: Miller, L.L., Mutschler Naismith, L., and Cloutier, P.F. Protein Catabolism in the Isolated Perfused Regenerating Rat Liver. Loss of Protein Catabolic Response to Glucagon. *Miami Winter Symposia* 2, pp. 516–529. North-Holland Publishing Co., Amsterdam, 1970.

In contrast to past emphasis on the role of enhanced nucleic acid and protein synthesis in regeneration of the experimentally hepatectomized rat liver, these studies were designed to examine the role of suppressed protein catabolism in favoring the net synthesis of hepatic proteins which characterizes the regenerating liver.

Adult male rats were partially hepatectomized (2/3 removal of the liver) and given L-leucine-1-^{14}C intraperitoneally to effect ^{14}C-labelling of hepatic proteins *in vivo*. Eighteen hours after injection of the ^{14}C-leucine, and 24 or 48 hr after hepatectomy, the livers were surgically removed, and perfused with heparinized oxygenated rat blood for periods of 6 or 7 hr. Minimal quantitative estimates of liver protein catabolism were made on the basis of net urea nitrogen produced and $^{14}CO_2$ evolved.

Diminished net urea production and $^{14}CO_2$ evolution by perfused 24 and 48 hr (but not by 9 day) regenerating livers support the hypothesis that diminished protein catabolism contributes significantly to the processes of net protein synthesis. Furthermore, the 50 to 100 percent increase in protein catabolism elicited by glucagon (1 μg/hr) in perfused normal livers, is totally suppressed in 24 and 48 hr regenerating livers even with doses of 10 μg/hr; this is further evidence of a profound alteration of liver protein catabolism in the regenerating liver.

Suppression of the protein catabolic effect of glucagon in 24 and 48 hr regenerating livers (without suppression of glycogenolysis) suggested that the regenerating liver is under an insulin like influence Although this view is also supported by demonstrated suppression of amino acid release in the perfused regenerating liver, anti-insulin antiserum, alone or with glucagon, was without effect on the regenerating liver.

* This paper is based on work performed under contract with the U.S. Atomic Energy Commission at the University of Rochester Atomic Energy Project and has been assigned Report No. UR-49-1310.

Of a number of liver enzymes potentially involved in overall control of protein catabolism, only glutaminase showed a major quantitative decrease in activity in the 24 and 48 hr regenerating liver.

1. Introduction

Although there is ample evidence for enhanced synthesis of nucleic acids and protein during liver regeneration [1], the possibility that impaired liver protein catabolism could contribute to the net gain of protein during regeneration has received relatively little attention. Thompson and Moss [2] observed a diminished total urinary urea output by rats for 3 days after subtotal hepatectomy and ascribed this to decreased food intake. Staib and Miller [3] used the technique of isolated liver perfusion to demonstrate urea production and decreased conversion of L-leucine-1-^{14}C labelled liver protein to $^{14}CO_2$ in 24 and 48 hr regenerating rat livers.

This report not only extends and confirms the observations of Staib and Miller [3] but also reveals that the 24 and 48 hr regenerating rat liver is uniquely insensitive to relatively massive doses of glucagon in failing to increase its protein catabolic rate estimated in terms of net urea production, and conversion of L-leucine-1-^{14}C labelled liver protein to $^{14}CO_2$; despite this, glucagon exerts its known glycogenolytic action in these livers.

Data are also presented to show that the 24 and 48 hr regenerating livers release significantly less free amino acids into the perfusion fluid than normal livers or 9 day regenerating livers.

Limited observations have been made on *in vitro* measurement of activities of a number of liver enzymes, activity of which may conceivably be involved in controlling the overall rate of protein catabolism in 24 and 48 hr regenerating rat liver; included were protease at both pH 3.5 and 9.0, alanine aminotransferase, glutamine-α-ketoisocaproic aminotransferase, glutaminase, glutamate dehydrogenase. Only glutaminase activity was found to be significantly decreased in the regenerating liver.

2. Experimental

2.1. *Animals*

Male rats of the Sprague-Dawley strain, obtained from either Holtzman or Charles River Farms, were used as blood and liver donors. Blood donors weighed from 300 to 500 g, while liver donors were somewhat smaller (200 to

400 g). To provide hepatectomized liver donors, partial hepatectomy (removal of 2/3 of liver) was carried out on fed rats according to the technique of Higgins and Anderson [4]. Immediately after partial hepatectomy 5 ml of Ringers solution was injected subcutaneously; no other post-operative care was given. All rats were maintained on Purina Checkers and water until 18 hr prior to perfusion at which time all food was removed from the cages. *In vivo* labelling of liver protein was achieved by intraperitoneal injection of liver donor rats with 2 ml (1.2 mg, 43.5 μCi) of L-leucine-1-^{14}C 18 hr before surgical removal of the liver for perfusion; here, better than 95 percent of ^{14}C-activity is present in liver protein [5].

2.2. *Perfusion methods*

Details of the operative procedure and of the perfusion apparatus used have been described elsewhere [5,6]. In the present experiments the perfusate volume was between 90 and 100 ml made up from whole heparinized blood and Ringers solution in a 2:1 ratio, to which 250 mg of glucose were added. Perfusions were run for either 6 or 7 hr during which time metabolic $^{14}CO_2$ was continuously collected and blood samples withdrawn from the reservoir at regular intervals. Glucagon, generously supplied by the Eli Lilly Co., was essentially free of insulin; it was dissolved in Ringers solution with minimal dilute sodium hydroxide (final pH 7.2 – 7.4). As indicated in figures for individual experiments, after a priming dose of either 3 or 10 μg of glucagon was added directly to the perfusate at the end of a 2½ or 3 hr base line period, a continuous infusion of glucagon at a rate of 3 or 10 μg per hour was started.

To explore the possibility that failure of glucagon to elicit increased protein catabolism was referable to the presence of endogenous insulin which is known to neutralize or prevent the protein catabolic effect of glucagon [7,8] in several experiments, 1 ml of guinea-pig anti-insulin serum was introduced into the perfusate after an initial 2 hr base line period. Pooled serum was obtained by bleeding guinea-pigs immunized by 4 weekly subcutaneous injections of 0.5–0.7 mg of glucagon-free insulin with Freund's adjuvant. One ml of this serum injected intravenously into fed rats resulted in a 2-fold increase in blood glucose level within 30 min; normal guinea-pig serum did not elicit hyperglycemia.

2.3. *Analytical methods*

A protein-free filtrate [9] was prepared immediately from each blood sample and aliquots of filtrate analyzed for glucose [10] and α-amino acid nitrogen [11]. Whole blood was analyzed for urea nitrogen content [12]. At the close of each perfusion the liver was weighed and immediately homogenized with 5 volumes of ice cold distilled water from which a protein-free filtrate was prepared and analyzed in the same manner used for the blood filtrates. In addition, the protein content of the liver homogenate was determined by the method of Lowry et al. [13].

Aliquots of whole blood, liver homogenate and bile were combusted by a modification of the method of Schöniger [14] and the $^{14}CO_2$ thus liberated absorbed in a mixture of ethanolamine and methyl cellosolve [15]. Assays for radioactivity were made in a Packard Tricarb liquid scintillation counter. Metabolic $^{14}CO_2$ was collected continuously and directly in 10 ml ethanolamine-methyl cellosolve, which was quantitatively replenished at 30 min intervals.

To compensate for differences in liver size, where appropriate, chemical results are expressed per 10 g of wet liver weight. For experiments in which proteins of perfused livers were priorly labelled with L-leucine-1-^{14}C, ^{14}C data are expressed as percent of total ^{14}C radioactivity recovered in a perfusion (sum-total of expired $^{14}CO_2$ and liver, bile and blood ^{14}C activity).

2.4. *Measurements of enzyme activity in regenerating liver*

For the measurements of enzyme activity liver removed at the time of hepatectomy or at 24 or 48 hr after hepatectomy was homogenized with ice cold distilled water for 3 min with a Teflon pestle homogenizer. Enzyme activities were compared on the basis of liver protein content as estimated by the method of Lowry et al. [13].

The methods for protease activity at pH 3.5 and 9.0 were essentially identical with those described by Koszalka and Miller [16] for muscle homogenates. The method of Rosen et al. [17] was used for alanine aminotransferase. For glutamine-α-ketoisocaproic aminotransferase, the method of Greenstein and Price [18] was used. The method of Sayre and Roberts [19] for glutaminase I was used. Glutamate dehydrogenase activity was estimated by the method of Bernt and Bergmeyer [20].

3. **Results**

3.1. *Protein catabolism in normal and regenerating livers*

The cumulative production of urea nitrogen in perfusions of normal and 24 and 48 hr regenerating livers is shown in fig. 1. In both types of regenerating livers urea production is depressed. It is perhaps of more physiological significance for a consideration of protein catabolism in the intact rat with a 24 or 48 hr regenerating liver to consider the actual observed urea nitrogen produc-

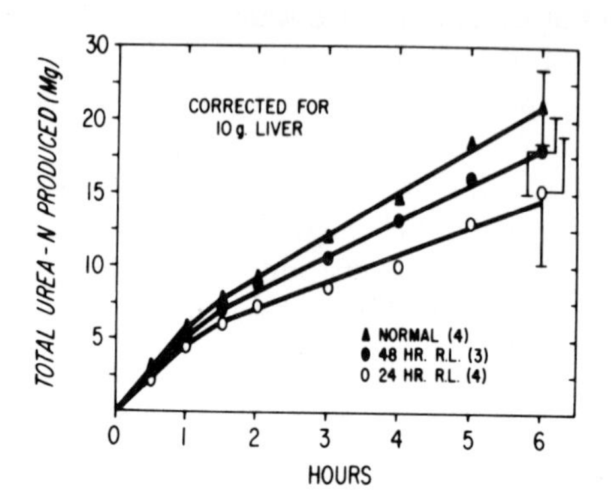

Fig. 1. Cumulative production of urea-*N* by perfused normal and regenerating livers. In this and all subsequent figures, brackets when present indicate extreme values for which averages are plotted.

tion. These data are shown in fig. 2 and indicate that these regenerating livers produce substantially less urea nitrogen than normal. The difference is most notable in the perfused 24 hr regenerating liver where on the average total urea nitrogen production is only 42 percent of normal. Although the data are not presented here, it is of interest that in 2 experiments with 9 day regenerating livers, urea production was found to be normal.

Fig. 3 depicts the percentage of ^{14}C-labelled liver protein expired as $^{14}CO_2$ in perfusions of both normal and regenerating livers. Previous studies with normal [5] and regenerating [3] L-leucine-1-^{14}C-labelled livers show that es-

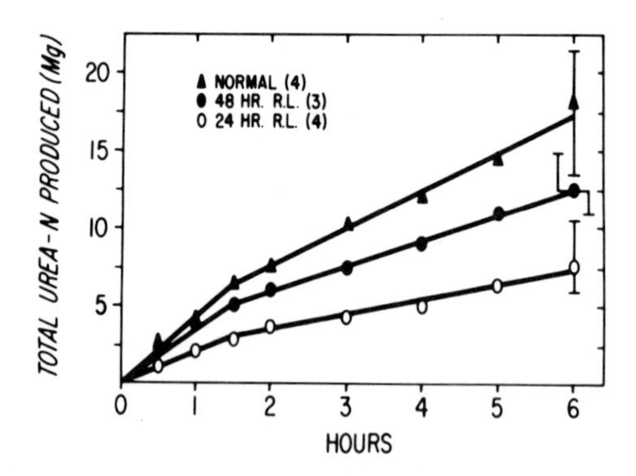

Fig. 2. Observed cumulative production (*not* corrected for 10 g liver weight) of urea-*N* by perfused normal and regenerating livers.

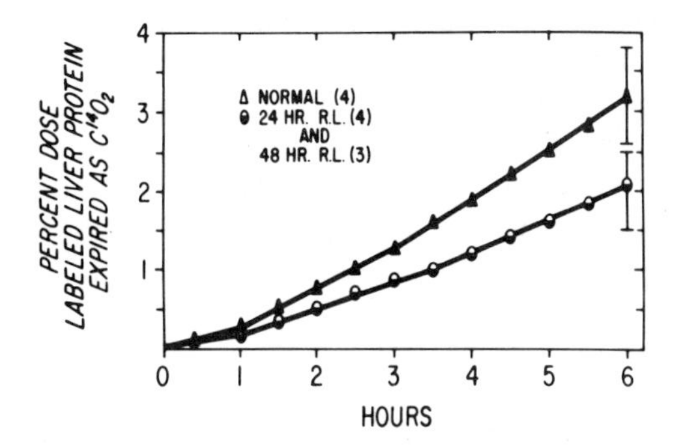

Fig. 3. Percentage of conversion to $^{14}CO_2$ of L-leucine-1-^{14}C labelled liver protein by perfused normal and regenerating livers. Data for 24 and 48 hr regenerating liver perfusions were identical and are, therefore, plotted together.

sentially all incorporated ^{14}C is in protein, and the amount of $^{14}CO_2$ produced affords a direct minimal measure of catabolism of labelled liver protein. As illustrated by this data both the rate and total amount of liver protein catabolized during perfusion is depressed in the regenerating liver. Production of $^{14}CO_2$ by perfused 9-day regenerating livers was normal in the two experiments carried out.

In the course of normal liver perfusions the α-amino acid nitrogen level of the perfusate has been shown to rise significantly especially after the first 3 to 4 hr of perfusion [7]. In perfusions with regenerating livers this phenomenon was not observed. Perfusate α-amino nitrogen levels, in mg per 100 ml, dropped

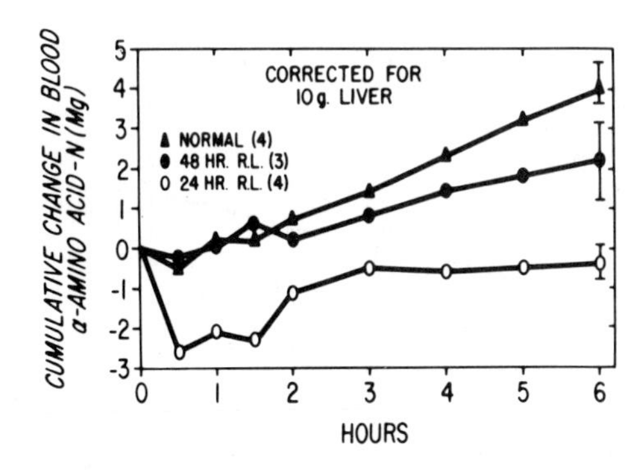

Fig. 4. Cumulative change in blood α-amino acid-*N* in normal and regenerating liver perfusions. Values less than that at zero time indicate a net loss and are plotted negatively.

precipitously and then gradually returned to initial values. When these results are plotted in terms of the cumulative change in perfusate α-amino nitrogen (fig. 4) it appears that there is a net removal of α-amino acid nitrogen from the perfusate in the case of the 24 hr regenerating liver. In the case of the 48 hr regenerating liver, while the level rises above the base line value, the rise is significantly less than that observed with normal livers. In this respect the regenerating liver resembles the normal liver under the continuous influence of insulin [7,8].

3.2. *The effect of glucagon on perfused regenerating livers*

As a consequence of the above findings it was of interest to determine whether glucagon would increase protein catabolism in perfused regenerating livers in the manner observed to occur consistently in normal livers [7,8]. Fig. 5 presents urea nitrogen production in perfusions in which glucagon was infused at a rate of 3 μg per hour and reveals that the powerful protein catabolic effect of this hormone seen in the normal liver (with a typical response shown in the top curve) is absent in the regenerating liver. Glucagon given at a rate of

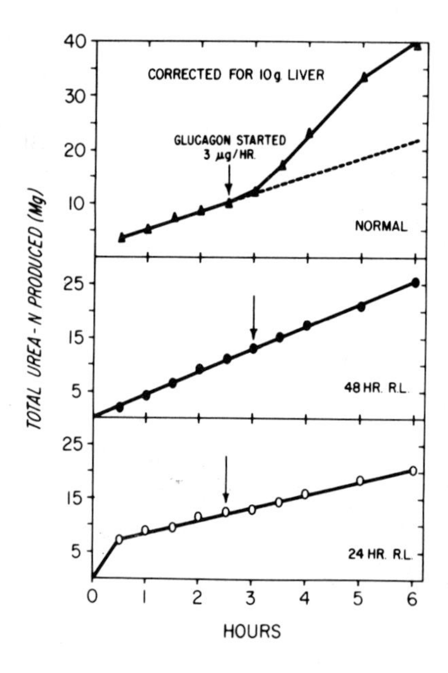

Fig. 5. Cumulative production of urea-*N* by perfused normal and regenerating livers. Continuous infusion of glucagon, 3 μg per hr, produces no increase in protein catabolism in regenerating livers. Each curve represents data from a single representative experiment.

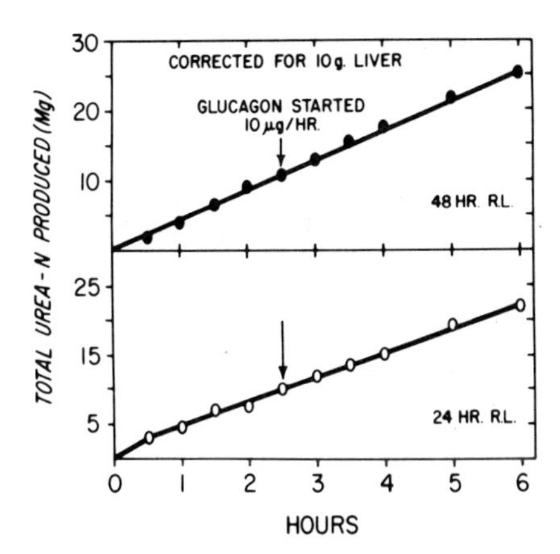

Fig. 6. Data from 2 representative experiments showing that continuous infusion of glucagon, 10 μg per hr, produces no increase in protein catabolism.

10 μg per hr was also without effect (fig. 6); nor was there any evidence for increased production of $^{14}CO_2$ (fig. 7). In contrast to the absence of a protein catabolic effect of even large doses of glucagon in regenerating livers, they remain responsive to the glycogenolytic action of glucagon as evidenced by a prompt rise in blood glucose levels (fig. 8). This is in contrast to the continued glucose disappearance curves from similar perfusions without glucagon.

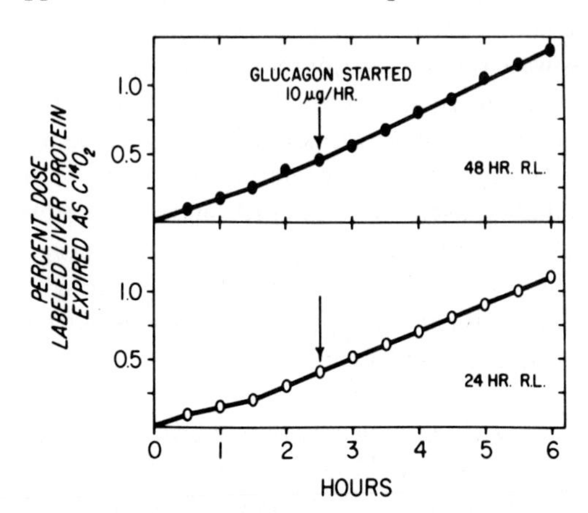

Fig. 7. Data from 2 liver perfusions showing that continuous infusion of glucagon, 10 μg per hr, is without effect in increasing conversion of leucine-^{14}C labelled liver protein to $^{14}CO_2$ in regenerating livers. (See fig. 6 for urea-*N* production by these livers).

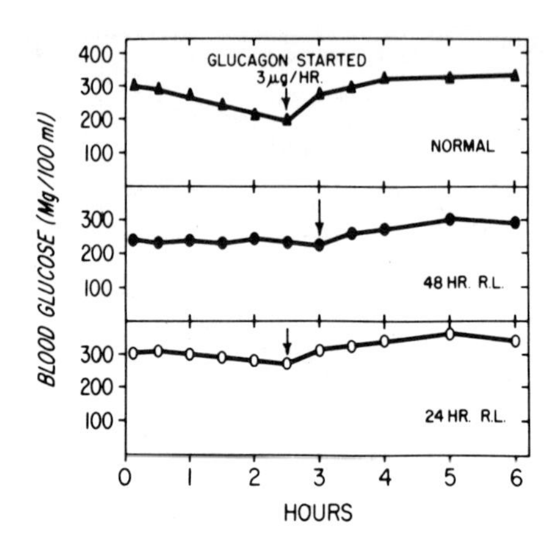

Fig. 8. Continuous infusion of glucagon, 3 μg per hr, causes glycogenolysis in both normal and regenerating livers. (These data are taken from experiments of fig, 5).

3.3. *The lack of effect of anti-insulin serum and glucagon on protein catabolism in regenerating livers*

Miller [7,8] has shown that insulin can completely inhibit glucagon-induced enhanced protein catabolism in the normal liver without abolishing its glycogenolytic action. The failure of the regenerating liver to respond to the protein catabolic action of glucagon while still responding to its glycogenolytic action, suggested the possibility that these livers were under an insulin-like influence. Perfusions were run with anti-insulin serum added to the perfusate at 2 hr with and without glucagon infusion. Neither anti-insulin serum alone, nor anti-insulin serum plus glucagon, infused at a rate of 10 μg per hr, increased urea nitrogen production by regenerating livers (fig. 9). Blood glucose levels and $^{14}CO_2$ production were no different from those observed in comparable control experiments. (Regenerating livers without glucagon served as controls for perfusions with anti-insulin serum and regenerating liver perfusions with glucagon served as controls for those perfusions in which both glucagon and anti-insulin serum were added.)

3.4. *α-Amino acid nitrogen levels in perfusate of regenerating livers*

It is of some interest to compare blood α-amino acid nitrogen levels at the end of 6 hr of perfusion in the various experimental groups already described. Table 1 shows that these levels are significantly lower than normal in perfusions of both 24 and 48 hr regenerating livers. In normal liver perfusions, continuous infusion of glucagon (3 μg per hr) produced a marked elevation of

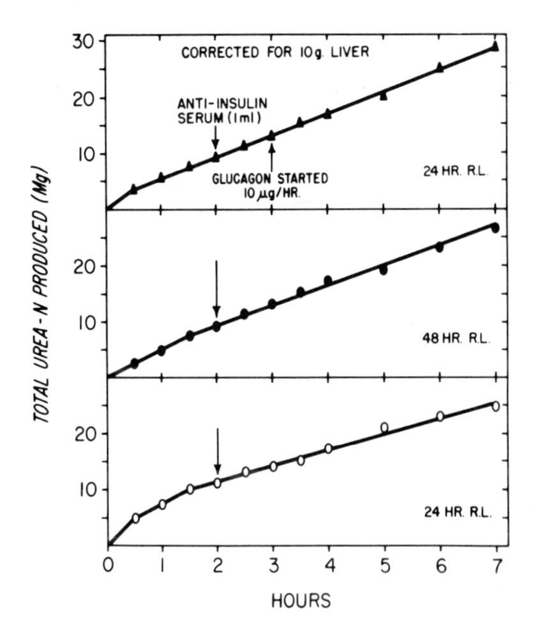

Fig. 9. Anti-insulin serum (1 ml total given at 2 hr) fails to increase urea-*N* production in regenerating livers. Glucagon, 10 μg per hr, given by continuous infusion starting at 3 hr, does not alter this response (upper-most curve).

blood amino acid level. The data suggest that both continuous infusion of glucagon and/or addition of anti-insulin serum bring about a small elevation in blood amino acid levels in regenerating liver perfusions. These results are, however, quantitatively much less pronounced than those found under similar conditions with the normal liver.

4. Discussion

The very nature of liver regeneration, that is hypertrophy and hyperplasia of the remaining tissue, implies that this process is associated with increased nucleic acid and protein synthesis. More direct evidence for this supposition has been obtained by numerous workers and has been reviewed by Bucher [1] in detail.

The concept of decreased protein catabolism as being an important aspect of the regenerative process has received relatively little attention. Thompson

Table 1
α-Amino acid nitrogen content of 6 hour blood samples in normal and regeneration liver perfusions.

Number of experiments	Liver donor	Treatment during perfusion	Average and range of blood α-amino-N levels (mg/100 ml)
4	Normal		9.8 (8.6 – 10.9)
1	Normal	Glucagon 3 μg at 2½ hr, 3 μg/hr 2½ – 6 hr.	13.7
1	Normal	Anti-insulin 1 ml at 2 hr; Glucagon 10 μg at 3 hr, 10 μg/hr 3 – 6 hr.	14.1
4	24 hr R.L.		5.8 (4.1 – 7.0)
3	24 hr R.L.	Glucagon 3 μg at 2½ hr, 3 μg/hr 2½ – 6 hr.	6.0 (5.0 – 7.5)
3	24 hr R.L.	Glucagon 10 μg at 2½ hr, 10 μg/hr 2½ – 6 hr.	7.5 (6.8 – 7.9)
4	24 hr R.L.	Anti-insulin 1 ml at 2 hr.	4.8 (3.0 – 6.2)
4	24 hr R.L.	Anti-insulin 1 ml at 2 hr; Glucagon 10 μg at 3 hr, 10 μg/hr 3–6 hr	6.8 (5.2 – 8.7)
3	48 hr R.L.		6.2 (4.9 – 7.1)
2	48 hr R.L.	Glucagon 3 μg at 3 hr, 3 μg/hr 3 – 6 hr.	6.9, 9.1
3	48 hr R.L.	Glucagon 10 μg at 2½ hr, 10 μg/hr 2½ – 6 hr.	7.1 (5.9 – 8.6)
3	48 hr R.L.	Anti-insulin 1 ml at 2 hr.	9.4 (8.1 – 11.3)
2	9 day R.L.		8.0, 8.0
1	9 day R.L.	Glucagon 3 μg at 3 hr, 3 μg/hr 3 – 6 hr.	12.7

and Moss [2] observed diminished urea production in rats for 3 days after partial hepatectomy, but concluded that this reflected decreased food intake. Staib and Miller [3] were the first to show that liver regeneration, at least in its early stages, is associated with decreased protein catabolism as demonstrated in the isolated perfused regenerating rat liver. The present studies confirm these findings in both the 24 and 48 hr regenerating liver and, furthermore, show that 9 days after partial hepatectomy, despite the fact that the liver has still not returned to its preoperative size, protein catabolism is apparently no longer suppressed.

If we assume that the same percentage of liver protein nitrogen is converted to urea nitrogen as liver protein carbon goes to $^{14}CO_2$, then it is a simple matter to estimate the catabolic contribution of liver protein to endogenously formed urea. Applying this method [5] to the present studies one finds that in a typical experiment with a normal liver, in the course of 6 hours, 3.15 percent of the initial total liver protein ^{14}C-radioactivity was expired as $^{14}CO_2$, while the total liver protein nitrogen content was 383 mg. Thus, the amount of protein nitrogen catabolized calculated to correspond to protein carbon catabolized equals (0.0315) (383) or 12.1 mg. Similarly during 6 hr of perfusion, (0.0207) (219) or 4.5 mg of liver protein nitrogen are catabolized by the 24 hr regenerating liver; (0.0213) (253) or 5.4 mg by the 48 hr regenerating liver; and (0.0401) (302) or 12.1 mg by the 9 day regenerating liver.

Although these figures indicate the magnitude by which liver protein catabolism is decreased in the regenerating liver, a more striking indication of suppressed protein catabolism is obtained from the experiments with glucagon. In the normal liver glucagon infused at the rate of 1 μg per hr enhances protein catabolism as evidenced by grossly increased urea production and by increased output of $^{14}CO_2$ from previously labelled liver protein; increasing doses of glucagon to 10 μg or even 20 μg per hr does not elicit a greater effect [7]. In the case of the 24 and 48 hr regenerating livers doses of glucagon as high as 10 μg per hr do not elicit a measurable increase in protein catabolism (figs 5 and 6). However glucagon causes some glycogenolysis in these regenerating livers as evidenced by the rise in blood glucose levels (fig. 8).

Both the failure of the regenerating liver to respond with a protein catabolic effect to massive doses of glucagon, and the decreased blood amino acid levels (table 1) suggest that the regenerating liver is under a pronounced insulin-like effect. However, a potent anti-insulin antiserum failed to modify protein catabolism, even when the antiserum was combined with glucagon infusion (fig. 9). One may speculate that the antibodies to insulin could not penetrate intracellularly, yet the small increase in perfusion blood amino acid levels suggests that the anti-insulin serum was somewhat effective, especially in the 48 hour regenerating liver.

The exact mechanism of the enhanced protein catabolism caused by glucagon in the isolated perfused normal rat liver is not known. Strong evidence has been presented [8] that it is only liver protein, rather than perfusate plasma protein, which is the source of the increased urea produced in response to glucagon, and that insulin can prevent or neutralize the protein catabolic effect of glucagon. Because unpublished experiments from our laboratory indicate that the apparent catabolic effect of glucagon in normal liver is not referrable to decreased liver protein resynthesis during otherwise normal turnover, or to increased oxidation of free amino acids, only the possibility of increased protein breakdown *per se* remains. If increased levels of cyclic 3′,5′-AMP are involved in mediating increased protein breakdown [21] in normal liver, then one may speculate that 24 or 48 hr regenerating liver either fails to produce increased cyclic 3′,5′-AMP or is unresponsive to it.

Although not presented here in detail, measurements of the following enzyme activities in homogenates of 24 hr regenerating livers were made: protease activity at pH 3.5 and pH 8.5, alanine aminotransferase, glutamine-α-keto-isovaleric aminotransferase, glutamate dehydrogenase, and glutaminase. Of these, only glutaminase activity was found significantly changed with a decrease in activity to about 50 percent of the preoperative level. Although minimizing hydrolysis of glutamine could favor purine nucleoside synthesis, it is not clear how it could contribute to decreased protein catabolism.

Whatever the mechanism for the apparent decrease in protein catabolism in 24 and 48 hr regenerating liver, it calls to mind evidence for decreased protein catabolism seen in precancerous livers of rats fed butter yellow [22] and the peculiarly constant protein catabolism in Walker carcinoma 256 cells in culture [23]. These observations suggest that diminished or unresponsive protein catabolism can contribute substantially to the net accretion of protein in neoplasia.

References

[1] N.L.R. Bucher, New England J. Med. 277 (1967) 686, 738; Intern. Rev. Cytol. 15 (1963) 245.
[2] J.F. Thompson and E.M. Moss, Proc. Soc. Exptl. Biol. Med. 89 (1955) 230.
[3] W. Staib and L.L. Miller, Biochem. Z. 339 (1964) 274.
[4] G.M. Higgins and R.M. Anderson, Arch. Pathol. 12 (1931) 186.
[5] M. Green and L.L. Miller, J. Biol. Chem. 235 (1960) 3202.
[6] L.L. Miller, C.G. Bly, M.L. Watson and W.F. Bale, J. Exptl. Med. 94 (1951) 431.
[7] L.L. Miller, in: Recent progress in hormone research, vol. 17, ed. G. Pincus (Academic Press, New York, 1961) p. 539.
[8] L.L. Miller, Federation Proc. 24 (1965) 3, 737.
[9] H.C. Hagedorn and B.N.Jensen, Biochem. Z. 135 (1923) 46.
[10] A.St.G. Huggett and D.A. Nixon, Lancet ii (1957) 368.
[11] E.C. Cocking and E.W. Yemm, Biochem. J. 58 (1954) xii.
[12] E.J. Conway and E. O'Malley, Biochem. J. 36 (1942) 655.
[13] O.H. Lowry, N.J. Rosebrough, A.L. Farr and R.J. Randall, J. Biol. Chem. 193 (1951) 265.
[14] R.J. Kelly, E.A. Peets, S. Gordon and D.A. Buyske, Anal. Biochem. 2 (1961) 267.

[15] H. Jeffay and J. Alvarez, Anal. Chem. 33 (1961) 612.
[16] T.R. Koszalka and L.L. Miller, J. Biol. Chem. 235 (1960) 665.
[17] F. Rosen, N.R. Roberts and C.A. Nichol, J. Biol. Chem. 234 (1959) 476.
[18] J.P. Greenstein and V.E. Price, J. Biol. Chem. 178 (1949) 695.
[19] F.W. Sayre and E. Roberts, J. Biol. Chem. 233 (1958) 1128.
[20] E. Bernt and H.U. Bergmeyer, in: Methods of Enzymatic Analysis (American Elsevier, New York, 1963) p. 384.
[21] L.A. Menahan and O. Wieland, European J. Biochem. 9 (1969) 55.
[22] W.T. Burke and L.L. Miller, Cancer Res. 16 (1956) 330.
[23] H.C. Jordan, L.L. Miller and P.A. Peters, Cancer Res. 19 (1959) 195.